T0142972

Studies in Computational Intelligence

Volume 524

Series Editor

Janusz Kacprzyk, Polish Academy of Sciences, Warsaw, Poland
e-mail: kacprzyk@ibspan.waw.pl

For further volumes:
http://www.springer.com/series/7092

About this Series

The series "Studies in Computational Intelligence" (SCI) publishes new developments and advances in the various areas of computational intelligence—quickly and with a high quality. The intent is to cover the theory, applications, and design methods of computational intelligence, as embedded in the fields of engineering, computer science, physics and life sciences, as well as the methodologies behind them. The series contains monographs, lecture notes and edited volumes in computational intelligence spanning the areas of neural networks, connectionist systems, genetic algorithms, evolutionary computation, artificial intelligence, cellular automata, self-organizing systems, soft computing, fuzzy systems, and hybrid intelligent systems. Of particular value to both the contributors and the readership are the short publication timeframe and the world-wide distribution, which enable both wide and rapid dissemination of research output.

Alejandro Peña-Ayala
Editor

Educational Data Mining

Applications and Trends

Editor
Alejandro Peña-Ayala
World Outreach Light to the Nations Ministries
Escuela Superior de Ingeniería Mecánica y Eléctrica, Zacatenco
Instituto Politécnico Nacional
Mexico City
Mexico

ISSN 1860-949X ISSN 1860-9503 (electronic)
ISBN 978-3-319-34499-7 ISBN 978-3-319-02738-8 (eBook)
DOI 10.1007/978-3-319-02738-8
Springer Cham Heidelberg New York Dordrecht London

Printed on acid-free paper

Springer is part of Springer Science+Business Media (www.springer.com)

Preface

Educational Data Mining (EDM) is a new discipline based on the Data Mining (DM) grounds (i.e., the baseline is composed of models, tasks, methods, and algorithms) to explore data from educational settings to find out descriptive patterns and predictions that characterize learners behaviors and achievements, domain knowledge content, assessments, educational functionalities, and applications.

This book introduces concepts, models, frameworks, tasks, methods, and algorithms, as well as tools and case studies of the EDM field. The chapters make up a sample of the work currently achieved in countries from the five continents, which illustrates the world labor of the EDM arena. According to the nature of the contributions accepted for this volume, four kinds of topics are identified as follows:

- *Profile* shapes a conceptual view of the EDM. It provides an introduction of the nature, purpose, components, processes, and applications. Through this section, readers are encouraged to: make an incursion in the EDM field, facilitate the extraction of source data to be mined, and acquire consciousness of the usefulness of this sort of approaches to support education policies.
- *Student Modeling* is an essential functionality of Computer-Based Educational Systems (CBES) to adapt their performance according to users needs. Most of the EDM approaches are oriented to characterize diverse student traits, such as: behavior, acquired domain knowledge, personality, and academic achievements by means of machine learning methods.
- *Assessment* evaluates learners' domain knowledge acquisition, skills development, and achieved outcomes, as well as reflection, inquiring, and sentiments are essential subjects to be taken into account by CBES. The purpose is to differentiate student proficiency at the finer grained level through static and dynamic testing, as well as online and offline assessment.
- *Trends* focus on some of the new demands for applying EDM, such as text mining and social networks analysis. Both targets represent challenges to cope with huge, dynamical, and heterogeneous information that new generations of students produce in their every day life. These paradigms represent new educational settings such as: ubiquitous-learning and educational networking.

This volume is the result of one year of effort, where more than 30 chapters were rigorous peer reviewed by a team of 60 reviewers. After several cycles of chapter submission, revision, and tuning based on the Springer quality principles, 16 works were approved, edited as chapters, and organized according to the prior four topics. So the Part I corresponds to *Profile* that includes Chaps. 1–3; the Part II represents *Student Modeling,* which embraces Chaps. 4–8; the Part III concerns *Assessment* and has Chaps. 9–12; the Part IV is related to *Trends* through Chaps. 13–16. A profile of the chapters is given next:

1. Chapter 1 provides a bibliographic review of studies made in the field of Educational Data Mining (EDM) to identify diverse aspects related to techniques and contributions in the field of computer-based learning. Authors pursue to facilitate the use and understanding of Data Mining (DM) techniques to help the educational specialists to develop EDM approaches.
2. Chapter 2 overcomes the lack of data preprocessing literature through the detailed exposition of the tasks involved to extract, clean, transform, and provide suitable data worthy to be mined. The work depicts educational environments and data they offer; as well as gives examples of Moodle data and tools.
3. Chapter 3 illustrates how EDM is able to support government policies devoted to enhance education. The work shapes the context of basic education and how the government aims at reforming the current practices of evaluation to academics and students. Several findings extracted from surveys are shown to highlight the opinion of the community and provide an initial diagnostic.
4. Chapter 4 presents the Student Knowledge Discovery Software, a tool to explore the factors having an impact on the student success based on student profiling. Authors deeply outline how to implement the software to help educational organizations to better understand knowledge discovery processes.
5. Chapter 5 explains how to automate the detection of student's personality and behavior in an educational game called Land Science. The work includes a model to learn vector space representations for various extracted features. Learner personality is detected by combining the features spaces from psycholinguistics and computational linguistics.
6. Chapter 6 attempts to predict student performance to better adjust educational materials and strategies throughout the learning process. Thus, authors design a multichannel decision fusion approach to estimate the overall student performance. Such an approach is based on the performance achieved in assignment categories.
7. Chapter 7 explores predictive modeling methods for identifying students who will most benefit from tutor interventions. Authors assert how the predictive capacity of diverse sources of data changes as the course progresses, as well as how a student's pattern of behavior changes during the course.
8. Chapter 8 predicts learner achievements by recording learner eye movements and mouse click counts. The findings claim: the most important eye metrics

that predict answers in reasoning questions include total fixation duration, number of mouse clicks, fixation count, and visit duration.

9. Chapter 9 focuses on coherence expressed in research protocols and thesis. Authors develop a coherence analyzer that employs Latent Semantic Analysis to mine domain knowledge. The analysis outcomes are used to grade students and provide online support with the aim at improving their writings.
10. Chapter 10 tailors an approach to automatically generate tests. It recognizes competence areas and matches the overall competence level of target students. The approach makes use of a concept map of programming competencies and a method for estimating the test item difficulty. The contribution is evaluated in a setting where its results are compared against a solution that randomly searches within the item space to find an adequate test.
11. Chapter 11 outlines methods oriented to support teachers' understanding of students' activities on Exploratory Learning Environments (ELE). The work includes an algorithm that intelligently recognizes student activities and visualization facilities for presenting these activities to teachers. The approach is evaluated using real data obtained from students using an ELE to solve six representative problems from introductory chemistry courses.
12. Chapter 12 adopts the concept of entropy from information theory to find the most dependent test items in student responses. The work defines a distance metric to estimate the amount of mutual independency between two items that is used to quantify how independent two items are in a test. The trials show: the approach for finding the best dependency tree is fast and scalable.
13. Chapter 13 proposes ReaderBench, an environment for assessing learner productions and supporting teachers. It applies text mining to perform: assessment of the reading materials, assignment of texts to learners, detection of reading strategies, and comprehension evaluation to fostering learner's self-regulation process. All of these tasks were subject of empirical validations.
14. Chapter 14 analyzes a data set consisting of student narrative comments that were collected using an online process. The approach uses category vectors to depict instructor traits and a domain-specific lexicon. Sentiment analysis is also used to detect and gauge attitudes embedded in comments about each category. The approach is useful to instructors and administrators, and is a vehicle to analyze student perceptions of teaching to feedback the educational process.
15. Chapter 15 introduces E-learning Web Miner, a tool that assists academics to discover student behavior profiles, models of how they collaborate, and their performance with the purpose of enhancing the teaching-learning process. The tool applies Social Network Analysis (SNA) and classification techniques.
16. Chapter 16 depicts an approach to assess the students' participation by the analysis of their interactions in social networks. It includes metrics for ranking and determining roles to analyze the student communications, the forming of groups, the role changes, and the interpretation of exchanged messages.

I express my gratitude to authors, reviewers, my assistant Leonor Adriana Cárdenas, the Springer editorial team, and the editors Dr. Thomas Ditzinger and Prof. Janusz Kacprzyk for their valuable collaboration to fulfill this work.

I also acknowledge the support given by the Consejo Nacional de Ciencia y Tecnología (CONACYT) and the Instituto Politécnico Nacional (IPN) of Mexico through the grants: CONACYT-SNI-36453, CONACYT 118862, CONACYT 118962-162727, IPN-SIP-20131093, IPN-COFAA-SIBE-DEBEC/647-391/2013.

Last but not least, I acknowledge the strength given by my Father, Brother Jesus, and Helper, as part of the research projects of World Outreach Light to the Nations Ministries (WOLNM).

August, 2013 Alejandro Peña-Ayala

Contents

Part I
Profile

Chapter 1
Which Contribution Does EDM Provide to Computer-Based Learning Environments?

Nabila Bousbia and Idriss Belamri

Abstract Educational Data Mining is a new growing research area that can be defined as the application of data mining techniques on raw data from educational systems in order to respond to the educational questions and problems, and also to discover the information hidden after this data. Over the last few years, the popularity of this field enhanced a large number of research studies that is difficult to surround and to identify the contribution of data mining techniques in educational systems. In fact, exploit and understand the raw data collected from educational systems can be "a gold mine" to help the designers and the users of these systems improving their performance and extracting useful information on the behaviors of students in the learning process. The use of data mining techniques in e-learning systems could be very interesting to resolve learning problems. Researchers' ambition is to respond to questions like: What can predict learners' success? Which scenario sequence is more efficient for a specific student? What are the student actions that indicate the learning progress? What are the characteristics of a learning environment allowing a better learning? etc. The current feedback allows detecting the usefulness of applying EDM on visualizing and describing the learning raw data. The predictions take also an interest, particularly the prediction of performance and learners' behaviors. The aim of this chapter is to establish a bibliographic review of the various studies made in the field of educational data mining (EDM) to identify the different aspects studied: the analyzed data, the objectives of these studies, the used techniques and the contribution of the application of these techniques in the field of computer based learning. The goal is not only to list the existing work but also to facilitate the use and the understanding of data mining techniques to help the educational field specialists to give their

N. Bousbia (✉) · I. Belamri
Laboratoire Méthode de Conception de Systèmes (LMCS), Ecole nationale Supérieure d'Informatique (ESI), BP 68M, 16309 Oued-Smar, Algiers, Algeria
e-mail: n_bousbia@esi.dz

I. Belamri
e-mail: i_belamri@esi.dz

A. Peña-Ayala (ed.), *Educational Data Mining*,
Studies in Computational Intelligence 524, DOI: 10.1007/978-3-319-02738-8_1,

feedback and to identify promoter research areas in this field to be exploited in the future.

Keywords EDM · Learner's behavior · Prediction of student's performance · Computer based learning environments (CBLE)

Abbreviations

CBLE	Computer based learning environment
DM	Data mining
EDM	Educational data mining
ITS	Intelligent tutoring system
KDD	Knowledge discovery in databases
KT	Knowledge tracing
LA	Learning analytics
LAK	Learning analytics and knowledge
LMS	Learning management system
NMF	Non-negative matrix factorization
SNA	Social network analysis

1.1 Introduction

EDM is an emerging discipline, with a suite of computational and psychological methods and research approaches for understanding how students learn, and the settings which they learn in [1].

Data of interest is not restricted to interactions of individual students with an educational system (e.g., navigation behavior, input to quizzes and interactive exercises) but might also include data from collaborating students (e.g., text chat), administrative data (e.g., school, school district, teacher), demographic data (e.g., gender, age, school grades), and data on student affect (e.g., motivation, emotional states) [2].

EDM can be applied to assess students' learning performance, to improve the learning process and guide students' learning, to provide feedback and adapt learning recommendations based on students' learning behaviors, to evaluate learning materials and courseware, to detect abnormal learning behaviors and problems, and to achieve a deeper understanding of educational phenomena [3].

For example, Ayesha et al. [4] described the use of k-means clustering algorithm to predict student's learning activities. Pal [5] used machine learning algorithm to find students which are likely to drop out their first year of engineering. Parack et al. [6] used multiple data mining algorithms for student profiling and grouping based on their academic records such as exam scores, term work grades, attendance and practical exams.

As the number of EDM studies found in the literature is growing considerably over the last few years, we aim in this chapter to establish a bibliographic review of these studies. Our goal is to discuss the data mining methods and tools used in computer based learning environments to analyze learners' behaviors and performance in order to facilitate the use and the understanding of data mining techniques to help the educational field specialists to give their feedback and to identify promoter research areas in this field to be exploited in the future.

Therefore, the remaining of the chapter is organized as follows: Sect. 1.2 is devoted to give a detailed view of the EDM field: definition, related areas, goals, methods, the analyzed data, process and the used tools. Section 1.3 presents some examples dealing with the two principal EDM applications: analyzing learners' behaviors and predicting learners' performance. We compare and discuss these examples according to their goals, the analyzed data and the used methods. We end the chapter with a conclusion in Sect. 1.4.

1.2 Educational Data Mining

1.2.1 Definition

Different definitions have been provided for the term 'Educational Data Mining' or EDM. Educational data mining is defined by the journal of educational data mining[1] and Baker [1] as "*an emerging discipline, concerned with developing methods for exploring the unique types of data that come from educational settings, and using those methods to better understand students, and the settings which they learn in*".

This definition does not mention data mining; open to exploring and developing other analytical methods that can be applied to educationally related data [7].

However, in [8] the authors precise that: "*EDM is both a learning science, as well as a rich application area for data mining, due to the growing availability of educational data. It enables data-driven decision making for improving the current educational practice and learning material*".

In the same way, Romero and Ventura [9, 10] define EDM as "*the application of data mining (DM) techniques to specific type of dataset that come from educational environments to address important educational questions*".

Although different in some details, these definitions share an emphasis on discovering knowledge based on educational data to improve educational systems. Note also that the definition of EDM is often confused with 'learning analytics' defined on the LAK (Learning Analytics and Knowledge) website as "*the measurement, collection, analysis and reporting of data about learners and their contexts, for purposes of understanding and optimizing learning and the environments in which it occurs*" [11].

[1] http://www.educationaldatamining.org/JEDM/ visited on August 6, 2013.

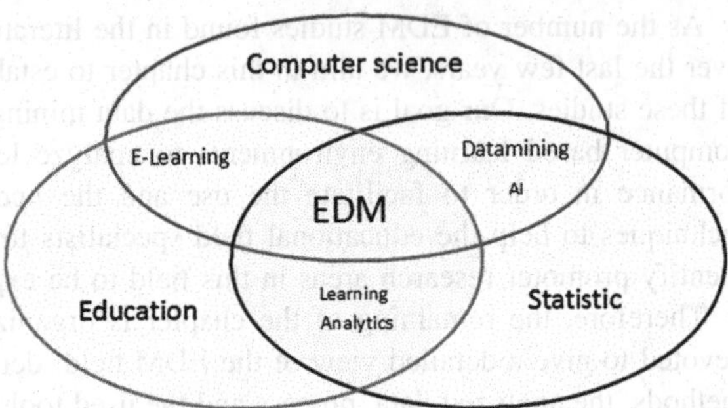

Fig. 1.1 Areas in relation with EDM [10]

Although there is no hard and fast distinction between these two fields, they have had somewhat different research histories and are developing as distinct research areas [12]. The objective of this chapter is not to draw up a comparative study between these two concepts (comparisons and details can be found in [10–12]).

However, we think worth mentioning that this field is the most related to the EDM field, as they share many goals and it is often difficult to differentiate if an application fits into one or the other of the two areas. The next subsection presents the related fields to EDM.

1.2.2 Areas in Relation to EDM

EDM can be drawn as the combination of three main areas (Fig. 1.1): computer science, education, and statistics. The intersection of those three areas also forms other subareas closely related to EDM such as learning analytics (LA), CBLE, DM and machine learning [10].

As an interdisciplinary area, EDM uses methods and applies techniques from statistics, machine learning, data mining, information retrieval, recommender systems, psycho-pedagogy, cognitive psychology, psychometrics, etc. The choice of which method or technique should be used depends on the addressed educational issue.

1.2.3 Objectives of the EDM

In the last several years, EDM has been applied to address a wide number of goals that are all parts of the general objective of improving learning [10]. Several studies [1, 8, 10, 12, 13] dress a list of these objectives.

Romero and Ventura [10] proposed to classify EDM objectives depending on the viewpoint of the final user (learner, educator, administrator, and researcher) and the problem to resolve:

- Learners. To support a learner's reflections on the situation, to provide adaptive feedback or recommendations to learners, to respond to student's needs, to improve learning performance, etc.
- Educators. To understand their students' learning processes and reflect on their own teaching methods, to improve teaching performance, to understand social, cognitive and behavioral aspects, etc.
- Researchers. To develop and compare data mining techniques to be able to recommend the most useful one for each specific educational task or problem, to evaluate learning effectiveness when using different settings and methods, etc.
- Administrators. To evaluate the best way to organize institutional resources (human and material) and their educational offer.

This view point clearly shows the benefit of EDM applications to the end user, but it is difficult to classify all EDM application goals according to these four actors, especially when an objective is related to more than one actor. That is why, based on the work of [1, 12–14] that focused on the related research goal of EDM applications, we distinguish between the following EDM general goals:

- Student modeling. User modeling in the educational domain incorporates such detailed information as students' characteristics or states such as knowledge, skills, motivation, satisfaction, meta-cognition, attitudes, experiences and learning progress, or certain types of problems that negatively impact their learning outcomes (making too many errors, misusing or under-using help, gaming the system, inefficiently exploring learning resources, etc.), affect, learning styles, and preferences. The common objective here is to create or improve a student model from usage information.
- Predicting students' performance and learning outcomes. The objective is to predict a student's final grades or other types of learning outcomes (such as retention in a degree program or future ability to learn), based on data from course activities. Examples of predicting student's performance can be found in Sect. 1.3.
- Generating recommendation. The objective is to recommend to students which content (or tasks or links) is the most appropriate for them at the current time [15].
- Analyzing learner's behavior. This takes on several forms: Applying educational data mining to answer questions in any of the three areas previously discussed (student models, Prediction, Generating recommendation). It is also used to group student according to their profile, and for adaptation and personalization purposes.
- Communicating to stakeholders. The objective is to help course administrators and educators in analyzing students' activities and usage information in courses. Macfayden and Dawson in [16] conducted a study that confirms that pedagogically meaningful information that is extracted from e-learning systems can be used to develop a customizable dashboard-like reporting tool for educators that will extract and visualize real-time data on student engagement and likelihood of

success. Romero et al. [17] provided feedback to help decision making for improving student learning and taking the appropriate proactive action. Other examples and case studies for this category of applications can be found in [14].

- Domain structure analysis. The objective is to determine domain structure and improving domain models that characterize the content to be learned and optimal instructional sequences, using the ability to predict the student's performance as a quality measure of a domain structure model. Performance on tests or within a learning environment is utilized for this goal.
- Maintaining and improving courses. It is related to the two previous goals. The objective here is to determine how to improve courses (contents, activities, links, etc.), using information (in particular) about student usage and learning.
- Studying the effects of different kinds of pedagogical support that can be provided by learning software. For example, Anaya and Boticario [18] proposed a method to analyze collaboration using machine learning techniques.
- Advancing scientific knowledge about learning and learners through building, discovering or improving models of the student, the domain, and the pedagogical support. For example, Siemens and Baker [19] developed and tested a scientific theory about improving learning technology, and formulated a new scientific hypothesis.

We note that these EDM objectives aim to improve several aspects of educational systems in general and CBLE in particular. In this specific context, the learner modeling is a key point to accomplish several goals and tasks (tutoring, adaptation, personalization, etc.). Indeed, the different objectives depend heavily on this first objective "Student modeling" which is often supplemented by the behavior analysis, and therefore, allows the prediction of performance, generating recommendation, providing administrators and educators the adequate information to maintain and improve the content and learning environments.

Thus, if the EDM facilitates the modeling, and thus achieve the objectives mentioned above, several treatments become easier in CBLE. To accomplish these goals, educational data mining researches use the categories of technical methods described below.

1.2.4 The Used Methods

To achieve the EDM objectives, the majority of traditional data mining techniques including but not limited to classification, clustering, and association analysis techniques have been applied successfully in the educational domain. Nevertheless, educational systems have special characteristics that require a different treatment of the mining problem [14]. That is why researchers involved in EDM apply not only data mining techniques, but also propose, develop and apply methods and techniques drawn from the variety of areas related to EDM (statistics, machine learning, text mining, web log analysis, psychometrics, etc.).

The most popular classification of these methods is the one proposed in Baker [1]: prediction, clustering, relationship mining, distillation for human judgment and discovery with models. Bienkowski et al. [12] then Romero and Ventura [10] extended this taxonomy. Based on these studies and those in [20, 21] we regroup these techniques into the following methods:

- Prediction. The goal is to develop a model which can infer a single aspect of the data (predicted variable) from some combination of other aspects of the data (predictor variables). Types of predictions methods are *classification* (when the predicted variable is a categorical value), *regression* (when the predicted variable is a continuous value), or *density estimation* (when the predicted value is a probability density function). An example of EDM application is predicting student's academic success [4] and behaviors [6].
- Clustering. Refers to finding instances that naturally group together and can be used to split a full dataset into categories. Typically, some kinds of distance measures are used to decide how similar instances are. Once a set of clusters has been determined, new instances can be classified by determining the closest cluster. In EDM, clustering can be used for grouping students based on their learning patterns or cognitive strategies [22].
- Relationship mining. Used for discovering relationships between variables in a dataset and encoding them as rules for later use. There are different types of relationship in mining techniques such as *association rule mining* (any relationships between variables), *sequential pattern mining* (temporal associations between variables), *correlation mining* (linear correlations between variables), and *causal data mining* (causal relationships between variables). In EDM, relationship mining is used to identify relationships between the students' online activities and the final marks [23] and to model learners' problem solving activity sequences [24].
- Distillation of data for human judgment. It is a technique that involves depicting data in a way that enables a human to quickly identify or classify features of the data. This approach uses summarization, visualization and interactive interfaces to highlight useful information and support decision-making. On the one hand, it is relatively easy to obtain descriptive statistics from educational data to obtain global data characteristics and summaries and reports on learner's behavior. On the other hand, information visualization and graphic techniques help to see, explore, and understand huge educational data at once. In [25] the visualization of sequences of student's activity helps to understand the patterns of learning environment use.
- Discovery with models. Its goal is to use a validated model of a phenomenon (using prediction, clustering, or knowledge engineering) as a component in further analysis such as prediction or relationship mining. It is used for example to identify the relationships between the student's behavior and characteristics [26].
- Outlier Detection. The goal of outlier detection is to discover data points that are significantly different than the rest of data. An outlier is a different observation (or measurement) that is usually larger or smaller than the other values in data.

In EDM, outlier detection can be used to detect deviations in the learner's or educator's actions or behaviors, irregular learning processes, and for detecting students with learning difficulties [27].

- Social Network Analysis. SNA or structural analysis, aims at studying relationships between individuals, instead of individual attributes or properties. SNA views social relationships in terms of network theory consisting of nodes (representing individual actors within the network) and connections or links (which represent relationships between the individuals, such as friendship, cooperative relations, etc.). In EDM, SNA can be used to interpret and analyze the structure and relations in collaborative tasks and interactions with communication tools [28].
- Process Mining. Its goal is to extract process related knowledge from event logs recorded by an information system to have a clear visual representation of the whole process. It consists of three subfields: conformance checking, model discovery, and model extension. In EDM, process mining can be used for reflecting students' behaviors in terms of their examination traces consisting of a sequence of course, grade, and timestamp triplets for each student [29].
- Text Mining. It is an extension of data mining to text that is focused on finding and extracting useful or interesting patterns, models, directions, trends, or rules from unstructured text documents such as HTML files, chat messages and emails. Text mining tasks include text categorization, text clustering, concept/entity extraction, production of granular taxonomies, sentiment analysis, document summarization, and entity relation modeling [10]. Text mining is used to analyze the content of discussion boards, forums, chats, Web pages, documents, etc. [3].
- Knowledge Tracing. KT is a popular method for estimating student mastery of skills that has been used in effective cognitive tutor systems. It uses both a cognitive model that maps a problem-solving item to the skills required, and logs of students' correct and incorrect answers as evidence of their knowledge on a particular skill. KT tracks student knowledge over time and it is parameterized by variables. There is an equivalent formulation of KT as a Bayesian network. In EDM, it is used for example for predicting student's behavior [30].
- Matrix Factorization. It is a decomposition of a matrix into a product of matrices. There are many matrix factorization techniques such as Non-negative Matrix Factorization (NMF). NMF consists of a matrix of positive numbers, as the product of two smaller matrices. For example, in the context of education, a matrix M that represents the observed examinee's test outcome data that can be decomposed into two matrices: Q that represents the Q-matrix of items and S that represents each student's mastery of skills [31]. Thai-Nghe et al. [32] used a matrix factorization model inspired from recommender systems to predict student performance.

We note here that an increasing number of techniques are used in EDM for the analysis of the different data produced in educational systems. The choice of which technique to use depends on the nature of the learning environment, the research objectives and the type of the available data. In what follows we discuss the type of the analyzed data.

1.2.5 The Analyzed Data

There are different analyzed data in EDM studies such as their objectives and techniques. We can distinguish these data according to the following features:

- Data availability:
  - Data already available recorded over the years in the institution databases (e.g. students' scores) or the log files of learning software.
  - Data generated during experiments within a research work.
  - Data available to researchers in benchmark repositories (PSL-Datashop[2], MULCE[3]).
- Collection sources:
  - Manual. Performed by a human observer that takes notes on the learning situation to evaluate the participants' activities.
  - Digital. Relies on the use of a hardware configuration that records the learner's activity. The result of such collection is a numerical trace that can be a log file, information stored in databases, audio or video records.
  - Mixed. Where both methods are used simultaneously.
- Learning environment [10]:
  - Traditional education. Primary, secondary, higher education, etc.
  - Computer-based education. Intelligent Tutoring System (ITS), Learning Management System (LMS), Adaptative Educational Hypermedia System (AEHS), Computer Supported Collaborative Learning (CSCL), serious games, test and quiz systems, etc.
- The educational described level [1, 9]:
  - The keystroke level, the answer level, the session level, the student level, the classroom level, the teacher level, and the school level.
- The type of data:
  - Qualitative or quantitative data.
  - Personal, administrative and/or demographic data (age, sex, etc.).
  - Answers to psychological questionnaires for measuring users' satisfaction, motivation, skills, cognitive features, etc.
  - Answers to questions and/or test scores of the academic system.
  - Individual interactions with the educational system: from fine grained actions such as mouse click, to high level ones such as number of attempts, the learner browsing pattern, etc.

---

[2] https://pslcdatashop.web.cmu.edu/ visited on August 6, 2013.

[3] http://mulce-pf.univ-fcomte.fr/PlateFormeMulce/visited on August 6, 2013.

- Social interaction (chat, sent messages, forum participation, etc.).
- Visual and facial reactions, etc.

We note here that the data are highly variable depending on the type of environment. In this chapter, we are interested in EDM applications on computer based education. In such systems, the collected data is often digital, and their size is often less important than traditional environments that have much bigger databases. However, several studies combine these two sources of data to give a complete view of the learner's behavior and performance. For instance, authors in [23] attempted to predict the success of students in the final exam based on their participation level in online forums. The fusion and the processing of these different types of data require several steps to implement the EDM process that we present in the following section.

1.2.6 Process of Applying the EDM

Romero and Ventura [10] and Sachin and Vijay [33] proposed a process of applying EDM close to the one of KDD (Knowledge discovery in databases) or other data mining application process (Fig. 1.2).

This process starts with collecting or choosing the data to study from the educational environment. The obtained raw data require cleaning and preprocessing (heterogeneous data fusion, treatment of missing and incorrect values, converting the data to an appropriate form, feature selection, etc.).

This phase often requires the use of some data mining techniques. That is why, and given its complexity some works try to eliminate this phase as [34] which provides a data model to structure data stored by Learning Management Systems,

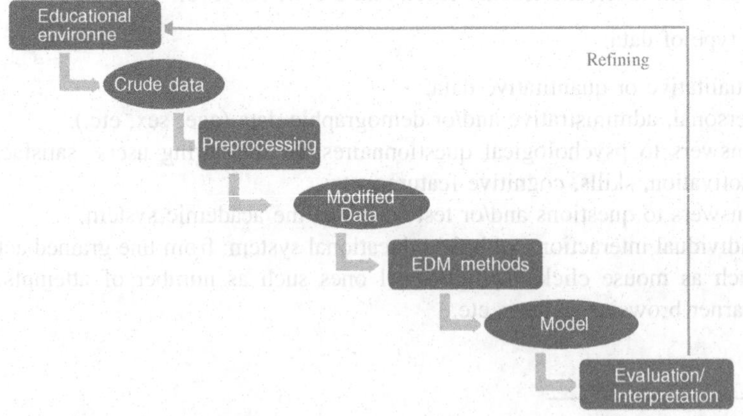

Fig. 1.2 The process of data mining application in educational data mining

and a tool that does the actual structure/export functionality, which they implemented for the Moodle LMS.

Once the data preprocessed, the appropriate EDM method/technique is applied. Finally, the last step is the interpretation and the assessment of the obtained results. To apply this process, which is often difficult given the heterogeneity of the data in the educational context, several tools are used.

1.2.7 Some Technological Tools Used in EDM

There are several tools and technologies used in the process of EDM not specifically designed for teaching and educational environments (Weka,[4] R,[5] etc.). However, in the last few years, a large number of data mining tools designed for educational purposes have been developed [10]. A summary of some of the most recent tools are presented in Table 1.1.

By analyzing these tools, we find that they are usually designed for computer-based educational systems. Moreover, apart from benchmark repositories (PSL-Datashop and MULCE), other tools are not re-used by other researchers of the EDM community.

This can be due to several reasons: their availability, the special format of the data to analyze, the difficulty of their deployment outside of their development environment or the ignorance of their existence. That is why; an effort should be made to make these tools available to the different learning actors (teachers, designers, administrators and researchers) to fulfill the different objectives and to analyze data from different environments.

Finally, now that we have an overview of the EDM field, we focus in the following on examples of its applications in computer-based educational systems. We particularly focus in behavior analysis and performance prediction and assessment.

1.3 Examples of EDM Applications in Computer-Based Learning Environments

In the last years, a wide number of EDM applications have been developed as seen in the previous sections. There are applications dealing with the assessment of students' learning performance, course adaptation and learning recommendations based on the student's learning behavior, evaluation of learning material and

[4] http://www.cs.waikato.ac.nz/ml/weka/ visited on August 6, 2013.

[5] http://www.r-project.org visited on August 6, 2013.

Table 1.1 Some tools for EDM applications

Tool	Goal	References
MMT tool	To facilitate the execution of all data mining steps in Moodle data	[2]
AAT	To analyze student's behavior in learning systems	[35]
E-learning web miner	To discover student's behavior profile in virtual course	[36]
SNAPP	To display the relation evolution between participants in discussion forums	[37]
EDM visualization tool	To visualize the process in which students solve procedural problems in logic	[38]
DRAL	To discover relevant e-activities for learners	[39]
IDLS	To analyze learner's browsing behavior and learning styles	[40]
eLAT	To explore and correlate learning object usage, user properties, user behavior, as well as assessment results based on graphical indicators	[41]
Meerkat-ED	To analyze interactions of students in asynchronous discussion forums of online courses	[28]
PSL-datashop	To store and analyze click-stream data, fine-grained longitudinal data generated by educational systems	[42]
MULCE		[43]

web-based courses, providing feedback to both teacher and students in e-learning courses, and detection of students' learning behaviors [21].

A review of these studies can be found in [9, 10, 12, 21, 44, 45]. Through these studies we noticed that the current mainstream EDM research is primarily focused on mining logs generated by the e-learning systems [13, 21]. We also found that the oldest and the most popular applications are the prediction of the student's performance and the analysis of learning behavior.

The term 'prediction' is generally used to characterize models (based on EDM techniques) designed for predicting new outcomes or scenarios based on new observations. Prediction is different from 'explanation', where the goal is to build models that explain underlying causal structure and to assess the explanatory power of such models [46]. This term is then linked to the study of the learner's behavior.

In the following, we present the most recent EDM studies from 2010 to 2013 related to these main objectives: learner's performance and behaviors in computer-based learning environments.

1.3.1 EDM Applications for Predicting and Evaluating Learning Performance

In this subsection, we analyze the current state of EDM research in learners' performance in CBLE. Table 1.2 summarizes some of the recent reviewed researches.

We note that the majority of the studied works applied EDM in LMS and ITS. In addition, as collaboration activities are often part of LMS, some studies treated collaborative data in LMS [23, 48, 56], while others [28, 55] analyzed collaboration usage data coming from a devoted environment.

The most tested LMS is Moodle. In [47] Jovanovica et al. applied classification models for predicting students' performance, and cluster models for grouping students based on their cognitive styles in Moodle. They developed a Moodle module that allows automatic extraction of data needed for educational data mining analysis and deploys models developed in this study. They indicate that the classification models helped teachers, students and business people, for early engaging with students who are likely to become excellent on a selected topic.

Furthermore, they indicate that clustering students based on cognitive styles and their overall performance enable better adaption of the learning materials with respect to their cognitive styles. Along the same lines, Falakmasir and Jafar [48] applied data mining methods (Feature Selection, decision trees) to the web usage records of students' activities in Moodle. As a result, they were able to identify and rank the students activities based on their impact on the performance of students in final exams/grades. Their findings suggest that students' participation in virtual classrooms had the greatest impact on their final grades.

Table 1.2 Some EDM applications for predicting the learner's performance in CBLE

References	CBLE	Data	Method	Goals
[47]	LMS: moodle	Data related to students' behaviors and cognitive style	Classification	To predict students' performance, and to cluster models for grouping students based on their cognitive styles
[48]	LMS: moodle	Web usage logs and final marks	Feature selection, classification (decision tree: C4.5 algorithm)	To identify and rank the students' activities based on their impact on the performance of students in final exams
[49]	LMS: moodle	Students' usage data in several courses	Statistical and classification methods (decision trees, rule and fuzzy rule induction methods, and neural network)	To predict the marks that university students will obtain in the final exam of a course
[23]	LMS: moodle	Students' usage data related to assignments and forums	Association rule mining algorithms	To evaluate the relation/influence between the on-line activities and the students' final marks
[46]	LMS: Sakai	Students' demographic data and Sakai log data of individual course events	Campbell's approach (factor analysis and logistic regression) classification, feature selection	To develop predictive models of students' success
[50]	Python tutor	Students' submissions and responses to satisfaction questionnaire	Statistics on the analyzed variables (means, average, etc.)	To generate personalized feedback and hints
[51]	Lecture capture system (LCS)	LCS log data combined with collected survey data	Statistics on the analyzed variables	To predict students' chance of passing exams
[32]	Computer-aided-tutoring systems	Log data of interactions (activities, success and progress indicators)	Matrix factorization	To predict student's performance
[52]	ITS: ASSiSTment tutor	Students' responses	Bayesian networks	To predict student's performance by modeling skills at a fine-grained level

(continued)

Table 1.2 (continued)

References	CBLE	Data	Method	Goals
[30]	ITS: ASSiSTment tutor	Student's first response to a question in the tutor	Bayesian knowledge tracing, performance factors analysis, correct first attempt rate, straightforward averaging, regression AdaBoost, neural networks, random forest (random decision trees)	To improve students' performance prediction with ensemble methods
[53]	ITS: ASSiSTment tutor	Data related to test scores	Clustering (spectral clustering, k-means) bootstrap aggregation ensemble method	To improve students' performance prediction and to introduce the utility of using spectral clustering
[54]	ITS: algebra and bridge to algebra	Students' answers to questions	Classification (neural network, KNN) and matrix factorization	To predict a student's ability to answer questions correctly based on historic results
[31]	Intelligent learning environments	Real data (responses to question items) and simulated data generated by probability matrix	Non-negative matrix factorization (NMF),visualization techniques	To improve the mapping of skills to question items in order to assess the student's mastery level
[55]	Forums manuel	Forum participation	Classification via clustering	To predict the student's final marks
[28]	Computer-supported collaborative learning (CSCL)	Interactions in asynchronous discussion forums	Social network analysis	To assess students' collaborations and course participation and facilitate fairer evaluation of these participation in online courses
[56]	A web-based discussion board	Asynchronous web discussion	Statistical techniques (variable analysis: ANOVA test, a post hoc least significant difference (LSD) analysis)	To evaluate to what degree the different types of web-based discussion affected students' language production performance

Romero et al. [49] fulfill trials and demonstrated how web usage mining can be applied in the Moodle e-learning system to predict the marks that university students will obtain in the final exam of a course. They also identified several avenues for using classification in educational settings: discovering student groups with similar characteristics, identifying learners with low motivations, proposing remedial actions, predicting and classifying students using intelligent tutoring systems. In the same way authors in [23] studied student's usage data from a Moodle system related to quizzes, assignments and forum activities to evaluate the relation/influence between the on-line activities and the final mark obtained by the students. They used several association rule mining algorithms. The discovered rules predict students' exam results (fail or pass) based on their frequent activities and can also help the instructor to detect infrequent students' behaviors/activities. In [46] Lauria et al. used another LMS: Sakai. They used demographic data and the LMS log data of individual course events to develop a predictive model of student success. They used many EDM methods (factor analysis and logistic regression, C4/5/C5.0 decision trees, support vector machine (SVM) classifiers, Bayesian network) to build data mining models that can help predict students' performance and take corrective actions in higher education institutions.

Regarding ITS studies, Dominguez et al. [50] created a system to generate personalized feedback and hints by mining the student data collected by Python Tutor, an online learning system. They found that students who used the hinting system achieved significantly better results than those who did not, and stayed active on the site longer. Gorissen et al. [51] analyzed the interactions of students with the recorded lectures using educational data mining techniques. They found discrepancies as well as similarities between students' verbal reports and actual usage as logged by the recorded lecture servers. The data suggests that students who do this have a significantly higher chance of passing the exams [3]. Thai-Nghe et al. [32] analyzed students' interactions log files to build success and progress indicators in order to predict students' performance using matrix factorization.

In [30, 52, 53] the authors carried out several experiments using data related to test scores and students' responses on the ASSiSTment tutor. They applied many EDM methods (classification, clustering, Knowledge Tracing, etc.) to improve student's performance prediction. Toescher and Jahrer [54] analyzed students answering questions from two ITS: Algebra and Bridge to Algebra. They used a set of collaborative filtering techniques adopted from the field of recommender systems (ex. matrix factorization), to predict a student's ability to answer questions correctly, based on historic results. Similarly, Desmarais [31] used Non-negative Matrix Factorization on students' scores to determine the skills required for a given question, and how strong different students are for these skills.

Regarding collaboration, López et al. [55] used classification and clustering to predict students' final marks from their participation in forums. In the same way, Rabbany et al. [28] analyzed students' interactions in forum asynchronous discussion of online courses using Social Network Analysis to facilitate fairer evaluation of students' participation in online courses. They also proposed

Meerkat-ED, a specific, practical and interactive toolbox for analyzing students' interactions in asynchronous discussion forums.

In [56] Chang et al. used a web-based discussion board provided by an online educational platform to analyze students' language production. They used statistical techniques (ANOVA test, least significant difference (LSD) analysis) to evaluate to what degree the different types of web-based discussion affected students' language production performance.

Regarding the analyzed data, the majority of these studies used students' question responses since their general goal was the prediction of learners' success in the final exam, based on their responses to previous tests [31, 52, 54] or previous attempts [30]. However, some studies also used interaction traces [32, 48], communication traces [28, 56], and responses to satisfaction questionnaire [50] or combined between several types of data such as in [23, 44, 48, 51]. Another approach that we found in [31] was to generate simulated data using a probability matrix in order to test several models.

We also notice that the data set size is very variable: from 27 [56] to 4,927 [51] participants producing data over several hours, weeks, months or years attending a big data set size (over 20 millions in [54]). This is related to the context of the study and the data origin: data collected in experiments or dataset already available and used in previous experiments (e.g. the ASSiSTment tutor). This second alternative facilitates the analysis by avoiding the collection step, often not very obvious. Moreover, it even allows testing several methods and environment as in [30].

We also note that in these studies, the used tools are often not mentioned or are DM tools (Pentaho[6] in [46], RapidMiner[7], R in [31]) except for Rabbany et al. [28] who proposed their own tool (Meerkat-ED). Concerning the used method, clustering and prediction (classification) are on the top of the implemented techniques. However, several studies addressed the use of other techniques such as text mining, sequential pattern, SNA and matrix factorization. Statistical methods as well are used in many studies not only during the treatment phase of the EDM process but also during the preprocessing step where it is often difficult to choice the adequate features to use among the available data. For instance, in [46] "Feature Selection" is used to select the relevant attributes to use.

Through this study, we note that the application of different techniques of EDM allowed to identify the learner's performance (usually measured as the success of the learner in the final exam), from simple data (previous results or question answers, participation in collaborative activities, productions, etc.).

This can be exploited to improve the learning systems in different ways. For instance, if the majority of learners have low performance on a resource, it could hint to the fact that the course resource and/or the learning material are inadequate and therefore should be changed and/or improved.

---

[6] http://www.pentaho.com visited on August 6, 2013.

[7] http://rapid-i.com visited on August 6, 2013.

Some reviewed studies have discussed some of these results that contribute to a better adaptation and personalization of CBLE: improving adaptation based on cognitive styles in [47], classifying the learners' activities according to their influence on the performance in [48], identifying the required skills for a learning resource in [31].

Other studies also discussed results contributing to help the educators in their tutoring and assistance task: identifying learners with little motivation as in [49], grouping students based their characteristics in [48, 49], detecting infrequent behaviors in [23], etc. Thus we think that these results should be used to improve and adapt the content and the organization of the learning materials in CBLE, and could be used to the advantage of all the actors involved in the learning process.

1.3.2 EDM Applications for Analyzing Learners' Behaviors

In this subsection, we analyze the current state of EDM researches for analyzing (identifying, explaining, etc.) learners' behaviors in computer-based learning environments. Table 1.3 summarizes some of the recent reviewed research.

Among the reviewed studies we found three that belongs to LMS. Krüger et al. [34] aimed to build a data model to ease analysis and mining of educational data. To experiment their model, they analyzed the data stored in the "Programming 1" course in the Moodle LMS to study learners' behaviors related to solving self-evaluation exercises using association rule. They found that as the semester progresses, less students solve them. Macfadyen and Dawson [16] made an analysis of LMS racking data from a Blackboard Vista-supported course. The goal was to explain the variation in students' final grades. Using regression, they found significant correlation between the students' final grades and their learning behaviors on the LMS, based on key variables such as the total number of discussion messages posted, and the number of assessments completed.

In [40] Bousbia et al. aimed to automatically identify the learner's behavior and learning style, based on navigation trace analysis in a web-based learning environment: the eFAD LMS. They defined four browsing behaviors using a decision tree and carried out experiments using statistical techniques and machine learning classifiers (C4.5 decision tree, KNN, Bayesian networks, and neural networks).

Learner's behavior is also studied in other types of CBLE. Peckham and McCalla [22] carried out an experiment in a learning environment designed to emulate hypermedia courses to identify patterns of students' behaviors in a reading comprehension task using EDM techniques (k-means clustering, and ANOVA test).

Desmarais and Lemieux [25] also aimed to better understand the patterns of use of a learning environment. They applied clustering and activity sequence visualization on gathered logs of learners' interactions in a self-regulated web based drill and practice learning environment.

Table 1.3 Some EDM applications for analyzing learners' behaviors in CBLE

References CBLE	Data	Method	Goals
[34] LMS: moodle	Students' attempts for doing an exercise	Association rule	Analyze how students use learning resources in a course
[16] LMS: blackboard	Log data related to asynchronous discussion and assessments	Regression	Explain the variation in students' final grades
[40] LMS: eFAD	Log of navigation interaction	Classification: k-means, neural network, naïve bayes	Analyze learner's browsing behaviors styles
[22] Learning environment designed to emulate hypermedia courses	Log data (events, actions)	K-means clustering, ANOVA	Identify students' behavior patterns in a reading comprehension task
[25] Self-regulated web based drill and practice environment	Logs of learners' interactions (answers to exercises, actions, events)	Clustering, classification, activity sequence, visualization	Understand the patterns of use of a learning environment
[3] A live video streaming environment	Students' online interaction	Text mining	Examine students' online interaction
[57] AutoTutor a dialogue tutor. The incredible machine TIM: a problem solving game), and Aplusix. A problem-solving based ITS	Study 1: standard pre-test intervention post-test, video of participants and of the content of their screen. Study 2 and 3: students' observation made by observers	Human judgments about the same set of six cognitive affective states, ANOVA test	Study the incidence, persistence, and impact of students' cognitive-affective states
[26] MetaTutor, an agent-based ITS	Trace data about tests	Clustering: expectation–maximization, sequence mining	Identify distinguishing patterns of behaviors
[58] Betty's brain: learning-by-teaching environment	Students' learning interactions as they teach an agent.	Sequence mining techniques, linear segmentation algorithm	Evaluate and compare students' learning behaviors

In [3] a live video streaming (LVS) system was used to study the students' patterns using data mining and text mining applied on data of online interaction. Bouchet et al. [26] analyzed students' characteristics and learning behaviors in MetaTutor, an agent-based ITS. They used clustering and sequence mining to distinguish patterns of behaviors. Similarly, Kinnebrew and Biswas [58] used sequence mining to identify learning behaviors in Betty's Brain, a learning-by-teaching environment.

In [57] Baker et al. carried out three studies in three CBLE: AutoTutor (a dialogue tutor), the incredible machine (TIM) (a problem solving game: a simulation environment), and Aplusix (a problem-solving based ITS). The studied data were pre-test–intervention–post-test, and video records of the participants and their computer screen in the first study, and observation made by observers related to cognitive affective states on the second and third studies. Using Human judgment and ANOVA test, the authors found that boredom was very persistent across learning environments and was associated with poorer learning and problem behaviors, such as gaming the system. Also, confusion and engaged concentration were the most common states within the three learning environments. These findings suggest that significant effort should be put into detecting and responding to boredom and confusion.

Throughout Table 1.3, we notice that all the reviewed studies used interaction traces, which are generally of low level (action/event) or specific to the analyzed activity (messages, reading task, etc.). These dataset are often structured in numerical attributes were task scores or statistics on log data (frequencies of actions, time spent in actions) are the most used. Moreover, the sample size used in these studies is less variable since almost all these works are based on experiments (from 28 in [22] to 148 participants in [26]). We note that the used tools for analysis are often not mentioned in these studies. The two mentioned ones are Weka in [26] and TraMiner-R in [25].

Regarding the used methods, clustering and classification still on the top as the majority of the presented works aim to identify common learning behaviors. Other methods were also used such as text mining, sequence mining, statistical methods (e.g. to calculate some variables) as well as Human judgments when the analyzed data referred to personal characteristics.

Thus, through the reviewed studies presented here, we find that EDM allows from low level traces to analyze and evaluate the student's behavior. This task is often a difficult one given the close relationships of the behavior to personal characteristics such as learning styles, emotions and its frequent changes according to the learner's state, the learning time, the type and the content of the learning materials, the learner's reaction to other actors, etc. Thus, behavioral analysis should be done in real time to provide a better feedback to teachers as well as learners in order to improve the tutoring and learning tracking tasks. This is still difficult even with the use of EDM techniques regarding the small size of the analyzed samples which does not allow the generalization of the obtained results that remain specific to the studied environments and the context of the carried out experiments.

However, even if the majority of research, such as those presented here, focus on the analysis of the past behaviors to explain a phenomena such as abandon, or evaluate the participation and the obtained results, their findings should be used to improve learning environments based on the students' behavioral patterns.

1.3.3 Discussion

The 25 reviewed studies presented in this section give an overview of the typical educational environment, data and methods used in EDM applications. LMS or generally online educational environment and ITS are the most exploited. This is probably due to their wide use in the educational environment, which facilitates the realization of experiments, which is often the data collection source of these studies.

The analyzed data are generally related to assessments (tests, quizzes, exams, etc.), fine grained online interaction (action, event) and also participation in collaboration activities. This data type is related to the type of the studied CBLE that provide such information in their database and log files, as well as the sighted objectives of these studies. However, some researches combined these data with the video recording of learners or the human manual observations during the learning sessions. This combination, although difficult to achieve, can refine the study especially in CBLE where there is less face to face interactions between the teacher and the learner.

Regarding EDM methods, the most used were prediction (classification, association rules, and regression) and clustering. This finding can be explained by the two objectives studied: learner's performance and behavior, and by the fact that these techniques are mature, widely known, tested and implemented in the DM used tools, and also provide satisfactory results even with small sample size, we often find here.

Other methods were also used, according to the analyzed data and the objective to exploit other techniques and improve the results. Note, however, that it is not easy to identify for a given CBLE type, a given type and size of data set, and a given goal, which is the best EDM technique to use. Certainly this information helps to establish a choice, but it does not limit or confirm that this is the best one.

This observation explains why in several studies several techniques were used to achieve the best results. Note that we did not discuss in this chapter the percentage of the obtained results, since they depend on the different context of the studied works (types of environment and the analyzed data, the student populations, the set parameters and hypothesis during the EDM process, etc.). Indeed, although the obtained results are generally satisfactory in their context, they remain is an experimental stage and cannot be generalized.

Finally, we note that although we focused on the study of EDM applications related to two main objectives, namely the prediction of performance and behavior analysis, the results of the presented research achieved other EDM objectives (student modeling, communicating to stakeholders, maintaining and improving courses, etc.). In addition, as the two studied objectives are closely related, we found studies dealing them both where EDM techniques were applied to explore the relationships between the learner's behavior and the learning performance, to improve the learning environment.

For example, in [59] learners' behaviors is used to predict the success or the failure of students without requiring the results of formal assessments. In [60] Bayer et al. focused on predicting drop-outs and school failures when students' data have been enriched with data derived from students' social behaviors.

We think that this last objective of analyzing the reasons of failure, drop-out and abandon is a promoter research area in the EDM field that should to be exploited in the future, especially for CBLE, where it is a common phenomenon. We also believe that the data collected from these environments should be enriched by other types of information such as demographic data, to provide a better explanation of the observed phenomena. An effort should also be provided to share the analyzed samples and provide significant benchmark to pass the experimental stage in order to generalize the established models and the results found in these studies to improve the learning environments.

For the same goal, it is required to improve EDM tools. In fact, although DM tools allow the analysis, they require some expertise to set the parameters and make the appropriate interpretation. It is therefore necessary that EDM have their own tools to make these techniques within the reach of teachers, and allow more advanced treatment combining multiple data sources, and proposing some methods according to the type of these data and the analyze goal. So we can imagine these tools included in learning environments to facilitate their access to the different learning actors.

1.4 Conclusions

In this chapter, we discussed the use of EDM in educational systems. We studied recent EDM applications (2010–2013) by taking into account: the educational system, the analyzed data, the used method for the analysis, the used tool, and the analysis goal, especially in computer based learning environments.

We noticed through this study that a large number of researches are interested today in the application of EDM in educational systems in general and in CBLE in particular, to exploit the available data or the one that can be collected in these environments to ensure their improvement through the various objectives. We can say that EDM introduces a major advantage, drawn from data mining and KDD fields, the one related to extract hidden information about learners and learning from recorded data.

We have reviewed, in some detail, recent research dealing with students' performance and behaviors. We found that the use of EDM methods helps the prediction of students' performance; especially final marks. It also helps to identify and explain usual and unusual learning behaviors that should facilitate the assistance of learners, and reduce the costs of educative personalization and adaptation processes.

However, these contributions have to go out of laboratories to be applied in the used educational systems in order to improve learning. Studies in this goal are initiated especially in traditional educational environments where the results of the application of EDM on existent data are used to improve the educational system [5].

We expect however, that this goal will be also applied in software educational systems, to find new ways to improve learning materials and reduce the abandon rate that is considerable in such environments. In fact, to make this area more mature, it is necessary that the established models in these studies could be tested in real environments for frequent use to affirm and exploit the found results to improve these environments.

A first step in this direction is the sharing and the reuse of the dataset through open data repositories and standard data formats to promote the exchange of data and models. It is also necessary to popularize the use of EDM through the popularization of tools targeted to the different learning actors for the analysis of educational data in a simple and intuitive way, while providing suggestion about methods to apply for a better result and facilitating the interpretation of these results. We think that it is also necessary to take into account the EDM process in the overall development process of the computer based learning environment to ensure a significant improvement.

References

1. Baker, R.S.J.d.: Data mining for education. In: McGaw, B., Peterson, P., Baker, E. (eds.) International Encyclopedia of Education, vol. 7, 3rd edn., pp. 112–118. Elsevier, Amsterdam (2010)
2. Pedraza-Perez, R., Romero, C., Ventura, S.: A java desktop tool for mining moodle data. In: Pechenizkiy, M., Calders, T., Conati, C., Ventura, S., Romero, C., Stamper, J. (eds.) Proceedings of 4th International Conference on Educational Data Mining, pp. 319–320. International Educational Data Mining Society, Eindhoven (2011)
3. He, W.: Examining students' online interaction in a live video streaming environment using data mining and text mining. Comput. Hum. Behav. **29**(1), 90–102 (2013)
4. Ayesha, S., Mustafa, T., Sattar, A., Khan, I.: Data mining model for higher education system. Eur. J. Sci. Res. **43**(1), 24–29 (2010)
5. Pal, S.: Mining educational data to reduce dropout rates of engineering students. Int. J. Inf. Eng. Electron. Bus. **2**(1), 1–7 (2012)
6. Parack, S., Zahid, Z., Merchant, F.: Application of data mining in educational databases for predicting academic trends and patterns. In: Proceedings of 2012 IEEE International Conference on Technology Enhanced Education, pp. 1–4. IEEE Press, Piscataway (2012)
7. Huebner, R.A.: A survey of educational data-mining research. Res. High. Educ. J. **19**, 1–13 (2013)
8. Calders, T., Pechenizkiy, M.: Introduction to the special section on educational data mining. ACM SIGKDD Explor. **13**(2), 3–6 (2011)

9. Romero, C., Ventura, S.: Educational data mining: a review of the state of the art. IEEE Trans. Syst. Man Cybern. Part C Appl. Rev. **40**(6), 601–618 (2010)
10. Romero, C., Ventura, S.: Data mining in education. Wiley Interdisc. Rev.: Data Min. Knowl. Discovery **3**(1), 12–27 (2013)
11. Chatti, M.A., Dyckhoff, A.L., Schroeder, U., Thüs, H.: A reference model for learning analytics. Int. J. Technol. Enhanced Learn. **4**(5–6), 318–331 (2012)
12. Bienkowski, M., Feng, M., Means, B.: Enhancing teaching and learning through educational data mining and learning analytics: an issue brief. US Department of Education, Office of Educational Technology, pp. 1–57 (2012)
13. Scheuer, O., McLaren, B.M.: Educational data mining. In: Seel, N.M. (eds.) Encyclopedia of the Sciences of Learning, pp. 1075–1079. Springer, US (2012)
14. Romero, C., Ventura, S., Pechenizkiy, M., Baker, R.S.J.d.: Introduction. In: Romero, C., Ventura, S., Pechenizkiy, M., Baker, R.S.J.d. (eds.) Handbook of Educational Data Mining, Chapman and Hall/CRC Data Mining and Knowledge Discovery Series, pp. 1–5. CRC Press, Boca Raton (2011)
15. Kotsiantis, S., Patriarcheas, K., Xenos, M.: A combinational incremental ensemble of classifiers as a technique for predicting students' performance in distance education. Knowl.-Based Syst. **23**(6), 529–535 (2010)
16. Macfayden, L.P., Dawson, S.: Mining LMS data to develop an ''early warning'' system for educators: a proof of concept. Comput. Educ. **54**(2), 588–599 (2010)
17. Romero, C., Zafra, A., Luna, J.M., Ventura, S.: Association rule mining using genetic programming to provide feedback to instructors from multiple-choice quiz data. Expert Syst. **30**(2), 162–172 (2013)
18. Anaya, A.R., Boticario, J.G.: Application of machine learning techniques to analyse student interactions and improve the collaboration process. Expert Syst. Appl. **38**, 1171–1181 (2011)
19. Siemens, G., Baker, R.S.J.d.: Learning analytics and educational data mining: towards communication and collaboration. In: Proceedings of 2nd International Conference on Learning Analytics and Knowledge, pp. 1–3. ACM, New York (2012)
20. Baker, R.J.D.F., Yacef, K.: The state of educational data mining in 2009: a review and future visions. J. Educ. Data Min. **1**(1), 3–17 (2009)
21. ALMazroui, Y.A.: A survey of data mining in the context of e-Learning. Int. J. Inf. Technol. Comput. Sci. **7**(3), 8–18 (2013)
22. Peckham, T., McCalla, G.: Mining student behavior patterns in reading comprehension tasks. In: Yacef, K., Zaïane, O., Hershkovitz, A., Yudelson, M., Stamper, J. (eds.) Proceedings of 5th International Conference on Educational Data Mining, pp. 87–94. International Educational Data Mining Society, Chania (2012)
23. Romero, C., Romero, J.R., Luna, J.M., Ventura, S.: Mining rare association rules from e-learning data. In: Baker, R.S.J.D., Merceron, A., Pavlik Jr., P.I. (eds.) Proceedings of 3rd International Conference on Educational Data Mining, pp. 171–180. International Educational Data Mining Society, Pittsburgh (2010)
24. Kock, M., Paramythis, A.: Activity sequence modeling and dynamic clustering for personalized e-learning. User Model. User-Adap. Inter. **21**(1–2), 51–97 (2011)
25. Desmarais, M.C., Lemieux, F.: Clustering and visualizing study state sequences. In: D'Mello, S.K., Calvo, R.A., Olney, A. (eds.) Proceedings of 6th International Conference on Educational Data Mining, pp. 224–227. International Educational Data Mining Society, Memphis (2013)
26. Bouchet, F., Azevedo, R., Kinnebrew, J.S., Biswas, G.: Identifying students' characteristic learning behaviors in an intelligent tutoring system fostering self regulated learning. In: Yacef, K., Zaïane, O., Hershkovitz, A., Yudelson, M., Stamper, J. (eds.) Proceedings of 5th International Conference on Educational Data Mining, pp. 65–72. International Educational Data Mining Society, Chania (2012)
27. Barahate. S.R.: Educational data mining as a trend of data mining in educational system. In: Proceedings of IJCA International Conference and Workshop on Emerging Trends in Technology, pp. 11–16 (2012)

28. Rabbany, R., Takaffoli, M., Zaïane, O.: Analyzing participation of students in online courses using social network analysis technique. In: Pechenizkiy, M., Calders, T., Conati, C., Ventura, S., Romero, C., Stamper, J. (eds.) Proceedings of 4th International Conference on Educational Data Mining, pp. 21–30. International Educational Data Mining Society, Eindhoven (2011)
29. Trčka, N., Pechenizkiy, M., Aalst W.v.d.: Process mining from educational data. In: Romero, C., Ventura, S., Pechenizkiy, M., Baker, R.S.J.d. (eds.) Proceedings of Handbook of Educational Data Mining, Chapman and Hall/CRC Data Mining and Knowledge Discovery Series, pp. 123–142. CRC Press, Boca Raton (2011)
30. Pardos, Z.A., Gowda, S.M., Baker, R.S.J.d., Heffernan, N.T.: The sum is greater than the parts: ensembling models of student knowledge in educational software. ACM SIGKDD Explor. **13**(2), 37–44 (2011)
31. Desmarais, M.C.: Mapping question items to skills with non-negative matrix factorization. ACM SIGKDD Explor. **13**(2), 30–36 (2011)
32. Thai-Nghe, N., Drumond, L., Krohn -Grimberghe, A., Schmidt-Thieme, L.: Recommender system for predicting student performance. Procedia Comput. Science **1**(2), 2811–2819 (2010)
33. Sachin, B.R., Vijay, S.M.: A survey and future vision of data mining in educational field. In: Proceedings of IEEE 2nd International Conference on Advanced Computing and Communication Technologies, pp. 96–100. ACM, New York (2012)
34. Krüger, A., Merceron, A., Wolf, B.: A data model to ease analysis and mining of educational data. In: Baker, R.S.J.D., Merceron, A., Pavlik Jr., P.I. (eds.) Proceedings of 3rd International Conference on Educational Data Mining, pp. 131–140. International Educational Data Mining Society, Pittsburgh (2010)
35. Graf, S., Ives, C., Rahman, N., Ferri, A.: AAT: a tool for accessing and analysing students' behaviour data in learning systems. In: Proceedings of 1st International Conference on Learning Analytics and Knowledge, pp. 174–179. ACM, New York (2011)
36. Zorrilla, M., Garcia-Saiz, D.: A service oriented architecture to provide data mining services for non-expert data miners. Decis. Support Syst. J. **55**(1), 399–411 (2013)
37. Bakharia, A., Dawson, S.: SNAPP: a bird's-eye view of temporal participant interaction. In: Proceedings of 1st International Conference on Learning Analytics and Knowledge, pp. 168–173. ACM, New York (2011)
38. Johnson, M., Barnes, T.: EDM visualization tool: watching students learn. In: Baker, R.S.J.D., Merceron, A., Pavlik Jr., P.I. (eds.) Proceedings of 3rd International Conference on Educational Data Mining, pp. 297–298. International EDM Society, Pittsburgh (2010)
39. Zafra, A., Romero, C., Ventura, S.: DRAL: a tool for discovering relevant e-activities for learners. Knowl. Inf. Syst. **36**(1), 211–250 (2013)
40. Bousbia, N., Rebaï, I., Labat, J.-M., Balla, A.: Learners' navigation behavior identification based on traces analysis. User Model. User-Adap. Inter. **20**(5), 455–494 (2010)
41. Dyckhoff, A.L., Zielke, D., Bültmann, M., Chatti, M.A., Schroeder, U.: Design and implementation of a learning analytics toolkit for teachers. Educ. Technol. Soc. **15**(3), 58–76 (2012)
42. Koedinger, K.R., Baker, R.S.J.d., Cunningham, K., Skogsholm, A., Leber, B., Stamper, J.: A data repository for the EDM community: the PSLC datashop. In: Romero, C., Ventura, S., Pechenizkiy, M., Baker, R.S.J.d. (eds.) Proceedings of Handbook of Educational Data Mining, Chapman and Hall/CRC Data Mining and Knowledge Discovery Series, pp. 43–55. CRC Press, Boca Raton (2011)
43. Reffay, C., Betbeder, M.-L., Chanier, T.: Multimodal learning and teaching corpora exchange: lessons learned in 5 years by the Mulce project. In: special issue on dataTEL: datasets and data supported learning in technology-enhanced learning. Int. J. Technol. Enhanced Learn. **4**(1–2), 11–30 (2012)
44. Kotsiantis, S.B.: Use of machine learning techniques for educational proposes: a decision support system for forecasting students' grades. Artif. Intell. Rev. **37**(4), 331–344 (2012)

45. Amershi, S., Conati, C.: Combining unsupervised and supervised classification to build user models for exploratory learning environments. J. Educ. Data Min. **1**(1), 18–71 (2009)
46. Lauria, E., Baron, J.: Mining Sakai to measure student performance: opportunities and challenges in academic. In: Proceedings of Enterprise Computing Community Conference (2011)
47. Jovanovica, M., Vukicevica, M., Milovanovica, M., Minovica, M.: Using data mining on student behavior and cognitive style data for improving e-learning systems: a case study. Int. J. Comput. Intell. Syst. **5**(3), 597–610 (2012)
48. Falakmasir, M., Jafar, H.: Using educational data mining methods to study the impact of virtual classroom in e-learning. In: Baker, R.S.J.D., Merceron, A., Pavlik Jr., P.I. (eds.) Proceedings of 3rd International Conference on Educational Data Mining, pp. 241–248. International Educational Data Mining Society, Pittsburgh (2010)
49. Romero, C., Espejo, P.G., Zafra, A., Romero, J.R., Ventura, S.: Web usage mining for predicting final marks of students that use moodle courses. Comput. Appl. Eng. Educ. J. **21**(1), 135–146 (2013)
50. Dominguez, A.K., Yacef, K., Curran, J.: Data mining to generate individualised feedback. In: Aleven, V., Kay, J., Mostow, J. (eds.) ITS 2010, Part II. LNCS, vol. 6095, pp. 303–305. Springer, Heidelberg (2010)
51. Gorissen, P., Bruggen, J., Jochems, W.: Usage reporting on recorded lectures using educational data mining. Int. J. Learn. Technol. **7**(1), 23–40 (2012)
52. Pardos, Z.A., Heffernan, N.T., Anderson, B.S., Heffernan, C.L.: Using fine-grained skill models to fit student performance with Bayesian networks. In: Romero, C., Ventura, S., Pechenizkiy, M., Baker, R.S.J.d. (eds.) Handbook of Educational Data Mining, Chapman and Hall/CRC Data Mining and Knowledge Discovery Series, pp. 417–426. CRC Press, Boca Raton (2011)
53. Trivedi, S., Pardos, Z.A., Sárközy, G.N., Heffernan, N.T.: Spectral clustering in educational data mining. In: Pechenizkiy, M., Calders, T., Conati, C., Ventura, S., Romero, C., Stamper, J. (eds.) Proceedings of 4th International Conference on Educational Data Mining, pp. 129–138. International Educational Data Mining Society, Eindhoven (2011)
54. Toescher, A., Jahrer, M.: Collaborative filtering applied to educational data mining. J. Mach. Learn. Res. (2010)
55. López, M.I., Luna, J.M., Romero, C., Ventura, S.: Classification via clustering for predicting final marks based on student participation in forums. In: Yacef, K., Zaïane, O., Hershkovitz, A., Yudelson, M., Stamper, J. (eds.) Proceedings of 5th International Conference on Educational Data Mining, pp. 148–151. International EDM Society, Chania (2012)
56. Chang, M.M., Lin, M.C., Tsai, M.J.: A study of enhanced structured web-based discussion in a foreign language learning class. Comput. Educ. **61**, 232–241 (2013)
57. Baker, R.S.Jd., D'Mello, S.K., Rodrigo, M.M.T., Graesser, A.C.: Better to be frustrated than bored: the incidence, persistence, and impact of learners' cognitive-affective states during interactions with three different computer-based learning environments. Int. J. Hum.-Comput. Stud. **68**(4), 223–241 (2010)
58. Kinnebrew, J.S., Biswas, G.: Identifying learning behaviors by contextualizing differential sequence mining with action features and performance evolution In: Yacef, K., Zaïane, O., Hershkovitz, A., Yudelson, M., Stamper, J. (eds.) Proceedings of 5th International Conference on Educational Data Mining, pp. 57–64. International EDM Society, Chania (2012)
59. McCuaig, J., Baldwin, J.: Identifying successful learners from interaction behaviour. In: Yacef, K., Zaïane, O., Hershkovitz, A., Yudelson, M., Stamper, J. (eds.) Proceedings of 5th International Conference on Educational Data Mining, pp. 160–163. International Educational Data Mining Society, Chania (2012)
60. Bayer, J., Bydzovska, H., Geryk, J., Obsıvac, T., Popelınsky, L.: Predicting dropout from social behaviour of students. In: Yacef, K., Zaïane, O., Hershkovitz, A., Yudelson, M., Stamper, J. (eds.) Proceedings of 5th International Conference on Educational Data Mining, pp. 103–109. International Educational Data Mining Society, Chania (2012)

Chapter 2
A Survey on Pre-Processing Educational Data

Cristóbal Romero, José Raúl Romero and Sebastián Ventura

Abstract Data pre-processing is the first step in any data mining process, being one of the most important but less studied tasks in educational data mining research. Pre-processing allows transforming the available raw educational data into a suitable format ready to be used by a data mining algorithm for solving a specific educational problem. However, most of the authors rarely describe this important step or only provide a few works focused on the pre-processing of data. In order to solve the lack of specific references about this topic, this paper specifically surveys the task of preparing educational data. Firstly, it describes different types of educational environments and the data they provide. Then, it shows the main tasks and issues in the pre-processing of educational data, Moodle data being mainly used in the examples. Next, it describes some general and specific pre-processing tools and finally, some conclusions and future research lines are outlined.

Keywords Educational data mining process · Data pre-processing · Data preparation · Data transformation

Abbreviations

AIHS Adaptive and intelligent hypermedia system
ARFF Attribute-relation File Format
CBE Computer-based education
CSV Comma-separated values

C. Romero (✉) · J. R. Romero · S. Ventura
Department of Computer Science and Numerical Analysis, University of Córdoba
Campus de Rabanales, Edificio C2-Albert Einstein, Córdoba, Spain
e-mail: cromero@uco.es

J. R. Romero
e-mail: jrromero@uco.es

S. Ventura
e-mail: sventura@uco.es

A. Peña-Ayala (ed.), *Educational Data Mining*,
Studies in Computational Intelligence 524, DOI: 10.1007/978-3-319-02738-8_2,

DM	Data mining
EDM	Educational data mining
HTML	Hypertext Markup language
ID	Identifier
IP	Internet Protocol
ITS	Intelligent tutoring system
KDD	Knowledge discovery in databases
LMS	Learning management system
MCQ	Multiple choice question
MIS	Management information system
MOOC	Massive Open Online Course
OLAP	Online Analytical Processing
SQL	Structured Query Language
WUM	Web Usage Mining
WWW	World Wide Web
XML	Extensible Markup Language

2.1 Introduction

Educational Data Mining (EDM) is a field that exploits Data Mining (DM) algorithms in different types of educational data in order to resolve educational research issues [1]. Data mining or Knowledge Discovery in Data-bases (KDD) is the automatic extraction of implicit and interesting patterns from large data collections [2]. The first step in the KDD process is the transformation of data into an appropriate form for the mining process, which is usually called data pre-processing in data mining systems [3]. It allows raw data to be transformed into a shape suitable for resolving a problem using a specific mining method, technique or algorithm [4]. In fact, the better raw data are pre-processed, the more useful information is possible to discover. However, the data pre-processing phase typically requires a significant amount of manual work, this phase coming to consume 60–90 % of the time, efforts and resources employed in the whole knowledge discovery process [5]. In particular, educational environments store a huge amount of potential mining data (raw, original or primary data) but often the data available to solve a problem are not in the most appropriate form (or abstraction), that is, the discovered models are not useful.

For example, obtaining a model with too many rules only containing very low level attributes would not be of interest to the instructor since it would not indicate how to improve the course. To resolve this difficulty it is necessary to pre-process data. It is often considered that once you have the correct transformation of the data (modified data), the problem is almost solved [6] and it is well known that the success of every data mining algorithm/technique and the resulting or discovered model/pattern are strongly dependent on the quality of the data used.

Data pre-processing in educational context is considered the most crucial phase in the whole educational data mining process [7], and it can take more than half of the total time spent in solving the data mining problem [3]. EDM users (such as an instructor, teacher, course administrator, academic staff, etc.) have to apply the most appropriate data pre-processing techniques for a particular data set and purpose. Thus, it is necessary that EDM users actively participate in the whole pre-processing process in order to select the pre-processing steps/tasks to be done and to decide how they should be ordered. Classical Web Usage Mining (WUM) pre-processing techniques, originally targeted at e-commerce, can be used in most cases, but new approaches more related to learning environments are required to reach interesting results [8].

There are some other special issues concerning educational data, e.g. data integration from multiple sources, integration of data with different granularities, etc. Thus, in the specific case of educational data, for example, the large number of attributes collected with information about each student can be reduced and summarized in a table for a better analysis with multi-relational analysis methods; attributes can be re-represented in binary representation whenever it is appropriate to allow association rule analysis; continuous attributes can be discretized to categorical attributes to improve the comprehensibility of data, etc. However, to our knowledge there are very few previous works exclusively focused on the pre-processing of educational data [7, 9, 10]. Therefore, in order to fill the gap of specific references about this important topic, this paper surveys the pre-processing task of educational data.

Our main goal is to survey different issues on data pre-processing to provide a guide or tutorial for educators and EDM practitioners.

Throughout this paper, Moodle is used as a coherent framework of pre-processing, and data extracted from Moodle learning management system [11] have served as case under study in most examples. The paper is organized as follows: Sect. 2.2 shows the different types of educational environments, whilst Sect. 2.3 discusses the different type of data they provide. Section 2.4 describes the main tasks and issues involved in the pre-processing of educational data. Section 2.5 lists most of the currently existing general and specific data pre-processing tools. Finally, some conclusions and further research are outlined in Sect. 2.6.

2.2 Types of Educational Environments

Traditional education or back-to-basics refers to long-established customs found in schools that society has traditionally deemed to be appropriate.

These environments are the most widely-used educational system, based mainly on face-to-face contact between educators and students that is organized through lectures, class discussion, small groups, individual seat work, etc. These systems gather information on student attendance, marks, curriculum goals, and individualized plan data. Also, educational institutions store many diverse and varied sources of information [12] such as administrative data in traditional databases (student's information, educator's information, class and schedule information, etc.). In conventional face-to-face classrooms, educators may attempt to enhance instruction by monitoring students' learning processes and analyzing their performance on paper and through observation. But with the increasing use of computers as educational tools, it is much easier for instructors to monitor and analyze students' behavior starting from their usage information.

Computer-Based Education (CBE) means using computers in education to provide guidance, to instruct or to manage instructions to the student. CBE systems were originally stand-alone educational applications that ran on a local computer without using artificial intelligence techniques. However, both the global use of Internet has led to today's plethora of new Web-based educational systems, together with artificial intelligence techniques has induced the emergence of new educational systems such as: learning management systems, intelligent tutoring systems, massive open online courses, etc. Each one of them provides very different data sources that have to be pre-processed in different ways depending on both the nature of available data and the specific problems and tasks to be resolved by DM techniques.

2.2.1 Learning Management Systems

Learning Management Systems (LMS) are a special type of Web-based educational platform for the administration, documentation, tracking, and reporting of training programs, classroom and online events, e-learning programs, and training content. They also offer a great variety of channels and workspaces to facilitate information-sharing and communication among all the participants in a course. Some examples of commercial LMSs are Blackboard and Virtual-U, while some examples of free LMS are Moodle, Ilias, Sakai and Claroline.

These systems accumulate massive log data with respect to students' activities and usually have built-in student tracking tools that enable the instructor to view statistical data [13]. They can record any student activities involved, such as reading, writing, taking tests, performing various tasks in real or virtual environments, and commenting on events with peers. LMSs normally also provide a relational database that stores all student information in different tables such as: personal user information (profile), academic results (grades), and the user's interaction data (reports).

2.2.2 Massive Open Online Courses

Massive Open Online Courses (MOOC) are growing substantially in numbers, and also in interest from the educational community [14]. MOOC is an online course aimed at large-scale interactive participation and open access via the Web that made it possible for anyone with an internet connection to enroll in free, university level courses. Some examples of MOOCs are Udacity, MITx, EdX, Coursera and Udemy. MOOCs store very similar student's usage information than LMSs but from thousands or hundreds of students. Thus, they also generate large amounts of data that makes necessary the use of data mining techniques to process and analyze it.

2.2.3 Intelligent Tutoring Systems

Intelligent Tutoring Systems (ITS) are systems that provide direct customized instruction or feedback to students. An ITS models student behavior and changes its mode of interaction with each student based on its individual model [15]. The ability of ITS to log and pool detailed, longitudinal interactions with large numbers of students can create huge educational data sets [16]. Although ITSs record all student-tutor interaction in log files or databases, there are some other data stores available within an ITS, for example, a domain model that incorporates a set of constraints relevant to the tutor's domain, a pedagogical data set that contains a set of problems and their answers, and a student model that stores information about each student with respect to all the constraints, satisfactions, and violations recorded.

2.2.4 Adaptive and Intelligent Hypermedia Systems

Adaptive and Intelligent Hypermedia Systems (AIHS) are one of the first and most popular kinds of adaptive hypermedia and provide an alternative to the traditional just-put-it-on-the-Web approach in the development of educational courseware [17]. They attempt to be more adaptive by building a model of the goals, preferences, and knowledge of each individual student and using this model throughout the interaction with the student in order to adapt to the needs of that student. The data coming from these systems is semantically richer and can lead to a more diagnostic analysis than data from traditional Web-based education systems [18]. In fact, the data available from AIHs are similar to ITS data; that is, AIHs store data about the domain model, student model and interaction log files (traditional Web log files or specific log files).

2.2.5 Test and Quiz Systems

Test and quiz systems are among the most widely used and well-developed tools in education. A test is an instrument consisting of a series of questions/items and other prompts for the purpose of gathering information from respondents.

The main goal of these systems is to measure the students' level of knowledge with respect to one or more concepts or subjects. There are different types of questions/items [19] such as: yes/no questions, multiple choice questions (MCQ), fill-in questions, open-ended answered questions, etc. Test systems store a great deal of information, such as questions, students' answers, calculated scores, and statistics.

2.2.6 *Other Types of Educational Systems*

There are also other types of educational environments, such as: educational game environments, virtual reality environments, ubiquitous computing environments, learning object repositories, wikis, forums, blogs, etc.

2.3 Types of Data

Most of the data provided by each of the above-mentioned educational environments are different, thus enabling different educational problems to be resolved using data mining techniques. In fact, they have conceptually different types of data that can be grouped in the next main types showed in Table 2.1.

2.3.1 *Relational Data*

Relational databases/data sets are one of the most commonly available and richest information repositories. A relational database is a collection of tables, and each is assigned a unique name. Each table consists of a set of attributes (columns or fields) and usually stores a large set of tuples (records or rows). Each tuple in a relational table represents an object identified by a unique key and described by a set of attribute values [2]. Relational data can be accessed by database queries

Table 2.1 Different types of data and DM techniques

Type of data	DM technique
Relational data	Relational data mining
Transactional data	Classification, clustering, association rule mining, etc.
Temporal, sequence and time series data	Sequential data mining
Text data	Text mining
Multimedia data	Multimedia data mining
World Wide Web data	Web content/structure/usage mining

Table 2.2 Some important Moodle database tables about student interaction

Name	Description
mdl_user	Information about all the users
mdl_user_students	Information about all students
mdl_log	Logs every user's action
mdl_assignement	Information about each assignment
mdl_assignment_submissions	Information about assignments submitted
mdl_forum	Information about all forums
mdl_forum_posts	Stores all posts to the forums
mdl_forum_discussions	Stores all forum discussions
mdl_message	Stores all the current messages
mdl_ message_reads	Stores all the read messages
mdl_quiz	Information about all quizzes
mdl_quiz_attempts	Stores various attempts at a quiz
mdl_quiz_grades	Stores the final quiz grade

written in a relational query language, such as Structured Query Language (SQL), or with the assistance of graphical user interfaces.

Moodle uses a relational database with a great number of tables (all their names start with mdl_ followed by a descriptive word) and the relationships between them. However, it is not necessary to take them all into account at a glance. For example, there are some tables called mdl_quiz_something. If the quiz module is the object of interest, then it is obviously necessary to understand these tables. But if the quiz module does not interest us, it can be ignored. The same is true for each activity module. Table 2.2 shows some examples of the most important Moodle tables from the point of student's usage information.

Relational data mining is the data mining technique used in relational databases. Unlike traditional data mining algorithms, which look for patterns in a single table (propositional patterns), relational data mining algorithms look for patterns among multiple tables (relational patterns). In fact, for most types of propositional patterns, there are corresponding relational patterns such as relational classification rules, relational regression trees, relational association rules, and so on.

In the area of EDM, relational data mining has been used, for example, to find association rules about student behavior in e-learning [20]. However, relational data are normally transformed into transaction data before the data mining is done [21].

2.3.2 Transactional Data

A transactional database/data set consists of a file/table where each record/row represents a transaction. A transaction typically includes a unique transaction identity number and a list of the items making up the transaction [2].

In our case, Moodle does not provide directly any transactional database or data set in itself. However, transactional data can be derived in Moodle, though it is not explicitly stored in its database. In fact, this chapter explains how to create a transactional summary table (Table 2.5 and Fig. 2.10) starting from some relational database tables (see Table 2.2) which contain student usage information on Moodle activities.

A great number of data mining methods can be applied over this type of data. In fact, most of the well-known and traditional data mining techniques, such as classification, clustering and association rule mining, work with this type of data. In fact, in the area of EDM, all these data mining techniques have been applied to Moodle student usage data to provide feedback to the instructor about how to improve both courses and student learning [21].

2.3.3 Temporal, Sequence and Time Series Data

Temporal, sequence and time series data consists of sequences of values or events changing with time [2]. A temporal database typically stores relational data that include time-related attributes. These attributes may involve several time-stamps, each one involving different semantics. A sequence database stores sequences of ordered events, with or without a concrete notion of time. A time-series database stores sequences of values or events obtained over repeated measurements of time (e.g., hourly, daily, weekly).

An example of a sequential database used by Moodle is a student's log. A log can be thought of as a list of a student's events, in which each line or record contains a time-stamp plus one or more fields that holds information about an activity at that instant. In particular, a Moodle log (see Fig. 2.1) [1] consists of the time and date it was accessed, the Internet Protocol (IP) address accessed from, the

Time	IP Address	Full name	Action	Information
Fri 15 January 2010, 12:49 PM	150.214.110.166		forum view forum	Foro de discusión sobre los Ejercicios de Reglas
Fri 15 January 2010, 12:05 PM	150.214.110.166		resource view	Ejercicios resueltos de Reglas
Fri 15 January 2010, 12:05 PM	150.214.110.166		resource view	Relacción de Ejercicios Reglas
Fri 15 January 2010, 12:05 PM	150.214.110.166		course view	Prácticas IA
Thu 14 January 2010, 07:43 PM	150.214.110.166		resource view	Ejercicios resueltos de Reglas
Thu 14 January 2010, 07:38 PM	150.214.110.166		resource view	Relacción de Ejercicios Reglas
Thu 14 January 2010, 07:38 PM	150.214.110.166		resource view	Relacción de Ejercicios Hechos
Thu 14 January 2010, 07:38 PM	150.214.110.166		assignment view	Entrega de la Relacción de Ejercicios de Hechos.
Thu 14 January 2010, 07:38 PM	150.214.110.166		upload upload	Relacion_Hechos.rar
Thu 14 January 2010, 07:38 PM	150.214.110.166		assignment upload	Entrega de la Relacción de Ejercicios de Hechos.
Thu 14 January 2010, 07:02 PM	150.214.110.166		assignment view	Entrega de la Relacción de Ejercicios de Reglas.
Thu 14 January 2010, 07:01 PM	150.214.110.166		assignment view	Entrega de la Relacción de Ejercicios de Hechos.
Thu 14 January 2010, 07:01 PM	150.214.110.166		assignment view all	
Thu 14 January 2010, 07:01 PM	150.214.110.166		assignment view	Entrega de la Relacción de Ejercicios de Hechos.
Thu 14 January 2010, 06:10 PM	150.214.110.166		resource view	Introducción a CLIPS
Thu 14 January 2010, 06:09 PM	150.214.110.166		resource view	Hechos
Thu 14 January 2010, 06:09 PM	150.214.110.166		course view	Prácticas IA
Thu 14 January 2010, 05:56 PM	150.214.110.166		forum view discussion	duda ejercicio 8
Thu 14 January 2010, 05:56 PM	150.214.110.166		forum view discussion	duda en ejercicio 8

Fig. 2.1 Example of Moodle log file

[1] The full name column covers the identification of subjects.

name of the student, each action (view, add, update and delete) performed in the different modules (forum, resource, assignment, etc.) and additional information about the action.

Sequential data mining, also known as sequential pattern mining, addresses the problem of discovering all frequent sequences in a given sequential database or data set [22]. In the area of EDM, sequential data mining algorithms can be used, for example, to recommend to a student which links are more appropriate to visit within an adaptive educational hypermedia system based on previous trails of students with similar characteristics [23].

2.3.4 Text Data

Text databases or document databases consist of large collections of documents from various sources, such as news articles, research papers, books, digital libraries, e-mail messages, chat and forum messages, and Web pages. Text databases may be highly unstructured, such as some Hypertext Markup Language (HTML) Web pages, or may be somewhat structured, that is, semi-structured, such as e-mail messages and eXtensible Markup Language (XML) Web pages.

Moodle provides a great amount of information in text format, such as: students' messages to forums, messages to chats and e-mails, and anything that students can read or write within the system.

Text mining or text data mining is roughly equivalent to text analytics [24], and can be defined as the application of data mining techniques to unstructured textual data. Typical text mining tasks include text categorization, text clustering, concept/entity extraction, production of granular taxonomies, sentiment analysis, document summarization, and entity relation modeling (i.e., learning relations between named entities). In the area of EDM, text data mining has been used, for example, to assess asynchronous discussion forums in order to evaluate the progress of a thread discussion [25].

2.3.5 Multimedia Data

Multimedia databases store image, audio and video data. Multimedia databases must support large objects, because data objects such as video can require gigabytes of storage. Specialized storage and search techniques are also required. Because video and audio data require real-time retrieval at a steady and predetermined rate in order to avoid picture or sound gaps and system buffer overflows, such data are referred to as continuous-media data. Multimedia information is ubiquitous and essential in many applications, and repositories of multimedia are numerous and extremely large.

Moodle also stores a great amount of multimedia data, for example, all the files uploaded by the instructors and the students. These files can be, for example, an

e-learning Uco. ▶ IA-ITIG(P) ▶ Files

Name	Size	Modified	Action
backupdata	8.8MB	18 May 2010, 07:02 PM	Rename
moddata	5.7MB	18 Jul 2007, 11:25 AM	Rename
EjerciciosHechosResueltos.pdf	15.8KB	18 Jul 2007, 11:57 AM	Rename
EjerciciosReglasResueltas.pdf	43KB	18 Jul 2007, 11:57 AM	Rename
IA-ITIG_P_Examen_Practicas_IA.pdf	7.2KB	18 Jul 2007, 11:19 AM	Rename
Introduccion_ia_items_32.jpg	21.2KB	18 Jul 2007, 11:19 AM	Rename
Introduccion_ia_items_33.jpg	34.5KB	18 Jul 2007, 11:19 AM	Rename
Introduccion_ia_items_34.jpg	29.6KB	18 Jul 2007, 11:19 AM	Rename
Introduccion_ia_items_35.jpg	39.3KB	18 Jul 2007, 11:19 AM	Rename

Fig. 2.2 Example of Moodle files

instructor's presentations (in Microsoft PowerPoint or PDF format, etc.), an instructor's images (in JGP or GIF format, etc.), a student's work and exercises (in Microsoft Word or PDF format, etc.), instructor's videos (in AVI, MOV or FLASH format), etc. All these files are stored in a Moodle data directory. Instructors can browse directly through all these files using Moodle files interface (see Fig. 2.2) or they can also download them to their own local disk in a backup ZIP file.

As seen in Fig. 2.2, Moodle data directory has a root directory (where all the files uploaded by the instructor are placed) and several default directories, such as the *Moddata* directory, which contains all the data submitted by the students, and the *Backupdata* directory, containing backup files of the entire course.

Multimedia data mining is a subfield of data mining that deals with an extraction of implicit knowledge, multimedia data relationships, or other patterns not explicitly stored in multimedia databases [26]. In EDM, for example, mining educational multimedia presentations has been used to establish explicit relationships among the data related to interactivity (links and actions) and to help predict interactive properties in multimedia presentations [27].

2.3.6 World Wide Web Data

World Wide Web (WWW) provides three main types of source data [28]:

- Content of Web pages. This usually consists of texts, graphics, videos and sound files, that is, text and multimedia data.
- Intra-page structure. Data that describe the organization of the content. Intra-page structure information includes the arrangement of various HTML or XML tags within a given page. The principal kind of inter-page structure information consists of hyper-links connecting one page to another.
- User usage data. Data that describe the patterns of Web page usage. Web-based systems record all the users' actions on Web logs, also known as click-streams records, which provide a raw tracking of the users' navigation on the site.

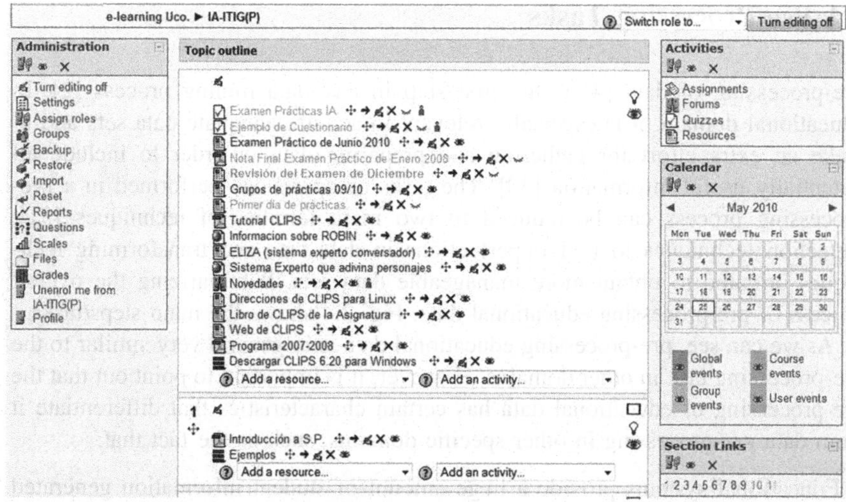

Fig. 2.3 Example of the main window of a Moodle course

In our case, Moodle has the same types of data sources as any other Web-based system. For example, user's usage data are stored in Moodle log files (see Fig. 2.1), some contents of Web pages are stored in the Moodle data directory (see Fig. 2.2), and some others, such as the Web page's text and the intra-page structure, can only be browsed and edited but not saved in files (see Fig. 2.3). This Fig. 2.3 shows the main windows of a Moodle course in editing mode. In the middle of this screen are the course activities and resources grouped into sections or blocks.

The instructor can manage this type of WWW data (contents and intra-page) by adding new resources and activities (using the "Add a resource" and "Add an activity" list boxes, respectively), modifying them, deleting them, hiding them and moving them (using the icons that appear below the text of the specific resource or activity).

Web mining [28] is the application of data mining techniques to extract knowledge from Web data. There are three main Web mining categories: Web content mining, which is the process of extracting useful information from the contents of Web documents; Web structure mining, which is the process of discovering structure information from the Web; and WUM, which is the discovery of meaningful patterns from data generated by client–server transactions on one or more Web localities. In EDM, there are many studies about applying Web mining techniques in educational environments. For example, different techniques such as clustering, classification, outlier detection, association rule mining, sequential data mining, text mining, etc. have been applied to Web educational data for different purposes [29].

2.4 Pre-Processing Tasks

Pre-processing of data [4] is the first step in any data mining process [2]. In educational domain, it is especially relevant to acquire adequate data sets and to make an extra effort for gathering and preparing data in order to include all potentially useful information [30]. The tasks or operations performed in a pre-processing process can be reduced to two main families of techniques [31]: Detection techniques to find imperfections in data sets and transforming techniques oriented to obtain more manageable data sets. Summarizing the overall process of pre-processing educational data, Fig. 2.4 shows the main steps/tasks.

As we can see, pre-processing educational data is in general very similar to the pre-processing task in other domains. However, it is important to point out that the pre-processing of educational data has certain characteristics that differentiate it from data pre-processing in other specific domains, such as the fact that:

- Educational systems provide a huge amount of student information generated daily from different sources of information (see Sects. 2.4.1 and 2.4.2).
- Normally, all the students do not complete all the activities, exercises, etc. In consequence, there is often missing and incomplete data (see Sect. 2.4.3).
- The user identification task is not normally necessary (see Sect. 2.4.4).
- There are usually a great number of attributes available about students and a lot of instances at different levels of granularity. So, it is necessary to use attribute selection and filtering tasks in order to select the most representative attributes and instances that can help to address a specific educational problem (see Sects. 2.4.5 and 2.4.6).
- Finally, some data transformation tasks, e.g. attribute discretization, can be normally applied for improving the comprehensibility of the data and the obtained models (see Sect. 2.4.7).

2.4.1 Data Gathering

Data gathering brings together all the available data, i.e. those that are critical to solve the data mining problem, into a set of instances. An instance can be defined as an individual, independent example of the concept to be learned by a machine learning scheme [32]. Numerous terms are used to describe data gathering and storing aspects such as data warehousing, data mart, central repository, meta-data, and others.

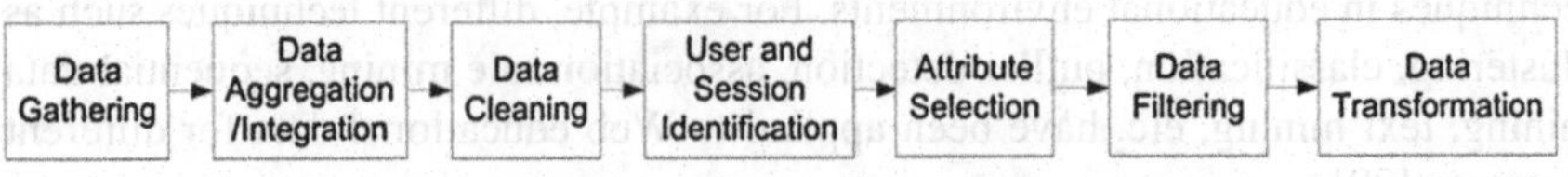

Fig. 2.4 Main pre-processing steps/tasks with educational data

Table 2.3 Examples of different data sources

Name	Description
Log file	Records all students-system interactions
Quiz/test	Stores information about quiz/test usage
Portfolio	Contains information about the students

Educational data are normally gathered from various sources (see Table 2.3) since they have been generated in different places at different times [33]: profile data that contains information about the students and instructors, content data and learning material, communication data that stores the information communicated between students, and activity data that records the students' learning process and the instructors' instruction activities.

Most of the learning tools and systems usually capture all students' fingertip actions (mouse and keyboard) in log files [34]. A typical log file usually takes the form of an ordered list of events occurring on the user interface of a software tool. It contains a record of the activity of one or more students, from the rather restrictive point of view of their fingertip actions. Intelligent tutoring systems commonly also record their interactions in the form of log files. Log files are easy to record, flexible in the information they capture and useful in debugging. Normally, Web server log files contain the access date and time, the IP address of the request, the method of the request and the name of the file requested [35]. However, log files generated by Moodle are a little different because they not only contain the access date and time and IP address, but also other more specific information such as the user name (full name of the student), action (module and specific action performed by the user), and additional information about the action (see Fig. 2.1).

On the other hand, quizzes and test data are stored and organized in a matrix in different ways. An example is the score matrix that is a collection of student scores for a set of questions [36]. This is a data matrix of student ratings (see Table 2.4) in which the first column is the student's name or ID (Identifier), and the first row shows the testing items. For example, in Table 2.4, R*ij* represents the score of item-*j* rating received by the student-*i*, in which the value 1 represent a correct answer, −0.5 if it is wrong, and 0 when not answered [37].

Table 2.4 Score matrix or data matrix of students' rates

Student	Item-1	Item-2	…	Item-j	…	Item-n
Student-1	0	1	…	1	…	−0.5
Student-2	1	1	…	1	…	1
…	…	…	…	…	…	…
Student-i	−0.5	0	…	Rij	…	0

Info | Reports | Preview | Edit Quiz

Overview | Regrade attempts | Item analysis

Item Analysis Table

Question text	Answer's text	% Correct Facility	SD	Disc. Index	Disc. Coeff.
En el sistema operativo LINUX,la combinación de teclas ctrl-c, produce el siguiente efecto: : **En el sistema operativo LINUX,la combinación de teclas ctrl-c, produce el siguiente efecto:**	Borra la línea completa.	64 %	0.464	0.87	0.82
	Detiene la ejecución de un programa.				
	Cierra el fichero.				
La orden mv pepe* "subdirectorio": : **La orden mv pepe* "subdirectorio":**	Dará error.	62 %	0.479	0.88	0.83
	Moverá cada fichero pepe* al correspondiente "subdirectorio".				
	Copiará cada fichero pepe* en el correspondiente "subdirectorio".				

Fig. 2.5 Example of Moodle item analysis

In our case, Moodle provides the score matrix when mark details are selected in the quiz results panel. Moodle also shows the full names of the students, and information about when they started and when they completed the quiz, the total time taken, as well as the final grade obtained together with the score for each question (score matrix). Quizzes can provide much more information, for example, the students' knowledge state can be determined from test question responses using a q-matrix. A q-matrix is the one that shows relationships between a set of observed variables (e.g. questions), and latent variables (concepts) that relate these observations [38]. In the context of education, for a given q-matrix Q, the value of Q (concept, question) represents the probability a student has of incorrectly answering the question due to the fact that he/she does not understand the concept involved. Learning management systems also provide some statistical information about quizzes. For example, Moodle has statistical quiz reports which provide item analysis (see Fig. 2.5). This table presents processed quiz data in a way suitable for analyzing and judging the performance of each question by way of assessment.

The statistical parameters used are calculated as explained by classical test theory (Facility Index or % Correct, Standard Deviation, Discrimination Index, Discrimination Coefficient). The teacher can see the most difficult and easiest questions for the students (% Correct Facility) as well as the most discriminating ones (Disc. Index and Disc. Coeff.). This information can also be downloaded in text-only or Excel formats in order to use a spreadsheet to chart and analyze it.

Another important educational data source is the portfolio. An e-portfolio can be seen as a type of learning record that provides actual evidence of achievement. An e-portfolio is a complete profile of a student that includes raw logged data and/or filled (predefined) templates; like traditional portfolios, it can facilitate the analysis of student learning behavior.

Portfolios can include all the records of students' activities during the learning process, such as their interaction with others, notes, assignments, test papers, personal work collections, their discussion content, online learning records and reports, etc. [39]. Learning portfolios can also include the students' learning path (routes used by students throughout the courses), preferred learning styles (approaches or ways of learning preferred by such groups as visual learners, auditory learners, kinesthetic learners, etc.), students' learning time (time used by students in each activity and/or the full course), course grade and difficulty, etc. [40].

Finally, it is important to highlight that software agents have been used to automatically capture students' interaction data. Although in general there are no differences between using an agent-based architecture or another type of architecture for logging, gathering and data analysis, agents can provide modularity, autonomy, persistence and social ability. A software agent or intelligent agent is a complex software entity capable of acting with a certain degree of autonomy in order to accomplish tasks on behalf of its user.

They have been used for extracting and evaluating log data from e-learning software and organizing that data in intelligent ways [41] to capture ITS data based on an agent communication standard [42], and for automatically recording useful information and organizing it into its corresponding tables in the database [43].

2.4.2 Data Aggregation/Integration

The goal of data aggregation/integration is to group together all the data from different sources [44]. The data can come from various sources, and so can be stored in different formats [30]. After the previous step of gathering all the required/desired data, the process of aggregation/integration can begin for combining data from multiple sources into a coherent recompilation, normally into a database. Aggregation and integration are different terms used to distinguish between the aggregations of the same type of data over multiple problems/sessions/students/classes/schools from the integration of different types of data about the same problem/session/student/class/school.

Educational systems normally provide several data sources that can be aggregated and/or integrated into one single database. Some of these data can be available for read in form of files, even when a certain part has to be transcribed manually from paper documents, because not all the useful information has been stored digitally [30]; this is the case of the attendance paper, in which all students sign at in-person classes. For example, Web log information can be used in conjunction with data from surveys, usability studies and other sources [45]; log files can be mixed with other inputs, such as student demographics and performance data and survey results [46].

Online learning environments normally store all the students' interactions not only in log files but also directly in databases [46]. And if this is not the case, during the pre-processing process, data for each individual student (profiles, logs, etc.) can be aggregated into a database [47, 48]. In a similar way, although ITSs use log files, it has been found that storing logging tutorial interactions directly into a properly designed and indexed database instead of using log files eliminates the need to parse them [49]. So, relational databases are more powerful than usual log text files and provide easier, more flexible and less bug-prone analyses [49].

In fact, most universities today have large and complex structure and activities (Multiversity) that are collected into one or several databases [50]. An example of a relational database is the Moodle database, which stores all the Moodle course information [11]. Moodle also provides different types of reports accessible from the Moodle interface. These reports are available to the course's instructor/s as an option in the Administration block and in some activities. All these reports can also be saved into files with .TXT, .ODS or.XLS format. Excel Pivot Tables have been also proposed to conduct a flexible analytical processing of Moodle usage data and gain valuable information [51]. A Pivot Table is a highly flexible contingency table that can be created from a large data set and offers the possibility to look at one section at a time.

Data warehouses have also been proposed for data gathering and integration [13]. A data warehouse schema can represent information about the structure and usage of the courses and can include several data sources, for example, the university Management Information System (MIS), Web server log files and LMS databases [52]. Data warehouses and data marts require concise, subject-oriented schemas that facilitate the analysis of data. For example, a star model (a modeling paradigm of educational data) of a data mart for a course of a LMS can contain a central access fact table that contains keys for each of the other tables, such as a time-dimension table, a user-dimension table and a learning-resource dimension table [53].

Finally, a multi-dimensional data cube structure has been used for carrying out an Online Analytical Processing (OLAP) operation on a database [54]. A data cube provides remarkable flexibility for manipulating data and viewing it from different perspectives. Building a Web log data cube allows the researcher to view and analyze Web log data from different angles, derive ratios and measure many different dimensions [55]. For example, in a study on Web-based education systems [56], student data were observed from three dimensional views (see Fig. 2.6): learning-behavior pattern dimension, student-personality characteristic dimension,

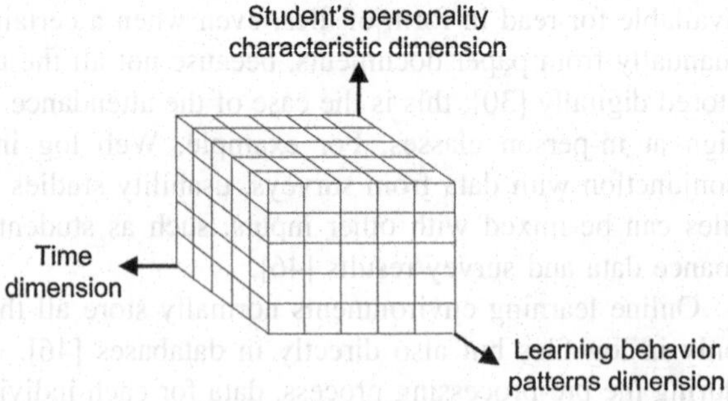

Fig. 2.6 Example of data *cube*

and time dimension. Each dimension was related to the so-called dimension table. One dimension may also describe a different level, for example, a time dimension may describe a level from a year, quarter, month, date and so on [56]. So, OLAP provides us with the possibility of analyzing different levels of aggregation, e.g. per day, per month, per duration of the course, or per student, per group, or the whole population, etc.

2.4.3 Data Cleaning

The data cleaning task consists of detecting erroneous or irrelevant data and discarding it [2]. The most common type of inaccuracies, such as missing data, outliers and inconsistent data [57], are described below.

Missing data are a common issue in the application of data analysis methods. In statistics, missing values occur when no value is stored for the variable in the current observation [58].

Some possible solutions are to use a global constant to fill in the missing value or to use a substitute value, like the attribute mean or the mode. For example, missing values have been replaced using linear interpolation of the previous and posterior 4 values for emotion detection in an educational scenario [59], or by determining what is the most probable value to fill in the missing value using regression [60]? A different and simple approach is to codify missing/unspecified values by mapping incomplete values [9], using for example the labels "?" (Missing) and "null" (unspecified).

In educational data, missing values usually appear when students have not completed or done all the activities in the course, or when we combine data from different sources and students have skipped some tasks [61]. For example, students who enroll in a course but do not actually participate, or those whose information or data are incomplete (missing data) [62]. In some extreme cases, in order to clean data and ensure their completeness, students who have all or almost all their values missed can be removed from data. For example, in e-learning courses some users only enter to a specific course one time (by error or in order to see one specific resource or to do an activity) but later they never come back to the course [21]. But normally, in the case that students show some missing values, whenever possible, these specific students may be contacted and asked (by the instructor) to complete the course, so that their information can be used and/or evaluated. When this is not possible, the missing information regarding students could be replaced by a predetermined value or label [9].

The elimination of noisy instances is one of the most difficult problems in data pre-processing. Frequently, there are samples in large data sets that do not comply with the general behavior of the data [63]. Such samples, which are significantly different from or inconsistent with the remaining set of data, are called outliers. Outliers can be caused by measurement error or they may be the result of inherent data variability, in which case the term "outlier" just refers to unlikely or

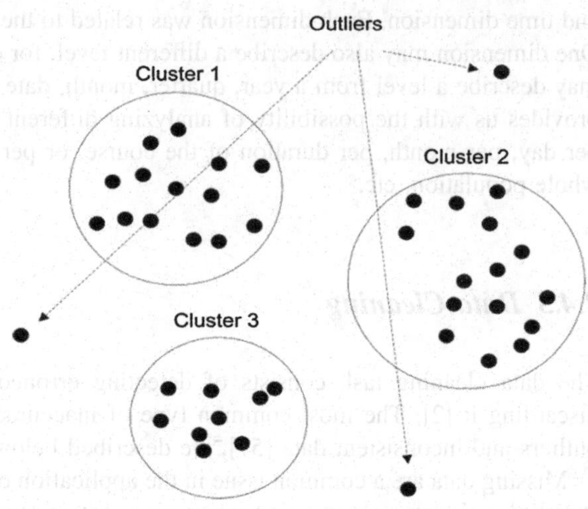

Fig. 2.7 Plot of *three* data clusters and some *outliers*

unexpected distribution, rather than outside a limit. For example, the value could be a typographical error, or it could be correct and represent real variability for the given attribute.

Many data mining algorithms try to minimize the influence of outliers in the final model, or eliminate them in the pre-processing phases. For example, in a 1-dimensional value a simple way to keep outliers from distorting results is to use medians instead of means [64]. This method has the added advantage of solving the prickly problem of distinguishing outliers from real values. It is also important to highlight that although outliers normally can be due to noise, in educational data they can be often true observations. For example, there are always exceptional students, who succeed with little effort or fail against all expectations. However, making the distinction between the outliers that should be accepted and those that should be rejected is not always easy. This requires knowledge of the domain in which the data was collected and depends on the aims of the analysis [65]. In this case, a relatively simple technique that can help to detect outliers is by visualizing data clusters, where values that fall outside the set of clusters may be considered outliers (see Fig. 2.7).

An example of data cleaning in a Web-based educational system [48] is to detect and eliminate both long periods of time between two actions carried out by the same student (longer than 10 min) and incomplete data (incompletely visited chapters, and unfinished tests and activities). Another similar example [61] shows that very high values were often recorded for attribute time because the student had left the computer without first exiting the exercise, concept or section. In order to address this problem, any times that exceeded a maximum established value (between 20 and 30 min is a common criterion) are considered noisy data, and this maximum value is assigned to any apparently erroneous data. A different approach is to use association rule mining for filtering data that match a predefined type of rule [66]. In this case, if results have the rule containing the conclusion *NO*, which indicates that students are not interested in the course, then the courses with the lowest students count are removed, as well as those students having the lowest course count.

Finally, inconsistent data appear when a data set or group of data is dramatically different from a similar data set (conflicting data set) for no apparent reason. In fact, some incorrect data may also result from inconsistencies in naming the conventions or data codes in use, or inconsistent formats for input fields, such as a date. For example, duplicate tuples can require data cleaning, e.g. Age = "42" and Birthday = "03/07/1997", shows discrepancy between duplicate records or oxymoron (self-contradiction).

2.4.4 User and Session Identification

One of the key issues in the pre-processing phase is to identify users. This step distinguishes the users as individuals, and as a result, different users are identified. This can generally be done in various ways, like through the use of IP addresses, cookies, and/or direct authentication (login/password).

Identifying users based on Web log files is not a straightforward problem, and so various methods have been developed [67]. On the other hand, user sessions also have to be identified. A session is a semi-permanent interactive information interchange between two or more communicating devices, for example, a login session is the period of activity between a user logging in and logging out.

Although user and session identification is not specific to education, it is especially relevant due to the longitudinal nature of student usage data. However, computer-based educational systems provide user authentication (identification by login and password). These logs include entries that identify the students/users who have logged on, thus identifying sessions a priori since users may also have to log out [68].

So it is not necessary to do the typical user identification task to identify sessions from logs, and session determination ceases to be a problem. In fact, all records can be sorted in an ascending order with the user ID as the primary key, and the event time as a secondary key [69]. After this sorting step, it is easy to identify user sessions by grouping contiguous records from one login record to the next one. Specifically, browsing records picked out between two successive login records are grouped into a browsing session, and an upper limit of the time interval between two successive clicks has to be set (from 15 to 45 min) in order to break the sequence of one student's click stream into sessions [35]. This value may result in increasing or decreasing the total number of identified sessions.

However, to our knowledge, there is no research on the relation between timeout of user session and its impact on quality of discovered knowledge [70]. It is also important to construct not only learning sessions but also learning episodes from logs [71] and tasks and activities [72]. A learning episode is a high level model of the student's learning task with information about the system situation at the beginning of the episode, the actions performed by the student, and the system situation at the end of the episode. A task is defined as a sequence of user interactions within one resource that ranging from passive reproductions of paper-based documents to sophisticated interactive resources [72]. An activity is defined as an interaction within the site and categorized to indicate whether the activity involved browsing or reading, or more interactive use.

Another noteworthy aspect to consider is that accessing to some information about users/students can be restricted due to privacy issues and special measures and permission may be required. It is also necessary to preserve student data anonymity/privacy but enabling that different pieces of information are linked to the same person without explicitly identifying but making sure that users can be de-coupled from their sessions if local, state or federal laws require it. Petersen [73] points to the importance of the de-identification of data before the data is made available for institutional use, including the option to retain unique identifiers for individuals in the data set, without identifying the actual identity of the individuals. A common solution for it consists in using a number randomly or incrementally generated, like a user ID or other kind of personal information, such as e-mail or an identification card instead of using someone's real name (see *id_student* attribute in Fig. 2.10). But, a better mechanism for assigning unique, disassociated IDs (from a specific name), may be required in some systems.

A different approach to protecting student identity consists in not revealing any private information in reports, tables, figures, etc. For example, notice that in some figures shown in this chapter (see Figs. 2.1 and 2.9), student names were blurred to protect their identity. Finally, whereas educational institutions have always had requirements to protect student and teacher privacy, new amendments to the existing regulations increase access to data for research and evaluation (including sharing across levels, such as from high school to college) while maintaining student privacy and parents' rights [74].

2.4.5 Attribute/Variable Selection

Feature selection and extraction chooses a subset of relevant attributes from all the available attributes [75]. This is also known as variable selection, feature reduction, attribute selection or variable subset selection. Choosing the right variables is one of the main tasks before applying data mining techniques [76], because the variables can be correlated or redundant. Consequently, the data must be pre-processed to select an appropriate subset of attributes and ignore irrelevant and redundant ones. An attribute may be redundant if it can be derived from another attribute or set of attributes. There may be redundancy, where certain features are correlated so that it is not necessary to include all of them in modeling; and interdependence, where two (or more) features together convey important information that is obscure if either of them is included on its own [77].

Attribute selection is very important in education because there could be a large number of attributes for learning schemes to handle in many practical situations [78] and this number of attributes can result in reducing the accuracy of a learning model due to overfitting problems. One solution to this issue is to select only the most important attributes/variables. For example, a ranking of several feature selection algorithms has been used for identifying which features or attributes have the greatest effect for predicting school failure [79]. A decision tree technique has been also used to choose the right variables (relevance analysis or feature selection) in educational data [76]. This method is used to obtain the most consistent variables by presenting them at various tree levels. Another solution is not to use irrelevant data that do not really provide any useful information to solve the problem. Some examples of well-known attributes that can be irrelevant are user password, student's e-mail, student's phone number, student's address, student's picture, etc.

Learning management systems, such as Moodle, store a huge amount of attributes/variables about courses, students and activities. So it is really relevant to select only a representative group of attributes in order to reduce the dimensionality of data. There are some proposals of indexes and metrics [80] in order to properly facilitate the evaluation of the course usage. However, even when there are many of these metrics in Web usage analysis for e-commerce, it is not the same situation in the case of e-learning. Then, these selected attributes can be stored all together in a new table comprising all the relevant information related to the students enrolled in the course [21].

For example, in the problem/case of predicting what it is the students' final performance in a course, starting from the usage information with Moodle, there is a lot of variables about the interaction between the students and Moodle system. Thus, it is necessary to select only the most related attributes with the student performance. Table 2.5 shows a list of the selected features/attributes for each student in a Moodle course, i.e. the fields of each summary record.

Table 2.5 Example of list of attributes selected per student in Moodle courses

Name	Description
id_student	Identification number of the student
id_course	Identification number of the course
num_sessions	Number of sessions
num_assigment	Number of assignments done
num_quiz	Number of quizzes taken
a_scr_quiz	Average score on quizzes
num_posts	Number of messages sent to the forum
num_read	Number of messages read on the forum
t_time	Total time used on Moodle
t_assignment	Total time used on assignments
t_quiz	Total time used on quizzes
t_forum	Total time used on forum
f_scr_course	Final score of the student obtained in the course

Finally, notice that some student-related variables or attributes might introduce a great degree of variance and this instability could represent a non-trait measure (i.e. a non- specific trait that a student has), denoting that this variable does not and should not describe the student [81]; for example, the consistency of students' behavior regarding the pace (also referred as speed or rate) of their actions (i.e. the number of logged actions divided by the session length in minutes) along the sessions of an online course. Nevertheless, some other variables (e.g. session length, response time, intensity of activity, preferred tasks) might have a great variance when they are repeatedly measured for the same student: this instability may also represent a non-trait measure.

2.4.6 Data Filtering

Data filtering selects a subset of representative data in order to convert large data sets into manageable (smaller) data sets [2]. Data filtering allows the huge amount of information available to be reduced. Data sets for analysis may contain hundreds of attributes, many of which may be irrelevant to the mining task or redundant. Some of the most common types of filtering techniques for educational data are the selection of data subsets relevant to the expected purpose, and the selection of the most convenient grain size to the research at hand.

Educational systems provide a huge quantity of information about all the events and activities performed by the students enrolled in courses. However, the instructor or educational research can be only interested on a certain subset of events, students or courses depending on the specific problem or task to be solved. For this reason filtering can be used to select only a specific subset of desired data [82]. These data can be filtered by defining the conditions of one or more attributes and removing the instances that violate them [83]. For example, Moodle allows log files to be filtered by course, participant, day, and activity.

A novel data preparation approach uses activity theory, which considers three levels of human activity, has been also used to pre-process data in order to get more interesting results to study what happens in a collaborative learning platform [10]. They propose to map the original data to a higher level, analysis-oriented representation by using activity theory.

A specific characteristic of data collection from the educational system is that there are different levels of granularity such as: keystroke level, answer level, session level, student level, classroom level, and school level [84]. Therefore, it is necessary to choose an appropriate level of granularity in order to only identify the variables that can be recorded at that specific level of granularity [83]. Logging data with multiple grain size facilitates viewing and analyzing data at different levels of detail [16]. Figure 2.8 shows various levels of granularity and amounts of data related to each level. It can be observed that a higher grain is related to a smaller amount of data and, on the other hand, a lower grain is related to a larger amount of data.

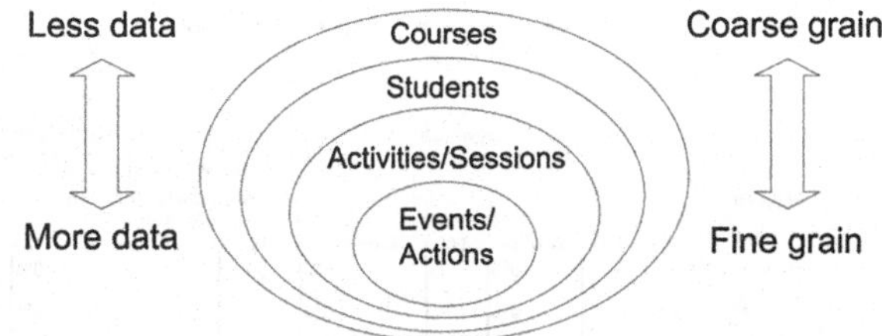

Fig. 2.8 Different levels of granularity and their relationship to the amount of data

The level at which events are logged constrains their analysis. For example, logging a mouse click by its *x* and *y* coordinates may help to analyze the student's motor skills. In contrast, logging the menu item selected may allow a student's reply to a multiple choice question to be scored. So, the level of granularity affects analysis, that is, the granularity of data should fit its intended analysis because the resulting temporary table would only contain attributes and transactions from students with respect to the level selected [61].

Sometimes, the raw learning logs collected by computer systems may be excessively detailed. To analyze these logs at the behavioral unit or grain size required by the educational research, we need to reformat the raw learning logs by systematically aggregating them and pass from low level events to high level learning actions [85]. Thus, it is necessary to define different abstractions of the log file data, such as:

- Event. It is a single action or interaction recorded on the log file.
- Session. It is a sequence of interactions of a user from a login to the last interaction.
- Task. A sequence of interactions of a user within one resource.
- Activity. A series of one or more interactions to achieve a particular outcome.

Moodle also provides several levels of data granularity. For example, as previously described, Moodle log reports provide fine grain information about all students' actions (see Fig. 2.1). In this case, there can be hundreds of instances or records or rows in the log file belonging to each student. However, Moodle also provides coarse grain information about students, for example by grades. Moodle grades show the grades of quizzes and other activities that students have done. In this case, there is only one instance or row for each student with columns for each activity. For example, Fig. 2.9[2] shows the grades of a course that has two quizzes (*Examen Practicas IA* and *Ejemplo de Cuestionario*) and two assignments (*Entrega de Relacción de Ejercicios de Hechos* and de *Ejercicios de Reglas*) evaluated using a scale from 0 to 10. The instructor can download the entire grade book as an .ODS, .XLS or .TXT file.

[2] The *student* column covers the identification of subjects.

e-learning Uco. ► IA-ITIG(P) ► Grades ► uncategorised

View Grades | Set Preferences | Set Categories | Set Weights | Set Grade Letters | Grade Exceptions

Download in ODS format | Download in Excel format | Download in text format

uncategorised Grades

Student	Examen Prácticas IA		Ejemplo de Cuestionario		Entrega de la Relacción de Ejercicios de Hechos.		Entrega de la Relacción de Ejercicios de Reglas.		Total Stats	
	10	Raw %	10	Raw %	10	Raw %	10	Raw %	40	Percent
	-	0%	-	0%	-	0%	-	0%	-	0%
	-	0%	-	0%	-	0%	-	0%	-	0%
	7.25	72.5%	2.3	23%	-	0%	-	0%	9.55	23.88%
	9.18	91.8%	8.9	89%	-	0%	-	0%	18.08	45.2%
	-	0%	-	0%	-	0%	-	0%	-	0%
	8.63	86.3%	10	100%	-	0%	-	0%	18.63	46.58%
	3.13	31.3%	-	0%	-	0%	-	0%	3.13	7.83%
	-	0%	-	0%	-	0%	-	0%	-	0%
	-	0%	-	0%	-	0%	-	0%	-	0%
	8.35	83.5%	1.3	13%	-	0%	-	0%	9.65	24.13%
	-	0%	-	0%	-	0%	-	0%	-	0%
	8.08	80.8%	7.9	79%	-	0%	-	0%	15.98	39.95%
	-	0%	-	0%	-	0%	-	0%	-	0%
	5.05	50.5%	8.9	89%	-	0%	-	0%	13.95	34.88%
	7.8	78%	8.9	89%	-	0%	-	0%	16.7	41.75%

Fig. 2.9 Example of Moodle grades

2.4.7 Data Transformation

Data transformation derives in new attributes from already available attributes [2]. Data transformation can facilitate a better interpretation of information. Some examples of transformation such as normalization, discretization, derivation, and format conversion, are described next.

Normalization is a data transformation technique where the attribute values are scaled within a specified range, usually from −1.0 to 1.0, or between 0.0 and 1.0. Within one feature there is often a great difference between maximum and minimum values, e.g. 0.01 and 1,000. Hence, normalization can be performed to scale the value magnitudes to low values. In this way, normalization may improve the accuracy and efficiency of the mining algorithms involving distance measurements [2].

Normalization also helps to prevent attributes with initially large ranges from outweighing attributes with initially smaller ranges. For example, one of the most important steps in data pre-processing for clustering is to standardize or normalize data in order to avoid obtaining clusters that are dominated by attributes with the largest amounts of variation [86].

There are many other methods for data normalization. However, in education, the most commonly used method is the *Min–max* normalization, which performs a linear transformation of the original data [87]. Suppose that *minA* and *maxA* are the minimum and maximum values of an attribute *A*, respectively. Then, the *Min–max* normalization maps a value, *v*, of *A* to *v'* in the range (*new minA*, *new maxA*) using the Eq. (2.1):

$$v' = \frac{v - \min_A}{\max_A - \min_A}(new\_\max_A - new\_\min_A) + new\_\min_A \quad (2.1)$$

Discretization divides the numerical data into categorical classes that are more user-friendly than precise magnitudes and ranges. It reduces the number of possible values of the continuous feature and it provides a much more comprehensible view of the data. Generally, discretization smooths out the effect of noise and enables simpler models, which are less prone to overfitting. It can be included as a reduction method that uses some data mining algorithms that do not work well with continuous attributes. For example, association rule mining algorithms usually work only with categorical data. A special type of discretization is the transformation of ordinal to binary representation, that is, from numbers denoting a position in a sequence to 0 or 1 value. This type of codification is used for example in frequent pattern mining research. Some discretization methods [2] are the following:

- Equal-width binning divides the range of possible values into N sub-ranges of the same size in which (2.2) For example, if our values are all between 0 and 100, 5 bins could be created as follows: [0–20], [20–40], [40–60], [60–80] and [80–100].

$$bin\_width = (\max value - \min value)/N \quad (2.2)$$

- Equal-frequency or equal-height binning divides the range of possible values into N bins, each of which holding the same number of instances.
- For example, there are the following 10 values: 5, 7, 12, 35, 65, 82, 84, 88, 90 and 95. Now, in order to create 5 bins, the range of values would be divided up so that each bin holds 2 values in the following way: [5, 7], [12, 35], [65, 82], [84, 88] and [90, 95].
- Manual discretization lets the user directly specify the cut-off points. A typical educational example of a manual discretization method is normally done with marks/scores. For example, if a range of values between 0.0 and 10.0 is applied, they could be transformed into the next four intervals and labels [21]:

$$\text{mark} = \begin{cases} FAIL: \ if\ value\ is < 5 \\ PASS: \ if\ value\ is \geq 5\ and < 7 \\ GOOD: \ if\ value\ is \geq 7\ and < 9 \\ EXCELLENT: \ if\ value\ is \geq 9 \end{cases} \quad (2.3)$$

Another example is the grade point average (normally a value between 0.0 and 4.0), which can also be translated into a letter grade, e.g. *A*, *B+*, *B*, *C+* and *C* [88]. A different approach is to use fuzzy intervals, in which fuzzy sets for grading are used instead of crisp intervals [89]. Finally, the most extreme but simplest discretization case is when attributes are binarized (0 or 1). Here, even if some information is lost, the resulting model can produce, for example, a more accurate classification [90].

Another technique of data transformation is the derivation, which enables to create new attributes starting from the previous ones. So, new attributes can be offshoots of other current attributes in a specific attribute derivation. In many cases, a new attribute needs to be derived from one (or more) of the attributes in a data set. The new attribute may result from a mathematical transformation of another attribute [83], such as the time difference attribute that could be converted to minutes instead of seconds. The most commonly used type of derivation performs some kinds of aggregation on another attribute. For example, when the constraints related to each attempt are grouped into an attribute of the attempt data set, then a "violated count" attribute is included as an attribute of the analysis data set. This attribute identifies the total number of violated constraints in each attempt [83].

A hash code has been also used as the encoding scheme for combining student information into a single hash number [9]. This hash number was simply created by multiplying each field with a distinct power of ten, in descendant order. Some other examples of attributes [91] derived from the information provided by an e-learning system is shown in Table 2.6.

A different approach is to enrich data (normally log files) by using expressions, annotations, labels, text replays, etc. However, it is important to note that data annotations and labeling are very labor-intensive tasks. Some other authors [92] propose making log files more expressive by overcoming a historical tendency to make log files cryptic in order to save file space. This change involves altering the representation of events in the log file and enriching the logged expressions so that more inferences can be drawn more easily. Other approach is to completely represent each event using English words, using English grammar and using standard log file forms [92].

Other authors [84] have also proposed using text replays as a method for generating labels or tags. Text replays produce data that can be more easily used by data mining algorithms. It is an example of distillation for human judgment that

Table 2.6 Example of derived attributes

Attribute	Description
UserId	A unique identifier per user
Performance	Percentage of correctly answered tests calculated as the number of correct tests divided by the total number of tests performed)
TimeReading	Time spent on pages (calculated as the total time spent on each page accessed) in a session
NoPages	The number of accessed pages
TimeTests	The time spent performing tests (calculated as the total time spent on each test)
Motivation	Engaged/disengaged

tries to make complex data understandable by humans to leverage their judgment. Text replays represent a segment of the student's behavior from the log files in a textual "prettily-printed" form. For example, the coder can see the time when each action has taken place, as well as the problem context, the input entered, the relevant skill and how the system assessed the action. Then the coder can choose one among a set of behavior categories and tags or can indicate that something has gone wrong. Other authors propose the use of hand-labeled data and the mapping of events to variables through intelligent tutors' data [16]. In this case, annotations about the level of students' engagement were made by an expert with tutoring experience to annotate sequences of students' actions with the label engaged or disengaged. A current approach proposes to create a grammar (i.e., a set of rules) to unambiguously combine some low-level entries into high level actions that correspond to functions provided at the user interface level of the Alice programming environment [93].

Finally, pre-processed data have to be transformed into the format required by the data mining algorithm or framework that will be used later. Therefore, data have to be exported to a specific format, such as the Weka's ARFF format (Attribute-Relation File Format) [32], Keel DAT format [94], Comma-separated values (CSV), XML, etc. Fig. 2.10 shows an example of a summary file in the ARFF format. Either the WekaTransform tool (http://sourceforge.net/projects/wekatransform/) or the Open DB Preprocess task in Weka Explorer (http://www.cs.waikato.ac.nz/ml/weka/) can be used in order to transform data directly from a database into ARFF format. Datapro4j (http://www.uco.es/grupos/kdis/datapro4j) can also be used to programmatically transform multiple data formats

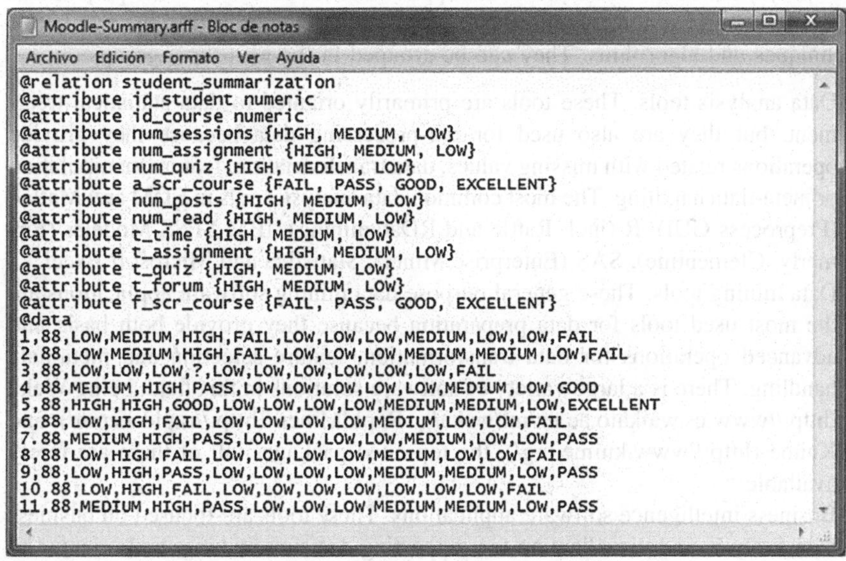


```
@relation student_summarization
@attribute id_student numeric
@attribute id_course numeric
@attribute num_sessions {HIGH, MEDIUM, LOW}
@attribute num_assignment {HIGH, MEDIUM, LOW}
@attribute num_quiz {HIGH, MEDIUM, LOW}
@attribute a_scr_course {FAIL, PASS, GOOD, EXCELLENT}
@attribute num_posts {HIGH, MEDIUM, LOW}
@attribute num_read {HIGH, MEDIUM, LOW}
@attribute t_time {HIGH, MEDIUM, LOW}
@attribute t_assignment {HIGH, MEDIUM, LOW}
@attribute t_quiz {HIGH, MEDIUM, LOW}
@attribute t_forum {HIGH, MEDIUM, LOW}
@attribute f_scr_course {FAIL, PASS, GOOD, EXCELLENT}
@data
1,88,LOW,MEDIUM,HIGH,FAIL,LOW,LOW,LOW,MEDIUM,LOW,LOW,FAIL
2,88,LOW,MEDIUM,HIGH,FAIL,LOW,LOW,LOW,MEDIUM,MEDIUM,LOW,FAIL
3,88,LOW,LOW,LOW,?,LOW,LOW,LOW,LOW,LOW,LOW,FAIL
4,88,MEDIUM,HIGH,PASS,LOW,LOW,LOW,LOW,LOW,MEDIUM,LOW,GOOD
5,88,HIGH,HIGH,GOOD,LOW,LOW,LOW,LOW,MEDIUM,MEDIUM,LOW,EXCELLENT
6,88,LOW,HIGH,FAIL,LOW,LOW,LOW,LOW,MEDIUM,LOW,LOW,FAIL
7.88,MEDIUM,HIGH,PASS,LOW,LOW,LOW,LOW,MEDIUM,LOW,LOW,PASS
8,88,LOW,HIGH,FAIL,LOW,LOW,LOW,LOW,MEDIUM,LOW,LOW,FAIL
9,88,LOW,HIGH,PASS,LOW,LOW,LOW,LOW,MEDIUM,MEDIUM,LOW,PASS
10,88,LOW,HIGH,FAIL,LOW,LOW,LOW,LOW,LOW,LOW,LOW,FAIL
11,88,MEDIUM,HIGH,PASS,LOW,LOW,LOW,MEDIUM,MEDIUM,LOW,PASS
```

Fig. 2.10 Example of Moodle summary ARFF file

using the Java language. Figure 2.10 is a snapshot of a Moodle summary file in ARFF format. Rows of the file represent a summary of all the online activities completed by the students and their final marks obtained in the Moodle course. Notice that the attributes defined in the file were already described in Table 2.5, although most of them are discretized.

The declaration section indicates that numeric attributes are specified by the keyword numeric, whereas nominal attributes are specified by the list of possible attribute values in curly brackets. Finally, the list containing all the instances is detailed right after the @data tag. Instances are listed in the comma-separated format, and a question mark represents a missing value.

2.5 Pre-Processing Tools

Data pre-processing phase requires strong efforts in the KDD process, which needs to be mitigated somehow with the use of software tools. These applications allow the EDM user to perform semi-automatic tasks for preparing data before accomplishing knowledge extraction. In general, these tools can be grouped in two categories: general purpose tools and specialized tools for pre-processing data.

2.5.1 General Purpose Data Pre-Processing Tools

Nowadays, the most mature software applications available for data preparation are general purpose tools, both in their scope of application and in the number of techniques and algorithms. They can be grouped in the next three groups:

- Data analysis tools. These tools are primarily oriented to data statistical treatment, but they are also used for pre-processing data because they provide operations related with missing values, data transformation, feature manipulation or meta-data handling. The most common data analysis tools in EDM are: Matlab (Preprocess GUI), R (incl. Rattle and RDatamining), IBM SPSS Modeler (formerly, Clementine), SAS (Enterprise Miner), Statistica and Microsoft Excel.
- Data mining tools. These general purpose data mining software applications are the most used tools for data preparation because they provide both basic and advanced operations for data transformation, feature selection and meta-data handling. There is a large variety of software solutions in this field, being Weka (http://www.cs.waikato.ac.nz/ml/weka/), RapidMiner (http://rapid-i.com/) and Knime (http://www.knime.org/), the top cited packages; all of them are freely available.
- Business intelligence software applications. These tools are focused on business data analysis and visualization for supporting decision making, and oriented to the treatment of large amounts of data. They provide advanced functionalities

for data exploration and transformation. Some examples are: Orange (http://orange.biolab.si), Angoss (http://www.angoss.com) and Amadea (http://alice-soft.com/html/prod_amadea_en.htm).

2.5.2 Specific Purpose Data Pre-Processing Tools

The number of tools exclusively devoted to data pre-processing is marginal. Although some specific tools have been developed for preparing data in any domain, as well as others especially well-suited for educational data, most of the currently existing tools are just prototypes providing restricted features or they are oriented to work only with a very specific type of data.

A sample of specific tool for data pre-processing is DataPreparator (http://www.datapreparator.com), a freely available proprietary and multi-platform software tool, that provides functionalities for data load (text files, relational databases, excel), data reduction (including attribute selection), data transformation (discretization, missing values, scaling, sorting) and outlier handling. In contrast to prior approaches, DMPML-TS [95] is a visual tool founded on the use of Data Mining Preparation Markup Language (DMPML), an XML notation designed to represent the data preparation phase of the KDD process in general domains. Its authors stand for the use of this type of representation, and claim that it facilitates data codification, cleaning and data transformation using XSLT transformations.

As for data pre-processing solutions in the field of EDM, we can also find some prototypical proposals in the literature. These tools are mainly oriented to the preparation of data extracted from log files in Web learning environments. A first proposal in the field of WUM was presented by Zaïne and Luo [96]. They highlighted the variety of Web log analysis tools available but specially tailored for e-business and, consequently, difficult to use by educators. Therefore, they proposed a novel tool based on a three-tier architecture for data gathering and pre-processing.

Its constraint-based approach allowed the educators to express restrictions and filters during the preparation phase, the patterns discovery phase, or the patterns evaluation phase. Having domain-specific filters during the first stages considerably reduces the search space and controls the performance and accuracy of the pattern extraction.

Similarly, Marquardt et al. [8] proposed an early tool prototype for the automation of the most typical tasks performed by the instructor in the pre-processing phase for the mining of data extracted from Web courses. Their tool, called WUM Prep, offered a set of scripts implementing classical WUM pre-processing techniques, i.e. data cleaning and filtering, user and session identification, path completion, data enrichment, and transaction identification.

Ceddia et al. [97] proposed Web Analysis Tool (WAT) to allow the educator to describe activities from sequences of Web site interactions that could be meaningful in the course context. The aim of this approach is to provide the teacher with an indication of how successful the educational Web course has been in assisting the students meet their objectives, based on the navigation path recorded. Even when it is also a prototype application, WAT provides a usable GUI and defines a process in two phases: activity definition, where the log files are preprocessed extracting those fields that are valuable for the activity and imposing some meaning to the selected interaction attributes; and activity extraction, where the file is processed according to the previous configuration, and the information about activities is extracted and shown to the educator.

More recently, Sael et al. [7] have proposed a specific pre-processing tool for e-learning platform using Moodle logs. It uses not only access log information but also SCORM activities in order to identify different levels of access to a course and thus to define episodes according to these levels.

Finally, EDM Workbench [98] is a specific tool that helps educational researchers with processing data from various sources (PLC shop, SQL tutor, Collaborative Learning System Database and Streamed Log Files). Though still in beta version, this application provides a GUI with some basic operations and algorithms specialized for EDM. It provides operations for collaborative labeling of log files, extraction of information for its subsequent use in machine learning and data transformation (e.g. random sampling, clipping, and a few others).

2.6 Conclusions

Nowadays, there are very few specific data pre-processing tools and so, EDM users normally use general software and DM tools for pre-processing. For example, database GUI administrator tools are used to data aggregation/integration, text editors or spreadsheets are used to manually eliminate some incomplete students' data, and DM tool are used for automatic attribute selection and filtering. However, most of the current data mining tools and general pre-processing tools are normally designed more for power and flexibility than for simplicity and thus, they do not suitably support pre-processing activities in the educational domain.

Most of the currently existing tools can be too complex for educators, EDM researchers and users who are not expert in data mining, since their features go well beyond the scope of what an educator may want to do. So, a very important future development will be the appearance of free EDM pre-processing tools and wizards in order to automate and facilitate all the pre-processing functions in an easy-to-use framework. In this way, the typical workload of the pre-processing phase could be significantly reduced by the automation of the most usual tasks.

On the other hand, there is currently only one public educational data repository, the PSLC DataShop [99] that provides a great number of educational data sets about ongoing courses. However, all this log data comes from ITSs, so it will be also useful to have in the future more public data sets available from other different types of educational environments such as AIHS, LMSs, MOOCs, etc. In this way, a wide range of educational benchmark data sets could be used directly without need to be pre-processed. Finally, the following main lessons have been learned with respect to the pre-processing of educational data:

- Pre-processing is always the necessary first step in any data mining process/application. This task is very important because the interestingness, usefulness and applicability of the obtained DM models highly depend on the quality of the used data.
- There are different types of educational environments that provide different type of data, and several pre-processing tasks can be applied. Although, this paper recommends the use of a specific flow or sequence in applying different pre-processing tasks/steps, some variations in the order of some tasks can also be possible. For example, data cleaning can be done later; attribute selection, data filtering and data transformation can be mixed or put in different order.
- Not all these pre-processing tasks/steps have to be applied in all the cases. That is, depending on the data and the specific issue to address, it might or might not be necessary to apply some of them. Examples are aggregation/integration (only if there are multiple sources), cleaning (only if there are erroneous, missing or incomplete data), user identification (only if the educational system does not provide user identification), data filtration and attribute selection (only if there is a huge amount of data and/or attributes respectively).
- Different types of techniques have also been used in each task/step, although there is no recipe or rule about which specific technique should be used in each pre-processing task. Therefore, the user will be the one in charge of selecting which one to apply each time depending on several issues, such as the specific characteristics of the data, the tools and algorithms available, and the final objective or data mining problem to be solved.

Acknowledgments This research is supported by projects of the Regional Government of Andalucía and the Ministry of Science and Technology, P08-TIC-3720 and TIN-2011-22408, respectively, and FEDER funds.

References

1. Romero, C., Ventura, S.: Data mining in education. WIREs Data Min. Knowl. Disc. **1**(3), 12–27 (2013)
2. Han, J., Kamber, M.: Data Mining: Concepts and Techniques. Morgan Kaufmann Publishers, San Francisco (2006)
3. Miksovsky, P., Matousek, K., Kouba, Z.: Data Pre-processing support for data mining. In: IEEE International Conference on Systems, Man and Cybernetics, pp. 208–212, Hammamet, Tunisia (2002)
4. Pyle, D.: Data Preparation for Data Mining. Morgan Kaufmann, San Francisco (1999)
5. Gonçalves, P.M., Barros, R.S.M., Vieria, D.C.L: On the use of data mining tools for data preparation in classification problems. In: 11th International Conference on Computer and Information Science, pp. 173–178, IEEE, Washington (2012)
6. Bohanec, M., Moyle, S., Wettschereck, D., Miksovsk, P.: A software architecture for data pre-processing using data mining and decision support models. In: ECML/PKDD'01 Workshop on Integrating Aspects of Data Mining, Decision Support and Meta-Learning, pp. 13–24 (2001)

7. Sael, N., Abdelaziz, A., Behja, H.: Investigating and advanced approach to data pre-processing in Moodle platform. Int. Rev. Comput. Softw. **7**(3), 977–982 (2012)
8. Marquardt, C.G., Becker, K., Ruiz, D.D.: A Pre-processing tool for web usage mining in the distance education Domain. In: International Database Engineering and Applications Symposium, pp. 78–87. IEEE Computer Society, Washington (2004)
9. Wettschereck, D.: Educational data pre-processing. In: ECML'02 Discovery Challenge Workshop, pp. 1–6. University of Helsinki, Helsinki (2002)
10. Simon, J.: Data preprocessing using a priori knowledge. In: D'Mello, S.K., Calvo, R.A., Olney, A. (eds.) 6th International Conference on Educational Data Mining, pp. 352–353. International Educational Data Mining Society, Memphis (2013)
11. Rice, W.H.: Moodle E-learning Course Development. A Complete Guide to Successful Learning Using Moodle. Packt publishing, Birmingham (2006)
12. Ma, Y., Liu, B., Wong, C., Yu, P., Lee, S.: Targeting the right students using data mining. In: Sixth ACM SIGKDD International Conference on Knowledge Discovery and Data mining, pp. 457–464. ACM, New York (2000)
13. Silva, D., Vieira, M.: Using data warehouse and data mining resources for ongoing assessment in distance learning. In: IEEE International Conference on Advanced Learning Technologies, pp. 40–45. IEEE Computer Society, Kazan (2002)
14. Clow, D.: MOOCs and the funnel of participation. In: Suthers, D., Verbert, K., Duval, E., Ochoa, X. (eds.) International Conference on Learning Analytics and Knowledge, pp. 185–189. ACM New York, NY (2013)
15. Anderson, J., Corbett, A., Koedinger, K.: Cognitive tutors. J. Learn. Sci. **4**(2), 67–207 (1995)
16. Mostow, J., Beck, J.: Some useful tactics to modify, map and mine data from intelligent tutors. J. Nat. Lang. Eng. **12**(2), 95–208 (2006)
17. Brusilovsky, P., Peylo, C.: Adaptive and intelligent web-based educational systems. Int. J. Artif. Intell. Educ. **13**(2–4), 159–172 (2003)
18. Merceron, A., Yacef, K.: Mining student data captured from a web-based tutoring tool: initial exploration and results. J. Interact. Learn. Res. **15**(4), 319–346 (2004)
19. Brusilovsky, P., Miller, P.: Web-based testing for distance education. In: De Bra, P., Leggett, J. (eds.) WebNet'99, World Conference of the WWW and Internet, pp. 149–154. AACE, Honolulu (1999)
20. Hanna, M.: Data mining in the e-learning domain. Campus-Wide Inf. Syst. **21**(1), 29–34 (2004)
21. Romero, C., Ventura, S., Salcines, E.: Data mining in course management systems: moodle case study and tutorial. Comput. Educ. **51**(1), 368–384 (2008)
22. Agrawal, R., Srikant, R.: Mining sequential patterns. In: Eleventh International Conference on Data Engineering, pp. 3–4. IEEE, Washington (1995)
23. Romero, C., Ventura, S., Zafra, A., De Bra, P.: Applying web usage mining for personalizing hyperlinks in web-based adaptive educational systems. Comput. Educ. **53**(3), 828–840 (2009)
24. Feldman, R., Sanger, J.: The Text Mining Handbook: Advanced Approaches in Analyzing Unstructured Data. Cambridge University Press, Cambridge (2006)
25. Dringus, L.P., Ellis, T.: Using data mining as a strategy for assessing asynchronous discussion forums. Comput. Educ. J. **45**(1), 141–160 (2005)
26. Petrushin, V., Khan, L. (eds.): Multimedia Data Mining and Knowledge Discovery. Springer, London (2007)
27. Bari, M., Lavoie, B.: Predicting interactive properties by mining educational multimedia presentations. In: International Conference on Information and Communications Technology, pp. 231–234. Bangladesh University of Engineering and Technology, Dhaka (2007)
28. Srivastava, J., Cooley, R., Deshpande, M., Tan, P.: Web usage mining: discovery and applications of usage patterns from web data. SIGKDD Explor. **1**(2), 12–23 (2000)
29. Romero, C., Ventura, S.: Educational data mining: a survey from 1995 to 2005. Expert Syst. Appl. **33**(1), 135–146 (2007)
30. Vranic, M., Pintar, D., Skocir, Z.: The use of data mining in education environment. In: 9th International Conference on Telecommunications, pp. 243–250. IEEE, Zagreb (2007)

31. Gibert, K., Izquierdo, J., Holmes, G., Athanasiadis, I., Comas, J., Sanchez, M.: On the role of pre and post processing in environmental data mining. In: Sánchez-Marré, M., Béjar, J., Comas, J., Rizzoli, A. E., Guariso, G. (eds.) iEMSs Fourth Biennial Meeting: International Congress on Environmental Modelling and Software (iEMSs 2008), pp. 1937–1958. International Environmental Modelling and Software Society, Barcelona (2008)
32. Witten, I.H., Frank, E.: Data Mining: Practical Machine Learning Tools and Techniques. Morgan Kaufmann, San Francisco (2005)
33. Zhu, F., Ip, H., Fok, A., Cao, J.: PeRES: A Personalized Recommendation Education System Based on Multi-Agents & SCORM. In: Leung, H., Li, F., Lau, R., Li, Q. (eds.) Advances in Web Based Learning—ICWL 2007. LNCS, vol. 4823, pp. 31–42. Springer, Heidelberg (2007)
34. Avouris, N., Komis, V., Fiotakis, G., Margaritis, M., Voyiatzaki, E.: Why logging of fingertip actions is not enough for analysis of learning activities. In: Workshop on Usage Analysis in Learning Systems, pp. 1–8. AIED Conference, Amsterdam (2005)
35. Chanchary, F.H., Haque, I., Khalid, M.S.: Web usage mining to evaluate the transfer of learning in a web-based learning environment. In: International Workshop on Knowledge Discovery and Data Mining, pp. 249–253. IEEE, Washington (2008)
36. Spacco, J., Winters, T., Payne, T.: Inferring use cases from unit testing. In: AAAI Workshop on Educational Data Mining, pp. 1–7, AAAI Press, New York (2006)
37. Zhang, L, Liu, X., Liu, X.: Personalized instructing recommendation system based on web mining. In: International Conference for Young Computer Scientists, pp. 2517–2521. IEEE Computer Society Washington (2008)
38. Barnes, T.: The Q-matrix method: mining student response data for knowledge. In: AAAI-2005 Workshop on Educational Data Mining, pp. 1–8, AAAI Press, Pittsburgh (2005)
39. Chen, C., Chen, M., Li, Y.: Mining key formative assessment rules based on learner profiles for web-based learning systems. In: Spector, J.M., Sampson D.G., Okamoto, T., Kinshuk, Cerri, S.A., Ueno, M., Kashihara, A. (eds.) IEEE International Conference on Advanced Learning Technologies, pp. 1–5. IEEE Computer Society, Los Alamitos (2007)
40. Wang, F.H.: A fuzzy neural network for item sequencing in personalized cognitive scaffolding with adaptive formative assessment. Expert Syst. Appl. J. **27**(1), 11–25 (2004)
41. Markham, S., Ceddia, J., Sheard, J., Burvill, C., Weir, J., Field, B.: Applying agent technology to evaluation tasks in e-learning environments. In: International Conference of the Exploring Educational Technologies, pp. 1–7. Monash University, Melbourne (2003)
42. Medvedeva, O., Chavan, G., Crowley, R.: A data collection framework for capturing its data based on an agent communication standard. In: 20th Annual Meeting of the American Association for Artificial Intelligence, pp. 23–30, AAAI, Pittsburgh (2005)
43. Shen, R., Han, P., Yang, F., Yang, Q., Huang, J.: Data mining and case-based reasoning for distance learning. J. Distance Educ. Technol. **1**(3), 46–58 (2003)
44. Lenzerini, M.: Data integration: a theoretical perspective. In: International Conference on ACM SIGMOD/PODS, pp. 233–246. ACM, New York (2002)
45. Ingram, A.: Using web server logs in evaluating instructional web sites. J. Educ. Technol. Syst. **28**(2), 137–157 (1999)
46. Peled, A., Rashty, D.: Logging for success: advancing the use of WWW logs to improve computer mediated distance learning. J. Educ. Comput. Res. **21**(4), 413–431 (1999)
47. Talavera, L., Gaudioso, E.: Mining student data to characterize similar behavior groups in unstructured collaboration spaces. In: Workshop on Artificial Intelligence in CSCL, pp. 17–23. Valencia (2004)
48. Romero, C., Ventura, S., Bra, P.D.: Knowledge discovery with genetic programming for providing feedback to courseware author. User modeling and user-adapted interaction. J. Personalization Res. **14**(5), 425–464 (2004)
49. Mostow, J., Beck, J.E.: Why, what, and how to log? Lessons from LISTEN. In: Barnes, T., Desmarais, M., Romero, R., Ventura, S. (eds.) 2nd International Conference on Educational Data Mining, pp. 269–278. International Educational Data Mining Society, Cordoba (2009)

50. Binli, S.: Research on data-preprocessing for construction of university information systems. In: International Conference on Computer Application and System Modeling, pp. 459–462. IEEE, Taiyuan (2010)
51. Dierenfeld, H., Merceron, A.: Learning analytics with excel pivot tables. In: Moodle Research Conference, pp. 115–121. University of Piraeus, Heraklion (2012)
52. Solodovnikova, D., Niedrite, L.: Using data warehouse resources for assessment of e-earning influence on university processes. In: Eder, J., Haav, H.M., Kalja, A., Penjam, J. (eds.) 9th East European Conference, ADBIS 2005. Advances in Databases and Information Systems. LNCS, vol. 3631, pp. 233-248. Springer, Heidelberg (2005)
53. Merceron, A., Yacef, K.: Directions to Enhance Learning Management Systems for Better Data Mining. Personal Communication (2010)
54. Yan, S., Li, Z.: Commercial decision system based on data warehouse and OLAP. Microelectron. Comput. **2**, 64–67 (2006)
55. Zorrilla, M.E., Menasalvas, E., Marin, D., Mora, E., Segovia, J.: Web usage mining project for improving web-based learning sites. In: Moreno-Díaz, R., Pichler, F., Quesada-Arencibia, A. (eds.) Computer Aided Systems Theory—EUROCAST 2005. LNCS, vol. 3643, pp. 205–210. Springer, Heidelberg (2005)
56. Yin, C., Luo, Q.: Personality mining system in e-learning by using improved association rules. In: International Conference on Machine Learning and Cybernetics, pp. 4130–4134. IEEE, Hong Kong (2007)
57. Heiner, C., Beck, J.E., Mostow, J.: Lessons on using ITS data to answer educational research questions. In: Lester, J.C., Vicari, R.S., Paraguaçu, F. (eds.) Intelligent Tutoring Systems, 7th International Conference, ITS 2004. LNCS, vol. 3220, pp. 1–9. Springer, Heidelberg (2004)
58. Rubin, D.B., Little, R.J.A.: Statistical Analysis with Missing Data. Wiley, New York (2002)
59. Salmeron-Majadas, S., Santos, O., Boticario, J.G., Cabestrero, R., Quiros, P.: Gathering emotional data from multiple sources. In: D'Mello, S.K., Calvo, R.A., Olney, A. (eds.) 6th International Conference on Educational Data Mining, pp. 404–405. International Educational Data Mining Society, Memphis (2013)
60. Shuangcheng, L., Ping, W.: Study on the data preprocessing of the questionarie based on the combined classification data mining model. In: International Conference on e-Learning, Enterprise Information Systems and E-Goverment, pp. 217–220. Las Vegas (2009)
61. García, E., Romero, C., Ventura, S., Castro, C.: An architecture for making recommendations to courseware authors using association rule mining and collaborative filtering. User Model. User-Adap. Inter. **19**(1–2), 99–132 (2009)
62. Huang, C., Lin, W., Wang, S., Wang, W.: Planning of educational training courses by data mining: using China Motor Corporation as an example. Expert Syst. Appl. J. **36**(3), 7199–7209 (2009)
63. Kantardzic, M.: Data Mining: Concepts, Models, Methods, and Algorithms. Wiley, New York (2003)
64. Beck, J.E.: Using learning decomposition to analyze student fluency development. In: Workshop on Educational Data Mining at the 8th International Conference on Intelligent Tutoring Systems, pp. 21–28. Jhongli (2006)
65. Redpath, R., Sheard, J.: Domain knowledge to support understanding and treatment of outliers. In: International Conference on Information and Automation, pp. 398–403. IEEE, Colombo (2005)
66. Sunita, S.B., Lobo, L.M.: Data preparation strategy in e-learning system using association rule algorithm. Int. J. Comput. Appl. **41**(3), 35–40 (2012)
67. Ivancsy, R., Juhasz, S.: Analysis of web user identification methods. World Acad. Sci. Eng. Technol. J. **34**, 338–345 (2007)
68. Rahkila, M., Karjalainen, M.: Evaluation of learning in computer based education using log systems. In: ASEE/IEEE Frontiers in Education Conference, pp. 16–21. IEEE, San Juan (1999)
69. Wang, F.H.: Content recommendation based on education-contextualized browsing events for web-based personalized learning. Educ. Technol. Soc. **11**(4), 94–112 (2008)

70. Munk, M., Drlík, M.: Impact of Different pre-processing tasks on effective identification of users' behavioral patterns in web-based educational system. Procedia Comput. Sci. **4**, 1640–1649 (2011)
71. Heraud, J.M., France, L., Mille, A.: Pixed: an ITS that guides students with the help of learners' interaction log. In: Lester, J.C., Vicari, R.S., Paraguaçu, F. (eds.) Intelligent Tutoring Systems, 7th International Conference, ITS 2004. LNCS, vol. 3220, pp. 57–64. Springer, Heidelberg (2004)
72. Sheard, J., Ceddia, J., Hurst, J., Tuovinen, J.: Inferring student learning behaviour from website interactions: a usage analysis. J. Educ. Inf. Technol. **8**(3), 245–266 (2003)
73. Petersen, R.J.: Policy dimensions of analytics in higher education. Educause Rev. **47**, 44–49 (2012)
74. Bienkowski, M., Feng, M., Means, B.: Enhancing Teaching and Learning Through Educational Data Mining and Learning Analytics: An Issue Brief. U.S. Department of Education, Office of Educational Technology, pp. 1–57 (2012)
75. Liu, H., Motoda, H.: Computational Methods of Feature Selection. Chapman & Hall/CRC, Boca Raton (2007)
76. Delavari, N., Phon-Amnuaisuk, S., Beikzadeh, M.: Data mining application in higher learning institutions. Inf. Educ. J. **7**(1), 31–54 (2008)
77. Kotsiantis, B., Kanellopoulos, D., Pintelas, P.: Data pre-processing for supervised learning. Int. J. Comput. Sci. **1**(2), 111–117 (2006)
78. Mihaescu, C., Burdescu, D.: Testing attribute selection algorithms for classification performance on real data. In: International IEEE Conference Intelligent Systems, pp. 581–586. IEEE, London (2006)
79. Márquez-Vera, C., Cano, A., Romero, C., Ventura, S.: Predicting student failure at school using genetic programming and different data mining approaches with high dimensional and imbalanced data. Appl. Intell. **38**(3), 315–330 (2013)
80. Wong, S.K., Nguyen, T.T., Chang, E., Jayaratnal, N.: Usability metrics for e-learning. In: Meersman, R., Tari, Z. (eds.) On the Move to Meaningful Internet Systems 2003: OTM 2003 Workshops, LNCS, vol. 2889, pp. 235–252. Springer, Heidelberg (2003)
81. Hershkovitz, A. Nachmias, R.: Consistency of students' pace in online learning. In: Barnes, T., Desmarais, M., Romero, R., Ventura, S. (eds.) 2nd International Conference on Educational Data Mining, pp. 71–80. International Educational Data Mining Society, Cordoba (2009)
82. Mor, E., Minguillón, J.: E-learning personalization based on itineraries and long-term navigational behavior. In: Thirteenth World Wide Web Conference, pp. 264–265. ACM, New York (2004)
83. Nilakant, K., Mitrovic, A.: Application of data mining in constraint based intelligent tutoring systems. In: International Conference on Artificial Intelligence in Education, pp. 896–898. Amsterdam (2005)
84. Baker, R., Carvalho, M.: A labeling student behavior faster and more precisely with text replays. In: Baker, R.S.J.d, Barnes, T., Beck, J.E. (eds.) 1st International Conference on Educational Data Mining, pp. 38–47. International Educational Data Mining Society, Montreal (2008)
85. Zhou, M., Xu, Y., Nesbit., J.C., Winne, P.H.: Sequential pattern analysis of learning logs: methodology and applications. In: Romero, C., Ventura, S., Pechenizkiy, M., Baker, R.S. J.D. (eds.) Handbook of Educational Data Mining, Chapman & Hall/CRC Data Mining and Knowledge Discovery Series, pp. 107–120. CRC Press, Boca Raton (2010)
86. Liu, B.: Web Data Mining: Exploring Hyperlinks, Contents and Usage Data. Springer, Heidelberg (2011)
87. Thai, D., Wu, H., Li, P.: A hybrid system: neural network with data mining in an e-learning environment. In: Jain, L., Howlett, R.J., Apolloni, B. (eds.) Knowledge-Based Intelligent Information and Engineering Systems, 11th International Conference, KES 2007, XVII Italian Workshop on Neural Networks. LNCS, vol. 4693, pp. 42–49. Springer, Heidelberg (2007)

88. Hien, N.T.N., Haddawy, P.: A decision support system for evaluating international student applications. In: Frontiers in Education Conference, pp. 1–6. IEEE, Piscataway (2007)
89. Kosheleva, O., Kreinovich, V., Longrpre, L.: Towards interval techniques for processing educational data. In: International Symposium on Scientific Computing, Computer Arithmetic and Validated Numerics, pp. 1–28. IEEE Computer Society, Washington (2006)
90. Hämäläinen, W., Vinni, M.: Classifiers for educational data mining. In: Romero, C., Ventura, S., Pechenizkiy, M., Baker, R.S.J.d. (eds.) Handbook of Educational Data Mining, Chapman & Hall/CRC Data Mining and Knowledge Discovery Series, pp. 57–71. CRC Press, Boca Raton (2010)
91. Cocea, M., Weibelzahl, S.: Can log files analysis estimate learners' level of motivation? In: Workshop week Lernen—Wissensentdeckung—Adaptivität, pp. 32–35. Hildesheim (2006)
92. Tanimoto, S.L.: Improving the prospects for educational data mining. In: Track on Educational Data Mining, at the Workshop on Data Mining for User Modeling, at the 11th International Conference on User Modeling, pp. 1–6. User Modeling Inc., Corfu (2007)
93. Werner, L., McDowell, C., Denner, J.: A first step in learning analytics: pre-processing low-level Alice logging data of middle school students. J. Educ. Data Min. (2013, in press)
94. Alcalá, J., Sanchez, L., García, S., Del Jesus, M.J., Ventura, S., Garrell, J.M., Otero, J., Romero, C., Bacardit, J., Rivas, V., Fernández, J.C., Herrera, F.: KEEL: a software tool to assess evolutionary algorithms to data mining problems. Soft. Comput. **13**(3), 307–318 (2009)
95. Gonçalves, P.M., Barros, R.S.M.: Automating data preprocessing with DMPML and KDDML. In: 10th IEEE/ACIS International Conference on Computer and Information Science, pp. 97–103. IEEE, Washington (2011)
96. Zaïne, O.R., Luo, J.: Towards evaluating learners' behaviour in a web-based distance learning environment. In: IEEE International Conference on Advanced Learning Technologies, pp. 357–360. Madison, WI (2001)
97. Ceddia, J., Sheard, J., Tibbery, G.: WAT: a tool for classifying learning activities from a log file. In: Ninth Australasian Computing Education Conference, pp. 11–17. Australian Computer Society, Darlinghurst (2007)
98. Rodrigo, M.T., Baker, R., McLaren, B.M., Jayme, A., Dy, T. : Development of a workbench to address the educational data mining bottleneck. In: Yacef, K., Zaïane, O., Hershkovitz, A., Yudelson, M., Stamper, J. (eds.) 5th International Conference on Educational Data Mining, pp. 152–155. International Educational Data Mining Society, Chania (2012)
99. Koedinger, K., Cunningham, K., Skogsholm, A., LEBER, B.: An open repository and analysis tools for fine-grained, longitudinal learner data. In: Baker, R.S.J.d, Barnes, T., Beck, J.E. (eds.) 1st International Conference on Educational Data Mining, pp. 157–166. International Educational Data Mining Society, Montreal (2008)

Chapter 3
How Educational Data Mining Empowers State Policies to Reform Education: The Mexican Case Study

Alejandro Peña-Ayala and Leonor Cárdenas

Abstract In this chapter we present a case study that illustrates how educational data mining (EDM) is able to support the implementation of government policies and assist the labor of public institutions. Specifically, we highlight the current educational reforms in Mexico and focus on one of its main goals: to enhance the education quality. In response, a valuable data source is mined to discover interesting findings what students think about education, family, teachers, and their surroundings. Thus, a brief description of the legal and social context is given, as well as a profile of the students opinions expressed in a national survey is shaped. Moreover, a framework to build an EDM approach is outlined and a sample of the mined results is stated. As a result of the findings generated by the EDM approach, an interpretation is provided to tailor a conceptual view of the observations made by students, as well as some initiatives to deal with the findings. The work concludes with an exposition of the reasons for presenting this kind of work, a comment on the research fulfilled, a viewpoint of the education in Mexico, and some suggestions to support State polices to enhance education.

Keywords Data mining · Knowledge discovery in databases · Educational data mining · Educational policies · Student opinions · Clustering · Association rules

Abbreviations

AIWBES Adaptive and intelligent web-based educational systems
CBIS Computer-based information systems

A. Peña-Ayala (✉)
WOLNM: Artificial Intelligence on Education Lab, 31 Julio 1859 No. 1099-B, Leyes Reforma Mexico City 09310, Mexico
e-mail: apenaa@wolnm.org; apenaa@ipn.mx

A. Peña-Ayala · L. Cárdenas
ESIME Zacatenco, Instituto Politécnico Nacional, Building Z-4, 2nd Floor, Lab 6, Miguel Othón de Mendizábal S/N, México, DF 07320, Mexico
e-mail: adriposgrado@gmail.com

A. Peña-Ayala (ed.), *Educational Data Mining*,
Studies in Computational Intelligence 524, DOI: 10.1007/978-3-319-02738-8_3,

CNTE	National Coordination of Workers of the Education
CPEUM	Politic Constitution of the Mexican United States
CV	Confidence value
DK	Domain knowledge
DM	Data mining
EDM	Educational data mining
EM	Expectation maximization
ENLACE	National Evaluation of the Academic Achievement of Scholar Centers
ES	Educational systems
EXCALE	Exams for the Quality and Educative Achievements
INEE	National Institute for Educative Evaluation
ITS	Intelligent tutoring systems
KDD	Knowledge discovery in databases
LCMS	Learning content management systems
LMS	Learning management systems
PISA	Program for International Student Assessment
SAV	Statistical Package for the Social Sciences data document
SEP	Secretary of Public Education
SNTE	National Union of Workers of the Education
SPS	Statistical Package for the Social Sciences syntax
SPSS	Statistical Package for the Social Sciences
SQL	Structured Query Language
TXT	Text data file
USA	United States of America
WBC	Web-based courses
XLSX	Excel extended

3.1 Introduction

Education is the ground to propel the progress of an individual, family, and society. In this context, the state of a nation is the rector of the education and practices. It mandates by the definition of constitutional articles and laws, as well as the founding of ministries and institutions to lead, fund, provide, and control education. The accomplishment of those responsibilities requires a responsible contribution of diverse actors, such as: politicians, members of Parliament, civil servants, principals, academics, labor union, and support staff, as well as students, parents, relatives, tutors, and classmates.

In addition to human assets, resource such as funding, facilities, logistics, supplies, and computer equipment are needed, as well as information, material, and technological resources. All of these contribute to build the setting to deliver education through classroom, open, and distance modalities. The education scope is extensive; it includes mandatory, professional, postgraduate, supplementary, and long-life categories. So, education is available for everybody at anytime, anyplace, and with anything.

Indeed, the educative labor is permanent, diverse, complex, demanding, controversial, and dynamic. It claims a legal driving force, realistic policies, generous funding, avant-garde programs, updated curricula, suitable contents and materials, useful artifacts, and appropriate facilities, as well as highly-trained and paid academics and specialized staff, motivated students, and committed families and tutors.

In an ideal landscape, *quality* is the reference to measure activities, behaviors, results, resources, and human assets. Subjects such as academic goals and programs, pedagogical and social criteria, students' profile and tracking-logs, performance and behavior records, psychometric and educational tests, domain knowledge (DK), skill assessments, mastery of competencies, the academics development and efficiency, and students' ratings are key sources to evaluate the quality of education.

However, the management of those subjects demands modern computer-based information systems (CBIS), which are able to automate the collection, representation, storing, processing, and exploitation of data derived from such assets, functions, factors, and sources. Furthermore, approaches for knowledge discovery in databases (KDD) are also needed to facilitate decision making for all the individuals involved in educational duties. In consequence, data mining (DM) emerges as a paradigm to scan huge data repositories for inferring relevant findings. Particularly, the DM specialty that deals with those specific requirements is: educational data mining (EDM).

EDM is a novel field that assists operative, tactical, and strategic educational activity. It is oriented to mine raw data extracted from conventional and online educational systems (ES). EDM offers proactive, real-time, and reactive support to ES. For instance, EDM aids these three stages when providing reliable knowledge to respectively: adapting teaching experiences prior to be delivered, offering suitable hints to students at problem solving, and facilitating the assessment of the acquired DK.

The work presented in this chapter shapes a sample of how EDM is an ally to deploy policies for reforming education. It highlights the current situation in Mexico, where the government reforms education by legal amendments and academic objectives, including quality enhancement. As a strategy, the students' opinion is considered a valuable feedback that is worthy to be taken into account. Thus, we build an EDM approach to mine a survey of students' opinions. As a result, diverse findings are discovered to reveal key issues that claim increasing education quality.

The exposition of the case study is organized as follows: The next three sections offer a profile of DM and EDM, a collection of related works, and the political, academic, and social context of the case study. Sections 3.5 to 3.7 introduce the test-bead and framework respectively used to extract data and build the EDM approach to discover findings, interpret, and diagnose students' opinions. The conclusion argues the reasons for including this kind of work as part of the book, sketches a general panorama of education in Mexico, and proposes EDM to support State policies.

3.2 Domain Study

The domain study of this work is made up of a pair of fields, DM and EDM. The first provides the grounds, logistics, and tools for KDD; whereas, the second is oriented to mine source data of educational settings and support academic functionalities carried out by ES. With the aim at tailoring the baseline of the domain study, a brief profile of DM and EDM is given in the following subsections.

3.2.1 A Glance at Data Mining

Essentially, CBIS are data processing systems oriented to produce information as the fundaments to back up knowledge [1]. A classic aim of CBIS is decisions-making, a routine task for humans and organizations. It demands essential data, useful information, and reliable knowledge to decide action courses that guide the behavior of people and organizations. Thus, efficient approaches composed of frameworks, models, methods, techniques, and tools are needed to support decisions-making practice.

Management information systems [2], decision support systems [3], data warehouses [4], online analytical processing [5], business intelligence [6], and KDD [7] are a sample of paradigms built to meet such needs, where DM is a new one [8].

DM is a field that mines information and extracts knowledge from large amounts of data [9]. DM finds out data patterns, uncovers hidden relationships, draws association rules, estimates unknown values to classify objects, composes sets of homogenous objects, and unveils findings that are not produced by classic CBIS.

The DM baseline is ground by diverse disciplines such as: statistic [10], probability [11], visualization [12], machine learning [13], and natural language [14]. They offer methods and algorithms that are useful to deal with DM models and tasks.

Fundamentally, descriptive and predictive DM models are designed. Descriptive models usually apply unsupervised learning functions to produce patterns that explain or generalize the intrinsic structure and relationships of data [15]. Predictive models frequently apply supervised learning functions to estimate unknown or future values of dependent variables based on the features of related independent variables [16].

A model is deployed by means of tasks that focus on similar problems to be solved. For instance, clustering [17], correlation analysis [18], association rules [19], and sequential pattern [20] implement descriptive models; whilst, classification [21], regression [22], and preferences [23] generate predictive models.

Some methods used to perform a task are the following: linear and logistic regression [24, 25], linear discriminant [26], support vector machines [27], Naïve Bayes [28], classification trees [29], rules induction [30], instances-based learning [31].

Once a method is chosen to fulfill a task, an algorithm is programmed to mine the source data [32]. A collection of popular algorithms is introduced as follows: Naïve Bayes tree [33], expectation maximization (EM) [34], C4.5 [35], CART [36], CN2 [37], Apriori [38], k-nearest neighbor [39], k-means [40], maximal frequent item set [41], randomized function K [42], independent choice logic [43], PageRank [44], and TwitterRank [45].

3.2.2 Educational Data Mining in a Nutshell

Education is a novel DM target of application for knowledge discovery, decisions-making, and recommendation [46]. EDM emerges as a paradigm oriented to design models, tasks, methods, and algorithms for exploring data from conventional, open, and distance educational settings [47]. EDM finds out patterns and makes predictions to depict learners' behaviors and achievements, DK, assessments, student support and ratings [48]. In order to shape an EDM profile, three surveys are stated as follows.

The first survey covers EDM works published from 1995 up to 2005 [49]. It contains 81 references, where only seven correspond to the 1990s. The review recognizes conventional and distance ES. The former is given in classrooms; whilst, the latter is delivered through web-based courses (WBC), adaptive and intelligent web-based educational systems (AIWBES), and learning content management systems (LCMS). According to three kinds of DM techniques, 36 citations are distributed for the three sorts of distance systems as follows: (a) 15 works of clustering, classification, and outlier detection implemented in WBC 3, AIWBES 9, LCMS 3; (b) 14 approaches of association rules and sequential pattern deployed in WBC 6, AIWBES 4, LCMS 4; (c) 7 text mining applications embedded in WBC 4, AIWBES 1, LCMS 2.

The second survey enhances the first adding publications from 2006 up to the start of 2010 [50]. Eight types of ES categorize 235 works as follows: 36 conventional, 54 web-based education, 29 learning management systems (LMS), 31 intelligent tutoring systems (ITS), 26 adaptive educational systems, 23 test-questionnaires, 14 text-contents, and 22 others. As for educational tasks, the works are gathered into eleven categories as follows: (a) analysis and visualization of data 35; (b) feedback provision 40; (c) recommenders 37; (d) students performance 76; (e) student modeling 28; (f) detecting student behaviors 23; (g) grouping students 26; (h) social network analysis 15; (i) concept maps 10; (j) constructing courseware 9; (k) planning and scheduling 11.

The third survey updates the chronicles of EDM evolution, due to it presents 240 works published from 2010 up to the first quarter of 2013 [51]. The sample is split into 222 approaches and 18 tools. The approaches are gathered according to six educational functionalities as follows: (a) student modeling 43; (b) student behavior modeling 48; (c) student performance modeling 46; (d) assessment 45; (e) student support and feedback 21; (f) curriculum, DK, sequencing, and teachers support 19. During the analysis of the collected approaches, a *profile* was designed to highlight educational and DM traits that characterize them. As result of the statistical and DM processes applied to the 222 profiles, the following counts are unveiled:

The most common ES are ITS, LMS, and conventional with 88, 20, and 20 cases; whilst, the most used ES instances are Algebra, ASSISTments, and Moodle with 20, 19, and 13 occurrences. As for the involved disciplines, probability, machine learning, and statistic provide the grounds for 181, 90, and 47 works; whereas, 60 % of the approaches correspond to predictive models and 40 % to descriptive. Concerning the tasks, classification, clustering, and regression are deployed by 102, 65, and 37 works; while Bayes theorem, decisions trees, and instance-based learning are taken into account by 48, 44, and 22 approaches. Regarding algorithms, the top four are: K-means, EM, J48, NaiveBayes, which are deployed by 19, 15, 15, and 15 approaches.

3.3 Related Works

One of the novel targets of EDM labor corresponds to students' opinions. It is devoted to support the students' right of expression! This ES functionality facilitates the students' feedback by means of collecting comments, suggestions, complains, requests, and evaluations about their teachers, family, facilities, resources, curricula, content, DK, and any other issue that students like to externalize. These opinions represent a valuable source that is worthy to be considered for estimating and improving the quality of education. In order to depict its nature, some works are stated next.

A well-known survey for gathering students' opinions is the Students Evaluations of Teaching [52]. The test is developed as a means of collecting data from students on their experience of learning at the individual subject or unit of study level. Another similar test is the Unit of Study Evaluation. It is designed to support aspects of the management and evaluation of coursework teaching at the University of Sidney [53].

An interesting approach is reported by Kim and Calvo [54], who detect learners' opinions (e.g., positive, negative…) and emotions (e.g., joy, surprise, anger, fear…). They apply category-based (e.g., latent semantic analysis and non-negative matrix factorization) and dimension-based emotion prediction models.

Opinions and emotions are inferred from textual and quantitative students' responses to a Unit of Study Evaluations test to provide a comprehensive understanding of student experience.

As for Champaign and Cohen [55], they explore the use of student annotations by allowing students to leave comments on learning objects they are using. Later, subsequent students could identify the annotations they find useful, which would then be intelligently shown to similar students. Authors develop a model for reasoning about which learning objects and annotations should be presented to future students.

Kopriska studies the effect of the stream, regular and advanced, on the student evaluation of teaching and course marks [56]. She identifies how the differences between the two streams (i.e., units of study demands and prior academic performance) affect the student evaluation of teaching and the units of study assessment results.

Barracosa and Antunes anticipate teachers' performance based on the analysis of pedagogical surveys filled by students [57]. Their approach pre-processes the surveys, applies sequential pattern mining to identify meta-patterns used for enriching the assessment of teachers. Thereby, they improve classification models accuracy.

Leong et al. [58] identify sentiments by mining short message service texts in teaching evaluation. Once the texts are read, parsed, and categorized, three models are built: base, corrected that adjusts for spelling errors, and sentiment which extends the corrected. An interestingness criterion selects the sentiment model from which the sentiments of the students towards the lecture are discerned.

In another vein, Gates et al. [59] maximize the value of student ratings by DM. They develop meaningful analysis of over one million student narrative comments collected through online process. Thus, they design a methodology to depict instructor traits and a domain-specific lexicon by positive and negative category vectors. Moreover, sentiment analysis is applied to detect and gauge attitudes expressed in comments about each category. The methodology is validated using three assessment approaches at the University of Mississippi.

3.4 Context

As the case study is concerned with educational reform in Mexico, specifically the enhancement of quality, it is pertinent to shape the study context. So, five subjects are introduced in this section. The first depicts the Mexican State as the main instance of representation and authority of the nation. The second identifies some actors and provides a sample of statistics to tailor the dimension of the educational community. The third summarizes the national assessments and the fourth shapes the educational reform that the current Mexican President heads to transform education and improve its quality. The last shows reactions of sectors that are essential to implement the reform.

3.4.1 The Mexican State

The Mexican United States, known as Mexico, is organized and regulated by a Mexican State. It is a Federation composed of territory, population, Federal Government, and laws. The Federation has thirty one states and one Federal District. The states are free and sovereign; and have their own political constitution and parliament. The Federal District is the site of the Federal Government and the country capital.

The territory is situated in North America, where the United States of America (USA) is at the north and Belize and Guatemala are located at the south. Its eastern coast corresponds to the Atlantic Ocean, and the western coast to the Pacific Ocean. The territory covers nearly 1,965,000 square kilometers.

As for the population, 112 millions inhabitants were counted as result of the last national census of population and housing taken in 2010. The rate of growth is 1.40 %, the population density per square kilometer is 57 [60]. The official estimation of population for age-ranges in 2010 is revealed in the fourth column of Table 3.1, where it is associated to four mandatory academic levels [61].

Concerning the Federal Government, constitutionally named Supreme Power of the Federation, it is based on a Presidential system, where the president is the Head of the State, as well as the Head of the Government. It embraces three Powers of the Union: Executive, Legislative, and Judicial. The Executive Power is headed by the President, who is assisted by State Secretaries in order to rule the nation. The Legislative Power is the Union Congress composed of two chambers: Senators and Deputies, which set the legal frame of the nation. The Judicial Power applies the laws to provide justice through the Supreme Court of Justice, magistrates' courts, and tribunals.

In another vein, the laws provide the legal framework to shape the live, values, principles, rights, duties, activities, and vision of the nation. The main referent is the Politic Constitution of the Mexican United States (CPEUM[1]) established in 1917. Article 40 asserts, "It is the will of the Mexican population to form a representative, democratic, and Federal Republic…" At that time, the Mexican constitution was the first constitution to include social rights. Particularly, the first three articles are consigned to acknowledge that the Mexican State must take care, provide, and guarantee human rights, public health, and education for all the members of society [62].

3.4.2 Educational Community

The educational community is made up of several sectors. However, six are the main protagonists of the case study stated in this chapter. One corresponds to

[1] Some acronyms maintain the initials written in Spanish to preserve their national identity.

Table 3.1 Mandatory academic levels, years of study, age-range, population, roll, covering

Level	Years	Age-range	Population	Roll	Covering (%)
Pre-primary	3	3–5	6,535,234	4,705,545	72
Primary	6	6–12	15,516,889	14,909,419	96
Middle school	3	13–15	6,570,144	6,167,424	94
High school	3	15–19	8,761,774	2,147,167	25

students, another to teachers, and one more to schools. The fourth concerns the Federal Government, the fifth the labor unions, and the last the public and private sectors.

Article 3 of the CPEUM asserts: All individuals have the right to receive education. The Mexican State will offer pre-primary, primary, middle, and high school education [62]. The four educational levels are mandatory; they are split into the academic years shown in the second column of Table 3.1. Generally, the students' age-range of each academic level is the one stated in the third column of Table 3.1. The official statistics for the roll of students in the scholar year of 2011–2012 is unveiled in the fifth column of Table 3.1 [63]; whereas the covering achieved for each academic level is identified at the right column of Table 3.1.

Regarding teachers, the academic assets assigned for the mandatory levels in the scholar year of 2011–2012 are given in column 2 of Table 3.2 [64]. It means, an academic teaches the average of students estimated in the column 4 of Table 3.2. The low average in middle and high schools is explained because: several teachers contribute to the academic development of students at those levels.

As for schools, the facilities available for the four mandatory levels during 2011–2012 are offered in column 3 of Table 3.2 [65]. Thus, the average school roll for each level is provided in column 5 of Table 3.2; whereas the average number of teachers assigned to schools in each level is stated in the last column of Table 3.2.

On the other hand, the Federal Government is responsible to define, guide, fund, administrate, provide, evaluate, and control education. The current presidential regime, which initiated its administration in December 1st 2012, ordered the Constitutional Reform in Education [66]. Thus, the Secretary of Public Education (SEP) supports the reform of the article 3 of the CPEUM [67]. As a consequence, the National System for Educative Evaluation is established under the coordination of the National Institute for Educative Evaluation (INEE) [68]. Therefore, the

Table 3.2 Mandatory academic levels, teachers schools, averages between them and students

Level	Teachers	Schools	Students/Teacher	Students/School	Teachers/School
Pre-primary	224,146	91,253	21	52	2.5
Primary	573,849	99,378	26	150	5.8
Middle school	388,769	36,563	16	169	10.6
High school	285,974	15,472	7.5	148	20.0

Deputies Chamber approved the General Law of Education [69] and the Law of the INEE [70].

Regarding academics, they are hired by the Federal Government to teach students in public schools and are organized into labor unions. The National Union of Education Workers (SNTE) [71] holds a membership of approximately 1.2 millions teachers [72]. It is organized into 54 sections along the Mexican territory, where some sections (e.g., 14, 18, 22) founded a fraction called the National Coordination of Education Workers (CNTE) [73], which holds a membership of tens of thousands in states such as: Oaxaca, Guerrero, and Michoacán.

As for public and private educational organizations concerned with education, some of them are: the Mexican Institute for the Competitiveness that founded an initiative to improve schools [74]; another is the Mexican Council for Educative Research, which is aimed to promote investigation in education [75]; Parents of Family Associations and Scholar Councils for Social Participation [76].

3.4.3 National Assessments

The assessment of education is mainly achieved in Mexico by the SEP and the INEE. The SEP uses several instruments, where one of them is called the National Evaluation of the Academic Achievement of Scholar Centers (ENLACE) [77]. As for the INEE [77], it evaluates students by means of diverse instruments such as the Program for International Student Assessment (PISA) and the Exams for the Quality and Educative Achievements (EXCALE) [78]. The PISA is a study well-known in the community that is promoted for measuring student success around the world by the Organization for Economic Co-operation and Development [79]. The EXCALE is a national assessment, whose profile is outlined in Sect. 3.5 because a sample is used as the source data of the case study.

An regards ENLACE, the SEP annually applies a national survey to assess the ratings in the Spanish language, mathematics, and a third subject (e.g., history, biology…) reached by students at primary, middle, and high school levels. During two days, the test is applied to primary schools students of the 3rd up to 6th years, as well as middle and high schools students of the 1st up to 3rd years. The evaluation is made from quantitative and qualitative perspectives. The former estimates the scoring of right answers (e.g., 50 and 90 items for Spanish language and math). Its average is 500 points, with a standard deviation of 100 and a range from 200 up to 800. The latter is based on the Item Response Theory to estimate the domain level [80].

The tests are applied, collected, validated, and processed to estimate outcomes at student, school, and state levels. The results are published and provided to students, parents, teachers, and school principals. A sample of the statistics produced in the scholar years 2012 and 2013 is outlined in Table 3.3, where the three states that reached the highest, middle, and lowest scores are identified, as well as the national average. Table 3.3 shows the national and state average reached in

Table 3.3 Results of the ENLACE test applied to students during the academic year 2012 for primary and middle levels, and 2013 for high school

Level/Nation/State	General	Primary school		Middle school		High school	
		Spanish	Math	Spanish	Math	Spanish	Math
Country	501	498	520	475	528	476	419
1 Federal district	555	580	583	504	531	497	427
2 Sonora	545	552	577	485	544	486	450
3 Tabasco	532	532	552	493	566	460	416
15 Yucatán	505	508	522	476	514	503	433
16 Quintana Roo	503	508	525	464	512	473	415
17 Sinaloa	501	493	522	471	525	473	439
30 Michoacán	466	480	501	401	456	459	411
31 Guerrero	430	404	427	432	501	432	394
32 Oaxaca	381	330	340	376	416	458	411

Spanish and math subjects for primary, middle and high school levels. The *id* edited at the left column reveals the ranking achieved by the state.

3.4.4 The Constitutional Reform in Education

Nowadays, Mexico is living a full transformation in strategic areas such as: education, energy, finances, fiscal, and labor. This synergy is leaded by the Executive Power and supported by the Legislative Power. As the first change, the Federal Government has ordered an educational reform. A briefing of the current progress is shaped as follows:

At the beginning of 2013, the President of Mexico submitted an initiative to reform article 3 of the CPEUM. Thus, the Union Congress discussed the amendment and modified the initiative. As a result, a decree to reform the article 3 was published on February 26, 2013 [81]. The decree reforms the fractions III, VII, and VIII of article 3 and fraction XXV of article 73. Fraction II is enhanced by the addition of the 3rd paragraph and item *d* of the 2nd paragraph. Fraction IX is included [68]. The essence of the reform for those subjects is given next [62]:

- Article 3, fraction III: The Executive Power will define the curricula of the pre-primary, primary, and middle school levels.
- Article 3, fraction VII. The universities and superior institutions which the law grants autonomy will hold the faculty and responsibility to govern themselves.
- Article 3, fraction VIII: The Union Congress will define the necessary laws to unify and coordinate the education in the whole country.
- Article 73, fraction XXV: The Union Congress holds the faculty to establish the Academic Professional Service.
- Article 3, fraction II, 3rd paragraph: Contribute to a better human coexistence and equality of rights without distinction of race, religion, groups, and sex.

- Article 3, fraction II, 2nd paragraph, item *d*: The criterion of the education will be quality based on the constant improvement and the maximization of the students' academic achievement.
- Article 3, fraction IX: In order to guarantee the quality of educative services, the National System for Educative Evaluation is established.

3.4.5 Community Reaction

As consequence of the educative reform, diverse reactions have been manifested by the community; some sectors support the reform, others are against it. A sample of the reactions produced by the Legislative Power, labor unions, students, and society is presented in this subsection in order to shape the current status of the reform.

Since the publication of the constitutional decree to reform education, the Deputies Chamber worked on outlining three complementary laws: General Law of Education, Law of the INEE, Law of Professional Academic Service. The first law rules the education provided by the Executive Power and private institutions [82]. The second defines the nature, purpose, and powers granted to the INEE in order to evaluate the quality and achievements of the education [83]. The third establishes the baseline for the professional development of academics, as well as the criteria, bases, and constraints of the obligatory evaluation for the entry, promotion, recognition, and permanence [84]. The first two laws were approved on August 22; but the third was postponed until September 1st due to political and social pressures exerted by the CNTE.

Even though most of the educational community sectors support the educative reform, the CNTE has presented a violent opposition. Since 1979 [85], the CNTE use to protest every year for improving their labor conditions, benefits, incomes, and diverse kinds of claims and interests. The CNTE members interrupt their academic labors during weeks or months; obstruct public streets, roads, airports, and highways, as well as the access to offices and commercial centers. They hijack public and political servants in their own work center and parliament. CNTE claims in such a way that members move from their states to the Federal District to stay for long periods to force rulers, deputies, senators, and state secretaries to modify the laws according their convenience or at least stop the legislative process. As for the most important complementary law, Professional Academic Service, demands the permanent development and progress of academics in order that they improve the education delivered in classrooms. However, CNTE rejects such commands and claims an ad-hoc law that guarantees they will preserve their "rights" and "benefits" based on an evaluation process that they themselves define and approve!

Students, parents of family, and members of the society are the main beneficiary of the educative reform. They are interested in the enhancement of the whole educational system, laws, labor conditions, facilities, curricula, pedagogical practices, educational content, and supplies. Moreover, they demand well-prepared and updated academics that are committed to perform their labor as well as possible because they love their teaching profession. In response to such demands, the SEP has extended the period up to 200 days for the academic year 2013–2014. The INEE is preparing to define, develop, implement, and apply the evaluation system that the Law of the INEE requires to assess and diagnose the educational status of students, teachers, schools, states, and the whole nation. Most of the teachers, affiliated to the SNTE started the new academic year from the first day: August 19, 2013; but members of the CNTE have not yet attended their schools. In consequence, it is a fact that: the states with the lowest academic achievement, shown in Table 3.3, are the ones where the CNTE is responsible for teaching the students of those states!

3.5 Source Data

This work shows how EDM empowers state policies aimed at improving the quality of education. Thus, the case study is demarcated into the reform of the CPEUM article 3, fraction IX and the Law of the INEE. These policies strengthen the role fulfilled by the INEE since its foundation in 2002 [86]. The INEE has been measuring the quality of education by many instruments, where EXCALE is the most used [78].

In consequence, a sample of the public databases offered by the INEE is used to mine and find out patterns. So, in this section the EXCALE database offered by INEE is described. Later on, a profile of the source data used to be mined is given. Afterwards, a framework for pre-processing and mining the source data is tailored. Finally, an exploratory analysis of the students' opinions to be mined is outlined.

3.5.1 EXCALE Databases

EXCALE assessment aims to evaluate DK, skills, achievements, ratings and opinions of students at pre-primary third year, primary third and sixth years, middle school third year, and high school third year. The surveys are concerned with opinions collected from students, teachers, and academics; as well as, DK evaluations about Spanish language, mathematics, biology, and other subjects [87].

The public data for those academic levels corresponds to surveys applied from 2004–2005 up to 2010–2011 academic years. However, the data concerning each academic level is not available for all the years of the prior range. For instance, data of middle school third year is only available for 2004–2005 and 2007–2008 years.

The single sample chosen to mine in this case study is found in the EXCALE database produced as a result of the assessment made to students of middle school third year in 2007–2008 [88]. It embraces information about the following kinds of data: DK assessments of Spanish language, mathematics, ethic, and biology; opinions given by students, teachers, and principals; documents to depict the survey and explain how to apply it. The source of student comments minded in this case study is the survey of students' opinions, which is introduced in the next subsection.

3.5.2 Source Data Students' Opinions

The survey about opinions given by students at middle school third year contains 88,198 records. Each row holds 180 items that are organized into three segments: (1) general: 19 items; (2) comments: 80 items; (3) statistics: 81 items. Only the first two segments are used for mining purposes. So, they are split into various categories.

The general segment holds five categories that contain several items: (1) student: 8 items (e.g., id, sex, age…); (2) school: 6 items (e.g., *id*, modality, classification, marginalization…); (3) teacher: 1 item, the *id*; (4) state: 2 items, *id* and name; (5) instrument: 2 items, *id* and variable. Regarding the comments segment, it embraces four categories that are split into subcategories. Such subcategories gather homogeneous items to characterize a specific subject of opinion, as Tables 3.4 and 3.5 outline.

Table 3.4 shows just one sample of question to illustrate the kind of items gathered by the subcategory. Thus, the third column states the *id* of the item and the respective question. As regards the value-instances available to respond to such questions, Table 3.5 identifies the integer value and its qualitative meaning that students choose to respond to a particular question. They reveal frequencies and specific close answers.

3.5.3 Framework

In order to mine the source data, a framework is designed to transform the raw data into a database. The framework carries out the five tasks sketched in Fig. 3.1. The first task picks the source data about students' opinions. The second task reads the raw data stored as a text (TXT) file and described as a Statistical Package by the Social Sciences (SPSS) [89] Syntax (SPS) file to transform them as a SPSS data document (SAV) file. Later, such a file is exported to an Excel extended (XLSX) file. The third task imports the XLSX file to create a Structured Query Language (SQL) [90] table. Next, statistical and DM processes are fulfilled. Finally, some findings are discovered and interpreted to shape a diagnostic of the study subject.

Table 3.4 Students' comments classified by category and subcategory to gather homogeneous items, where one of them is illustrated through its respective question

Category—subcategory	Number of items	A sample of the items Example: *id* question
Student	12	
behavior	6	16: Do you focus your attention at class?
development	2	12: What you learn helps you daily?
economic help	2	18: Do you fulfill a job to receive incomes?
personal	5	26: Do you drink alcoholic beverages?
metacognition	2	13: When you do not understand something in class, do you seek additional information in other sources?
Study	15	
development	4	22: Did you study the primary in just one school?
surroundings	8	31: Did a peer bring a gun to school in this year?
institution	2	19: What is the kind of your primary school?
Family	26	
surroundings	2	04: With which parent do you live?
economic help	2	51: Does your family receive an economical support?
emigration	4	79: Are you thinking of going to another country?
resources	7	39: How many books are there in your home?
social	5	01: Which language did you learn first?
family help	6	05: Are your parents aware of your marks?
Teacher	22	
behavior	5	53: How often do your teachers miss their class?
performance	1	61: Do you understand what your teacher teaches?
teaching style	16	63: Does your teacher advice how to correct your faults?

Table 3.5 Value-instances for the questions illustrated in Table 3.4

Item	Value instances
16	1: never; 2: rarely; 3: frequently; 4: quite often; 8: multiple; 9: null
12	1: never; 2: rarely; 3: frequently; 4: quite often; 8: multiple; 9: null
18	0: never; 1: one day per week; 2: two–three days per week; 4: four-more days per week; 8: multiple; 9: null
26	1: never; 2: rarely; 3: frequently; 4: quite often; 8: multiple; 9: null
13	1: never; 2: rarely; 3: frequently; 4: quite often; 8: multiple; 9: null
22	0: no; 1: yes; 8: multiple; 9: null
31	1: never; 2: rarely; 3: frequently; 4: quite often; 8: multiple; 9: null
19	0: public; 1: private; 8: multiple; 9: null
04	0: none parent; 1: father; 2: mother; 3: father & mother; 4: multiple; 9: null
51	0: no; 1: yes; 8: multiple; 9: null
79	0: ignore; 1: I am not going to do; 2: I likely do; 3: yes, some day I will leave to other country: multiple; 9: null
39	0: nothing; 1: up to 10 L; 2: up to 25 L; 3: up to 50 L; 4: up to 100 L
01	1: Spanish; 2: local language; 3: foreign language; 8: multiple; 9: null
05	1: never; 2: rarely; 3: frequently; 4: quite often; 8: multiple; 9: null
53	0: quite often; 1: frequently; 2: rarely; 3: never; 8: multiple; 9: null
61	1: quite often; 2: frequently; 3: rarely; 4: never; 8: multiple; 9: null
63	0: never; 1: rarely; 2: frequently; 3: quite often; 8: multiple; 9: null

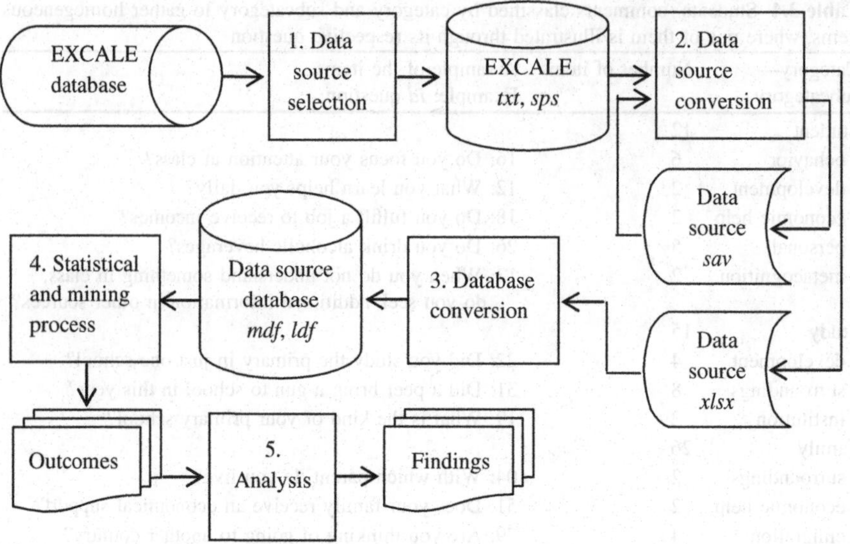

Fig. 3.1 Framework to pre-process and mine source data of the EXCALE students' opinions

3.5.4 Exploration Analysis

Based on the data source and its information characteristics, an exploratory analysis is accomplished to determine the topics to be mined. The analysis is made according to the segments, categories, and subcategories previously identified. These references guide the definition of a population of 114 cases that represents meaningful findings to highlight the students' opinions based on the study context.

The general segment focuses on two kinds of cases to mine one category and the relationship between two categories. The former defines three cases: (1) student; (2) school; (3) state. The latter depicts case number 4 for the school-state relationship.

Concerning the comments segment, twenty one kinds of cases are considered to accumulate 110 specific cases. The kinds of cases and their specific instances are presented in the following relation, where the cases are identified by the progressive *id*. The first kind embraces four cases to mine one category of opinions and the second identifies one case to mine the relationship between two categories. The kinds 3–6 correspond to cases where one subcategory is mined; whilst, 7–10 mine the relationship between two subcategories of the same category. The kinds 11–19 study the relationship between subcategories that pertain to different categories. The kinds 20–21 mine the relationship between one category of the general segment and one subcategory of the comments segment.

In total, 114 cases are designed to be explored by means of an EDM approach that mines the already described source data. The EDM approach and the outcomes are summarized in Sect. 3.6, whereas the findings are discussed in Sect. 3.7.

1. One category: (5) student; (6) study; (7) family; (8) teacher.
2. Two categories: (9) student–teacher.
3. One subcategory of the *student* category: (10) behavior; (11) development; (12) economic help; (13) personal; (14) metacognition.
4. One subcategory of the *study* category: (15) development; (16) surroundings; (17) institution.
5. One subcategory of the *family* category: (18) surroundings; (19) economic help; (20) emigration; (21) resources; (22) social; (23) family help.
6. One subcategory of the *teacher* category: (24) behavior; (25) performance; (26) teaching style.
7. Two subcategories of *student* category: (27) behavior–development; (28) behavior–economic help; (29) behavior–personal; (30) behavior–metacognition; (31) development–economic help; (32) development–personal; (33) development–metacognition; (34) economic help–personal.
8. Two subcategories of *study* category: (35) development–surroundings; (36) development–institution; (37) surroundings–institution.
9. Two subcategories of the *family* category: (38) surroundings–economic help; (39) surroundings–emigration; (40) surroundings–resources; (41) surroundings–social; (42) surroundings–family help; (43) economic help–emigration; (44) economic help-resources; (45) economic help-social; (46) emigration–resources; (47) emigration–family help; (48) resources–social; (49) resources–family help; (50) social–family help.
10. Two subcategories of the *teacher* category: (51) behavior–performance; (52) behavior–teaching style; (53) performance–teaching style.
11. *Student behavior* subcategory with: (54) study development; (55) study surroundings; (56) study institution; (57) family surroundings; (58) family resources; (59) study family help; (60) teacher behavior; (61) teacher performance; (62) teacher teaching style.
12. *Student development*: (63) study development; (64) family surroundings; (65) family economic help; (66) family emigration; (67) family resources; (68) family/family help.[2]
13. *Student economic help*: (69) study development; (70) family surroundings; (71) family economic help; (72) family emigration; (73) family resources.
14. *Student personal*: (74) study development; (75) study surroundings; (76) study institution; (77) family surroundings; (78) family economic help; (79) family emigration; (80) family/family help; (81) teacher behavior.
15. *Student metacognition*: (82) study development; (83) study surroundings; (84) study institution; (85) family emigration; (86) teacher behavior; (87) teacher performance; (88) teacher teaching style.
16. *Study development*: (89) family surroundings; (90) family economic help; (91) family emigration; (92) family resources; (93) family social; (94) family/

[2] The symbol/separates the category from its subcategory.

family help; (95) teacher behavior; (96) teacher performance; (97) teacher teaching style.

17. *Study surroundings*: (98) teacher behavior.
18. *Study institution*: (99) family resources; (100) teacher behavior; (101) teacher teaching style.
19. *Family emigration*: (102) teacher behavior; (103) teacher teaching style.
20. *General* segment, *student* category with *comments* segment: (104) student behavior; (105) student economic help; (106) student personal; (107) student metacognition; (108) study development; (109) study surroundings; (110) family/family help; (111) teacher performance.
21. *General* segment, *school* category with *opinion* segment: (112) study institution; (113) family emigration; (114) family resources.

3.6 Educational Data Mining Approach

The EDM approach is built and exploited as part of the fourth task that composes the framework to mine students' opinions. In order to fulfill the DM processes, the Weka[3] software [91] is used to develop the approach by a workstation. Its design takes into account the source data nature, its taxonomy (i.e., the conceptual organization made up of segments, categories, and subcategories to homogeneously organize 180 items) and the 114 cases earlier defined as a result of the exploratory analysis. The approach is progressively built through a cycle oriented to define parameters, execute DM processes, and analyze results.

The loop ends when a meaningful outcome is reached. Next, the results are interpreted and new mining processes are made. The description of the EDM approach and the results are organized as follows: Firstly, an essential mining process is developed and exploited for the 114 cases that compose the *population*. Secondly, the EDM approach is extended to discover additional knowledge for a *sample* of 32 cases.

3.6.1 Essential Mining

An essential mining process is carried out through an EDM approach whose DM baseline is characterized by means of the following traits: Machine learning discipline is chosen to provide the grounds to mine data [13]. A descriptive DM model is tailored to produce homogeneous groups of instances [15]. Clustering task is considered to solve the problem to gather similar instances [17]. The

[3] Waikato Environment for Knowledge Analysis.

instances-based learning method is applied to scan the pre-processed source data [31]. *k*-means algorithm is picked from the repository offered by Weka to mine the source data [40].

With the aim at illustrating the development cycle achieved for the EDM approach, one sample of the outcomes generated by Weka for the case 96 is given in Fig. 3.2. This shows the screen produced by Weka to output the results of the application according to the DM baseline, whose traits were introduced like parameters.

The main window of Fig. 3.2 describes five clusters created to gather several viewpoints of the students' opinions according to five items. The first column identifies the number of the item (e.g., 22–25, and 65). The second to the seventh columns inform the "average" value estimated by the DM process for each item with the purpose to collect similar instances (i.e., they are approximated values of the grouped instances). The heading of such columns exhibits the cluster *id* and the number of instances that they gather. The second column corresponds to the universe of 88,198 records; whilst, the others represent the five generated clusters.

Once a series of results are produced for a specific case, they are analyzed, some findings are discovered, and the outcomes interpreted. A sample of these deliverables produced for the most relevant cases taken from the population is presented as follows. Firstly, the findings are organized according the nature-context they

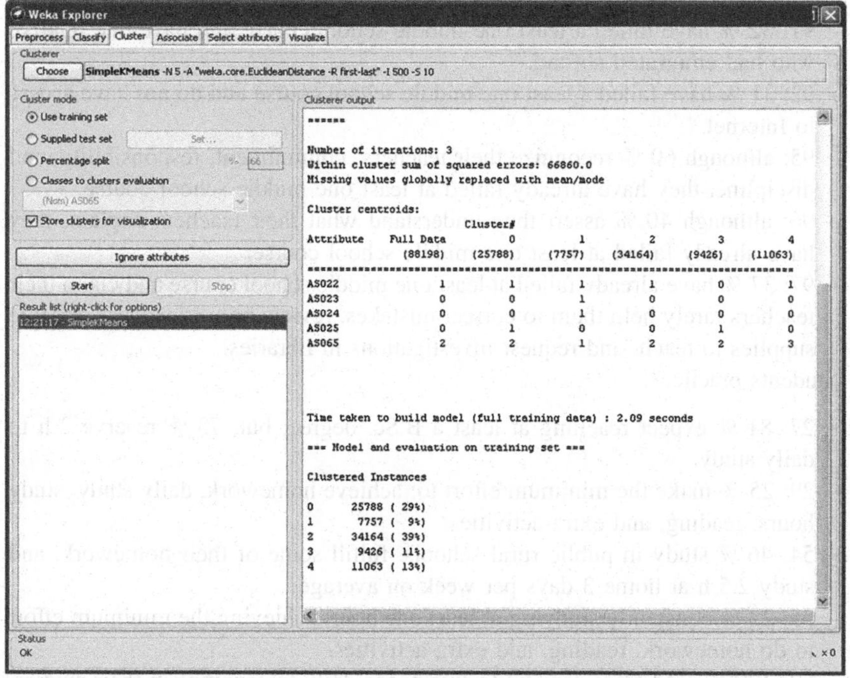

Fig. 3.2 Clusters produced for the case 96 to unveil what students think about teaching

represent. Secondly, the *id* of the alluded case precedes the concise description of the finding, which declares the percentage of the involved students. Finally, an interpretation of such findings is pointed out in Sect. 3.7.1.

- Academic context:
  - 1: student population is made up of 44,908 males and 43,290 women. One cluster gathers 95 % of males who are 14.6 years old on average, and the other contains 98 % of women that are 15.3 years old on average.
  - 17: 85 % study in public schools and 15 % in private; 55 % study in urban schools and 45 % study in rural or community schools.
  - 37: 9 % study in private urban schools where strict discipline is observed.
  - 19: 37 % lack of any kind of health service.
- Students achievements:
  - 89: although 9 % have not failed a course in the middle school all of them failed the first year of primary school. But, 55 % who have failed at least one middle school course did not fail any primary school year.
  - 90: 9 % have failed a least one primary school year and have one relative who abandoned school due to economic issues.
  - 94: in spite of the full scholar support and help given by their parents, 60 % had failed one primary school year or at least one middle school course.
  - 108: 43 % are male students who have failed at least one middle school course and 9 % are women who failed one primary school year.
  - 91: 32 % have failed a least one middle school course and have one relative who had emigrated abroad.
  - 92: 31 % have failed a least one middle school course and do not have access to Internet.
  - 95: although 60 % recognize their teachers' commitment, responsibility, and discipline, they have already failed at least one middle school course.
  - 96: although 40 % assert they understand what their teachers explain, they have already failed at least one middle school course.
  - 97: 37 % have already failed at least one middle school course and claim their teachers rarely help them to correct mistakes, review homework, use suitable supplies to teach, and request investigations in libraries.
- Students practices:
  - 27: 81 % expect reaching at least a B.Sc. degree; but, 73 % reserve 2 h to daily study.
  - 29: 25 % make the minimum effort to: achieve homework, daily study, study hours, reading, and extra-activities.
  - 54: 46 % study in public rural schools, fulfill some of their homework, and study 2.5 h at home 3 days per week on average.
  - 55: 22 % constantly studying at least one hour, achieving the minimum effort to do homework, reading, and extra-activities.
  - 60: 100 % read from zero up to five pages daily to accomplish their studies.

- Students metacognitive and attitude habits:
  - 14: 67 % rarely apply self-regulation (i.e., they to seek content of topics they do not understand or are interested in by their own initiative an effort).
  - 86: 16 % frequently apply self-regulation and recognize a responsible behavior of their teachers that they usually demonstrate in class.
  - 107: 6 % are 15.3 years old women who always apply self-regulation.
- Students economic, grants, and work issues:
  - 12: 22 % have to work at least two days per week to get some income; whereas 30 % receive an economical scholarship.
  - 31: although 80 % do not receive a grant, they aim to get at least a B.Sc. degree; but 60 % of the students who receive a grant conforming to only reach high school level.
  - 73: 7 % work two days per week and their family receive social support, but they lack commodities at home (e.g., phone, Internet access, car…).
  - 105: 5 % are 15.3 years old women who in spite of receiving an economical scholarship have to work at least two days per week to get additional income.
- Students academic aims:
  - 11: 94 %, claim: what they learn at school is useful for their daily life. 62 % expect to develop postgraduate studies. However, 6 % are not convinced of their studies' usefulness and their goal is to reach only up to high school level.
  - 49: 70 % of their parents expect their children to achieve postgraduate studies. However, 19 % of their parents expect them to reach only up to high school.
  - 59: although the parents of the 9 % attended primary school, they fully support and help their children to develop their studies and expect them to reach post-graduate studies.
  - 68: 100 % reveal similar academic aims as their parents!
  - 85: 25 % receive the support and help of their parents and apply self-regulation; where all of them expect the children earn a postgraduate diploma.
- Students claims about violence:
  - 16: 80 % report violent acts in classroom and 91 % complain of bullying.
- Students addictions:
  - 13: no student consumes drugs, 8 % smoke, and 23 % drink alcohol; both addictions have been practiced since they were 14 years old. Moreover, students who smoke also drink alcohol.
  - 34: 3 % work two days per week, receive a grant, drink alcohol, and smoke.
  - 76: 7 % study in public middle school and drink alcohol and smoke.
  - 78: 73 % do not drink alcohol nor smoke.
  - 80: 5 % drink alcohol and smoke, and consider emigrating abroad.
- Family context:
  - 18: 19 % live only with their mother.

- 38: 12 % live with their mother, help her in domestic duties, but do not receive social support from the government.
- 22: parents of the 80 % have menial work and 10 % hold a B.Sc. degree.
- 42: 36 % of their fathers hold at least a B.Sc. degree and 24 % of their mothers at least achieved high school.
- 45: 51 % have fathers who earned at least a B.Sc. degree and mothers that perform housework.
- 41: 13 % live with people who speak a native language.

• Family academic support given to their children:

- 23: although the fathers of all students are interested in their marks, only 8 % of them help their children to develop the homework.

• Family facilities provided to their children:

- 21: 28 % have Internet access at home.
- 40: 6 % live with their mother and enjoy commodities, such as: books, private health service, telephone line, Internet access, and home facilities.
- 44: 19 % have less than 11 books and lack of commodities (e.g., phone, Internet access, car…).
- 46: 11 % have up to 50 books and commodities (e.g., phone, Internet…).
- 48: 18 % have fathers who work as menial laborers and mothers that do housework, their homes lack commodities and have less than 11 books.

• Family emigration influence on their children:

- 39: the fathers of 15 % had emigrated abroad and students would like to emigrate too; but, 17 % who have no family abroad aim at staying in this country.
- 43: although the family of 12 % receives social assistance, they have relatives that had abandoned school and emigrated abroad.
- 72: 8 % work two days per week, receive a grant, their father had emigrated, and they are also considering emigrating in the future.

• Students favorable opinions of their teachers:

- 52: 23 % recognize their teachers' commitment, responsibility, and discipline.
- 53: 18 % admit their teachers often help to correct mistakes, review homework, use suitable accessories to teach, and request investigations in libraries.
- 101: 59 % acknowledge their teachers commitment, responsibility, and discipline, as well as recognize their teachers frequently help them to correct mistakes, review homework, and use suitable accessories to teach…

• Students claims of their teachers:

- 25: 24 % frequently do not understand what teachers explain.
- 26: 24 % claim their teachers rarely help to correct mistakes, review homework, use suitable accessories to teach…
- 101: 100 % manifest their teachers rarely request to investigate in libraries.
- 102: 14 % complain their teachers start the class late.

3.6.2 Supplementary Mining

Once the essential mining process has been performed for the population of 114 cases, a *sample* of 32 representative cases is organized to achieve supplementary mining. Thus, the EDM approach is extended according to the following attributes: Machine learning discipline shapes the baseline of the application [13]. Descriptive DM model guides the organization of homogeneous groups of instances [15]. Association rules task defines relationships between trait values to gather sets of similar instances [19]. Rules induction method is applied to exploit source data [31]. The Weka version of the Apriori algorithm is chosen to mine the students' opinions [38].

In order to illustrate the cycle to build the second EDM approach, a sample of the results produced by Weka for case 47 is given in Fig. 3.3. The mining criterion used by Weka corresponds to the parameters that represent the EDM baseline. Figure 3.3 shows ten rules that gather instances whose items satisfy specific values. The case uses ten items (e.g., 5, 6, 7...) to depict students' opinions of the support given to them by their parents, their academic aims, and emigration of relatives.

When the antecedent items are instantiated by specific values, then the consequent items hold certain values. The association is ground on the instances,

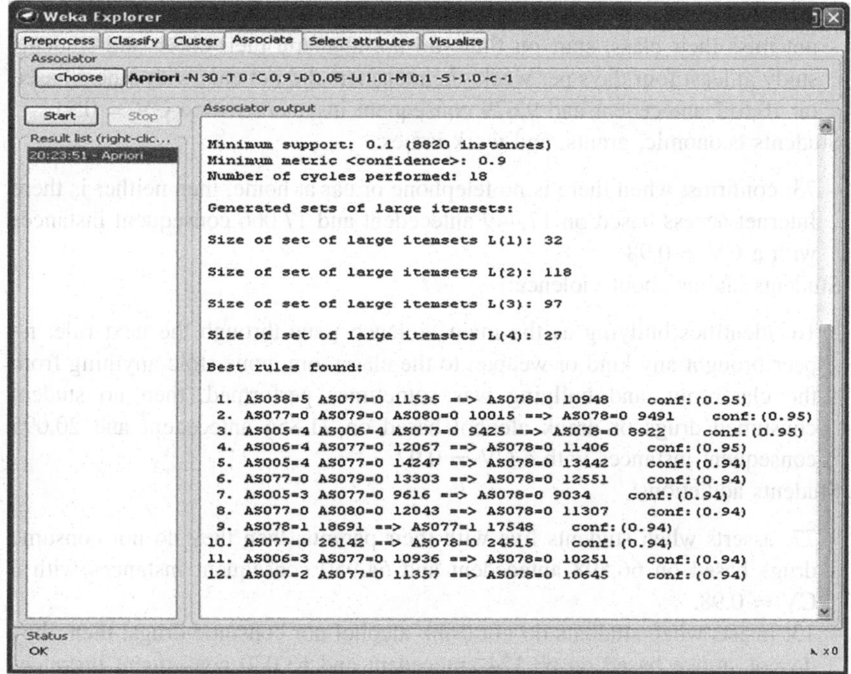

Fig. 3.3 Clusters produced for case 47 to reveal family and personal academic aims

where antecedent and consequent occur with a confidence value (CV) ranged between 0 and 1. The CV is the result of dividing the consequent between the antecedent occurrences. After the generation of several results for a case, they are analyzed to discover findings and interpret them. An example of the rules generated for interesting cases taken from the sample are given in this subsection. The findings are organized according to the criterion used in Sect. 3.6.1, but their interpretation is outlined in Sect. 3.7.2.

- Academic context:
  - 17: reveals: when schools are public and urban, then they have one teacher assigned for each academic grade based on 11,637 antecedent and 10,566 consequent instances with a CV = 0.91.
- Students achievements:
  - 89: confirms: in spite students failing a course in the middle school, they did not fail any year of primary school based on 23,193 antecedent and 23,045 consequent instances with a CV = 0.99.
  - 96: supposes: when students at middle school quite frequently understand what their teachers explain and have not failed a course; then they had not also failed a primary school year based on 21,725 antecedent and 19,725 consequent instances with a CV = 0.91.
- Students practices:
  - 60: recognizes students and teachers' behaviors depicted as: when teachers do not miss their class, start on time, do not abandon their class, and students study at least four days per week, then teachers did not chat with others based on 10,611 antecedent and 9,679 consequent instances with a CV = 0.91.
- Students economic, grants, and work issues:
  - 73: confirms: when there is no telephone or car at home, then neither is there Internet access based on 17,449 antecedent and 17,066 consequent instances with a CV = 0.98.
- Students claims about violence:
  - 16: identifies bullying as the main violence issue through the next rule: no peer brought any kind of weapon to the classroom, none stole anything from the classroom, and bullying was sometimes performed, then no student consumed drugs or drank alcohol based on 21,856 antecedent and 20,095 consequent instances with a CV = 0.92.
- Students addictions:
  - 77: asserts when students live with their parents, then they do not consume drugs based on 66,598 antecedent and 64,950 consequent instances with a CV = 0.98.
  - 13: insists when students do not drink alcohol nor consume drugs; then, they do not smoke based on 63,124 antecedent and 63,030 consequent instances with a CV = 1.0.

- Family context:
  - 22: unveils: even though the father woks as a menial laborer and the mother does housework, the spoken language is Spanish at home based on 24,476 antecedent and 23,027 consequent instances with a CV = 0.94.
  - 41: shows: when the student live with their parents, then the spoken language is Spanish at home based on 65,598 antecedent and 64,213 consequent instances with a CV = 0.96.
- Family academic support given to their children:
  - 47: asserts: when parents remain at home in Mexico, then they expect their children to reach postgraduate studies based on 11,535 antecedent and 10,948 consequent instances with a CV = 0.95.
- Family facilities provided to their children:
  - 40: denounces: even though some students live with their parents and they lack a telephone, they do not have Internet access based on 13,043 antecedent and 12,822 consequent instances with a CV = 0.98.
  - 58: asserts: when there are less than 11 books and there is no telephone at home, then there is not access to Internet based on 11,523 antecedent and 11,187 consequent instances with a CV = 0.97.
- Family emigration influence on their children:
  - 43: reveals: when students talk about emigration, they quite frequently think about their fathers experience overseas based on 26,142 antecedent and 24,515 consequent instances with a CV = 0.94.
- Students favorable opinions of their teachers:
  - 8: depicts how teachers treat them: when teachers demand maximum effort and provide confidence to their students, then they encourage their students to continue their academic development based on 55,612 antecedent and 55,144 consequent instances with a CV = 0.97.
  - 24: depicts teachers' behavior such as: when teachers do not miss their class, start their lecture on time, and do not abandon the classroom, then they do not distract talking with others during the lecture based on 10,611 antecedent and 9,679 consequent instances with a CV = 0.91.
- Students claims of their teachers:
  - 26: denounces a contradiction: even though teachers do not request homework to be carried out in libraries, they encourage students to continue their academic development based on 44,841 antecedent and 42,345 consequent instances with a CV = 0.94.

3.7 Discussion

Once the EDM approach was built and used to mine the population of 114 cases and extended to discover more knowledge of a sample of 32 cases, an interpretation of the findings and a diagnostic are given in this section. The exposition follows the organization applied in Sect. 3.6 and uses the case id. Thus, a pair of subsections is outlined to declare the interpretation that corresponds to basic and supplementary mining. What is more, the subjects that identify the nature-context of the findings are taken into account for heading the statement of the respective interpretations. In addition, a third subsection is reserved to sketch a diagnostic of the revelations denounced by the students' feedback, findings, and interpretations.

3.7.1 Interpretations of the Basic Findings

As a result of the findings analysis unveiled in Sect. 3.6.1, we provide one interpretation, of several that could be inferred, to highlight the relevance of the case developed by the basic EDM approach. Thus, the next interpretations uncover a given situation or request a specific demand that must be considered by government, politics, academic authorities, teachers, and parents to improve the quality of education.

- Academic context:
  - 1: unveils that boys are younger than girls in middle school 3rd year; thus, a sexual education program is needed to protect girls from unwanted pregnancy.
  - 17: uncovers the need to improve facilities in rural or communitarian towns.
  - 19: claims the enhancing of the scope given by medical systems.
  - 37: advices public schools to emulate discipline measure of private schools.
- Students achievements:
  - 89, 90, and 108: observe the need to review the curricula of the 1st year primary school.
  - 91: investigate how family issues bias on student failings.
  - 92: facilitate accessories and commodities to reinforce student learning.
  - 94, 95, and 96: discover unknown causes that influence student failings in spite of the positive support provided by family and teachers.
  - 97: insists encouraging teachers to improve their lectures and support delivered to students.
- Students practices:
  - 27, 29, 54, and 55: motivates students to increase the time and effort to study.
  - 60: proposes programs to stimulate reading.

- Students metacognitive and attitude habits:
  - 14 and 107: propose motivating and training students to apply self-regulation.
  - 86: notes that teachers' behavior stimulates students' self-regulation learning.
- Students economic, grants, and work issues:
  - 12, 73, and 105: consider providing grants to students, especially for those who work at least two days per week.
  - 31: entail benefited students to accomplish higher academic goals.
- Students academic aims:
  - 11, 49, 59, 68, and 85: demand enhancing the scope and capacity of universities to satisfy the aims of developing graduate and postgraduate studies.
- Students claims about violence:
  - 16: requests campaigns against violence and respect of human rights.
- Students addictions:
  - 13, 76, and 80: urge preventive campaigns to avoid consuming cigarettes and alcohol.
  - 34: depicts boys that are like adults with economical needs and addictions.
  - 78: reward and disseminate the exemplary attitude of this sort of students.
- Family context:
  - 18: demonstrates the heavy responsibility carried by many women.
  - 22: asks for academic and training programs for adults.
  - 38: encourages government to provide social support for single mothers.
  - 41: promotes bilingual education.
  - 42: facilitates academic development for women.
  - 45: requests supplementary programs for the development of women.
- Family academic support given to their children:
  - 23: encourages the coexistence between parents and children.
  - 40 and 48: disseminate the exemplary attitude of this kind of parents.
- Family facilities provided to their children:
  - 21: exposes the poor availability of Internet access throughout the country.
  - 40: admires the effort made by single mothers to support their children.
  - 44 and 48: consider providing social support to parents of students who work at least two days per week.
  - 46: pursues to increment the number of students that have commodities.
- Family emigration influence on their children:
  - 39 and 72: warn the influence that emigration of family members exerts on students' plan for leaving, as well as the failing on courses at middle school.

  - 43: highlights extreme instances of economical needs.

- Students favorable opinions of their teachers:
  - 52, 53, and 101: award teachers' exemplary behavior.
- Students claims of their teachers:
  - 25: stimulates improving teaching practices.
  - 26 and 101: motive teachers to help students, use accessories, promote investigation in libraries.
  - 102: inform students' feedback for teachers to trigger reflection.

3.7.2 Interpretation of Supplementary Mining

In this subsection an interpretation of the findings discovered by the supplementary mining is stated. The purpose is to complement the knowledge of the situation uncovered by the survey of students' opinions. Therefore, the next interpretations follow the thematic organization used to classify the nature-context of the findings presented in Sect. 3.6.2.

- Academic context:
  - 17: proposes: teachers should be specialized and overload of duties avoided.
- Students achievements:
  - 89: aims at finding causes that explain why adolescents fail middle school courses even though they did not fail any primary school year.
  - 96: acknowledges how teachers' efforts and style to explain their lectures influence students' achievements.
- Students practices:
  - 60: finds out a correlation between the committed behavior of students and teachers.
- Students economic, grants, and work issues:
  - 73: provides Internet access for free to students who lack commodities.
- Students claims about violence:
  - 16: recognizes bullying as the most relevant violence manifestation that happens in classrooms.
- Students addictions:
  - 77: values the presence of parents that influence their children to avoid consuming drugs.
  - 13: reinforces preventive health programs to avoid addictions.

- Family context:
  - 22: explains that: even though the job of the parents is modest, the Spanish is the language spoken at home.
  - 41: reinforces the preponderance of the Spanish language in family surroundings.
- Family academic support given to their children:
  - 47: expresses how the presence of parents at home motives their children to reach a high education level, such as postgraduate studies.
- Family facilities provided to their children:
  - 40 and 58: claim to facilitate Internet access throughout the country.
- Family emigration influence on their children:
  - 40: denounces that fathers are the usual relatives who emigrate.
- Students favorable opinions of their teachers:
  - 8: recognizes the teachers' manner used to treat students as: demanding, encouraging, and providing confidence.
  - 24: recognizes the teachers' behavior demonstrated in classroom as: formal, consistent, and congruent.
- Students claims of their teachers:
  - 26: motivates teachers to stimulate the investigation of contents in libraries.

3.7.3 A Diagnostic of Students Opinions

A diagnostic of what students opine, the findings produced by the EDM approach, and their interpretations is outlined in this subsection. Where, essential and supplementary findings and interpretations are gathered according to the topics early defined. Thus, the diagnostic represents a pragmatic viewpoint that the educative community could consider to define, develop, and implement educational polices.

In regards to academic context, it is necessary to provide funds to maintain and increment rural and community schools in order that they improve their facilities and train teachers to specialize and enhance education. Particularly, states with poor economy, such as Oaxaca, Guerrero, and Michoacán. Moreover, sexual education is needed to avoid students interrupting their academic development and abandoning school. Furthermore, health programs should be provided to students whose parents lack official and private medical assistance.

As for students' achievements, the primary school first year represents an obstacle for students, who some tend to fail. However, most of them, do not fail again during the rest of the primary and during middle school. It is paradoxical that even though students receive the support of their parents, teachers, and schools, as well as understand what academics teach, some of them have failed middle school courses. So, causes that exert on students failing is a target for investigation, such as Internet usefulness, parents influence, emigration, lack of commodities, teacher styles.

On the other hand, students' practices reveal a minimal effort made by students outside classrooms; specially the reading of books and investigation in libraries. Therefore, motivational programs, as well as accessories and facilities, are needed to encourage students to increase the time scheduled for extra-labors. In addition, the example and support demonstrated by parents and teachers are decisive in the behavior of students. In fact, when teachers are demanding, students exert themselves. On the contrary, most of the parents are interested in being providers of commodities instead of helpers who assist their children to do homework and prepare for exams.

Concerning students' metacognitive and attitude habits, few students exercise self-initiative and self-regulation to improve their learning. As result, motivational and metacognitive topics should be delivered to students to reinforce their attitudes and cognitive activity. What is more, academics must be trained to develop motivational and metacognitive practices in class with the aim that their pupils acquire experience and develop their own skills and strategies to make their efforts more profitable.

Regarding students' economy, grants, and work issues, the amount provided by scholarships is not enough to avoid some students working at least two days per week to obtain extra-income, acquire or hire commodities, and get Internet access. The worst case is: some students who receive a grant are resigned to reach only up to high school level. Thus, a revision of the budget, criteria, and effects of the scholarships is needed to accomplish the goals of supporting and encouraging students to avoid labor distractions, continue their studies, and have the essential commodities to study.

In another vein, students' academic aims assert most students are convinced of the usefulness of their studies in their daily life. Surprisingly, a high percentage of students and their parents aim to earn a postgraduate diploma! Particularly, parents who only reached primary school wish their children to accomplish at least a M. Sc. degree! In contrast, a small proportion of students and their parents consider that it is more than enough to complete high school. These aims in Mexico are very hard to be satisfied, due to there is not enough capacity of public and private schools to satisfy the educational demands at high school, college, and postgraduate levels. For example, Table 3.1 reveals that only 25 % of the high school demand is met. In consequence, government polices should be focused on extending the scope.

Students' claims about violence recognize that bullying is the most common practice that happens in classrooms. Even though, the Federal Government has stated laws and polices to combat such a practice, it still happens and is increasing. Therefore, academics and staff should respectively attack this practice and offer psychological assistance to prevent and correct its occurrence.

With respect to students' addictions, most of the students deny consuming alcohol and smoking. However, the age of fourteen is transcendental, because students that practice these addictions recognize they started at this age. Therefore, preventive and reactive programs to combat addictions should be permanently applied at schools in order to warn students and provide physical and mental treatments to avoid that they engage in such habits. Moreover, the coexistence with their parents at home is vital to impede consuming drugs.

As well as the family context, most of the students live with their parents, and the remainder live with their mothers who take on a heavy responsibility. Usually, fathers achieved higher academic levels than mothers. What is more, fathers perform menial works, whilst mothers carry out housework. Thus, the Federal Government measures should be intensified to extend their scope and enhance the opportunities to develop women, support single mothers, and strengthen social programs. In rural communities where native languages are spoken, bilingual education must be provided to preserve local values and customs, due to students only speaking Spanish.

Family academic support provided to their children is mainly focused on providing commodities and requesting marks, but are not committing in spending time to help children to develop homework and extra-curricular activities. So, social programs should be spread to stimulate coexistence and the pursuit of common goals, where very often both parents and children coincide in their academic aims.

As for family facilities given to students, while parents try to provide domestic facilities, some of them are not able to provide the necessary support, such as Internet. Thus, a public program to access free Internet is demanded to meet such lacks.

In relation to family emigration influence on their children, fathers travel abroad to work, obtain incomes, and deliver money to maintain their family. Unfortunately, this is a serious social and economical issue in Mexico. Approximately 33.6 millions of people whose origin is Mexican live in USA [92]. Such a situation biases some students who have failed middle school courses and considered emigrating in the future too. Most of them receive grants and have to work at least two days per week because the family revenue is not enough. Therefore, the critical conditions of these students should be analyzed in order to improve the social assistance to their families.

As regards students' favorable opinions of their teachers, many recognize their academics' commitment, responsibility, and discipline. Moreover, some of them acknowledge the support provided to develop their academic activities in class. Furthermore, many teachers encourage students to continue their academic development. Stimulation programs that offer promotion and bonuses could be designed to encourage teachers who perform their academic labor with excellence.

Concerning students' claims of their academics stands up that they do not understand what their teachers explain. What is more, students complain their teachers do not help them to review homework, correct mistakes, request investigations, and start the class on time. So, teachers require programs designed for motivating, updating, stimulating, and rewarding the improvement of their performance, teaching style, use of accessories, and the manner in which they deal with their students; as well as criteria and rules for penalizing recurrent behaviors that have not been corrected after several opportunities to be amended.

3.8 Conclusions

Even though this book is dedicated to disseminate the scientific and technological work being carried out in the EDM arena, researchers must be aware that the final purpose of their outcomes is to serve society. Moreover, scientists and technocrats are constrained by the legal rules that lead funding, activity, application, and diffusion of the deliverables and findings. Therefore, this kind of labor should not be alien to the constitutional grounds, scope of the laws, aim of the government initiatives, as well as social needs, interests, and reactions. When, the scientific community takes into account such a reality, the work and its products are coherent with reality. Thereby, the research efforts will be useful to improve the quality of life for human beings.

Corresponding with such a vision, in this chapter we have summarized a case study that actually afflicts Mexican society and challenges the Federal Government, the improvement of education. We have introduced the Mexican State, as well as the Powers of the Union. A profile of the educational community, a sample of statistics, and a briefing of the academic achievements were outlined. The constitutional reform in education proposed by the Mexican Presidency and the laws decreed by Parliament were presented. Furthermore, a sample of the social reactions has been stated, where the irrational opposition of a small number of teachers attempts to hinder the consummation of initiatives conceived to enhance the quality of education in Mexico.

However, this book concerns the EDM arena. In order to highlight the contribution of this chapter, we illustrated how the development of an EDM approach is useful to support government initiatives to enhance education. In consequence, a description of DM and EDM fields has been given. The characteristics of the data source that records the students' opinions collected in a national survey have been explained. In addition, a framework for pre-processing and mining was sketched, as well as the results, findings, and interpretations of the achieved mining were revealed. Thus, a panorama of the students' feedback and proposal are presented as follows.

Usually, students' opinions are not listened to by academics, principals, public servants, and politicians. Therefore, the main target of education is silent and without the right of expression. The prevalent interests are the labor conditions of the academics, the Federal Government initiatives, and the budget available for maintaining the educational system. However, national surveys about the students' feedback, as the one introduced in this chapter, and the analysis of their opinions, as expressed in this work, represent valuable sources that are worthy to be known.

The data, findings, interpretations, and diagnostic earlier stated reveal a vision of life and the aspirations of a better future for the generations in progress; where, many of them are considering postgraduate studies. The support given by their families is an essential factor to provide, help, and motive students to pursue their dreams. The role played by academics is vital as the person who facilitates DK acquisition, skills development, learning experiences, and the establishment of principles that rule life.

Parents and academics, as human beings, are not perfect, but should be encouraged to improve the support, assistance, coexistence, and love for the students. As for the Federal Government, social policies are needed to extend and enhance the provision of public health services, grants for scholars, free Internet access, preventive and reactive assistance against addictions and violence, as well as awarding outstanding academic results, performance, and behaviors of students, teachers, and parents.

As part of future work to be fulfilled, the promotion of EDM virtues is strategic to participate as a technological agent that supports State policies to improve education. The design, development, and deployment of EDM approaches to mine educational data are considered as an instrument to implement the Educative Reform,

As a final comment that highlights the case study: by the end of this book's edition, nearly 1.3 million children that live in Oaxaca State have not had classes because their teachers, approximately 13,500 members of the CNTE [93], have been protesting in Mexico City instead of working in their schools.

Acknowledgments The first author gives testimony of the strength given by his Father, Brother Jesus, and Helper, as part of the research projects of World Outreach Light to the Nations Ministries (WOLNM). Moreover, a mention is given to Mr. Lawrence Whitehill Waterson, a native British English speaker who tunes the manuscript, as well as Aldo Ramírez Arellano for his edition support. In addition, a special mention is given to the Instituto Nacional para la Evaluación de la Educación for the publication of the source data used in this case study for exclusively research purposes. Finally, this research holds a partial support from grants given by: CONACYT-SNI-36453, CONACYT 118862, CONACYT 118962-162727, IPN-PIFI/20131093-28/1398, CONACYT 360532/289763. and IPN-COFAA-SIBE DEBEC/647-391/2013.

References

1. Vlahos, G.E., Ferratt, T.W., Knoepfle, G.: The use of computer-based information systems by German managers to support decision making. J. Inf. Manag **41**(6), 763–779 (2004)
2. Lucey, T.: Management Information, 9th edn. Thomson, London (2005)
3. Power, D.J., Sharda, R.: Decision support systems. In: Shimon, N.Y. (ed.) Springer Handbook of Automation, pp. 1539–1546. Springer, Heidelberg (2009)
4. Mazón, J.N., Trujillo, J.: An MDA approach for the development of data warehouses. J. Decis. Support Syst. **45**(1), 41–58 (2008)
5. Alkharouf, N.W., Curtis-Jamison, D., Matthews, B.F.: Online analytical processing: A fast and effective data mining tool for gene expression databases. J. Biomed. Biotechnol. **2005**(2), 181–188 (2005)
6. Turban, E., Sharda, R., Delen, D.: Decision Support and Business Intelligence Systems, 9th edn. Prentice Hall, New Jersey (2010)
7. Wille, R.: Why can concept lattices support knowledge discovery in databases? J. Exp. Theor. Artif. Intell. **14**(2–3), 81–92 (2002)
8. Dunham, M.H.: Data Mining Introductory and Advanced Topics. Pearson Education, Harlow (2006)
9. Han, J., Kamber, M., Pei, J.: Data Mining Concepts and Techniques, 2nd edn. Morgan Kaufmann, California (2006)

10. Hill, T., Lewicki, P.: Statistics: Methods and Applications: A Comprehensive Reference for Science, Industry, and Data Mining. StatSoft, Oklahoma (2006)
11. Karegar, M., Isazadeh, A., Fartash, F., Saderi, T., Navin, A.H.: Data-mining by probability-based patterns. In: Luzar-stiffler, V. (ed.) 30th International Conference on Information Technology Interfaces, pp. 353–360. IEEE, Cavtat (2008)
12. Soukup, T., Davidson, I.: Visual Data Mining: Techniques and Tools for Data Visualization and Mining. Wiley, New Jersey (2002)
13. Witten, I.H., Frank, E., Hall, M.A.: Data Mining: Practical Machine Learning Tools and Techniques, 3rd edn. Morgan Kaufmann, California (2011)
14. McCarthy, P.M., Boonthum-Denecke, C.H. (eds.): Applied Natural Language: Identification, Investigation and Resolution. Information Science Reference, Pennsylvania (2011)
15. Peng, Y., Kou, G., Shi, Y., Chen, Z.: A descriptive framework for the field of data mining and knowledge discovery. Int. J. Inf. Technol. Decis. Making **7**(4), 639–682 (2008)
16. Hand, D.J., Mannila, H., Smyth, P.: Principles of Data Mining. MIT Press, Massachusetts (2001)
17. Berkhin, P.: Survey of clustering data mining techniques. In: Kogan, J., Nicholas, C., Teboulle, M. (eds.) Grouping Multidimensional Data, pp. 25–71. Springer, Heidelberg (2006)
18. Hardoon, D.R., Shawe-Taylor, J., Szedmak, S.: Canonical correlation analysis: An overview with application to learning methods. J. Neural Comput. **16**(12), 2639–2664 (2004)
19. Hong, T.P., Lin, K.Y., Wang, S.L.: Fuzzy data mining for interesting generalized association rules. J. Fuzzy Sets syst. **138**(2), 255–269 (2003)
20. Pinto, H., Han, J., Pei, J., Wang, K., Chen, Q., Umeshwar, D.: Multidimensional sequential pattern mining. In: Paques, H., Liu, L., Grossman, D. (eds.) 10th International Conference on Information and Knowledge Management, pp. 81–88. ACM, New York (2001)
21. Chau, M., Cheng, R., Kao, B., Ng, J.: Uncertain data mining: An example in clustering location data. In: Ng, W.K., Kitsuregawa, M., Li, J., Chang, K. (eds.) Advances in Knowledge Discovery and Data Mining, LNCS, vol. 3918, pp. 199–204. Springer, Heidelberg (2006)
22. Wu, Z., Li, C.H.: L0-constrained regression for data mining. In: Ng, W.K., Kitsuregawa, M., Li, J., Chang, K. (eds.) Advances in Knowledge Discovery and Data Mining, LNCS, vol. 4426, pp. 981–988. Springer, Heidelberg (2007)
23. Corteza, P., Cerdeirab, A., Almeidab, F., Matos, T., Reisa, J.: Modeling wine preferences by data mining from physicochemical properties. J. Decis. Support Syst. **47**(4), 547–553 (2009)
24. Du, W., Han, Y.S., Chen, S.: Privacy-preserving multivariate statistical analysis: Linear regression and classification. In: 4th SIAM International Conference on Data Mining, pp. 222–233. Society for Industrial and Applied Mathematics, Lake Buena Vista (2004)
25. Komarek, P.: Logistic Regression for Data Mining and High-dimensional Classification. Carnegie Mellon University, Pennsylvania (2004)
26. Pang, S., Ozawa, S., Kasabov, N.: Incremental linear discriminant analysis for classification of data streams. J. IEEE Trans. Syst. Man Cybern. B Cybern. **35**(5), 905–914 (2005)
27. Wang, L. (ed.): Support Vector Machines: Theory and Applications. Springer, Heidelberg (2005)
28. Glick, M., Klon, A.E., Acklin, P., Davies, J.W.: Enrichment of extremely noisy high-throughput screening data using a naive Bayes classifier. J. Biomol. Screen. **9**(1), 32–36 (2004)
29. Rogan, J., Miller, J., Stow, D., Franklin, J., Levien, L., Fisher, C.: Land-cover change monitoring with classification trees using Landsat TM and ancillary data. J. Photogram. Eng. Remote Sens. **69**(7), 793–804 (2003)
30. Alves, R.T., Delgado, M.R., Lopes, H.S., Freitas, A.A.: An artificial immune system for fuzzy-rule induction in data mining. In: Yao, X., Burke, E.K., Lozano, J.A., Smith, J., Merelo-Guervós, J.J., Bullinaria, J.A., Rowe, M.J.E., Tiňo, P., Kabán, A., Schwefel, H.P. (eds.) Parallel Problem Solving from Nature-PPSN VIII, LNCS, vol. 3342, pp. 1011–1020. Springer, Heidelberg (2004)

31. Brighton, H., Mellish, C.: Advances in instance selection for instance-based learning algorithms. J. Data Min. Knowl. Disc **6**, 153–172 (2002)
32. Wu, X., Kumar, V., Quinlan, J.S., Ghosh, J., Yang, Q., Motoda, H., McLachlan, G.J., Ng, A., Liu, B., Yu, P.S., Zhou, Z.H., Steinbach, M., Hand, D.J., Steinberg, D.: Top 10 algorithms in data mining. J. Knowl. Inf. Syst. **14**(1), 1–37 (2008)
33. Williams, N., Zander, S., Armitage, G.: A preliminary performance comparison of five machine learning algorithms for practical IP traffic flow classification. ACM SIGCOMM Comput. Commun. Rev. **36**(5), 5–16 (2006)
34. McLachlan, G.J., Krishnan, T.: The EM Algorithm and Extension, 2nd edn. Wiley, New Jersey (2007)
35. Yao, Z., Liu, P., Lei, L.,Yin, Y.: R-C4.5 decision tree model and its applications to health care dataset. In: International Conference on Services Systems and Services Management, vol. 2, pp. 1099–1103. IEEE, Chongqing (2005)
36. Waheed, T., Bonnell, R.B., Prasher, S.O., Paulet, E.: Measuring performance in precision agriculture: CART—a decision tree approach. J. Agric. Water Manag. **84**(1–2), 173–185 (2006)
37. Lavrac, N., Kavsek, B., Flach, P., Todorovski, L.: Subgroup discovery with CN2-SD. J. Mach. Learn. Res. **5**, 153–188 (2004)
38. Lazcorreta, E., Botella, F., Fernández-Caballero, A.: Towards personalized recommendation by two-step modified Apriori data mining algorithm. J. Expert Syst. Appl. **35**(3), 1422–1429 (2008)
39. Zhang, M.L., Zhou, Z.H.: ML-KNN: A lazy learning approach to multi-label learning. J. Pattern Recognit. **40**(7), 2038–2048 (2007)
40. Jain, A.K.: Data clustering: 50 years beyond K-means. J. Pattern Recognit. Lett. **31**(8), 651–666 (2010)
41. Burdick, D., Calimlim, M., Flannick, J., Johannes, G., Tomi, Y.: MAFIA: A maximal frequent item set algorithm. IEEE Trans. Knowl. Data Eng. **17**(11), 1490–1504 (2005)
42. Dwork, C.: Differential privacy: A survey of results. In: Agrawal, M., Zu, D., Duan, Z., Li, A. (eds.) Theory and Applications of Models of Computation, LNAI, vol. 4978, pp. 1–19. Springer, Heidelberg (2008)
43. Lamma, E., Mello, P., Riguzzi, F., Storari, S.: Applying inductive logic programming to process mining. In: Blockeel, H., Ramon, J., Shavlik, J., Tadepalli, P. (eds.) Inductive Logic Programming, LNCS, vol. 4894, pp. 132–146. Springer, Heidelberg (2008)
44. Evans, M.P.: Analysing Google rankings through search engine optimization data. J. Internet Res. **17**(1), 21–37 (2007)
45. Weng, J., Lim, E.P., Jiang, J., He, Q.: Twitter rank: Finding topic-sensitive influential twitters. In: 3rd International Conference on Web Search and Data Mining, pp. 261–270 (2010)
46. Vialardi-Sacin, C., Bravo-Agapito, J., Shafti, L., Ortigosa, A.: Recommendation in higher education using data mining techniques. In: Barnes, T., Desmarais, M., Romero, R., Ventura, S. (eds.) 2nd International Conference on Educational Data Mining, pp. 190–199. International Educational Data Mining Society, Cordoba (2009)
47. Anjewierden, A., Kollöffel, B., Hulshof, C.: Towards educational data mining: Using data mining methods for automated chat analysis to understand and support inquiry learning processes. In: International Workshop on Applying Data Mining in e-Learning, EC-TEL 2007, pp. 23–32. EC-TEL, Crete (2007)
48. Luan, J.: Data mining and its applications in higher education. J. New Dir. Inst. Res. **113**, 17–36 (2002)
49. Romero, C., Ventura, S.: Educational data mining: A survey from 1995 to 2005. Expert Syst. Appl. **33**(1), 135–146 (2007)
50. Romero, C., Ventura, S.: Educational data mining: A review of the state of the art. IEEE Trans. Syst. Man Cybern. C Appl. Rev. **40**(6), 601–618 (2010)
51. Peña-Ayala, A.: Educational data mining: A survey and a data mining-based analysis of recent works. Expert systems with applications 41(4), doi:10.1016/j.eswa.2013.08.042 (2014)

52. Marsh, H.W., Roche, L.A.: Effects of grading leniency and low workload on students' evaluations of teaching: Popular myth, bias, validity, or innocent bystanders? J. Educ. Psychol. **92**(1), 202–228 (2000)
53. About the Students Evaluations of Teaching (USE): Institute for Teaching and Learning, University of Sidney. http://www.itl.usyd.edu.au/use/
54. Kim, S.M., Calvo, R.A.: Sentiment analysis in student experiences of learning. In: Baker, R.S.J.D., Merceron, A., Pavlik Jr, P.I. (eds.) 3rd International Conference on Educational Data Mining, pp. 111–120. International Educational Data Mining Society, Pittsburgh (2010)
55. Champaign, J., Cohen, R.: An annotations approach to peer tutoring. In: Baker, R.S.J.D., Merceron, A., Pavlik Jr, P.I. (eds.) 3rd International Conference on Educational Data Mining, pp. 231–240. International Educational Data Mining Society, Pittsburgh (2010)
56. Koprinska, I.: Mining assessment and teaching evaluation data of regular and advanced stream students. In: Pechenizkiy, M., Calders, T., Conati, C., Ventura, S., Romero, C., Stamper, J. (eds.) 4th International Conference on Educational Data Mining, pp. 359–360. International Educational Data Mining Society, Eindhoven (2011)
57. Barracosa, J., Antunes, C.: Anticipating teachers' performance. In: KDD 2011 Workshop: Knowledge Discovery in Educational Data, pp. 77–82. San Diego (2011)
58. Leong, C.K., Lee, Y.H., Mark, W.K.: Mining sentiments in SMS texts for teaching evaluation. Expert Syst. Appl. **39**(3), 2584–2589 (2012)
59. Gates, K., Wilkins, D., Conlon, S., Mossing, S., Eftink, M.: Maximizing the value of student ratings through data mining. In: Peña-Ayala, A. (ed.) Educational Data Mining: Applications and Trends. Studies in Computational Intelligence. Springer, Heidelberg (2013, in press)
60. Instituto Nacional de Estadística y Geografía: Population in 2010. http://www.inegi.org.mx/
61. Instituto Nacional de Estadística y Geografía: Population in 2010 organized by age ranges from 3 up to 19 year old. http://www3.inegi.org.mx/sistemas/sisept/default.aspx?t=medu01&s=est&c=21778
62. Constitución Política de los Estados Unidos Mexicanos: Constitution published by the Official Diary of the Federation on 5 Feb 1917, Deputy Chamber of the Honorarium Congress of the Union, Mexico. http://www.diputados.gob.mx/LeyesBiblio/pdf/1.pdf
63. Instituto Nacional de Estadística y Geografía: Students roll in 2011–2012 organized by academic level from pre-primary up to high-school. http://www3.inegi.org.mx/sistemas/sisept/default.aspx?t=medu17&s=est&c=21788
64. Instituto Nacional de Estadística y Geografía: Teachers inventory in 2011–2012 organized by academic level from pre-primary up to high-school. http://www3.inegi.org.mx/sistemas/sisept/default.aspx?t=medu03&s=est&c=21819
65. Instituto Nacional de Estadística y Geografía: Schools inventory in 2011–2012 organized by academic level from pre-primary up to high-school. http://www3.inegi.org.mx/sistemas/sisept/default.aspx?t=medu57&s=est&c=21815
66. Presidencia de la República de los Estados Unidos Mexicanos: Educational Reform News. http://www.presidencia.gob.mx/reforma-educativa-construyendo-la-gran-obra-colectiva-que-es-mexico/
67. Secretaría de Educación Pública: Constitutional Reform in Educative Subject. http://www.sep.gob.mx/
68. Diario Oficial de la Federación: Educational reform decree, addition of the fraction IX to the article 3 of the Politic Constitution of the Mexican United States. http://www.sep.gob.mx/work/models/sep1/pdf/promulgacion_dof_26_02_13.pdf
69. Cámara de Diputados del Honorable Congreso de la Unión: Bulletin number 1937 to approve the verdict to found the General Law of Education. http://www3.diputados.gob.mx/camara/005_comunicacion/a_boletines/2013_2013/agosto_agosto/22_22/1936_aprueba_camara_de_diputados_reformas_a_la_ley_general_de_educacion
70. Cámara de Diputados del Honorable Congreso de la Unión: Bulletin number 1937 to approve the verdict to found the Law of the National Institute for Educative Evaluation. http://www3.diputados.gob.mx/camara/005_comunicacion/a_boletines/2013_2013/agosto_agosto/

22_22/1937_aprueban_diputados_dictamen_que_crea_la_ley_del_instituto_nacional_para_la_evaluacion_de_la_educacion
71. Sindicato Nacional de Trabajadores de la Educación: Workers of the Education. http://www.snte.org.mx/index2.php
72. El Universal Newspaper: SNTE surveys Workers of the Education. http://www.eluniversal.com.mx/primera/38857.html
73. Coordinadora Nacional de Trabajadores de la Educación: National Coordination of Workers of the Education. http://cntrabajadoresdelaeducacion.blogspot.mx/
74. Instituto Mexicano para la Competitividad: Initiative improve your School. http://www.mejoratuescuela.org
75. Consejo Mexicano de Investigación Educativa: Home page. https://www.comie.org.mx
76. Secretaría de Educación Pública: Parents of Family Associations and Scholar Councils for Social Participation. http://csep.sepdf.gob.mx/wcsep/default.asp?id=5
77. Secretaría de Educación Pública: National Evaluation of the Academic Achievement of Scholar Centers. http://www.enlace.sep.gob.mx/que_es_enlace/
78. Instituto Nacional para la Evaluación de la Educación. Exams for the Quality and Educative Achievements. http://www.inee.edu.mx/index.php/bases-de-datos/bases-de-datos-excale/marcos-de-referencia
79. Organization for Economic Co-operation and Development: Program for International Student Assessment. http://www.oecd.org/pisa/
80. Fayers, P.: Item response theory for psychologists. Qual. Life Res. **13**(3), 715–716 (2004)
81. Diario Oficial de la Federación: Decree to reform the article 3 of the Politic Constitution of the Mexican United States. http://www.dof.gob.mx/nota_detalle.php?codigo=5288919&fecha=26/02/2013
82. Cámara de Diputados del Honorable Congreso de la Unión: General Law of Education. http://www.diputados.gob.mx/LeyesBiblio/pdf/137.pdf
83. El Universal Newspaper: Law of the National Institute for Educative Evaluation. http://www.dstjal.com.mx/est92/images/docs/biblioteca/Ley_Evaluacion.pdf
84. El Universal Newspaper: Law of Professional Academic Service. http://www.dstjal.com.mx/est92/images/docs/biblioteca/leyservicioprofesionaldocente.pdf
85. Noticiario de José Cárdenas Informa: Radio Fórmula 103.3 FM. CNTE is founded 1979, radio news provided on 27 Aug 2013 at 19:20. http://www.radioformula.com.mx/estaciones.asp?idEs=tele&iPos=2
86. Instituto Nacional para la Evaluación de la Educación: Home page. http://www.inee.edu.mx/
87. Instituto Nacional para la Evaluación de la Educación: Databases. http://www.inee.edu.mx/index.php/bases-de-datos
88. Instituto Nacional para la Evaluación de la Educación: EXCALE, middle-school third grade evaluation made in 2007–2008. http://www.inee.edu.mx/index.php/bases-de-datos/bases-de-datos-excale/excale-09-ciclo-2007-2008
89. Pallant, J.: SPSS Survival Manual: A Step by Step Guide to Data Analysis Using SPSS. McGraw-Hill International, Berkshire (2010)
90. Skulschus, M., Tittel, J., Wiederstein, M.: MS SQL Server 2012 (4): Data Mining, Analyse und Multivariate Verfahren. Comelio GmbH, Munchen (2013)
91. Hall, M., Frank, E., Holmes, G., Pfahringer, B., Reutemann, P., Witten, I.H.: The WEKA data mining software: An update. SIGKDD Explor **11**(1), 10–18 (2009)
92. El Universal Newspaper: Census reports 33.6 millions of Mexicans living in USA. http://archivo.e-consulta.com/2013/index.php/nacionales/item/reporta-censo-336-millones-de-mexicanos-en-estados-unidos
93. Noticiario de Joaquín López Dóriga: Radio Fórmula 103.3 FM. 1.3 millions of students without classes, radio news provided on 26 Aug 2013 at 14:00. http://www.radioformula.com.mx/estaciones.asp?idEs=103-3-fm&iPos=0

Part II
Student Modeling

Chapter 4
Modeling Student Performance in Higher Education Using Data Mining

Huseyin Guruler and Ayhan Istanbullu

Abstract Identifying students' behavior in university is a great concern to the higher education managements (Kumar and Uma, Eur J Sci Res 34(4):526–534). This chapter proposes a new educational technology system for use in Knowledge Discovery Processes (KDP). We introduce the educational data mining (EDM) software and present the outcome of a test on university data to explore the factors having an impact on the success of the students based on student profiling. In our software system all the tasks involved in the KDP are realized together. The advantage of this approach is to have access to all the functionalities of the Structured Query Language (SQL) Server and the Analysis Services through a single developed software item, which is specific to the needs of a higher education institution. This model (Guruler et al., Comput Educ 55(1):247–254) aims to help educational organizations to better understand the KDPs, and provides a roadmap to follow while executing whole knowledge projects, which are non-trivial, involve multiple stages, possibly several iterations.

Keywords Educational data mining · Educational technology system and architectures · Student relationship management · Knowledge discovery software · Decision tree

H. Guruler (✉)
Department of Information Systems Engineering, Technology Faculty, Mugla Sitki Kocman University, 48000 Kötekli, Mugla, Turkey
e-mail: hguruler@mu.edu.tr

A. Istanbullu
Department of Computer Engineering, Engineering and Architecture Faculty, Balikesir University, 10145 Cagiş, Balikesir, Turkey
e-mail: iayhan@balikesir.edu.tr

A. Peña-Ayala (ed.), *Educational Data Mining*,
Studies in Computational Intelligence 524, DOI: 10.1007/978-3-319-02738-8_4,

Abbereviations

CM	Correlation matrices
DBMS	Database management system
DM	Data mining
DT	Decision tree
DTS	Data transformation services
EDM	Educational data mining
GPA	Grade point average
KDD	Knowledge discovery in databases
KDP	Knowledge discovery process
MDAC	Microsoft data access components
MDT	Microsoft decision tree
OLAP	On-line analytical processing
PDCA	Plan-do-check-act
SKDS	Student knowledge discovery software
SRM	Student relationship management
SQL	Structured query language

4.1 Introduction

Appropriate decisions can be made by effectively analyzing and managing the growing volume of data. Gaining information from business data started with data collection in the 1960s; this type of data collection answered questions related to the past. In the 1980s, with the development of relational databases, data access methods were introduced. In the 1990s, data warehousing and decision support systems were created based on multi-dimensional databases and On-line Analytical Processing (OLAP). Today, data mining (DM) produces a particular enumeration of patterns in data. This should be understandable and usable by the business end user. To accomplish this, there is a typical data-driven business process consisting of multiple stages between multiple servers and data extracts, preprocessing, and conversions with advanced algorithms, multi-processor computers and massive databases [1].

DM is a new data-oriented technology, which is able to discover valuable interactions in human activities using computer implementations. For this purpose, an automated-process to uncover trends, patterns, and relationships from accumulated electronic traces is used to collect the data [2].

Recently, knowledge discovery in databases (KDD) methodologies have been used to enhance and evaluate higher education tasks [3]. This process, contributes to the enhancement of the quality of a higher educational system by evaluating student data. Analyzing and manipulating the existing data with respect to predefined goals provide high quality, student-specific, and student-centered education for higher education institutions. Thus, DM promises better ways to produce higher quality in education, and greater satisfaction for student [4]. Moreover, Web-based systems routinely collect vast quantities of data on user patterns, and DM methods can be applied to these databases. Newly developed web-based educational technologies, also offer researchers unique opportunities to evaluate the factors affecting students' learning capacity which is an important element of their academic success [5].

Another fundamental role of universities is to raise the quality of education, as well as producing and disseminating information. Recorded data in universities contains valuable information regarding students, which is usually used for official procedures such as producing transcripts. In fact, this data could also be used in academic guidance of students using a separate discovery investigation to extract information relevant to the individual student's progress [6]. Additionally, competitive advantages could be obtained by identification of the students' demands through the available data. In this direction, some models have been proposed and implemented. One of them demonstrates how DM can be utilized in a higher educational system to improve the efficiency and effectiveness of the traditional processes [7]. The other model was combined with a deterministic model to analyze the students' results over the 2 or 3 semesters in the academic year in a private educational institution [8].

In the increasing commercialized education environment, higher educational institutions need to become more efficient, provide a better quality service to deliver exceptional student experience [9]. Moreover students and their parents want an education that is tailored to their needs. Student Relationship Management (SRM) is one of the responses to these demands [10]. SRM can be described as a proactive management system which creates a single, holistic view of each student by bringing together different elements of data from various sources such as; academic departments, student services and independent systems such as finance and accommodation [11]. SRM is valuable when data is scattered across an institution, in different departments, in various file formats. It is designed to impact on every connection in the student lifecycle and integrates with an institutions current projects and systems, avoiding duplication and ensuring a fluid, step-change in student management.

This smarter student management uses predictive analytics that considers the mix of very different metrics on students and from this data can be confidently predicted their potential failure or success. The results can trigger action to bring proactive support to the learner and help remove the factors that lead to failure. Integrated profiles, analytics and tools to increase the quantity and quality of admissions across the institution, furthermore, the success rate of the establishment increases [11].

There are several approaches to KDP. A chapter in the book [12] describes the KDP, presents models, and explains why and how these could be used for a successful DM project. In the context of DM, Crisp-DM model is considered to be a significant standard, however, it is highly recommended in a technical project report [13] that following a structured plan-do-check-act (PDCA) cycle in DM applications to achieve an optimized quality and success. The PDCA cycle as an approach to change and problem solving is very much at the heart of Deming's quality-driven philosophy [14]. The four phases in the PDCA Cycle are:

- Plan: Identifying and analyzing the problem.
- Do: Developing and testing a potential solution.
- Check: Measuring how effective the test solution was, and analyzing whether it could be improved in any way.
- Act: Implementing the improved solution fully.

Crisp-DM can be combined with the PDCA cycle in this study as presented in Fig. 4.1 in which, the framework of the PDCA cycle has eight stages, which captures all the facets of the DM tasks [15]. The major stages are: problem identification; gathering and selection of data; data preprocessing for missing, duplicate or erroneous information; selection of appropriate learning algorithms; preparation and processing of data; construction and evaluation of the models; interpretation of the discovered knowledge; and finally, taking action.

The cycle starts with the plan stage where a precise business objective and the related business problem or opportunity, is defined and the application domain is demarcated. This stage is very important since it determines the scope of the project. Then comes the do stage that includes the stages of the KDP which tends to be highly iterative and interactive [12].

The target data set is selected from a large database. After selecting the target data set, cleaning, preprocessing and reduction create the appropriate data set for further transformation and combination. After choosing the functions of DM and the DM algorithm(s) follow next the DM process is initiated. The outcome of this whole process is the discovered knowledge that is interpreted and evaluated for the business client.

This knowledge consists of the relationships and patterns found in the data which becomes the input for the check stage where the analyses are completed to assess whether the knowledge is applicable to the scope of the project. If the results are interesting and satisfactory, last stage is to act on the results by implementing the solution suggested by the KDD results. This cycle is continuous, as new or related problems arise over time the cycle becomes continuous and each time the mechanism will try to find solutions with the help of DM tools [16].

This chapter proposes a new EDM system 'Student Knowledge Discovery Software (SKDS)', introduces its architecture for use in KDPs and presents the outcome of a test on university data to explore the factors having an impact on the success of the students based on student profiling [17, 18]. SKDS is a specific system that integrates the EDM process with the database management system (DBMS) [19]. Although there are different commercial software applications available that are adapted to DM [20], our approach has some major advantages for educational institutions. First, the data analysis realizes where it is generated therefore; the analysis can be easily repeated on new inserted data. Secondly, the software can use the functions of the SQL Server and the Analysis Services, which

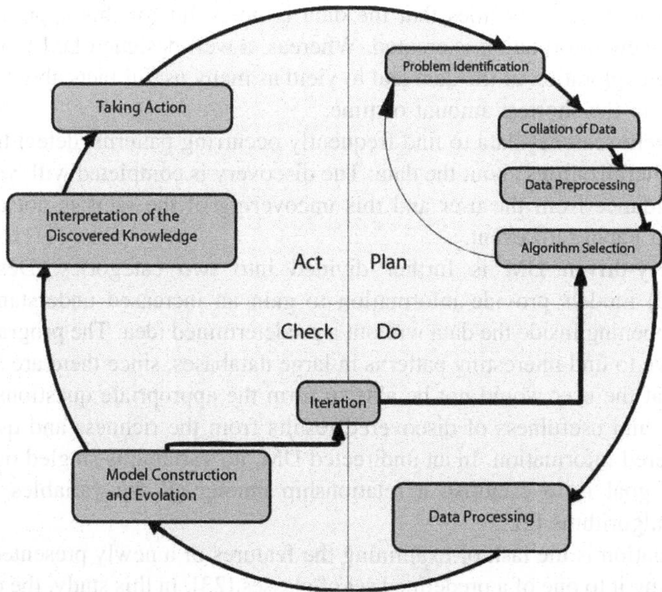

Fig. 4.1 The framework of PDCA cycle for DM

have programmed DM techniques. Thus, the user can perform tasks without any need for complicated Structured Query Language (SQL) statements. Finally, the specifically designed user interface makes it possible to follow the EDM process and to perform the database management activities in an easy and orderly manner. The end-user of this software must have fundamental skills for computer and database operations with the knowledge of EDM. The results can be assessed by the decision-makers, such as advisors and administrators.

The rest of the chapter is organized as follows: Sect. 4.2 presents the basic concepts of decision tree (DT) classification analysis. Section 4.3 introduces the system overview, user interface and architecture of SKDS. Section 4.4 gives notations related to our performed study. Section 4.5 presents the results of our EDM model. In Sect. 4.6 the conclusion enumerates the advantages and validates the proposed approach.

4.2 Background

There are basically two types of DM. The user creates an explicit or implicit hypothesis about the data in the "verification-driven" DM. Limited by the hypothesis, a query concerning the data is conducted and the results of this query are examined. If the result is positive, the process ends otherwise, a new query is formulated and the process iterates until the resulting data either verifies the hypothesis or the user decides that the data is not valid for this approach [21]. Thus, little new information is created. Whereas, a well-designed DM tool is able to build the exploration of the data and to yield as many useful facts about the data as possible in the shortest amount of time.

The system searches data to find frequently occurring patterns, detect trends or produce generalizations about the data. The discovery is completed with very little (or no) guidance from the user and this uncovering of the facts is not a consequence of a haphazard event.

Discovery-driven DM is further divided into two categories. Descriptive (undirected) models provide information to gain an increased understanding of what is happening inside the data without a predetermined idea. The program takes the initiative to find interesting patterns in large databases, since there are so many patterns that the user would not be able to form the appropriate questions to ask. The power and usefulness of discovered results from the richness and quality of the discovered information. In an undirected DM, no variable is singled out as the target, the goal is to establish a relationship among all the variables such as clustering algorithms [22].

Classification is the task of examining the features of a newly presented object and assigning it to one of a predefined set of classes [23]. In this study, the decision tree (DT) classification was chosen since this structure offers the ability to easily generate rules, provide understandable models, and achieve a high level of integration with information technology processes because it requires little preprocessing of data [24].

4.2.1 The Decision Tree Classification Model

The DT classification is a supervised learning method that constructs a tree from a set of examples. It creates classification models by examining already classified data from a historical database and inductively finding a predictive pattern. This pattern can be used both to understand the existing data and predict how new instances will behave. It is a predictive model viewed as a tree consisting of decision nodes, branches and leaves. A decision node specifies a test to be carried out, which branches are to be supplied without losing any data.

The split decision is made at the node "in the moment", it is never revisited and also univariate. In addition, all splits are made sequentially, so each split is dependent on its predecessor. Thus, all future splits are dependent on the first split, which means the final solution could be very dissimilar if a different first split had been made. Each branch of the tree is a possible answer to the classification question and will lead either to another decision node or to the bottom of the tree, called a leaf node. The leaves are the partitions of the data set with their classification. The DT process starts at the root node and moves to each subsequent node until a leaf node is reached [25].

From a business perspective, DTs can be viewed as creating a segmentation of the original data set to predict some important piece of information (each segment would be one of the leaves of the tree). The predictive segments are similar with respect to the information being predicted and contain a description of the characteristics that define the predictive segment. Thus, although the DTs and algorithms may be complex, the results are easy-to-understand [26].

There are several major advantages as well as disadvantages in using DTs. The most important advantage of DTs are generating understandable roles no matter how complicated the inputs are. It is generally easy to follow any one path through the tree, so explaining the decisions along the way is also easy.

The computation cost for each split is minimal. In practice, algorithms tend to produce DTs with a low branching factor with simple tests at each node, so the tree does not grow too large and these tests translate into simple boolean and integer operations that are fast and inexpensive. Using DTs, the field, which is the best at splitting the training records, can be singled out for analysis. This will enable the user to determine which variable mostly influences their data. However, when there are a large number of factors affecting data it might be very difficult to determine specific factors; therefore DTs are not suited for numbers covering large intervals [27].

4.2.2 The Decision Tree Mechanism

DTs are built using recursive partitioning which is an iterative process of splitting the data up into partitions. Initially, the algorithm seeks to create a tree that works as perfectly as possible on all the available data, but this does not usually work. The process starts with a training set consisting of pre-classified records. In order to build a tree that distinguishes the classes, the best possible question to ask at each branch point of the tree has to be found. The goal is for the leaves of the tree to be as homogeneous as possible with respect to the prediction value. The diversity measure is calculated for the two partitions, and the best split is that with the largest decrease in diversity. After the tree has been grown to certain size, the algorithm has to check if the model overfits the data which it does by a cross validation approach. The tree size can be controlled via stopping rules limiting growth [28].

The quality of a tree depends on both its size and the classification accuracy [29]. The method first chooses a subset of the training examples to form a DT. If the tree does not give the correct answer for all the objects, a selection of the exceptions is added to the window and the process continues until the correct decision set is found.

The eventual outcome is a tree in which each leaf carries a class name, and each interior node specifies an attribute with a branch corresponding to each possible value of that attribute. Entropy is a measure commonly used in information theory. The higher the entropy of an attribute, the more uncertainty there is with respect to its outcomes. Thus, we would want to select attributes in order of increasing entropy, where the root node of our tree would correspond to the attribute with the lowest entropy value. More information about the methodology and related measures of DTs can be found in [30, 31].

4.3 System Overview, Software Interface and Architecture

In this EDM application, the Microsoft Windows Server and the Microsoft SQL Server were used as an operating system and a relational DBMS, respectively. In addition, the Analysis Services in the SQL Server were used to create and validate DM models, the Microsoft Data Access Components (MDAC) were used to access the data, Angoss DM consumer controls were used to display the models and validation results. Furthermore, the SKDS was developed using the programming technologies (plugins, controls and tools) of the Microsoft Visual Basic.

While implementing some tasks related to the EDM process, the SQL Server tools were called upon for example, in the transformation stage where the Data Transformation Services (DTS) import/export wizard accesses the data set and then transforms the column values. Moreover, the DTS creates predictions based on the DM model and performs actions according to the results.

In the solution development phase of the study, a task sharing mechanism between the SQL Server and the Analysis Services has been developed in order to implement the tasks involved in every individual EDM stage. In fact, each of the tasks can be separately implemented either on the SQL server or the Analysis Services. In order to perform all the tasks together, SKDS was developed. SKDS was specifically designed for use with student demographic data. The purpose of SKDS is:

- To set an example for the EDM process.
- To access all the tasks from a single program.
- To access all functions of the SQL Server and the Analysis Services by means of programming techniques. In this way, the user can perform tasks without any need for complicated SQL statements.

The SKDS user interface is shown in Fig. 4.2. SKDS consists of three main sections; the database connection, data preparation and model development. In each section a button represents a task and takes the user either to a form or to a wizard related to this task. Since it is important to undertake the tasks in order of precedence, the interface was designed to indicate this order. Forms accessed through the user interface make it possible to follow the KDP from the perspective of the PDCA cycle and to perform the database management activities in an easy and orderly manner.

Figure 4.3 presents the SKDS working principle as a block diagram. This shows that the user (computer science professional familiar with the principles of knowledge discovery and student data that is to be used) who will perform the DM can access the forms directly (table management, sampling, cleaning, research and exploring, splitting, modeling, control and validating) or indirectly (transformation) for eight different tasks.

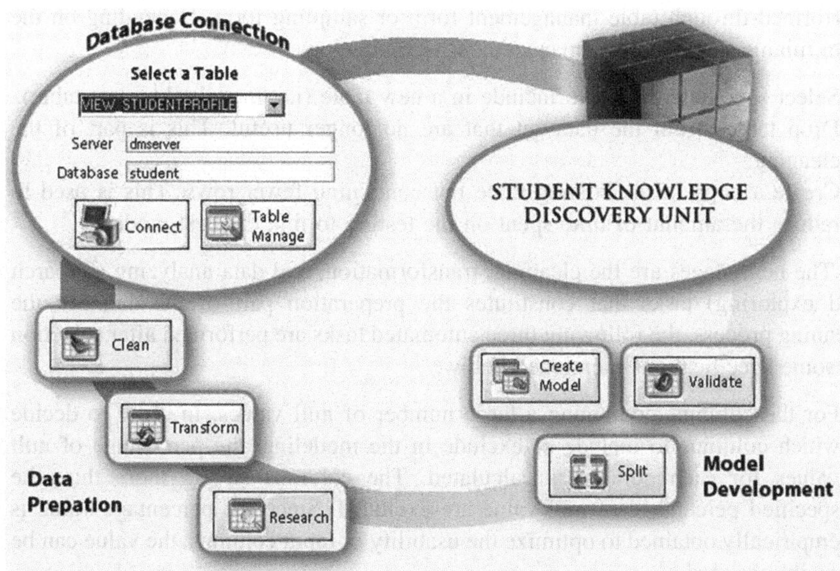

Fig. 4.2 SKDS user interfaces

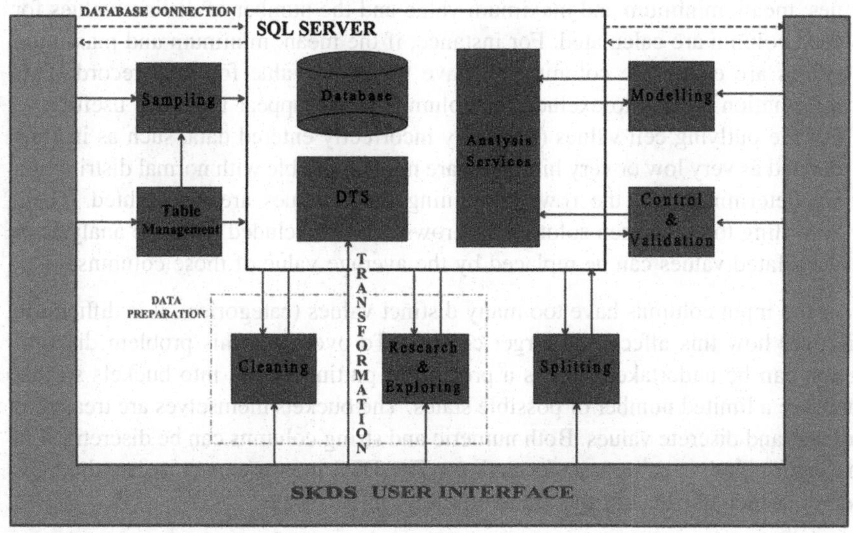

Fig. 4.3 SKDS architecture

In SKDS, first a database connection is made to the data set that contains the data on which an investigation is to be carried out. After the connection is established, the various editing and backup activities listed below can be performed through table management form or sampling form. Depending on the data mining task the table management techniques are:

- Select specific columns to include in a new table (i.e. model table, test table).
- Drop tables from the data set that are no longer useful. This is part of the cleaning.
- Create a copy of an existing table but containing fewer rows. This is used to reduce the amount of time spent on the testing to find the best model.

The next stages are the cleaning, transformation, and data analyzing (research and exploring) tasks that constitutes the preparation part of the data. In the cleaning process, the following three automated tasks are performed after selection of some specifications mentioned below.

- For the columns containing a large number of null values; in order to decide which columns to include or exclude in the modeling, the percentage of null values for each column is calculated. The columns having more than the specified percentage of null value are excluded. Since the percentage value is empirically obtained to optimize the usability of input columns, the value can be easily changed.
- For the columns that have one (i.e. same content), a few (distinct values in a group with same content or a spread of neutral status) or too many (close to the number of records such as; students' address and phone numbers) distinct values; mean, minimum and maximum value and the number of distinct values for each column are calculated. For instance, if the mean, minimum and maximum values are equal, the column will have the same value for each record. This information is used to exclude the columns which appear not to be useful.
- For the outlying cell values (generally incorrectly entered data, such as income defined as very low or very high) that are not compatible with normal distribution are determined and the rows containing these values are highlighted. Then, according to the chosen solution, the rows can be excluded from the analysis or the related values can be replaced by the average value of those columns.

If the input columns have too many distinct values (categories), it is difficult to discover how this affects the target column. To overcome this problem discretization can be undertaken; this is a process of putting values into buckets so that there are a limited number of possible states. The buckets themselves are treated as ordered and discrete values. Both numeric and string columns can be discretized in the input columns using transformations. The DTS import/export assists the SQL Server, which carries out this transformation in two ways:

- The input columns having a large number of possibilities for the values are categorized exhaustively into a limited number of categories such as category 1, category 2, …, category k.
- In some cases in which the variables that may have some effect on target column are not in a single column or not directly available the transformations on these input columns can be used to create new useful columns.

The last procedure of data preparation is the data analyzing, before splitting the data, consists of the research and explore. On these tasks the columns, which are expected to further affect the result are preferred. In the research task of this stage, correlation matrices (CM) were used to find columns that only had numeric values.

These correlations indicate the degrees of relationship between the target columns and the potential predictor input columns. The specified correlation value (i.e. ± 0.01) is accepted as the lowest limit in the CMs, so the correlations of the columns with the target columns below this value were ignored for the DM models. During the exploration, the other part of this stage, both numeric and non-numeric columns are shown in histograms to determine visually the reliability or usability of the columns.

In the model development section; the data set was split into training and test data sets. Splitting the data allows the user to create a model and to test this model using data from the same source. Two new training and test tables are created.

Then the main table's rows were allocated to the new training and test tables (i.e. 70 and 30 % of all data set records respectively), by random distribution. SKDS calculates the percentages of the positive and negative values in consultation with the target column (e.g. in Model I: grade point average (GPA) value below 2.0 is negative, above 2.0 positive for each record). A model is then formed using the training data set. Finally, the validity of the model is checked using test data set during the validation process. A lift chart method was chosen for the evaluation of the efficiency of the models. To accomplish all these processes, SKDS benefited from the SQL Server, which is mainly utilized in the data base management activities, and the Analysis Services, which are primarily utilized in modeling and validating.

4.4 Case Study: Modeling Student Performance

This study aims to reveal individual student characteristics that are associated with academic success using a DT classification technique. Each student is categorized as either successful or unsuccessful according to their GPAs. The stages of the study are given below.

4.4.1 Data Description

This study uses the demographic data of students enrolled in the faculty of Economics and Social Sciences of Mugla Sitki Kocman University. This faculty was chosen for this study since the departments are very similar furthermore; it is the oldest in the university and has the most students.

The data used in the discovery process was mostly obtained from when the students registered at the university. The data consisted of; information required by the state such as city of residence and date of birth, high school information including matriculation certificate, education type and knowledge of foreign language); Turkish university entrance exam score and university placement information, socio-economic status of the student's family and student's academic standing in terms of the semester based GPA scores from their university department.

This study was conducted in the University of Mugla Sitki Kocman, Turkey. In this university, student's academic standing is calculated in the form of the cumulative GPA taking into account all the courses they have taken over the whole degree program. After each course was completed, the student is given a letter grade for which there is a point equivalent (AA = 4.0, BA = 3.5, BB = 3.0, CB = 2.5, CC = 2.0, DC = 1.5, DD = 1.0, FD = 0.5 and FF = 0.0). The GPA is the average of all the grades accumulated over the courses taken within a specific period of education. To be awarded a bachelor degree, a student must obtain at least DD from each course and have a GPA of at least 2.00 to graduate. Honors degrees are given to students with a cumulative GPA from 3.00 to 3.49 and for high honors from 3.50 to 4.00.

4.4.2 Data Preparation

The student data in this study came from many different data files in multiple databases; in the university departments, faculty, central registration system, and archives. Thirteen tables related to the scope of this study were selected from the databases given above. The SQL Server DBMS was implemented at the university registration office. Using SQL queries, the target data set was formed from six separate relational tables, which contained records reflecting academic, demographic, identification, undergraduate and graduate course information. These related tables consisting of a total of 111 columns with 6,470 records were combined in a single view. Afterwards, this target data set is subject to further data cleaning and pre-processing. Unnecessary attributes which are irrelevant for the proposed model are omitted. Thus, handling missing data fields and accounting for known changes is completed, see Fig. 4.4.

The GPAs of the students are generally better indicators because they are reliable and ultimately objective, numeric and accredited to measure academic success in education. In the study, it was found that the greatest correlation coefficient was obtained when comparing the correlations of each column to the other columns of KDD data in the correlation matrices. Thus, the column containing the GPAs was used by the authors as the target column to establish the models. The other columns in the same data set were used as input for the DT models.

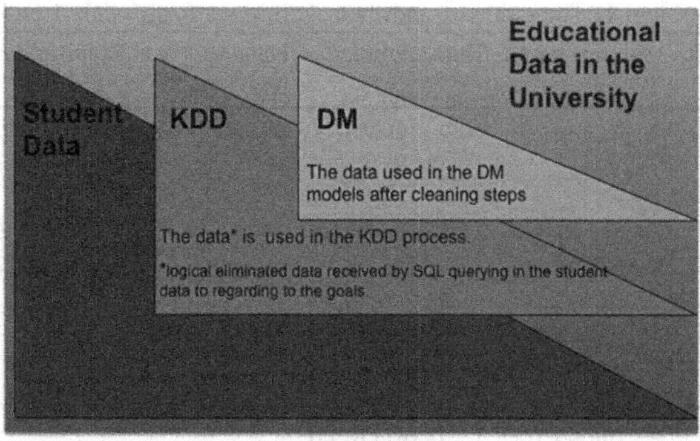

Fig. 4.4 Data selection in the EDM

4.4.3 Analyzer Model

In this study, the classification of the students was undertaken according to their individual success characteristics and DTs were chosen for the model as they assist in producing more understandable results.

Microsoft Decision Trees (MDTs) in the Microsoft SQL Server Analysis Services were applied to create DT models [19]. Originally MDT is a probabilistic classification tree algorithm that is an improvement over the ID3 DT algorithm with some of the add-ons. The basis of the MDT algorithm is introduced in [32].

4.5 Discussion of Results

This section presents the results of the discovery process performed on the student data. Table 4.1 shows the columns and their definitions in the DT models. In the first DT model, the columns affecting the target column, in the order of importance, were YEARECNO and GRANTPTF, which slightly affected the target. The first separations happened in the YEARECNO values: 5, 4, 1 and 2, respectively. This is not a surprising prediction because students have to complete their education over a period of 4–7 years with a requirement of cumulative GPA ≥2.00. Thus, the DT models refined as a result of PDCA cycle and YEARECNO were removed from the model to obtain more interesting and hidden results which is a function of DM.

After this operation, the columns that affected the target column are given in the first DT model in Fig. 4.5. Here, the most affective columns used to predict the target column are LANGPREP and REGTYPE. Since LANGPREP is English and REGTYPE is different from the normal type such as transferred from another university, these are seen to be influential on student success in DT. Students transferring from another university must go through adaptation training for a year and must be successful to a certain extent.

Table 4.1 The columns and their definitions for Models 1 and 2

No	Column name	Data included	Data-type	No	Column name	Data included	Data-type
1	KEY	Key column (1, 2, 3, …)	Single	13	LANGPREP	Foreign language to be learned in prep-school	Varchar
2	TARGET_A[a] (target for model I)	If GPA is between 2.0 and 4.0, 1, else 0	Integer	14	PROGTYPE[b]	Department of university	Varchar
3	TARGET_B[b] (target for model II)	If GPA is between 3.0 and 4.0, 1 else 0	Integer	15	HIGHGRAD1[b]	Finishing high school with best average (yes/no)	Integer
4	REQPREP	Request for preparatory school (yes/no)	Boolean	16	SEMRECNO[b]	Number of semesters attended in university	Integer
5	MILITARY	Military service status (completed or not completed)	Boolean	17	SEMCOUNT	Total of semesters to be attended	Integer
6	GENDER	Gender (M/F)	Boolean	18	YEARECNO[a]	The number of years spent at school	Integer
7	GRANTPTF	Grants for tuition fees (receiving/not receiving)	Boolean	19	FMINCOME	Monthly income of family	Single
8	EDUCTYPE	Type of education (day/evening classes)	Boolean	20	CITY2	Region in which the student was born	Varchar
9	DEPTNAME	Department name	Varchar	21	LIVECITY2	Region in which student currently lives	Varchar
10	IDCODE	Department and type of education	Varchar	22	PREFERNO	Order of preference of university attended according to student choice	Integer

(continued)

Table 4.1 (continued)

No	Column name	Data included	Data-type	No	Column name	Data included	Data-type
11	REGTYPE	Type of school registration	Varchar	23	PREFERNO2	Order of preference of university location according to student choice (five categories grouped)	Integer
12	TYPEHIGH	Type of high school	Varchar	24	FMINCOME2[b]	Monthly income of family (four categories grouped)	Integer

[a] only included in the 1st model
[b] only included in the 2nd model

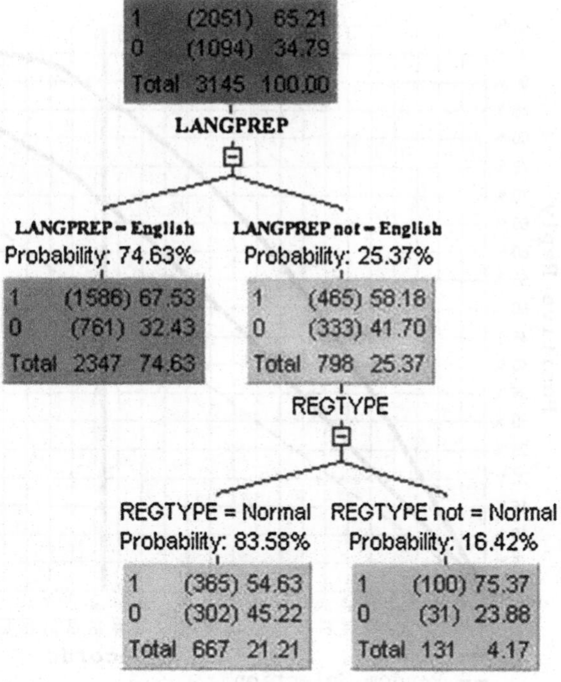

Fig. 4.5 Graphical display of the DTs for Model I

Fig. 4.6 Graphical display of the DTs for Model II

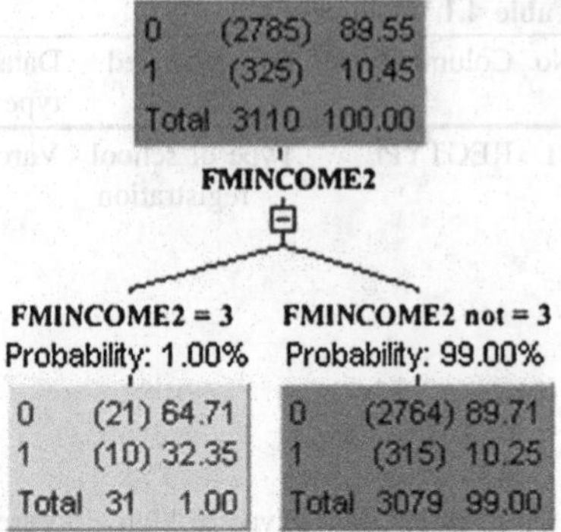

Figure 4.6 shows the resulting DT from Model II. The DT algorithm selected FMINCOME2 as the most important factor that determines the split on the data. The column FMINCOME2 contains categorized data on the monthly incomes of the students' family with the value 3 corresponding to the middle level income (over the lowest rate at which an employer can legally pay an employee; usually

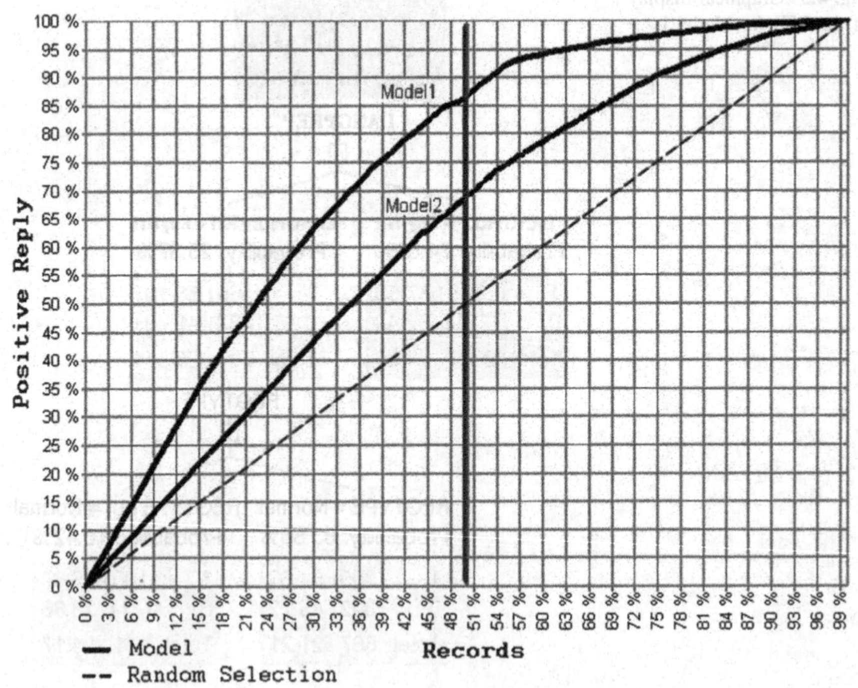

Fig. 4.7 Lift graphics [33] for the models (Model I predicts that the students will have a GPA that is greater or equal to 2.0, whereas Model II predicts that the students will have a GPA that is greater or equal to 3.0)

expressed as pay per month) group. Thus, it is concluded that a high level of success is more correlated with the middle level income.

Lift graphics, which specify the validation of the models are given in Fig. 4.7. When trained models are employed, the percentage of total positive responses in all records was 87 % for Model I and 68 % for Model II. Accordingly, the lift value is 87/50 = 1.74 for Model I and 68/50 = 1.36 for Model II. These results indicate that these models are able to confidently predict the outcome. The reason for the lower prediction potential of Model II is that Model II has less positive values in the target column in comparison with Model I. Thus, a small positive ratio makes it difficult to establish relationships during the training of the model. Model I had a 65 % positive value ratio and for Model II, this ratio was 11 %.

4.6 Conclusions

Besides having a very important role in knowledge production and dissemination, universities offer educational support services for students. Leading universities should discover new ways to base their decision making process in the educational domain on sound business analysis thus providing the best service for their customers; the students. In order to achieve customer satisfaction there needs to be a high student achievement. This can be attained by good guidance and support thus meeting the academic demands of the students during their university education. Analyzing and manipulating existing data with respect to pre-defined goals brings a competitive advantage for higher education institutions in providing high quality, student-specific and student-centered education.

This study aimed to evaluate and develop data-driven approach to the improvement of the performance of university students using a new developed educational technology system, SKDS, using DM methods. With this EDM system, all the tasks involved in the KDP are performed collectively. Our approach has some major advantages for educational institutes. First, the data analysis is realized where it is generated thus, this analysis can be easily repeated on new inserted data. Second, software can use the functions of the SQL server and the Analysis Services, which have essential DM models [34].

In this way, the user can perform tasks without any need for complicated SQL statements. SKDS were designed by the authors considering the needs of the used data and the problem investigated. So, the specifically designed user interface, makes it possible to follow the EDM process and perform the database management activities in an easy and orderly manner. This study may help other researchers working on the integration of specific EDM processes into the DBMS of educational institutions.

In evaluating student performance, the DT classification technique was used since it can produce rules in the tree structure, provide simple and easy-to-understand models and operations that can be carried out even with minimal information preparation. Therefore, DTs can easily be integrated with information technologies and render high level of automation possible. The classifications attempt to discover which demographic data has the most impact on student GPA. In the current study two classification models were obtained limited to determining the profiles of students whose GPAs ranged from 2.0 to 4.0 and the second group with GPAs of 3.0–4.0.

In the first model, the types of registration to the university and in the second model, the monthly income of the family were found to be the greatest factors affecting the target. In checking the performance of the models, lift graphics were used. According to the lift graphics, values 1.74 for Model I and 1.36 for Model II were found which shows that the models have a prediction capability.

Missing data in some of the columns in the dataset had a direct impact on the success of the system. Therefore, more accurate predictions about student success can be made when the amount of data and the number of variables is increased. DTs handle non-numeric data very well. This ability to accept categorical data minimizes the amount of data transformations. Even though DTs are good at classifying data, they alone may not be sufficient for discovery.

In terms of future work, since SKDS can connect to the SQL Server and the Analysis Services, which include more DM algorithms, an extension of the current study based on different analysis parameters could show different perspectives of a student's performance and progress through their university career.

References

1. Abdous, M., He, W., Yen, C.J.: Using data mining for predicting relationships between online question theme and final grade. J. Educ. Technol. Soc. **15**(3), 77–88 (2012)
2. Campagni R., Merlini D., Sprugnoli R.: Analyzing paths in a student database. In: Yacef, K., Zaiane, O., Hershkovitz, A., Yudelson, M., Stamper, J. (eds.) 5th International Conference on Educational Data Mining, pp. 208–209. International Educational Data Mining Society, Chania (2012)
3. Oyelade, O.J., Oladipupo, O.O., Obagbuwa, I.C.: Application of k-means clustering algorithm for prediction of students' academic performance. Int. J. Comput. Sci. Inf. Secur. **7**(1), 292–295 (2010)
4. Scheuer, O., McLaren, B.M.: Educational DM. In: Seel, N.M. (ed.) Encyclopedia of the Sciences of Learning. Springer, New York (2011)
5. Bidgoli B.M., Kashy D.A., Kortemeyer G., Punch W.F.: Predicting student performance: an application of DM methods with an educational web-based system. In: 33rd ASEE/IEEE Frontiers in Education Conference, pp. 13–18. IEEE, Boulder (2003)
6. Romero, C., Ventura, S.: Educational data mining: a review of the state of the art. IEEE Trans. Syst. Man Cybern. Part C **40**(6), 601–618 (2010)
7. Delavari, N., Amnuaisuk, S.P., Beikzadeh, M.R.: DM application in higher learning institutions. Inform. Educ. **7**(1), 31–54 (2008)
8. Vialardi, C., Chue, J., Peche, J.P., Alvarado, G., Vinatea, B., Estrella, J., Ortigosa, A.: A data mining approach to guide students through the enrollment process based on academic performance. User Model. User-Adap. Inter. **21**(1–2), 217–248 (2011)
9. Kumar, N.V.A., Uma, G.V.: Improving academic performance of students by applying data mining technique. Eur. J. Sci. Res. **34**(4), 526–534 (2009)

10. IBM Case Study, Hamilton County Department of Education: Improving student performance and school effectiveness with predictive analytics. http.//www.ibm.com/analytics/us/en/case-studies
11. IBM Case Study, Seton Hall University: Social media marketing analytics helps engage incoming prospects and increase enrollment yield http://www.ibm.com/analytics/us/en/case-studies
12. Cios, K.J., Swiniarski, R.W., Pedrycz, W., Kurgan, L.A.: Data Mining: A Knowledge Discovery Approach, pp. 9–24. Springer, New York (2007)
13. Meints, M., Möller, J.: Privacy preserving data mining: a process centric view from a European perspective. Report of the project FIDIS (Future of Identity in the Information Society), http://www.fidis.net/fileadmin/journal/issues/1-2007/Privacy_Preserving_Data_Mining.pdf
14. Jalili, M., Rezaie, K.: Quality principles deployment to achieve strategic results. Int. J. Bus. Excellence **3**(2), 226–259 (2010)
15. Maimon, O., Rokach, L.: Introduction to knowledge discovery and data mining. In: Maimon, O., Rokach, L. (eds.) Data Mining and Knowledge Discovery Handbook, pp. 1–15. Springer, New York (2010)
16. Micić, Ž., Micić, M., Blagojević, M.: ICT Innovations at the platform of standardisation for knowledge quality in PDCA. Comput. Stand. Interfaces **36**(1), 231–243 (2013)
17. Guruler, H., Istanbullu, A., Karahasan, M.: A new student performance analyzing system using knowledge discovery in higher educational databases. Comput. Educ. **55**(1), 247–254 (2010)
18. Guruler, H., Karahasan, M., Istanbullu, A.: Determining profile of university students: a case study on Mugla University databases. Mugla Univ. J. Soc. Sci. **18**, 27–37 (2007)
19. Larson, B., English, D., Purington, P.: Delivering Business Intelligence with Microsoft SQL Server 2012. McGraw-Hill, New York (2012)
20. The Cyber Security and Information Systems Information Analysis Center (CSIAC): A Comparison of Leading DM Tools. https://sw.thecsiac.com/databases/url/key/222/225
21. Chikalov, I., Lozin, V., Lozina, I., Moshkov, M., Nguyen, H.S., Skowron, A., Zielosko, B.: Logical analysis of data: theory, methodology and applications. In: Chikalov, I., Lozin, V., Lozina, I., Moshkov, M., Nguyen, H.S., Skowron, A., Zielosko, B. (eds.) Three Approaches to Data Analysis, pp. 147–192. Springer, Heidelberg (2013)
22. Khan, D.M., Mohamudally, N., Babajee, D.K.R.: A unified theoretical framework for data mining. Procedia Comput. Sci. **17**, 104–113 (2013)
23. Özekes, S.: Classification and prediction in data mining with neural networks. Istanbul Univ. J. Electr. Electron. Eng. **3**(1), 707–712 (2012)
24. Cano, A., Zafra, A., Ventura, S.: An interpretable classification rule mining algorithm. Inf. Sci. **240**, 1–20 (2013)
25. Guan, H.: A new data mining approach combining with extension transformation of extenics. In: Deng, W. (ed.) Future Control and Automation, vol. 173, pp. 199–205. LNEESpringer, Heidelberg (2012)
26. Lakshmi, T.M., Martin, A., Begum, R.M., Venkatesan, V.P.: An analysis on performance of decision tree algorithms using student's qualitative data. Int. J. Mod. Educ. Comput. Sci. **5**(5), 18–27 (2013)
27. Lin, C.F., Yeh, Y.C., Hung, Y.H., Chang, R.I.: Data mining for providing a personalized learning path in creativity: an application of decision trees. Comput. Educ. **68**, 199–210 (2013)
28. James, G., Witten, D., Hastie, T., Tibshirani, R.: Tree-based methods. In: Casella, G., Fienberg, S., Olkin, I. (eds.) An Introduction to Statistical Learning, vol. 41, pp. 303–335. Springer, New York (2013)
29. Yang, H., Fong, S.: Optimized very fast decision tree with balanced classification accuracy and compact tree size. In: 3rd International Conference on Data Mining and Intelligent Information Technology Applications, pp. 57–64. IEEE, Coloane (2011)
30. López-Chau, A., Cervantes, J., López-García, L., García Lamont, F.: Fisher's Decision Tree. Expert Systems with Applications **40**(16), 6283–6291 (2013)

31. Aggarwal, C.C., Zhai, C.: A survey of text classification algorithms. In: Aggarwal, C.C., Zhai, C.X. (eds.) Mining Text Data, pp. 163–222. Springer, New York (2012)
32. Microsoft Decision Trees Algorithm Technical Reference http://msdn.microsoft.com/en-us/library/cc645868.aspx
33. Prati, R.C., Batista, G.E.A.P.A., Monard, M.C.: A survey on graphical methods for classification predictive performance evaluation. IEEE Trans. Knowl. Data Eng. **23**(11), 1601–1618 (2011)
34. Data Mining Algorithms (Analysis Services - Data Mining) http://msdn.microsoft.com/en-us/library/ms175595.aspx

Chapter 5
Using Data Mining Techniques to Detect the Personality of Players in an Educational Game

Fazel Keshtkar, Candice Burkett, Haiying Li and Arthur C. Graesser

Abstract One of the goals of Educational Data Mining is to develop the methods for student modeling based on educational data, such as; chat conversation, class discussion, etc. On the other hand, individual behavior and personality play a major role in Intelligent Tutoring Systems (ITS) and Educational Data Mining (EDM). Thus, to develop a user adaptable system, the student's behaviors that occurring during interaction has huge impact EDM and ITS. In this chapter, we introduce a novel data mining techniques and natural language processing approaches for automated detection student's personality and behaviors in an educational game (Land Science) where students act as interns in an urban planning firm and discuss in groups their ideas. In order to apply this framework, input excerpts must be classified into one of six possible personality classes. We applied this personality classification method using machine learning algorithms, such as: Naive Bayes, Support Vector Machine (SVM) and Decision Tree.

Keywords Personality · Classification · Conversation · Larry's Rose framework · Natural language processing · Educational data

F. Keshtkar (✉) · C. Burkett · H. Li · A. C. Graesser
Southeast Missouri State University, DH 021F, Mail Stop 5950, USA
e-mail: fkeshtkar@semo.edu

C. Burkett
e-mail: cburkett@memphis.edu

H. Li
e-mail: hli5@memphis.edu

A. C. Graesser
e-mail: a-graesser@memphis.edu

A. Peña-Ayala (ed.), *Educational Data Mining*,
Studies in Computational Intelligence 524, DOI: 10.1007/978-3-319-02738-8_5,

Abbreviations

CBLE Computer based learning environment
CRF Conditional random field
EDM Educational data mining
ITS Intelligent tutoring system
LIWC Linguistic inquiry and word count
NPC Non-player characters
SVM Support vector machine

5.1 Introduction

Interpersonal conversation is not an easy task. During conversation in educational games, ITS, or chat interaction, the students may have different ideas from the others. Because they may affect by different moods or personality when they listen or say something. On the other hand, students might have different personality characters, i.e., to be cooperative, leading, aggressive, or dependent. For all these reason, we believe personality traits should be considered in computer-based learning environments (CBLE) such as educational game and intelligent tutoring systems. For example, attitudes toward computers can be related to personality types such that those displaying higher scores on neuroticism may have greater computer related anxiety. Furthermore, it is known that it is important to take individual differences into account during learning in CBLE. For example, ITS are known for their ability to simulate effective human tutoring methods as well as take into account the individual needs of learners [1].

Although the efforts to classify personality traits can be a particularly useful endeavour, the detection of personality and/or behavior in conversation using natural language, as it turns out, is a rather difficult task. For example, in serious games in which communication occurs in chat rooms, players may discuss different ideas than others they are chatting with during conversation. Likely, they are also exposed to or affected by the different personalities or moods of other players during communication. On the other hand, players may demonstrate various personality characteristics (such as those related to helping, leading, or aggression) that may result in varied behavioral indicators within conversation.

This chapter aims to investigate, how chat interactions from student log data can be used to determine a student model to classify personality. In results, it turned out that we developed a supervised learning model based on annotated data to automatically detect the students' personality based on their chat interaction in an educational game. The purpose of this research is to identify personality traits of students in textual excerpts in an Educational Game in order to develop an automatic classier that determines the personality characteristics of a student based on their discourse in game. This automatic classier will then be implemented within the ITS module.

This research is divided into two parts; Manual Annotation, and Automatic personality detection. We also aim to answer the important questions: (a) how different types of student's behavior impact other students learning in different ways; (b) how variations (such as human computer interaction) in an ITS and educational game impact students behavior.

Moreover, in this chapter we present a dataset that we have annotated containing personality excerpts based on Leary's Rose framework (Competitive, Dependent, Leading, Helping, Aggressive, Withdrawn). By this, we have presented that the detection of personality behavior is more efficient than that of human judges. Consequently, we have presented three automated methods to personality detection, based on understanding from research in natural language processing (NLP), machine learning, and psychology. We explore that text classification based on n-gram (Unigrams and Bigrams) is the best particular detection approach. We also examined a combination method such as Linguistic Inquiry and Word Count (LIWC) and subjective lexicons features.

In the first task, we performed a coding scheme based on Leary's framework personality dimension by human judges. Therefore, we annotated the personality characteristics of students and their chat interactions from log data set. Furthermore, we have analyzed a random of 200 student's textual excerpts from the chat our annotated data set to test our automated personality detection performance. Two human judges manually annotated this subset of excerpts see Sect. 5.4.

In the second task we develop a supervised method, using data mining, NLP, and machine learning algorithm, to detect the personality of students. We have used machine learning algorithms (i.e., SVM, J48, Naive Bays) for classification. We also used Weka and other NLP tools (i.e., Standford Parser [2], and OpenNLP [3]) to develop this automated system. Our model for classification personality explained in this chapter are performed using a tenfold cross validation method under its default setting in Weka [4]. We reported: Accuracy, Precision, Recall and F-Measure.

We observed that our automated classifier approaches out performed human judges annotation with accuracy of 83 %. We analyzed our results with ANOVA method. It is used to test the difference in LIWC component scores among six types of personality: competitive, leading, between dependent, withdrawn, helping, and aggressive.

The remainder of this chapter is organized as follows. The Sect. 5.2 covers the literature review and the previous works in personality related to education data and student modeling. In the Sect. 5.3 we introduce the Leary's Rose Frame work. The Sect. 5.4 presents the annotation scheme and human annotation for data set. Section 5.5 presents our model, the main functionality of our system for automatic personality classification. Section 5.6 presents the experiences and results. In Sect. 5.7, we illustrate discussion and analysis of our results. Finally, this chapter ends with conclusion and future works in Sect. 5.8.

5.2 Literature Review

5.2.1 Personality in Computer-Based Learning Environments

There are numerous reasons personality traits should be considered in CBLE. For example, even at a very basic level, attitudes toward computers can be related to personality types such that those displaying higher scores on neuroticism may have greater computer related anxiety [1]. Also, it is useful to consider differences in students or group dynamics into account during learning in CBLE. ITS are good examples to measure the ability of students against human tutoring methods as well as needs of learners [1]. This task should not be taken lightly, however, as for both human tutor and ITS, it is difficult to accurately assess both the cognitive and emotional states of individual learners. Similarly, it is a rather complex process to categorize personality traits solely from natural language user input in CBLE.

5.2.2 Emotion Detection Using Leary's Rose Frameboard

Researchers have had some success on the deLearyous gaming project [5]. To our knowledge, this is the only research that has been done specifically on the automatic classification of sentences based on Leary's Rose for emotion detection. DeLearyous researchers described a methodology for a serious gaming project which aims at developing an environment in which users can improve their communication skills by interacting with a virtual character in written natural language (Dutch). In order to apply Leary's framework, they classified the input sentences into one of four possible "emotion" classes (above, below, opposed, together). They applied several machine learning algorithms SVM, Naive Bayes, and Conditional Random Field (CRF) to obtain the classification performance. For this, they used different features set from their dataset (unigrams, lemma trigrams and dependency structures). They obtained 52.5 % accuracy, around 25 % over the baseline. The researchers noted, however, that the manually annotated sentences used to compile their training set were labeled by one human annotator and thus may have been susceptible to issues with reliability.

5.2.3 Automatic Detection of Personality

In other research [6, 7] found that identification of personality (Big Five in speech) by automatic analysis performed better than the baseline. Their analysis confirms previous findings linking language and personality and also reveals many new linguistic and prosodic markers. However, there was a limitation in their method in that speech recognition, such as prosodic features.

In addition, there has been other research conducted in order to let a machine learner determine the appropriate sentiment/emotion class. For instance, [8] and [9] attempted to classify LiveJournal posts according to their mood using SVM trained with frequency features (word counts, POScounts), length-related features (length of posts/sentences, etc.), semantic orientation features (using WordNet to calculate the distance of each word to a set of manually classified keywords) and special symbols (emoticons).

5.2.4 Personality and Student Behavior

Gore et al. investigated the relation between personality and organizational citizenship behaviors in student populations [10]. They tested the hypothesis that conscientiousness, agreeableness, and neuroticism predict unique variance in academic citizenship attitudes. They studied 270 college students who completed an online questionnaire assessing their personality and academic citizenship attitudes.

They claimed that results confirmed the hypothesis. In another study they also found that academic citizenship attitudes mediate the association between personality and citizenship behavior. Their results showed that general conscientiousness was associated with citizenship behavior, but academic conscientiousness attitudes mediated this association.

5.2.5 The Relationship Between Personality Traits and Information Competency

Song and Kwon examined differences between Korean and American cultures in terms of the relationships between Big Five personality traits and information competency [11]. In their research, Korean (n = 245) and American (n = 185) college students completed the NEO-Five Factor Inventory and the Information Competency Scale. Their results showed both similarities and differences between the two culture groups.

They showed that Conscientiousness and openness to experience significantly predicted information competency in both Korean and American students. On the other hand, they conducted that the influence of extroversion was significant only for American students. This result happened due to the high value placed on extroversion in American culture [11].

5.2.6 Personality Traits and Learning Style in Academic Performance

In [12], Furnham conducted various tests soon after students arriving at university on the Big Five Personality Traits [13]. The first study (N = 178) showed Conscientiousness and General intelligence to be the only significant predictor of overall first year grade accounting for 11 % of the variance.

The second study (N = 93) showed that ability and non-ability factors differed in terms of their predictive validity depending on the exams taken. Individual difference factors account for around 10 % of the variance in college examination success [12].

5.2.7 A Neural Network Model for Human Personality

In this research, Read et al. [14], presented a neural network model that aims to bridge the historical gap between dynamic and structural approaches to personality. The model integrates work on the structure of the trait lexicon, the neurobiology of personality, temperament, goal-based models of personality, and an evolutionary analysis of motives. It is organized in terms of two overarching motivational systems, an approach and an avoidance system, as well as a general des-inhibition and constraint system. Each overarching motivational system influences more specific motives.

Traits are modeled in terms of differences in the sensitivities of the motivational systems, the baseline activation of specific motives, and inhibitory strength. The result is a motive-based neural network model of personality based on research about the structure and neurobiology of human personality. The model provides an account of personality dynamics and person situation interactions and suggests how dynamic processing approaches and dispositional, structural approaches can be integrated in a common framework [15].

5.2.8 Relationships Between Academic Motivation and Personality Among the Students

Relationships between personality and academic motivation were examined using 451 first-year college students [16]. In this research, multiple regressions compared three types of intrinsic motivation, three types of extrinsic motivation and motivation to five personality factors. Results indicated that those who were intrinsically motivated to attend college tended to be extroverted, agreeable, conscientious, and open to new experiences; although these trends varied depending on the specific type of intrinsic motivation.

Those who lacked motivation tended to be extroverted, agreeable, conscientious, and neurotic; depending on the type of extrinsic motivation. Those who lacked motivation tended to be disagreeable and careless. These results suggest that students with different personality characteristics have different reasons for pursuing college degrees and different academic priorities [16].

5.2.9 Relation Between Learning from Errors and Personality

This research focused on the relationship between negative emotionality and learning from errors [17]. Specifically, negative emotionality was expected to impair learning from errors by decreasing motivation to learn. Perceived managerial intolerance of errors was hypothesized to increase negative emotionality, whereas emotional stability was proposed to decrease negative emotionality. All the hypotheses were tested in a laboratory simulation.

Contrary to the prediction, a positive association was found between negative emotionality and motivation to learn. The effects of perceived managerial intolerance of errors and emotional stability on negative emotionality were as predicted. Moreover, exploratory data analysis were conducted at the level of specific negative emotions and revealed differentiated effects of specific negative emotions on learning from errors [17].

5.2.10 Academic Achievement and Big Five Model

Poropat [18] reported a meta-analysis of personality-academic performance relationships, based on the 5-factor model, in which cumulative sample sizes ranged to over 70,000. Most analyzed studies came from the tertiary level of education, but there were similar aggregate samples from secondary and tertiary education. There was a comparatively smaller sample derived from studies at the primary level. Academic performance was found to correlate significantly with Agreeableness, Consciousness, and Openness. Where tested, correlations between Conscientiousness and academic performance were largely independent of intelligence.

When secondary academic performance was controlled for, Conscientiousness added as much to the prediction of tertiary academic performance as did intelligence. Strong evidence was found for moderators of correlations. Academic level (primary, secondary, or tertiary), average age of participant, and the interaction between academic level and age significantly moderated correlations with academic performance. Possible explanations for these moderator effects are discussed, and recommendations for future research are provided [14].

5.2.11 The Big Five Personality, Learning Styles, and Academic Achievement

Personality and learning styles are both likely to play significant roles in influencing academic achievement [19]. College students (308 undergraduates) completed the Five Factor Inventory and the Inventory of Learning Processes and reported their grade point average. Two of the Big Five traits, conscientiousness and agreeableness, were positively related with all four learning styles (synthesis analysis, methodical study, fact retention, and elaborative processing), whereas neuroticism was negatively related with all four learning styles.

In addition, extraversion and openness were positively related with elaborative processing. The Big Five together explained 14 % of the variance in grade point average (GPA), and learning styles explained an additional 3 %, suggesting that both personality traits and learning styles contribute to academic performance. Further, the relationship between openness and GPA was mediated by reflective learning styles (synthesis-analysis and elaborative processing). These latter results suggest that being intellectually curious fully enhances academic performance when students combine this scholarly interest with thoughtful information processing. Implications of these results are discussed in the context of teaching techniques and curriculum design [19].

5.2.12 Using Personality and Cognitive Ability to Predict Academic Achievement

Beaujean et al. [20], conducted a study on the relationship between cognitive ability, personality, and academic achievement in post-secondary students, using latent variable models. By testing both simple and complex relationships, they found that cognitive ability and personality predicted reading achievement independently, but that they interact when predicting math achievement, at least in the Conscientiousness and Openness to Experience domains [20].

5.3 Leary's Interpersonal Frame Board

Leary's Interpersonal Circumplex (or Leary's Rose Frame Board) has been used by researchers for decades as a foundation for categorizing personality through the discourse [21]. The Circumflex defines characteristics according to two dimensions: the above-below axis represents variation from dominant (above) to submissive (below) whereas the opposed-together axis represents variations of cooperation from accommodating (together) to opposition (opposed) (See Fig. 5.1). Based on these two dimensions, the Rose can easily be separated into four quadrants and then further split into eight different categories (See Table 5.1 for examples) [6].

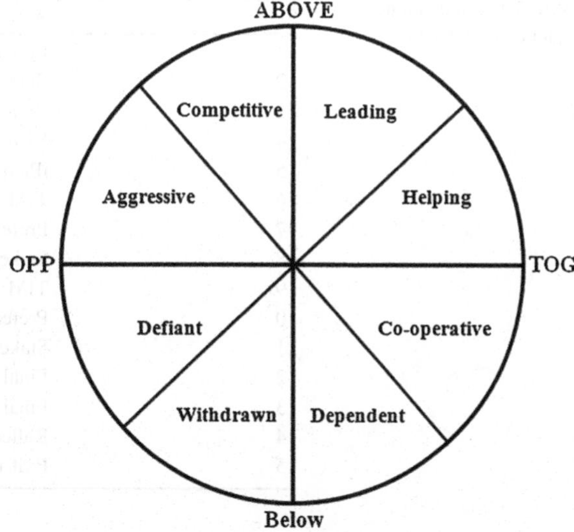

Fig. 5.1 Leary's interpersonal circumplex (Leary's Rose). *OPP* opposite, *TOG* together

Table 5.1 Leary's Rose categories examples from land science game

Statement	Leary category	Leary quadrant
Finish your task now so we can move on	Leading	Above-together
How can I help you with that?	Helping	Above-together
My plan is better than your plan	Competitive	Above-opposed
That idea is stupid. It will never work	Aggressive	Above-opposed
Sure, we can work together on this project	Cooperative	Below-together
What should I do now?	Dependent	Below-together
Sorry, never mind, I'm not thinking	Withdrawn	Below-opposed
No. I am not going to do that	Defiant	Below-opposed

5.3.1 Land Science Game

Land Science is a serious game created by researchers at the University of Wisconsin-Madison that has been designed to simulate a regional planning practicum experience for students [11, 22–24]. During the 10-hour game, students play the role of interns at a fictitious regional planning firm (called Regional Design Associates).

Where they make land use decisions in order to meet the desires of virtual stakeholders who are represented by Non-Player Characters (NPC). Students are split into groups and progress through a total of 15 stages (all these stages are shown in Table 5.2) of the game in which they complete a variety of activities including a virtual site visit of the community of interest in which students familiarize themselves with the history and ecology of the area as well as the desires of different stakeholder groups.

Table 5.2 Land science sequence of activities

Stage #	Activity
1	Intake interview
2	Staff page
3	Request for proposals
4	Virtual site visit and site assessment
5	iPlan
6	TIM 1
7	Preference survey 1
8	Stakeholder assessment 1
9	TIM 2
10	Preference survey 2
11	Stakeholder assessment 2
12	Final plan (individual)
13	Final proposal (individual)
14	Reflection
15	Exit interview

In addition, students get feedback from the stakeholders, and use a custom designed Geographic Information System (iPlan) to create a regional design plan. Throughout the game players communicate with other members of their planning team as well as a mentor (i.e., an adult who is representing a professional planner with the fictitious planning firm) through the use of a chat feature that is embedded in the game.

5.3.2 *Participants and Data Set Construction*

Participants included 12 middle school students who played the epistemic game Land Science as a part of an enrichment program at the Mass Audubon Society in Massachusetts. As previously mentioned, players in the game communicated with both other players and mentors using a chat feature embedded in the interface. For the purposes of detecting the personality of players, we only analyzed the players' chat excluding mentors' chat. Annotation was done using the coding scheme (further discussed under Human Annotation in Sect. 5.4) that was developed by the researchers based on the Timothy Leary's Interpersonal Circumplex Model [21]. The researchers selected 1,000 excerpts (average = 4.8 words) to be analyzed. For our purposes, an excerpt was defined as a turn of speech that was taken by the student.

On the other words, one excerpt occurred each time a student typed something and clicked "send" or hit "enter" in the chat function. The excerpts were selected from a larger set of 3,227 excerpts, so approximately 31 % of the player excerpts were randomly used in the analyzed data set. We have used the distribution for all stages for selecting data set. Our model is illustrated in Fig. 5.2 and in the following sections we describe the components of this model.

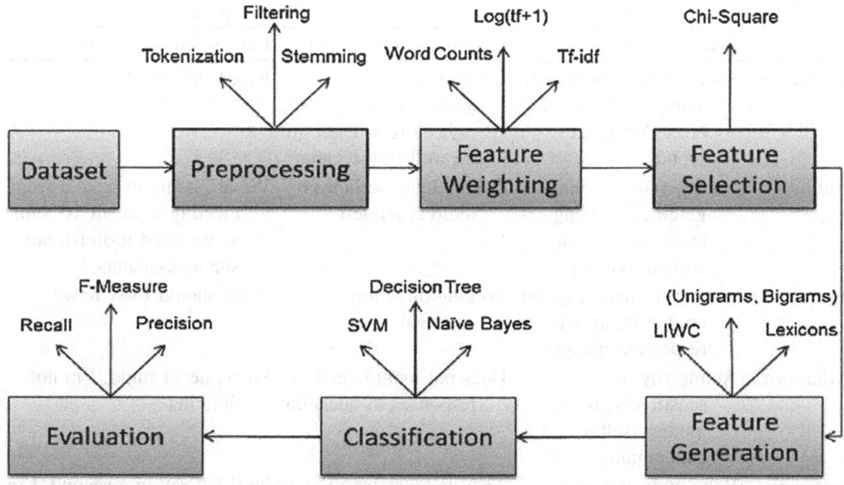

Fig. 5.2 Architecture of automated classification

5.4 Annotation Scheme

As previously described, the researchers developed a coding scheme based on Leary's Interpersonal Circumflex, which focused on 6 categories from all 4 quadrants of Leary's Rose: Competitive, Leading, Dependent, Withdrawn, Aggressive/Defiant and Helping/Cooperative.

Regarding the current annotation scheme we combined Leary's original categories of Aggressive and Defiant because in our data set there was little differentiation between these two categories. Similarly, we also combined Leary's original categories of Helping and Cooperative. Definitions and examples of each of these categories are included in Table 5.3.

5.4.1 Human Annotation

Using this coding scheme, two trained researchers annotated the data set of 1,000 excerpts. The first series of training required the human annotators to independently code 200 excerpts randomly selected from the Land Science corpus. The Kappa statistic was computed to assess inter-rater reliability on this set and agreement was fair (0.33). Following this, the annotators discussed and refined any issues regarding the coding scheme and then annotated a new set of randomly selected excerpts. The Kappa statistic was computed to assess inter-rater reliability on the second training set and agreement was substantial (0.69). Results indicated increased reliability and thus completed the training of the human annotators. Once the two annotators were trained they independently annotated a set of 1,000 excerpts.

Table 5.3 Annotation scheme; category definitions and examples from land science

Category	Leary definition	Additional information	Land science example
Competitive	Narcissistic, competing, acting confidently, boast, brag, act proud	Competitive with another orby indicating a desire to do well in the game	Beat team Eva!!
Leading	Managerial, directing, guiding, advising, teaching, ordering around, bossing	Can include explicit or indirect request	We are going to have a team meeting in about 10 min, so we need to finish our site assessments
Dependent	Asking for help, depend on, act in an over respecting manner	Seeking direction or approval	What should I say now?
Withdrawn	Acting shy or sensitively, being modest, self-condemning	Does not include lack of responses to question	Sorry, never mind. I'm not thinking
Aggressive/defiant	Rebellious actions, complaining, wariness, being skeptical	Also includes taking a strong stance and passing the blame on to someone else	No!! I'll say in a second. I'm on something else
Helping/cooperative	Takes responsibility, helping, offering, giving, agree, co-operate, compromise	Includes working together as a group or participating in group activity	If you want me to look at your plan I can.

Table 5.4 Inter-rater reliability (Kappa) for 1,000 coded excerpts

Personality category	Inter-rater reliability (kappa)
Competitive	0.82
Leading	0.65
Dependent	0.83
Withdrawn	0.77
Helping/cooperative	0.58
Aggressive/defiant	0.65
Neutral	0.60
Overall average	0.70

Overall, personality category agreement between the two annotators on the set of 1,000 excerpts was substantial (Kappa = 0.70). As shown in Table 5.4, agreement is substantial for the Competitive, Dependent and Withdrawn categories, and is moderate for the Leading, Helping/Cooperative, Aggressive/Defiant and Neutral categories. The two human annotators agreed on the personality category for a total number of 1,523 excerpts (see Table 5.5). Of those agreements, the largest percentage of excerpts is Neutral (35.78 %), indicating that the annotators agreed that there was no evidence of a personality category represented.

Table 5.5 Number of instances agreed present for each personality category

Personality category	Number of instances
Competitive	277
Leading	134
Dependent	343
Withdrawn	46
Helping/cooperative	141
Aggressive/defiant	37
Neutral	545
Total	1,523

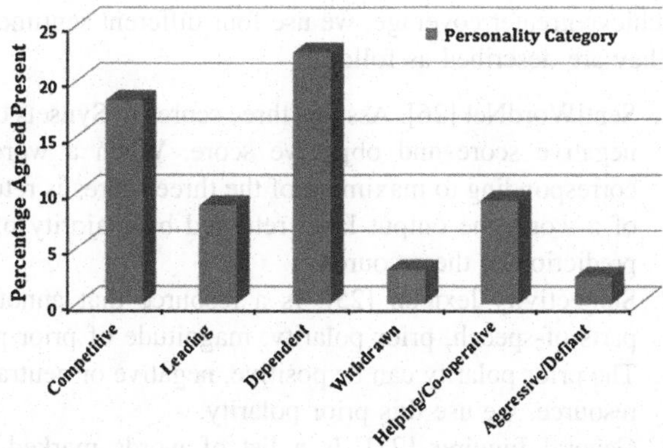

Fig. 5.3 Percentage of excerpts agreed in each personality category

Regarding excerpts for which there is an agreement that a personality category present, as seen in Fig. 5.3, the largest percentage is Dependent (22.52 %) followed by Competitive (18.19 %) and Helping/Cooperative (9.26 %). The least represented personality categories are Leading (8.80 %), Withdrawn (3.02 %) and Aggressive/Defiant (2.43 %).

5.5 Model

To our knowledge, the only research has been done specifically on the automatic classification of sentences based on Learys Rose for emotion detection is done by [25]. They described a methodology for a serious gaming project, deLearyous, which aims at developing an environment in which users can improve their communication skills by interacting with a virtual character in (Dutch) written natural language. In order to apply this framework, they classified the input sentences into one of four possible "emotion" classes (above, below, opp, tog, see Fig. 5.1).

They applied several machine learning algorithms, such as SVM, Naïve Bayes, Conditional Random field to obtain the calcification performance. For this, they used different features set from their dataset (unigrams, lemma trigrams and dependency structures). They obtained 52.5 % accuracy around 25 % over the baseline. In contrast, in our method we use Leary's Rose framework to detect personality rather than emotion.

5.5.1 Lexicon Resources

Sentiment-based lexical resources annotate words/concepts with polarity. To achieve greater coverage, we use four different sentiment-based lexical resources. They are described as follows.

1. SentiWordNet [26]. Assigns three scores to Synsets of WordNet: positive score, negative score and objective score. When a word is looked up, the label corresponding to maximum of the three scores is returned. For multiple synsets of a word, the output label returned by majority of the Synsets becomes the prediction of the resource.
2. Subjectivity lexicon [25]. Is a resource that annotates words with tags like parts-of-speech, prior polarity, magnitude of prior polarity (weak/strong), etc. The prior polarity can be positive, negative or neutral. For prediction using this resource, we use this prior polarity.
3. General Inquirer [27]. Is a list of words marked as positive, negative and neutral. We use these labels to use Inquirer resource for our prediction.
4. Taboada [28]. It is a word-list that gives a count of collocations with positive and negative seed words. A word closer to a positive seed word is predicted to be positive and vice versa.

5.5.2 Feature Extraction

From this dataset we extracted a wide range of different features. The sentences were first parsed with Stanford POS Tagger, an English language parser [2], which allowed us to extract linguistic information such as word tokens, lemmas, part-of-speech tags, syntactic functions and dependency structures.

The actual feature vectors were then generated on the basis of this linguistic information by using a "bag of n-grams" approach, i.e. by constructing n-grams (unigrams, bigrams and trigrams) of each feature type (e.g. n-grams of word tokens, n-grams of part-of-speech tags…) and by counting for each n-gram in the training data how many times it occurs in the current instance. In addition to these n-gram counts, we also included punctuation counts, average word length and average sentence length.

Sentiment Score Feature. Based on predictions of individual traits, we compute the Sentiment prediction for each trait with respect to a keyword in form of percentage of positive, negative and objective content. This is on the basis of predictions by each resource by weighting them according to their accuracies. These weights have been assigned to each resource based on experimental results. For each resource, the following scores are determined (see Eqs. 5.1, 5.2, 5.3).

$$PositiveScore(s) = \sum_{i=0}^{i=n} PiW_{Pi} \tag{5.1}$$

$$NegativeScore(s) = \sum_{i=0}^{i=n} NiW_{Ni} \tag{5.2}$$

$$ObjectiveScore(s) = \sum_{i=0}^{i=n} OiW_{Oi} \tag{5.3}$$

where, PositiveScore(s) = Positive score for each excerpt s; NegativeScore(s) = Negative score for each excerpt *s*; ObjectiveScore(s) = Objective score for each excerpt s; n = Number of resources used for prediction; Pi, Ni, Oi = Positive, Negative, and Objective count of excerpt predicted respectively using resource i; WPi,WNi,WOi = Weights for respective classes derived for each resource *i*.

5.5.3 The Linguistic Inquiry and Word Count Features

We extracted features derived from the Linguistic Inquiry and Word Count (LIWC) output. Specifically, LIWC counts and groups the number of instances of nearly 4,500 keywords into 80 psychologically meaningful dimensions. We create one feature for each of the 80 LIWC dimensions summarized under the following four categories:

- Linguistic processes: Functional aspects of text (e.g., the average number of words per sentence, the rate of misspelling, swearing, etc.)
- Psychological processes: Includes all social, emotional, cognitive, perceptual and biological processes, as well as anything related to time or space.
- Personal concerns: Any references to work, leisure, money, religion, etc.
- Spoken categories: Primarily filler and agreement words.

For each instance, we calculate the ratio of words in each category from the LIWC toolkit [18], as these features are correlated with the personality dimensions (as shown in Table 5.6). Indeed, the LIWC2007 software used in our experiments subsumes most of the features introduced in other work. Thus, we focus our psycholinguistic approach to personality detection on LIWC-based features.

Table 5.6 LIWC features [18]

Feature category	Features included
Standard counts	*Word count* Words per sentence, type/token ratio, words captured, words longer than 6 letters, negations, assents, articles, prepositions, numbers, pronouns: 1st person singular, 1st person plural, total 1st person, total 2nd person, total 3rd person
Psychological processes	*Affective or emotional processes* Positive emotions, positive feelings, optimism and energy, negative emotions, anxiety or fear, anger, sadness, cognitive processes: causation, insight, discrepancy, inhibition, tentative, certainty, sensory and perceptual processes: seeing, hearing, feeling, social processes: communication, other references to people, friends, family, humans
Relativity	*Time* Past tense verb, present tense verb, future tense verb, Space: up, down, inclusive, exclusive, motion
Personal concerns	*Occupation* School, work and job, achievement, leisure activity: home, sports, television and movies, music, money and financial issues, metaphysical issues: religion, death, physical states and functions, body states and symptoms, sexuality, eating and drinking, sleeping, grooming
Other dimensions	*Punctuation* period, comma, colon, semi-colon, question, exclamation, dash, quote, apostrophe, parenthesis, other, Swear words, non-fluencies, fillers

For each instance, we calculate the ratio of words in each category from the LIWC toolkit [18], as these features are correlated with the personality dimensions [18]. These features and their categories are shown in below.

5.5.4 Automated Approaches to Personality Classification

We explain three automated approaches to classify detecting personality behavior, each of which utilizes classifiers trained on the dataset of Sect. 5.3.2. The features employed by each strategy are described here.

Psycholinguistic Personality Detection. The Linguistic Inquiry and Word Count (LIWC) software [18] is a popular automated text analysis tool used widely in the social sciences. It has been used to detect personality traits [6], to study tutoring dynamics [29], and, most relevantly, to analyze personality detection [6].

Since LIWC software does not include a text classifier, we create features derived from the LIWC output. In particular, LIWC counts and groups the number of instances of nearly 4,500 keywords into 80 psychologically meaningful dimensions. We construct one feature for each of the 80 LIWC dimensions, which can be summarized under the four categories that explained in Sect. 5.3. Indeed, the LIWC2007 software used in our experiments subsumes most of the features introduced in other work. Thus, we focus our psycholinguistic approach to personality detection on LIWC-based features.

5.5.5 Classification Method

Naive Bayes Classifier Provides a simple approach and it is a classifier as a form of Bayesian network and it leans on two simple assumptions. First, it assumes that the predictive attributes are conditionally independent given the class. Then, it posits that no hidden or latent attributes influence the prediction process [30]. For a document X, with label class c, the Naive Bayes classifier gives us the following decision rules (see Eqs. 5.4 and 5.5) [30]:

$$P(C = c|X = x) = \frac{p(C = c)\,p(X = x|C = c)}{p(X = x)}, \tag{5.4}$$

where

$$P(X = x|C = x) = \prod_{i}^{n} P(Xi = xi|C = c) \tag{5.5}$$

We use John and Langley [30] Naïve Bayes classifier in Weka [4] to train our Naive Bayes models on all three approaches and feature sets described above, namely LIWC, lexicons, Unigrams, Bigrams. We also evaluate every combination of these features, but for brevity include only UNIGRAMS + BIGRAMS, which performs best with tenfold cross validation on the corresponding dataset.

Support Vector Machine. We also train SVM classifiers, which find a high-dimensional separating hyper-plane between two groups of data. To simplify feature analysis in Sect. 5.5, we restrict our evaluation to linear SVM, which learn a weight vector *w* and bias term *b*, such that a document x can be classified by (5.6):

$$y = sign\ (\vec{w}.\vec{x}) + b \tag{5.6}$$

We use SMO [31] to train our SVM models on all three approaches and feature sets described above: LIWC, LEXICONS, UNIGRAMS, and BIGRAMS. We also evaluate every combination of these features, but for shortness include only LIWC + BIGRAMS, and LEXICON + BIGRAMS which performs best.

Decision Trees. We use J48, an open source Java implementation of the C4.5 algorithm in Weka [4] data mining tool to train our dataset for decision trees classifier. We evaluate approaches on all combination of feature set, but we consider the features which performed best (UNIGRAMS + BIGRAMS, UNIGRAMS + LIWC). Our classification experiments are carried out with tenfold cross validation on the corresponding dataset. A sample of the results achieved by the three methods is stated in Table 5.7.

Table 5.7 Automated classifier performance for three approaches based on tenfold cross- validation experiments

Approach	Features	Acc. (%)	COM			DEP		
			P	R	F	P	R	F
LEXICAL	Lexicons$_{j48}$	61.95	0.67	0.66	0.67	0.56	0.66	0.61
LIWC	Liwc$_{j48}$	59.30	0.57	0.62	0.60	0.64	0.67	0.65
Method	Unigrams$_{svm}$	60.54	0.74	0.70	0.72	0.64	0.63	0.63
	Bigrams$_{svm}$	70.40	0.92	0.65	0.76	0.92	0.75	0.82
	Liwc + bigrams$_{\text{svm}}$	77.47	0.90	0.75	0.82	0.93	0.78	0.85
	Lexicons + bigrams$_{\text{svm}}$	<u>83.71</u>	0.96	0.80	0.87	0.96	0.84	*0.90*
	Bigrams$_{nb}$	65.02	0.04	0.87	0.62	0.87	0.77	0.50
	Unigrams + bigrams$_{nb}$	60.53	0.77	0.60	0.67	0.72	0.68	0.50
	Unigrams + bigrams$_{j48}$	62.78	0.83	0.67	0.74	0.83	0.71	0.46
	Unigrams + liwc$_{j48}$	74.0	0.86	0.80	0.83	0.81	0.77	0.63
Approach	Features	Acc. (%)	LEA			WIT		
			P	R	F	P	R	F
LEXICAL	Lexicons$_{j48}$	61.95	0.61	0.55	0.58	0.56	0.54	0.55
LIWC	Liwc$_{j48}$	59.30	0.52	0.40	0.45	0.62	0.42	0.50
Method	Unigrams$_{svm}$	60.54	0.50	0.32	0.50	0.83	0.39	0.53
	Bigrams$_{svm}$	70.40	0.79	0.40	0.52	1	0.44	0.62
	Liwc + bigrams$_{\text{svm}}$	77.47	0.95	0.64	0.77	0.93	0.54	0.68
	Lexicons + bigrams$_{\text{svm}}$	<u>83.71</u>	0.98	0.76	0.86	1	0.74	0.85
	Bigrams$_{nb}$	65.02	0.50	0.21	0.3	0.80	0.44	0.57
	Unigrams + bigrams$_{nb}$	60.53	0.50	0.53	0.51	1	0.39	0.54
	Unigrams + bigrams$_{j48}$	62.78	0.46	0.43	0.45	0.82	0.47	0.60
	Unigrams + liwc$_{j48}$	74.0	0.63	0.64	0.63	0.85	0.78	0.81
Approach	Features	Acc. (%)	COP			AGG		
			P	R	F	P	R	F
LEXICAL	Lexicons$_{j48}$	61.95	0.66	0.60	0.63	0.25	0.21	0.22
LIWC	Liwc$_{j48}$	59.30	0.60	0.62	0.61	0.50	0.53	0.51
Method	Unigrams$_{svm}$	60.54	0.52	0.74	0.61	0.17	0.33	0.22
	Bigrams$_{svm}$	70.40	0.50	1	0.66	1	0.17	0.29
	Liwc + bigrams$_{\text{svm}}$	77.47	0.54	0.95	0.69	1	0.2	0.33
	Lexicons + bigrams$_{\text{svm}}$	<u>83.71</u>	0.98	0.76	0.86	*1*	*0.32*	*0.48*
	Bigrams$_{nb}$	65.02	0.46	0.96	0.62	1	0.16	0.28
	Unigrams + bigrams$_{nb}$	60.53	0.40	0.74	0.52	0.67	0.5	0.57
	Unigrams + bigrams$_{j48}$	62.78	0.42	0.84	0.56	0.26	0.22	0.24
	Unigrams + liwc$_{j48}$	74.0	0.63	0.75	0.69	0.50	0.68	0.57

Reported Accuracy, (*P*) precision, (*R*) recall and (*F*) measure

5.6 Experience and Results

5.6.1 Classification Results

The model for classification personality strategies explained in Sect. 5.5 are per-formed using a tenfold cross validation method under its default setting in Weka [4]. The parameters for model are chosen for each test fold based on standard cross validation experiments on the training dataset. All folds are chosen so that each includes all instances from six classes; therefore, learned classifiers are always measured on dataset from unseen instances.

Table 5.8 shows the results of the top scores that we managed to achieve with each of the three classifiers over three approaches. We also use the combination of features and learner parameters that were determined to give the best accuracy by the classifiers. "Approach" column shows the model that have been tested, the "features" column indicates the types of features that have been used, the rest of columns indicates the results based on Accuracy, Precision, Recall, and F-measure (Acc., P, R, F) for all six classes. We observe that our automated approaches outperformed human judges (Kappa) and baseline for most of feature sets. The statistical baseline for these six classes classification problem, considering the slight imbalances in the class distribution, is 30 %. However there is an exception such as Recall for "aggressive" which is not significant.

We can argue on this due to low number of instances in this class. However, this is expected given that human judges often focus on unreliable cues to aggressive utterances. We observe that our automated approaches outperformed

Table 5.8 Top 15 highest weighted features learned by BIGRAMS + LEXICONSsvm and LIWCsvm. The results show for binary classification of "helping, aggressive" and "leading, dependent"

BIGRAMS + LEXICONSsvm	LIWCsvm
Helping, aggressive	Leading, dependent
Always want	Six letters
Didn't seem	Pronoun
Don t	Personal pronoun
For me	I
Is quite	We
It is	You
Need to	She/he
No need	They
People don	Impersonal pronouns
Quite deadly	Article
Really that	Verb
Seem to	Auxiliary verbs
Slow down	Past tense
Speaking Spanish	Present tense

human judges (Kappa) and baseline for most of feature sets. The statistical baseline for these six classes classification problem, considering the slight imbalances in the class distribution, is 30 %. However there is an exception such as Recall for "aggressive" which is not significant. We can argue on this due to low number of instances in this class. However, this is expected given that human judges often focus on unreliable cues to aggressive utterances.

If we look at the confusion matrix in Table 5.9; firstly, we note that most of the aggressive instances (8) classified as "helping" personality. Many other classes considered as "helping" as well. We figured out, this happened due to human judge's evaluation, because the judges considered many small responses such as: OK, Yep, Thanks, Cool, etc. as "helping" class. Secondly, as it shown in Table 5.1 the number of instances in "aggressive" class is low. We found out that the players are not often aggressive during chat conversation. It might be due to their work environment in that they are supervised by a human mentor during the game.

Interestingly, the psycholinguistic approach (LIWCj48) performs almost 30 % more accurately than baseline rather than SVM or NB. Also J48 perform higher than SVM and NB on lexical subjective scores features. Overall, all the standard text categorization approaches proposed in Sect. 5.5 perform between 9 and 53 % more accurately than baseline. However, best performance overall is achieved by combining features from these two approaches. Particularly, the combined model LEXICONS + BIGRAMSSVMis 83.71 % accurate at personality classification.

Surprisingly, models trained only on UNIGRAMSsvm(60.54 %), the simplest n-gram feature set, outperform LIWC (non-text classification) approaches, and models trained on BIGRAMSnb(65.02 %) perform even better. This suggests that a universal set of feature such as psycholinguistic keyword personality (i.e., LIWC) cannot be the best model for personality detection, and a context-sensitive approach (e.g., BIGRAMS) might be necessary to achieve state-of-the-art personality detection performance.

To better understand the models learned by these automated approaches, we report in Table 5.8 the top 15 highest weighted features for two pair classes (Helping, Aggressive and Leading, Dependent) as learned by BIGRAMS + LEXICONSsvm and LIWCsvm. From BIGRAMS + LEXICONSsvm approach we have chosen classifier for classes "Helping" (with highest F-measure) and "Aggressive" (lowest F-measure), for LIWCsvm approach we have chosen classifier for classes "Leading, Dependent" with similar reason.

We note that player with "Helping" personality behavior tend to use somehow similar language with "Aggressive" players; in particular, "need to" and "no need", the former one can be consider as "Helping" behavior and later one can be regarded as "Aggressive" attitude. Accordingly, in term of global features such as psycholinguistic features (LIWC), "Leading" and "Dependent" players tend to use similar pronouns(personal or impersonal) (i.e.; i, we, you, she/he, they). Finally, when we look at Confusion Matrix (Table 5.9), it turns out that all misclassified instances from "Aggressive" class fall into "Helping" class and similarly almost 75 % of misclassified instances in "Leading" class are classified as "Dependent" class.

Table 5.9 The confusion matrix performed by SVM classifiers approach over BIGRAMS and subjective lexicon features

a	b	c	d	e	f	Classified as
130	1	2	1	37	0	a = competitive
2	155	1	0	47	0	b = dependent
5	4	61	0	26	0	c = leading
0	0	0	14	12	0	d = withdrawn
2	1	0	0	146	0	e = helping
0	0	0	0	8	2	f = aggressive

5.7 Discussion and Analysis

5.7.1 Personality Trait Tracking Analysis

An additional aim of the current study is to explore the consistency of personality characteristics displayed by individual participants across the various stages of the game. In order to do this we randomly selected two participants and charted their-coded personality traits throughout the game.

For the purposes of the current results we focused only on three of the most prevalent personality categories overall (Competitive, Leading and Dependent). Figs. 5.4 and 5.5 display the personalities displayed by these two players (referred to as Player A and B) for each of the 15 stages of the game (numbered 0–14). First, it is important to note that both players exhibited different personalities during different stages of the game. More specifically, Player A (see Fig. 5.4) demonstrated a variety of noticeable trend for the first few stages of the game. However, there was a drastic increase in Dependent statements in stage 6 followed by an increase in Leading statements in stage 7. Competitive statements then become the most dominant for most of the final stages of the game.

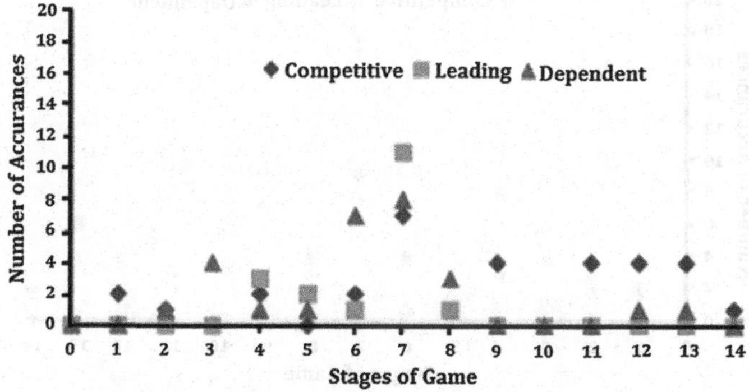

Fig. 5.4 Player A personality characteristics displayed for each stage of the game

In addition, Player B (See Fig. 5.5) exhibited a variety of personality characteristics throughout the game. For example, Dependent statements dominated 6 of the first 9 stages of the game with a drastic increase in Stage 7. However, like Player A, Competitive statements were most prevalent for the final 6 stages of the game.

Based on the above results, the changes that occur in Stage 7 of the game seem to be especially relevant. These changes highlight that players may be altering their statements based both on the demands of the game as well as the personalities exhibited by other players in the group dynamic. Specifically with the above examples, notice that during Stage 7 of the game Player A had a drastic increase in Leading statements while Player B had a drastic increase in Dependent statements. It is possible that there may be something about the task associated with Stage 7 that encourages a group dynamic in which some players become more dependent while others become more directive.

Overall, results indicate substantial agreement between two trained human annotators. Regarding coded personality categories Leading, Dependent, Helping/Cooperative and Competitive are the four most commonly present categories, whereas, Withdrawn and Aggressive/Defiant statements are less prevalent. Furthermore, players demonstrate different personality characteristics depending on the stage of the game and, likely, the dynamics of the group.

5.7.2 ANOVA Analysis

One way analysis of variance (One-way ANOVA) is used when two or more groups are compared with their mean scores on one continuous variable, also called the independent variable. A one way analysis of variance (ANOVA) will tell people whether these groups differ.

Consequently, post hoc comparisons will help to test which groups are significantly different from one another. One-way between-groups ANOVA was used

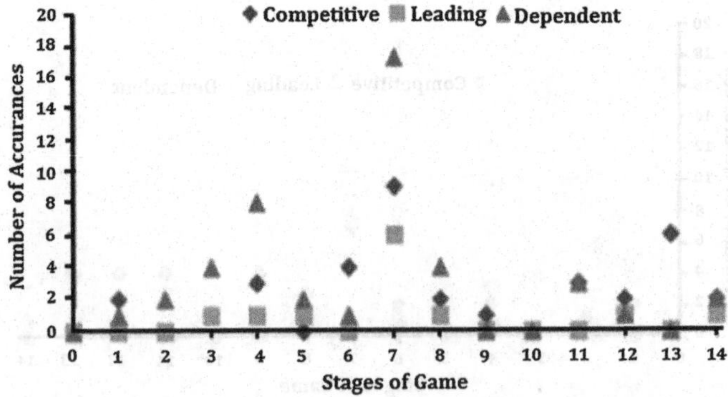

Fig. 5.5 Player B personality characteristics displayed for each stage of the game

to test the difference in LIWC component scores among six types of personality: competitive, leading, dependent, withdrawn, helping, and aggressive. The type of personality is one factor and normalized LIWC components related to psychological features are the dependent variables: Argumentation (Persuasion), Achievement, and Negative Valence [32]. Table 5.10 shows the ANOVA results that each LIWC component scores differed significantly across the three types of personality:

- Argumentation, F (5, 521) = 5.12, $p < .001$
- Achievement, F (5, 521) = 7.26, $p < .001$
- Narrative, F (5, 521) = 67.87, $p < .001$
- Negative Valence, F (5, 521) = 55.45, $p < .001$; and
- Embodiment, F (5, 521) = 12.35, $p < .001$.

Tamhane post hoc tests comparisons of the six groups indicate for LIWC component, Argumentation, the dependent personality (M = 3.05, 95 % CI [2.58, 3.52]) gave significantly higher score than leading personality (M = 1.72, 95 % CI [1.04, 2.40], $p = .025$), and helping type (M = 1.27, 95 % CI [0.57, 1.96], $p = .001$). Comparisons between the other groups were not statistically significant at $p < .05$.

The results indicated that the players with dependent personality tended to use significantly more argumentation, in other words, more cognitive words than leading and helping personality. In terms of LIWC component, Achievement, the competitive type (M = 2.04, 95 % CI [1.69, 2.40]) was significantly higher than leading personality (M = 0.29, 95 % CI [−0.19, 0.72], $p < .001$), dependent (M = 0.87, 95 % CI [0.41, 1.32], $p = .001$), and withdrawn (M = −0.45, 95 % CI [−1.75, 0.86], $p = .013$).

Moreover, helping (M = 1.51, 95 % CI [0.83, 2.20]) was significantly higher than leading personality (M = 0.29, 95 % CI [−0.19, 0.72], $p = .043$). These findings showed competitive personality tended to use significantly more achievement words compared to leading, dependent and withdrawn personality.

For LIWC component Negative Valence, withdrawn (M = 10.43, 95 % CI [4.23, 16.62]) was significantly higher than competitive (M = −0.55, 95 % CI

Table 5.10 ANOVA results of LIWC psychological features with personality type as the factor

Personality type	Groups	*df*	*F*	η	*p*
Argumentation	Between groups	5	5.116	0.047	0.000
	Within groups	521			
	Total	526			
Achievement	Between groups	5	7.261	0.065	0.000
	Within groups	521			
	Total	526			
Negative valence	Between groups	5	5.447	0.347	0.000
	Within groups	521			
	Total	526			

[−0.93, −0.17], $p = .022$), leading (M = −0.85, 95 % CI [−1.26, −0.45], $p = .018$), and dependent (M = 0.61, 95 % CI [−1.03, −0.19], $p = .021$); and helping (M = 6.54, 95 % CI [4.98, 8.09], was significantly higher than competitive (M = −0.55, 95 % CI [−0.93, −0.17], $p < .001$), leading (M = −0.85, 95 % CI [−1.26, −0.45], $p < .001$), dependent (M = −0.61, 95 % CI [−1.03, −0.19], $p < .001$), and aggressive (M = 1.32, 95 % CI −0.43, 3.07], $p < .001$).

The aforementioned findings showed that withdrawn and helping personality tended to express more negative emotions than competitive, leading, and dependent. Moreover, helping also used more negative emotion words than aggressive.

5.8 Conclusion and Future Research

In this chapter we have developed a dataset containing personality excerpts based on Leary's Rose Frameboard. By this, we have developed automatic personality detection that shows are more efficient than that of human judges. Consequently, we have presented three automated methods to personality detection, based on understanding from research in natural language processing, machine learning, and psychology characteristic.

We conducted that while text classification based on n-gram (UNIGRAMS, BIGRAMS) is the best particular detection approach, a combination-method such as LIWC and Subjective Lexicons features along with n-gram features can achieve better performance.

Eventually, we have done several notable contributions. Particularly, our results indicate to take into account both the context, such as BIGRAMS, rather than precisely using a global set of personality indications (e.g., LIWC and Subjective Lexicons). We have also reported results based on the feature weights that show the difficulties confronted by judges in annotating the dataset. Finally, we have found a possible connection between personality behavior by players, such "Helping and Aggressive" and "Dependent & Leading", based on BIGRAMSs and LIWC similarities.

For future work, we want to include an extended experiment of the methods pro-posed in current research to sentiment analysis, opinion mining, as well as emotion detection in other domains. Also, we want to extend the method in this work to apply in Big-Five personality detection. It will help us to not only detect the player's behaviors but also to detect introvert and extrovert players and a focus on approaches with POS features might be useful.

Acknowledgments This work was funded by the National Science Foundation (DRK-12-0918409). Any opinions, findings, and conclusions or recommendations expressed in this material are those of the authors and do not necessarily reflect the views of these funding agencies, cooperating institutions, or other individuals.

References

1. D'Mello, S., Onley, A., Person, N.: Mining collaborative patterns in tutorial dialogues. J. Educ. Data Mining. **2**(1), 1–37 (2010)
2. Toutanova, K., Klein, D., Manning, C., Singer, Y.: Feature-rich part-of-speech tagging with a cyclic dependency network. In: Proceedings of Conference of the North American Chapter of the Association for Computational Linguistics on Human Language Technology, pp. 173–188. Association for Computational Linguistics, Stroudsburg (2003)
3. Wilcock, G.: Text annotation with OpenNLP and Uima. In: 17th Nordic Conference of Computational Linguistics, pp. 7–8. University of Southern Denmark, Odense (2009)
4. Hall, M., Frank, E., Holmes, G., Pfahringer, B., Reutemann, P., Witten, I.H.: The WEKA data mining software: an update. SIGKDD Explor. Newsl. **11**(1), 10–18 (2009)
5. Vaassen, F., Daelemans, W.: Emotion classification in a serious game for training communication skills. In: 20th Meeting of Computational Linguistics in the Netherlands, pp. 155–168. Utrecht Institute of Linguistics, Utrecht (2010)
6. Mairesse, F., Walker, M., Mehl, M., Moore, R.: Using linguistic cues for the automatic recognition of personality in conversation and text. J. Artif. Intell. Res. **30**(1), 457–500 (2007)
7. Mairesse, F., Walker, M.: Automatic recognition of personality in conversation. In: Human Language Technology Conference of the North American Chapter of the ACL, pp. 85–88. Association for Computational Linguistics, Stroudsburg (2006)
8. Pennebaker, J.W., Chung, C.K., Ireland, M., Gonzales, A., Booth, R.J.: Linguistic Inquiry and Word Count (LIWC). Lawrence Erlbaum Associates, Mahwah (2007)
9. Keshtkar, F., Inkpen, D.: Using sentiment orientation features for mood classification in blogs. In: International Conference on Natural Language Processing and Knowledge Engineering, pp. 1–6. IEEE Press, New York (2009)
10. Gore, J., Kiefner, A., Combs, K.: Personality traits that predict academic citizenship behavior. J. Appl. Soc. Psychol. **42**(10), 2433–2456 (2012)
11. Song, H., Kwon, N.: The relationship between personality traits and information competency in Korean and American students. Int.J.Soc.Behav.Pers. **40**(7), 1153–1162 (2012)
12. Furnham, A., Chamorro-Premuzic, T., McDougall, F.: Learning style, personality traits and intelligence as predictors of college academic performance. Learn. Individ. Differ. **10**(3), 117–128 (2012)
13. Costa, P.T., McCrae, R.R.: Four ways five factors are basic. Pers. Individ. Differ. **13**(6), 653–665 (1992)
14. Read, S.J., Monroe, B.M., Brownstein, A.L., Yang, Y., Chopra, G., Miller, L.C.: A neural network model of the structure and dynamics of human personality. Psychol. Rev. **117**(1), 61–92 (2010)
15. Shaffer, D.W., Chesler, N., Arastoopour, G., D'Angelo, C.: Nephrotex: teaching first year students how to think like engineers. In: Course, Curriculum, and Laboratory Improvement PI Conference, Poster. NSF, Washington (2011)
16. Clark, M.H., Schroth, C.A.: Examining relationships between academic motivation and personality among college students. Learn Individ. Differ. **20**(1), 19–24 (2010)
17. Zhao, B.: Learning from errors: the role of context, emotion, and personality. J. Organ. Behav. **32**(3), 435–463 (2011)
18. Poropat, A.E.: A meta-analysis of the five-factor model of personality and academic performance. Psychol. Bull. **135**(2), 322–338 (2009)
19. Komarraju, M., Karau, S.J., Schmeck, R.R., Avdic, A.: The big five personality traits, learning styles, and academic achievement. Pers. Individ. Differ. **51**(4), 472–477 (2011)
20. Beaujean, A.A., Firmin, M.W., Attai, S., Johnson, C.B., Firmin, R.L., Mena, K.E.: Using personality and cognitive ability to predict academic achievement in a young adult sample. Pers. Individ. Differ. **51**(6), 709–714 (2011)
21. Leary, T.: The Interpersonal Diagnosis of Personality. Wiley, Hoboken (1957)

22. Bagley, E.A.S.: Stop talking and type: mentoring in a virtual and face-to-face environmental education environment. PhD thesis, University of Wisconsin-Madison (2011)
23. Shafer, D., Hatfield, D., Svarovsky, G., Nash, P., Nulty, A., Bagley, E., Franke, K., Rupp, A., Mislevy, R.: Epistemic network analysis: a prototype for 21st century assessment of learning. Int. J. Learn. Media **1**(2), 33–53 (2009)
24. D'Angelo, C., Arastoopour, G., Chesler, N., Shaffer, D.: Collaborating in a virtual engineering internship. In: Spada, H., Stahl, G., Miyake, N., Law, N. (eds.) Connecting Computer-Supported Collaborative Learning to Policy and Practice, pp. 626–630. International Society of the Learning Sciences, Hong Kong (2011)
25. Wiebe, J., Wilson, T.: Learning to disambiguate potentially subjective expressions. In: 6th Conference on Natural Language Learning, pp. 112–118. Association for Computational Linguistics, Stroudsburg (2002)
26. Esuli, A., Sebastian, F.: Sentiword-net: a publicly available lexical resource for opinion mining. In: 5th Conference on Language Resources and Evaluation, pp. 417–422. European Language Resources Association, Paris (2006)
27. Stone, P., Dunphy, D., Smith, M., Ogilvie, D.: The General Inquirer: A Computer Approach to Content Analysis. MIT Press, Cambridge (1966)
28. Taboada, M., Grieve, J.: Analyzing appraisal automatically. In: AAAI Spring Symposium on Exploring Attitude and Affect in Text: Theories and Applications, pp. 158–161. Association for the Advancement of Artificial Intelligence, Menlo Park (2004)
29. Cade, W.L., Lehman, B.A., Olney, A.: An exploration of off topic conversation. In: Human Language Technologies: The 2010 Annual Conference of the North American Chapter of the Association for Computational Linguistics, pp. 669–672. Association for Computational Linguistics, Stroudsburg (2010)
30. John, G.H., Langley, P.: Estimating continuous distributions in bayesian classifiers. In: Besnard, P., Hanks, S. (eds.) Eleventh Conference on Uncertainty in Artificial Intelligence, pp. 338–345. Morgan Kaufmann Publishers, San Francisco (1995)
31. Keerthi, S., Shevade, S., Bhattacharyya, C., Murthy, K.: Improvements to platt's SMO algorithm for SVM classifier design. Neural Comput. **13**(3), 637–649 (2001)
32. Li, H., Cai, Z., Graesser, A., Duan, Y.: A comparative study on English and chinese word uses with LIWC. In: Youngblood, G.M., McCarthy, P.M. (eds.) 25th International Florida Artificial Intelligence Research Society Conference, pp. 238–243. AAAI Press, Palo Alto (2012)

Chapter 6
Students' Performance Prediction Using Multi-Channel Decision Fusion

H. Moradi, S. Abbas Moradi and L. Kashani

Abstract A teacher or an artificial instructor, embedded in an intelligent tutoring system, is interested in predicting the performance of his/her students to better adjust the educational materials and strategies throughout the learning process. In this chapter, a multi-channel decision fusion approach, based on using the performance in "assignment categories", such as homework assignments, is introduced to determine the overall performance of a student. In the proposed approach, the data gathered are used to determine four classes of "expert", "good", "average", and "weak" performance levels. This classification is conducted on both overall performance and the performance in assignment categories. Then, a mapping from the performances in "assignment categories" is learned, and is used to predict the overall performance. The main advantage of the proposed approach is in its capability to estimate students' performance after a few assignments. Consequently, it can help the instructors better manage their class and adjust educational materials to prevent underachievement.

Keywords Learning analytics · Student performance prediction · Educational data mining · Decision fusion · Assignment categories

---

H. Moradi (✉) · S. A. Moradi
School of ECE, University of Tehran, North Kargar St,
P.O. Box 14395-515 Tehran, Iran
e-mail: moradih@ut.ac.ir

S. A. Moradi
e-mail: s.a.moradi@ut.ac.ir

L. Kashani
Department of Psychology and Special Education, University of Tehran,
Jalal Al-e-Ahmad Avenue, P.O. Box 11455/6456 Tehran, Iran
e-mail: lkashani@ut.ac.ir

A. Peña-Ayala (ed.), *Educational Data Mining*,
Studies in Computational Intelligence 524, DOI: 10.1007/978-3-319-02738-8_6,

Abbreviations

BKT	Bayesian knowledge tracing
CART	Classification and regression tree
CHAID	Chi-square automatic interaction detection
EDM	Educational data mining
ITS	Intelligent tutoring systems
LMS	Learning management system

6.1 Introduction

Performance prediction is very important in conducting in-time interventions for the students, whose performances are not at the expected level, especially the students at-risk[1] for poor academic performance [1]. The detection of the at-risk college students becomes highly important in the critical first year [2]. The first year failing students are more likely to fail throughout the rest of that academic program [1]. It should also be noted that the typical performance evaluation at the middle of a learning process, i.e. the midterm exams, is too late for detection of the at-risk students [3].

According to Tinto's Student Integration Model [2], student's academic performance is one of the factors influencing student's integration in academic life, which impacts the student's decision to stay in the program of study (retention) or to leave the program of the study (attrition). Since retention is a formidable issue in any educational institute, academic performance and its prediction is of significant importance.

From another point of view, performance prediction is also necessary for an instructor to be able to plan the instructions personalized according to the learners' zone of proximal development [4]. According to Vygotsky's zone of proximal development, any given learner has a level of learning without any guidance; the learner also has another level of performance, in which the student cannot complete tasks unaided, but can perform well with guidance. The difference between these two levels indicates any learner's zone of proximal development, which is very crucial for the instructor to find and plan the instructional materials and strategies accordingly. In other words, the performance prediction helps an instructor to determine whether or not a student falls in his/her zone of proximal development. Then the teacher can adjust the teaching materials and strategies to the learners' potentials. Furthermore, the instructions can be planned to be more challenging than the current level of knowledge, but still within the possible range of potential capabilities and competencies.

[1] At-risk students are the ones who do not experience success at school, more likely to fail academically, and may drop out. It is important to detect them and help them as early as possible to avoid future failures and eventual drop out.

The professional instructors have a natural talent or have built an expertise over the time to predict the performance of their students as soon as possible. Unfortunately, this expertise does not come easily and it is not acquired in a short time; it cannot be easily transferred to pre service or in-service instructors, and is not available for e-Learning systems. Consequently, it is important to develop methods and algorithms that can help teachers or artificial instructors to better predict the performance of their students as early as possible. That has been the drive behind many research studies in the area of performance prediction and skill estimation [5–11].

The development of such methods, to predict the performance of students, can help a teacher in the form of a program, such as a module for a spread sheet or a standalone program, to process the assignments' grades. The prediction made by the program can be used by the instructor to monitor his/her students' progress, and decide on his/her next educational activities accordingly. On the other hand, these methods can be implemented as a module for a Learning Management System (LMS) to automatically retrieve the assignments' grades from the database, and predict the performance level. The results would be accessible to the instructor for planning his/her activities. Similarly, an artificial tutor in an Intelligent Tutoring Systems (ITS) automatically uses these methods to predict the performance level of its learner(s) and adjust the educational materials accordingly.

From another point of view, such a method can be used in traditional, semi-online,[2] and online learning environments to predict the learners' performance, and adjust the learning materials accordingly. The experiments reported in this chapter are conducted in a semi-online setup, in which the students receive the lectures in a face-to-face format while they receive the materials and complete quizzes through an LMS.

In the subsequent sections, the importance of performance prediction and the related research is further examined. Furthermore, the data mining approaches in performance prediction has been reviewed. In Sect. 6.3, the multi-channel decision fusion performance prediction is introduced. Section 6.4 is dedicated to experimental setups and the performance of the multi-channel decision fusion approach. Discussion on the results and future works are explained in Sects. 6.5 and 6.6.

6.2 Student Modeling

Student modeling module is one of the 4 major modules in an ITS, including the student module, the expert module, the tutoring module, and the user interface module. These intelligent educational systems gain much of their power from having the student model that describes the learner's proficiencies at various

[2] Also referred as the mixed methods.

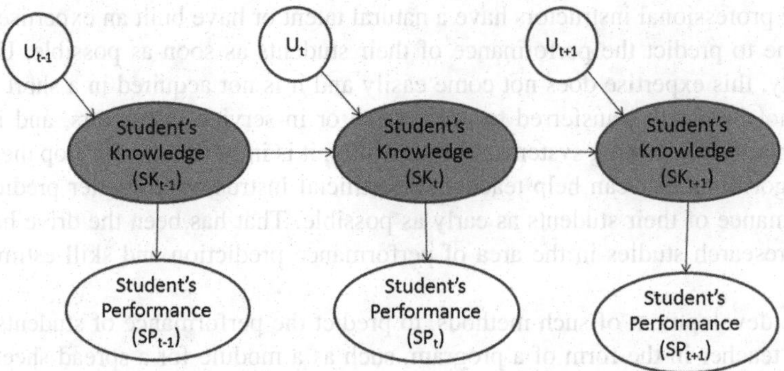

Fig. 6.1 The dynamic Bayesian Network that characterizes the evolution of a student's knowledge based on the previous knowledge (SK_{t-1}) and the learning activities (U_t). The student's performance is the measurable outcome of the knowledge. The unwanted effects during the learning process can be considered as noise for the learning activity

aspects of the domain to be learned. A student's model [12] represents the student's long term characteristics, such as personality, and short term characteristics, such as motivation and emotion. These characteristics can be used to diagnose errors and mistakes in a learner's knowledge, to determine learners' misconceptions, to predict a possible learner's reaction to the learning materials, and to evaluate learner's performance.

One of a learner's characteristics is his/her current knowledge, and the change in his/her knowledge, which is the most significant issue in any learning process. Several widely used models, such as overlay model, differential model, perturbation model, and constraint-based model, are used to model learners' knowledge [13]. However, since knowledge is not directly measurable, performance is used as a measurable outcome of the knowledge at any given time. That is why evaluation tools are used throughout the human history to indirectly measure the knowledge level of a person. To model the relationship between the knowledge and performance evaluation, Bayesian Networks can be used (Fig. 6.1).

The learning process, i.e. the change in the knowledge, can be modeled using a Bayesian Network based on Markov assumption, in which the current knowledge of a learner is shaped based on his/her prior knowledge and the learning activities (Fig. 6.1). In this model, the performance evaluation is the measurable outcome of the knowledge directly affected by knowledge. The knowledge at time $t-1$ can be acquired through previous studies or through interaction with friends or the environment. The change in the knowledge represents the probability of acquiring new knowledge, i.e. changing the level of the knowledge, after facing new learning materials (Eq. 6.1).

$$p\left(SK_t | SK_{t-1}, U_t\right) \tag{6.1}$$

In which SK_{t-1} and SK_t are learner's knowledge at times $t-1$ and t accordingly. U_t represents the learning materials in the given learning period. The unmodeled or undesired learning can be represented as noise for U_t. In other words, a student's knowledge can be estimated by his/her previous knowledge and the learning activity. To improve this knowledge estimation, the Bayesian filtering approach can be used (Eqs. 6.2 and 6.3), in which the estimated belief in a knowledge level is the summation of all possible previous knowledge levels that may reach the estimated knowledge level (Eq. 6.2), i.e.:

$$\bar{bel}(SK_t) = \sum_{SK_{t-1}} p\,(SK_t|SK_{t-1}, U_t)bel(\,SK_{t-1}) \tag{6.2}$$

In which $\bar{bel}(SK_t)$ is the estimated student's knowledge at time t and $bel(SK_{t-1})$ is the belief in the previous learner's knowledge. An effective, useful, and yet simple partitioning of the performance level is to classify it into four categories of weak, average, good, and excellent knowledge levels. In other words, the stereotype model is used to represent the knowledge level of a learner into four stereotypes, i.e. stereotype one as weak learners, stereotype two as average learners, stereotype three as good learners, and stereotype four as excellent learners. Obviously, it is possible to change the stereotype models based on the domain knowledge if needed, such as the case that has been conducted in knowledge analysis studied in PeRSIVA [14] in which 8 stereotypes are defined. The estimated belief in the knowledge can be corrected (modified) by observing the knowledge through performance evaluation (Eq. 6.3).

$$bel(SK_t) = \eta p\,(SK_t|SK_{t-1}, U_t)\;\bar{bel}(SK_t) \tag{6.3}$$

In other words, by having the learning materials and activities, it is possible to guess the extent to which the students have met the learning objectives. By measuring the performance, we can correct the guess, and determine the real learning acquired. It should be noted that determining how much a learning material or activity can affect learners' knowledge is an important issue. This is what talented and experienced educators have empirically learned.

On the other hand, the probability of a performance assuming a given knowledge (Eq. 6.4) depends on many factors such as the probability of slip or guess [11] and the performance evaluation materials.

$$p\,(SP_t|SK_t) \tag{6.4}$$

Since the extent to which a student has met the learning goals in the process of learning is the measure of successfulness of an educational system, it becomes important to accurately determine the knowledge level of a learner.

Although the above model can help to better evaluate the current knowledge of a learner, the early prediction of his/her knowledge remains an important issue in order to improve the learning process and the educational system. The following sections address this issue in more detail.

6.3 Performance Prediction

One of a learner's characteristics is his/her current knowledge, which is evaluated using his/her performance in an assignment or examination. From the learner's point of view, the performance evaluation can be formative [15], such that it would give appropriate feedback to the learner on his/her current state, or summative [15] that summarizes the knowledge of the learner at the end of the learning period. From the system's point of view, or an instructor's point of view, this evaluation can be used in diagnostic, predictive and evaluative forms to improve the learning process.

Due to the importance of the performance prediction, which allows on the fly adjustment to the learning materials in improving the learning process outcomes, this chapter is devoted to performance prediction using assignment categories. In other words, based on the current performance, i.e. SP_t, in different assignments the most accurate predicted performances for the midterm, i.e. $SP_{t=midterm}$, and for the final exam, i.e. $SP_{t=final}$, are needed.

In the following, the position of the performance prediction in a general student modeling module is explained. Later, its position in an ITS and the data mining approaches used for performance prediction are explained. A comprehensive list of research studies using Educational Data Mining (EDM) has been prepared by Romero and Ventura [16].

6.3.1 Performance Prediction in ITS

Although performance prediction is useful for all learning environments, from traditional face-to-face learning environments, to e-Learning and ITS, nonetheless, it becomes more valuable for e-Learning and ITS, since it tries to compensate the shortcomings caused by the lack of face-to-face interaction of a human expert.

It should be noted that assessing a learner's knowledge, especially through performance evaluation, is difficult because (1) part of the learner's proficiency evaluation comes from visual observation which is not available in online systems, in ITS, and even in large traditional classrooms since the instructor is incapable of having direct interaction with most of the learners, (2) due to low reliability and validity of most available testing instruments, the learner's performance in an exam or quiz may not be a perfect reflection of the learner's knowledge and proficiency in a field, and (3) the fact that the state of the learner's knowledge changes over time.

Additionally, the information gathered through human–computer interaction might not clearly and/or uniquely represent real world situation. Consequently, it is important to design a data mining system capable of correctly and efficiently process the data to predict the performance.

It should also be noted that the predicted performance can be represented in the typical scale format, i.e. 0–4 or 0–20 depending on the educational setup, or it can be represented in a few major classes such as weak, normal, good, and excellent. The later one is more appropriate since all the predictions are not 100 % accurate, and providing the predicted performance in the scale might create unnecessary expectations.

6.3.2 Data Mining Approaches for Prediction

Based on the fact that the performance, in any measurable setup such as exams or homework assignments, is a representation of the knowledge of a learner, a wide range of research studies have been performed to model and estimate performance. Decision Trees, classification methods and Bayesian Networks [13, 16, 17] are from the most widely used approaches. For instance, TELEOS is a learning environment in which Bayesian network is used to diagnose the student's knowledge state [18].

A set of research studies using Data Mining approaches have focused on considering the possibility of answering questions correctly/incorrectly, i.e. slip and guess, by chance [11]. The effects of slip and guess have also been considered in the Bayesian Knowledge-Tracing (BKT) approach [14].

Classification and regression trees [19, 20] are typical machine-learning methods for constructing prediction models from data. The models are obtained by recursively partitioning the data space and fitting a simple prediction model within each partition.

C4.5 and CART [20] (Classification and Regression Tree) are two classification tree algorithms that follow the general recursive tree building approach. C4.5 uses entropy for its impurity function, whereas CART uses a generalization of the binomial variance called the Gini index. These approaches first grow an overly large tree, and then prune it to a smaller size to minimize an estimate of the misclassification error. By default, CART employs tenfold cross validation, whereas C4.5 uses a heuristic formula to estimate error rates. CART estimates the dependent variable, while C4.5 estimates the class to which dependent variable belongs. In the proposed multi-channel decision fusion approach, in which the nearest neighbor approach is employed for classification, CART is used.

Chi-squared Automatic Interaction Detection (CHAID) [21] employs yet another strategy. If the input is ordered, its data values in the node are split into 10 intervals and one child node is assigned to each interval. If the input is unordered, one child node is assigned to each value of the input. Then, CHAID uses significance tests and Bonferroni corrections to try to iteratively merge pairs of child nodes. This approach has two consequences. First, a few nodes may be split into more than two child nodes. Second, considering the sequential nature of the tests and the inaccuracy in the grading, the method is biased toward selecting variables with fewer distinct values.

Due to the fuzzy nature of the human performance, fuzzy approaches have been widely used in the field [22, 23]. At the end, it should be mentioned that Pardos et al. [24] showed that an ensemble of different approaches could result in better predicting a student's knowledge level.

6.4 Multi-Channel Decision Fusion Performance Prediction

To predict performance in a course or in a learning setup, available grades from different assignments such as homework, lab assignments, projects, and online quizzes, which we will refer to them as assignment categories, can be used. As mentioned earlier, one approach to performance evaluation and/or prediction is to use all available grades from different assignments to determine the performance level, and to predict the performance in the midterm or final exam (Fig. 6.2a). However, this approach does not directly consider the importance and impact of each assignment category in the prediction and evaluation of the performance.

This becomes important when assignments are from different categories and each has its own characteristics and importance. For instance, the performance level that is represented by the homework assignment category, in which the students have adequate time to think, consult friends, or the teaching assistants, is completely different from the performance level that in class quizzes represent, in which the students have limited time and should analyze and answer the questions individually. Figure 6.2b represents a multi-channel decision fusion approach, in which the performance in each assignment category is determined, and those performance levels are used to determine the overall performance.

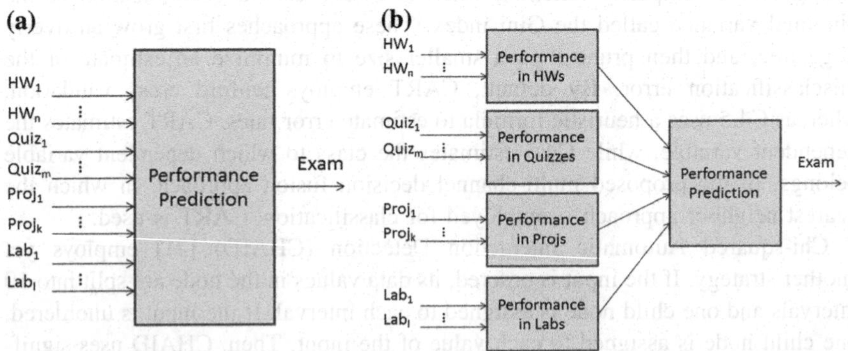

Fig. 6.2 Different approaches in performance prediction. **a** One shot performance prediction, in which the performances in assignments are used to predict the performance in the exam. **b** The multi-channel decision fusion for performance prediction in which each assignment category has a performance level, based on the related assignments, and the overall performance level is determined using only the performance in assignment categories

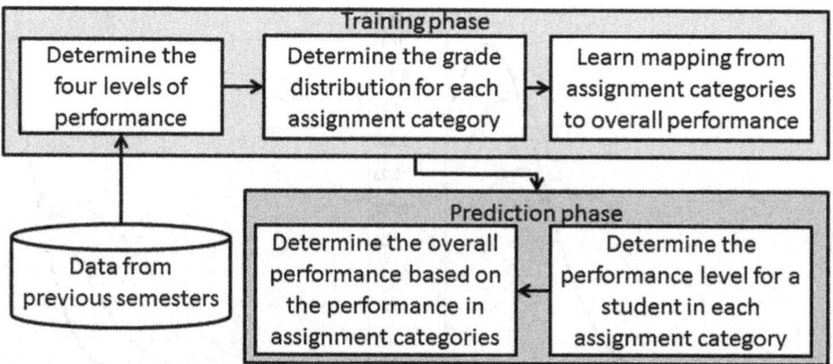

Fig. 6.3 The proposed multi-channel decision fusion approach consists of two phases, the training phase and the estimation phase. In the training phase, the data from previous semesters is used to create the mapping, from assignment categories to overall performance, which can be used in the current or future semesters

In order to conduct the performance prediction, a training phase is needed to learn the relationship between the performance in assignment categories and the overall performance, which is normally conducted based on the data provided from the previous runs of a course. In the case of the multi-channel decision fusion performance prediction, this training phase should be performed in order to determine the performance in assignment categories, the overall performance level, and the mapping between the two, i.e. from the performance in assignment categories to the overall performance. The overall training and estimation phases are shown in Fig. 6.3. In the following subsections, these three are explained.

6.4.1 Determining the Performance Level in Assignment Categories

As shown in Fig. 6.3, the first step in the training phase is to classify students into a few groups, which normally consists of four groups of expert, good, average and weak. Each group is represented by a normal distribution, i.e. $N(\mu, \sigma^2)$, with μ representing the mean, and σ^2 representing the variance. These groups are determined based on the data collected from previous runs of the course, which includes all the grades and final performance of the students in the course.

This classification can be made manually by an instructor or an expert, or by using intelligent methods such as K-means. The number of groups can also be determined manually or detected automatically. It does worth mentioning that the automatic approach normally results into more accurate classification than the fixed human classification. In the human classification, typically the mean for the expert performance level is set for 18.5 and its minimum is set for 17 in a scale of 20. However, in an automatic classification, 18.2 were determined to be the

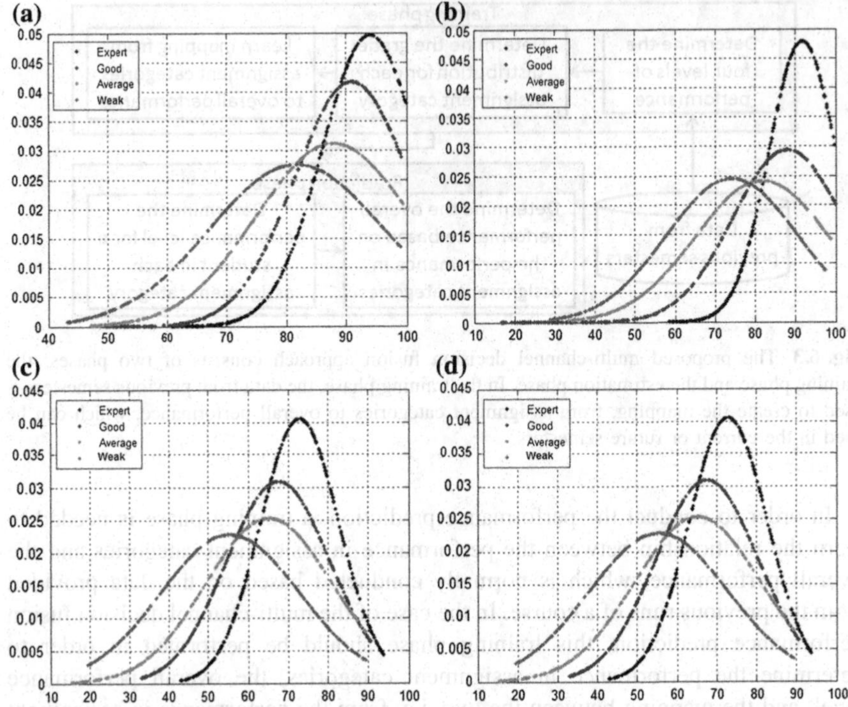

Fig. 6.4 The grade distribution for each assignment category in the four performance levels: **a** lab assignments, **b** homework, **c** quizzes, and **d** projects. The performance distributions are represented using Gaussian distribution (Table 6.3), in which the horizontal axes are the grades out of 100

mean and 17.40 as the minimum for the expert class. This happens to be more realistic based on the specific grades achieved in a specific course and was verified by an expert. Also the standard deviation is determined more accurately and can be updated over time.

After classifying the students into four performance levels, the grade distribution for each assignment category is determined (Fig. 6.4). By analyzing Fig. 6.4 several interesting points which are consistent with the reality can be observed:

1. The impact of plagiarism: There is no significant difference between groups of "average" and "good" learners in online quizzes. Perhaps it could be due to the fact that the learners collaborated together in answering the questions, which was confirmed by our indirect observations.[3] Plagiarism also shows its impact in homework assignments. Although it was possible to detect plagiarism in

[3] A survey was performed to validate this observation which has been explained at the end of Section 6.5.

homework assignments, it is not as easy as detecting plagiarism in programming projects, for which MOSS[4] is used to detect similar codes and the students had to orally deliver the projects. In the case of lab assignments, the fixed structure of the lab assignments does not allow adequate discrepancy between performance levels.
2. The impact of interaction between close performance level groups: It is fairly visible that in case of collaboration and/or plagiarism, the learners tend to interact with peers in their own performance group or in a performance group close to them. For instance, the learners in the "weak" group tend to interact and get help from the learners in the "average" group. That is why their scores are closer to each other. Similarly, the learners in the "average" group tend to get help from the learners in the "good" performance level group. This phenomenon is more visible between the top three groups rather than the weak group. The reason could be that the weak learners, who mostly constitute at-risk students, tend to lose their motivation for better achievement, while the learners in average and good performance level groups still have hope to gain better grades through cheating and collaboration with their stronger peers.

Although the above two facts reduce the effectiveness of assignment categories in predicting the overall performance level, especially for lab and quiz assignments, these still provide clear distinction between "weak" and other groups, especially the "excellent" group.

That is actually the main reason that multi-channel decision fusion provides better prediction since these features/characteristics of the assignment categories can be clearly considered in the prediction.

6.4.2 Determining Overall Performance Levels

The overall performance is typically determined based on the result in the final assessments, i.e. final exams in an educational system. On the other hand, midterm exams are used to provide feedback to both learners and instructors such that they can adjust their activities for better results. These methods of performance assessment are chosen as the ground truth of the performance level since:

1. These assessment methods suffer less from noise since these are conducted in a controlled setup, in which there is a lower possibility of plagiarism.
2. These assessments show the sole understanding and knowledge of a learner, since learners should answer questions on their own. In other words, in case of homework, they cannot get help from others or hide behind the performance of others, in case of team projects.

[4] http://moss.stanford.edu.

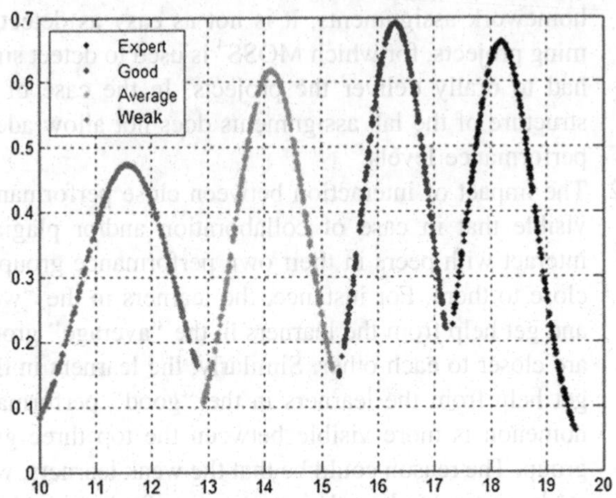

Fig. 6.5 The clustering of the students into four distinctive performance levels

3. They represent the intermediate and final performance levels, one as a formative assessment and the other as a summative assessment.
4. These are accepted assessments approaches for performance evaluation.

Consequently, to determine the overall performance, the midterm and final grades from previous semesters are classified into the four groups. Similar to the approaches considered to determine the performance level in assignment categories, the four groups of performance level in midterm and final exams can be determined using manual or automatic classification.

Figure 6.5 shows the clustering performed on the final exam for a course using K-means. In this specific course, the four performance levels have clear distinction from each other. A comparison between Figs. 6.4 and 6.5 shows how midterm and final exams have stronger differentiating capability between the performance level groups, compared to assignment categories. This could be due to the noise involved in the assignment categories so it cannot be a clear representation of the learner's knowledge.

6.4.3 Mapping from the Performance in Assignment Categories to Overall Performance

Based on the discussion in the previous subsection, the mapping from the performance in assignment categories to overall performance is based on the midterm and final exam grades. To develop the mapping between the performance level in assignment categories and exam grades, different data mining approaches can be used. The simplest approach can be linear regression. Other possible approaches can be CHAID and CART. Even an ensemble of different methods can be used to take advantage of the strength of each method. A study on a specific course shows that linear regression has the best performance among these three, which would be discussed in the experimental results section.

After selecting the right method for mapping between the assignment categories and the overall performance, the training data is used to train the map. In the prediction phase, the mapping is used to find the overall performance of a learner based on the available grades in the assignment categories.

The nearest neighbor, i.e. the performance level with the closest average performance to the performances determined in the training phase, is considered as the performance level of the learner. The nearest neighbor is determined by Euclidean distance between the means of two groups. In the case that the distances between the means are too close, the distance between variances is also considered.

In the last step, the accuracy of the mapping between the performance levels in the assignment categories and the final performance level is measured using the current midterm and final exams. Then, the mapping is updated to reduce the mapping error.

6.4.4 The Characteristics of Assignment Categories

To better understand the importance of multi-channel decision fusion performance prediction, it is necessary to compare the assignment categories based on the features that have impact on evaluating the knowledge level.

The following list (Table 6.1) is a set of important features proposed in this chapter. Other features may be added in the future. The importance of a feature in

Table 6.1 Typical assignment categories and their features

	Plagiarism	Plagiarism detection	Time limit	Discriminatory	Legitimate help	Slip and guess	Team work
e-Homework	+	+	−	+	0	−	−
Homework	+	−	−	+	+	−	−
Coding Projects	+	+	−	+	+	−	−
Hardware projects	0	0	−	+	+	−	−
Online quiz	+	−	−	0	−	+	−
In-class quiz	−	−	+	0	−	+	−
Laboratory	0	0	0	−	+	−	−
e-Essay	+	+	−	+	+	−	−
Essay	+	−	−	+	+	−	−
Take home exam	+	−	0	+	−	−	−
Exams	−	−	+	+	−	0	−
Team projects	0	+	0	0	+	−	+

any of the assignment categories is shown by "−", when it is not important, "0" when it is neutral, and "+" when it is important.

- Plagiarism: This feature shows the possibility of cheating in a given assignment. For example, it is harder to plagiarize in essay exams, compared to coding projects.
- Detecting Plagiarism: This feature shows the capability of detecting plagiarized materials. For instance, hard copy assignments are difficult to be checked against others while checking electronics assignments is easier.
- Time limit: This feature shows if a learner can be under pressure due to the limited time for completing the assignment.
- Discriminatory: Which shows that the assignment can clearly discriminate between different knowledge levels or not. For instance, laboratory assignments are too structured and less discriminatory since the answers are normally fixed.
- Legitimate help: Which shows if asking for help and guidance is allowed for a given assignment or not. It shows if the instructed materials have been completely absorbed, or the learner still needs guidance to make use of his/her knowledge.
- Slip or guess: That shows if slip and guess can easily happen in a given assignment. For instance, in coding projects, guessing is hardly possible, while in quizzes, assuming multiple choice or true/false quizzes, it is easy.
- Team work: This refers to the fact that in the assignments completed in groups, it is difficult to evaluate the contribution shares, i.e. the knowledge level and the effort of each team member.

Table 6.1 shows the importance of each of the above features in a set of typical assignment categories. If a new assignment is designed, its features can be compared to the listed features in this table to decide whether it should be considered as a new category or it can be included into an existing category.

6.5 Experimental Results and Discussion

In order to evaluate the proposed method, 387 students who took the "Introduction to Computers and Programming" course at the school of ECE, University of Tehran, in the fall semesters of 2009–2010, have been selected. The majority of these students are freshmen, taking this course at the university level for the first time and in the first semester after entering college. Since these students have to pass the national entrance exam to enter the school, most of the students are among the top 1,000 students in the country.

The course includes four assignment categories, i.e. online quizzes, laboratory assignments, homework, and projects with 5, 15, 5 and 15 % of the total grade respectively. The course is conducted in the combined traditional face-to-face and online format, in which Moodle is used as LMS to deliver quizzes, slides

Table 6.2 Extracted skill levels from K-means classification

	Min	Max	Mean	SD	Total
Expert	17.40	19.5	18.20	0.55	92
Good	15.30	17.3	16.35	0.56	81
Average	12.90	15.2	14.12	0.63	80
Weak	10.00	12.8	11.53	0.81	51

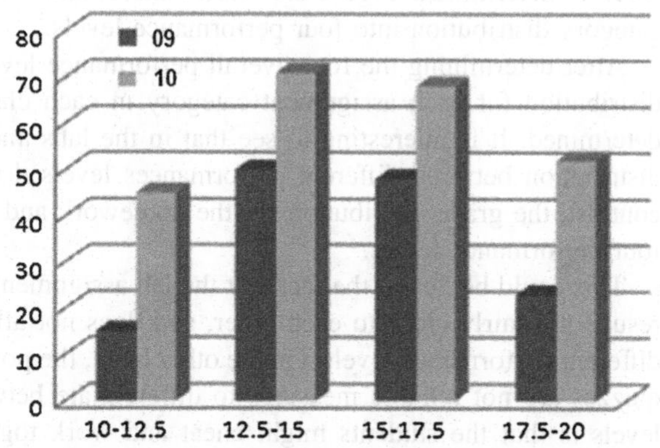

Fig. 6.6 The distribution of the grades in the four categories of expert, good, average and weak. The percentages of the grades in each range are shown for two different semesters, i.e. fall 2009 and fall 2010. The grades are out of 20

and readings, assignments, grades, and to provide online collaborative features, such as discussion forums and news. Consequently, the LMS contains all the grades and data about the course. 80 % of the data is used for training and classification and 20 % is used for testing. The data processing and normalizations are performed using Weka.[5]

Table 6.2 shows automatic data classification that has distinguished four different performance levels, i.e. expert, good, average, and weak, which matches human intuition. 30.16 % of the students have been categorized with expert performance level, 26.6 as Good, 26.2 as average and 16.7 % as weak (Fig. 6.6). Total represents the total number of students in each performance level. It should be noted that the number of students below 10 were very small, and did not constitute a group with adequate data.

Figure 6.5 shows four clear levels of performance existing in the course. It should be noted that k-means is used to perform clustering in this step. We have observed that k-means clusters in four or five groups based on the given data in the class. Thus, k-means is setup to get four groups to have a fixed set of groups for all the data sets, i.e. grades in different assignment categories. Also, this automatic approach has more accurate clustering than the fixed human clustering since the learning setup would be slightly/greatly different from semester to semester. This difference could be based on the individual differences between students, instructors teaching the course, changes in the course materials and assignments,

[5] Waikato Environment for Knowledge Analysis.

and the variations in the teaching assistants helping with the course. In the human clustering, normally the mean for an expert is set for 18.5 and its minimum is set for 17. However, in automatic clustering, 18.2 is determined to be the mean and 17.40 as the minimum for the expert class. Also the standard deviation is determined more accurately, and can be updated over time.

After determining the four overall performance levels in the course, the grades' normal distribution for each assignment category in each class of performance level is determined. Figure 6.4 shows the results of analyzing the assignment category distribution into four performance levels.

After determining the four overall performance levels in the course, the grade distribution for each assignment category in each class of performance level is determined. It is interesting to see that in the labs and online quizzes, the grade distribution between different performances levels do not differ significantly. In contrast, the grade distribution for the homework and projects differ between the four performance levels.

This could be due to the fact that the lab assignments are so systematic that the results are fairly close to each other, and does not allow differentiation between different performance levels. On the other hand, the possible reason that the online quizzes are not reliable measures to differentiate between different performance levels is that the students might cheat and work together to answer the online quizzes. It should be noted that although the labs and online quizzes are not good measures for differentiation among all the performance levels, however, they can be used to differentiate between expert and weak performance levels.

This can be justified based on the fact that the students who work together to answer quizzes are within the group or groups who feel closer together to collaborate with each other. For instance, weak and average students may work together to answer an online quiz, while students in good and expert groups tend to work together. Table 6.3 shows the distribution of the grades in each assignment category in the four performance levels.

In this step, the distributions of students' grades are calculated for these four assignment categories. Then each performance level that has the closest mean and standard deviation from the distribution of a student's grades is considered as the performance level of that specific student.

The estimation of the final performance through the assignment categories can be done using different methods. As mentioned in the previous section, three methods have been used and compared to each other for this estimation. The

Table 6.3 Calculated grade range for the learning objects

	Lab		Quiz		Homework		Project	
	μ	σ	μ	σ	μ	σ	μ	σ
Expert	94.2	7.7	73.7	9.7	92.0	8.4	97.2	5.3
Good	91.0	8.5	68.3	13.0	88.4	12.2	87.8	14.3
Average	88.1	12.2	64.8	15.9	81.6	16.2	75.9	20.5
Weak	82.1	14.6	54.2	18.4	74.7	17.1	61.7	23.2

Table 6.4 The comparison of different methods for performance level classification. It is clear that, over all, regression has better results than the other two

	CHAID	CART	Regression	CHAID	CART	Regression
Estimated	*Was Expert*			*Was Good*		
Expert	64.3	82.1	96.6	19.2	42.3	26.9
Good	35.7	14.3	3.4	46.2	42.3	73.1
Average	00.0	3.6	00.0	34.6	15.4	00.0
Weak	00.0	00.0	00.0	00.0	00.0	00.0
Estimated	*Was average*			*Was weak*		
Expert	25.0	15.4	0.0	0.0	0.0	00.0
Good	50.0	15.4	15.4	6.3	6.3	00.0
Average	33.3	7.7	38.5	6.3	00.0	12.5
Weak	00.0	61.5	61.5	87.5	93.7	87.5

results are shown in Table 6.4 An advantage of CART and CHAID is that the results are shown in a hierarchical tree, and the results can be analyzed easier. Figure 6.7 shows a branch of the tree generated by CHAID and Figs. 6.8 and 6.9 show the result of CART and CHAID trees, respectively.

In this branch, the root node consists of all samples. At this point, it is possible to predict the midterm grade with 76.5 % accuracy. At the second level, the algorithms try to come up with a range, i.e. 95.6–98, in the LAB grades to improve the prediction. The accuracy has dropped to 72.3 %. Including the homework grades increased the accuracy to 78 % at the 3rd level and the projects could increase the accuracy to 92.9 %. It should be noted that at the root level, no assignment category is considered and the classification is done based on the raw assignment grades. Thus, it clearly shows the advantage of multi-channel decision fusion approach to the basic approach with 16 % increase in the prediction rate.

As shown in Table 6.4, CHAID classified the students with 64.3, 46.2, 33.3 and 87.5 % accuracy in expert, good, average, and weak classes, respectively. As it can be seen, CHAID performs better in good class than CART, while CART performs better in expert and weak. Both approaches were not very successful in classifying students in the good group. As it can be seen in the table, regression approach outperforms the other two approaches. Consequently, linear regression is more suitable for performance level estimation than the other two approaches.

Prediction of a student's final performance level, as soon as possible, is very important in adjusting the course materials. Also, as mentioned earlier, this is crucial for helping at-risk students, especially in the first year of college or university [1]. Consequently, the lower the number of assignment categories needed to effectively predict the performance of a student, more suitable the item to be used for performance level prediction. That is why the results in the "Introduction to Computers and Programming" course has been analyzed to determine the best learning objects for performance level prediction. The result shows that the performance of a student can be predicted by using two to three homework or project grades. Using three to four grades can determine the performance level with high confidence.

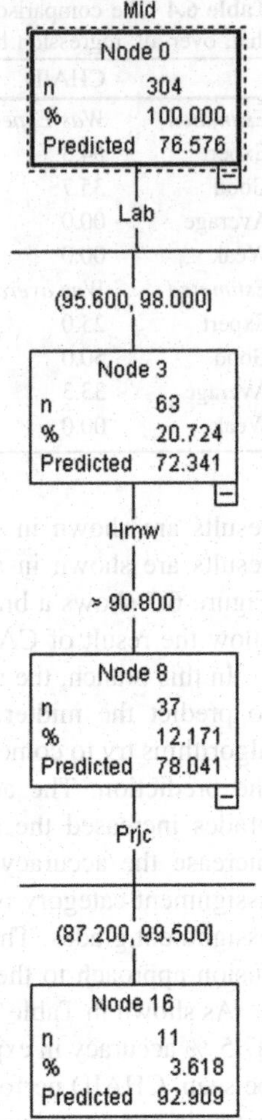

Fig. 6.7 The figure shows a branch of the tree generated by CHAID which is shown in Fig. 6.8. In this example, it is shown how the midterm prediction for a skill level is conducted based on lab, then HW, followed by the project. "*n*" represents the number of students predicted, "*%*" the percentage of the students predicted in this group and "predicted" represents the accuracy of prediction

As mentioned earlier, the grades for quizzes and lab assignments do not have clear differentiation power between all four performance levels and cannot be used for this purpose. Furthermore, four to six grades are needed to be able conduct performance level classification. Consequently, those would not be used for the purpose of early classification.

To further evaluate the effectiveness of the proposed approach, the approach was conducted in the Artificial Intelligence course, at the school of electrical and computer engineering, University of Tehran, with 60 registered students. Two assignment categories have been used in this course. Since the study was only performed in one semester and the data from other semesters was not available, both training and testing were performed on the same data from one semester. Consequently, to train the system, beside the assignment categories, the midterm results are used as the performance level ground truth. Also, 80 % of the data is

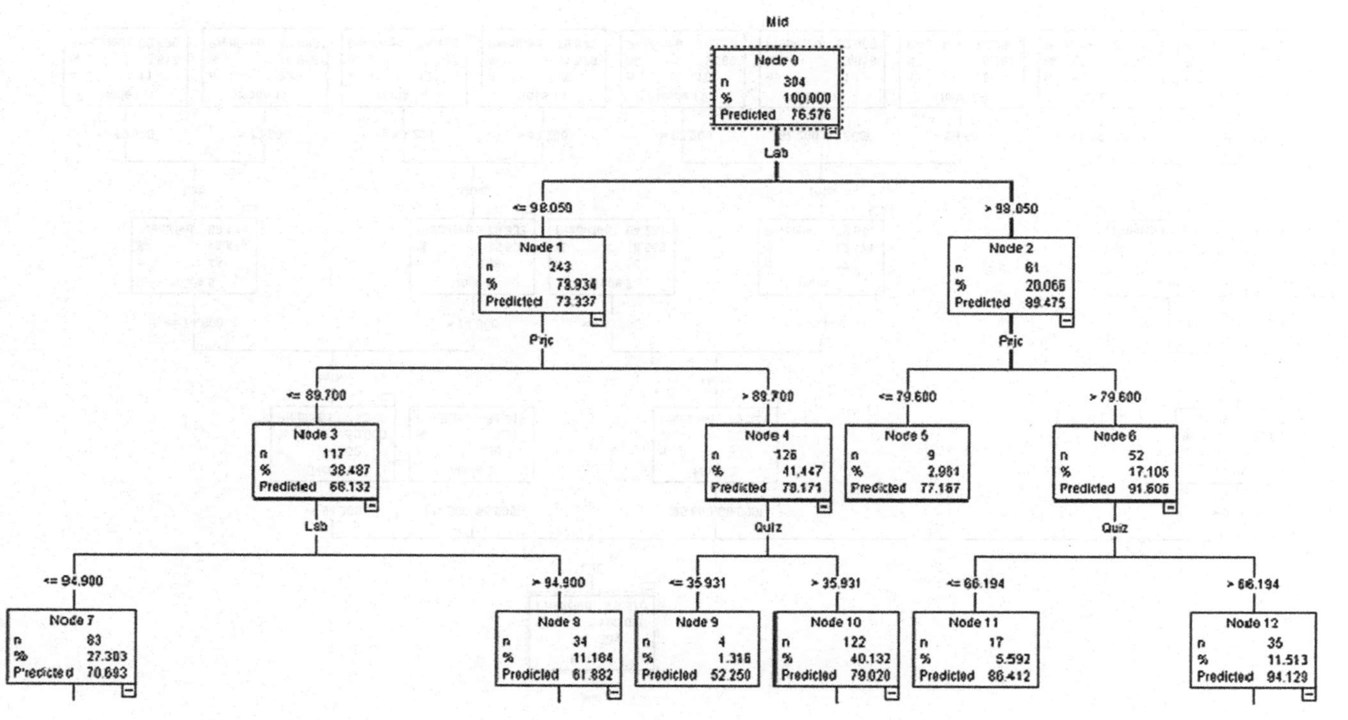

Fig. 6.8 Four levels of CART decision tree

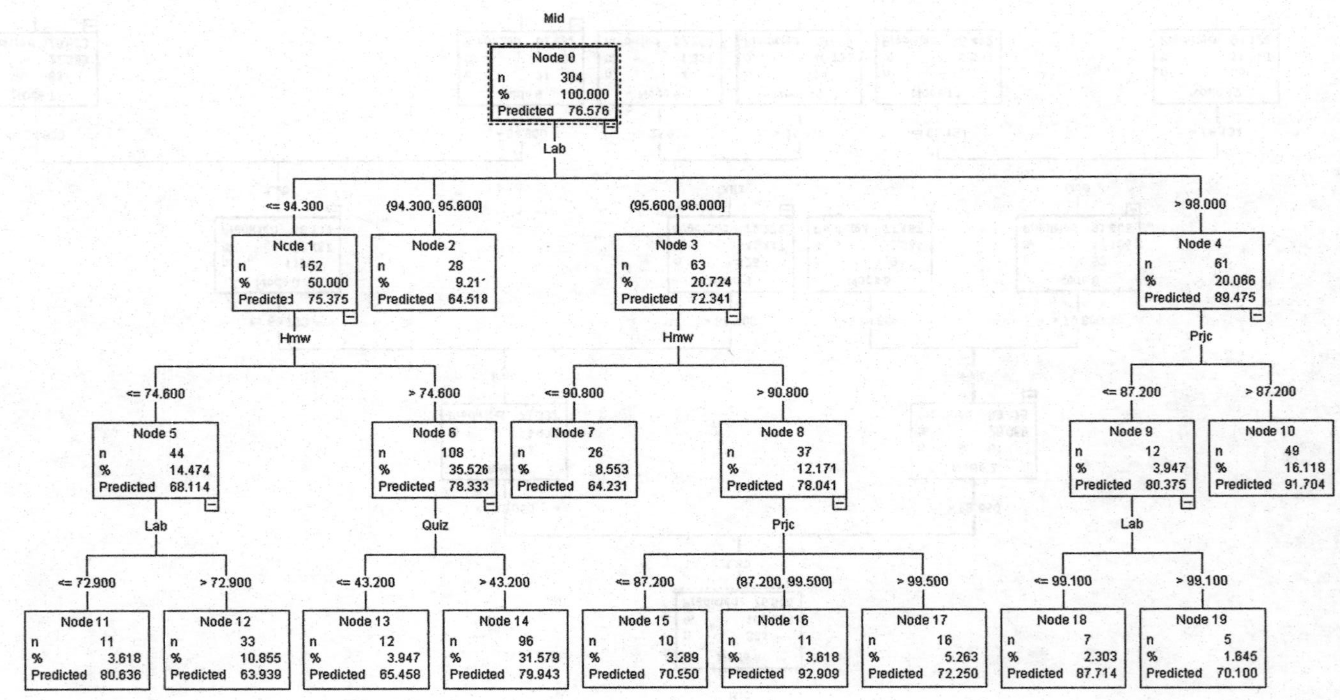

Fig. 6.9 CHAID decision tree

Table 6.5 The distribution of the grades in the Artificial Intelligence course in three groups

	Homework		CA		Midterm		Final		Total	
	μ	σ	μ	σ	μ	σ	μ	σ	μ	σ
Expert	84.67	6.62	85.91	3.73	78.40	6.64	73.00	7.87	15.93	0.80
Good	72.88	7.79	78.31	10.17	66.53	5.22	63.41	7.17	13.94	0.75
Average	44.31	9.22	53.33	20.18	57.22	14.06	50.00	10.63	11.17	1.00

used as the training set and 20 % as the test set. Since the students normally perform well in this course, k-means clustered the students in three groups based on their performances, i.e. excellent, good, and average. Interestingly, the importance of the homework assignment category in performance prediction was lower than the importance of the projects.

This could be due to the fact that the students could easily cheat in homework, while it was harder to plagiarize in projects. Furthermore, the result shows 75 % accuracy in predicting the midterm exam grades only by having two homework grades and one project grade. This shows that this approach is effective even in a new course with limited assignment categories, in order to predict the performance level of a learner. Table 6.5 shows the distribution of the grades in the Artificial Intelligence course, in which the grades are classified into three groups.

It should be mentioned that in this course, the same conclusion, i.e. regression outperforms CART and CHAID, has been made. Regression correctly predicted the performance level of 12 students out of the 16 test cases. Meanwhile, CART only predicted five cases correctly, and CHAID could not make the decision tree in two performance levels.

The reason behind the better performance of linear regression over CART and CHAID can be in the fact that linear regression does not ignore the possible correlation between the variables, i.e. the assignment categories in this problem. However, CART and CHAID consider each variable separately to make the classification. The results show that the correlation between these variables is not negligible, and should be considered.

As mentioned earlier, it was observed that the interaction between students in groups may reduce the differentiation between the groups, specially the average and good groups which are at the middle of the group spectrum. To validate this observation, the students were asked to give the name of three classmates with whom they have interacted the most in completing their assignments. Eighty-three students responded to this survey. The average grade of each set of students, whom interacted with each other were, used to classify them to their nearest performance level group. The standard deviation of these set of students is 1.96, confirming the fact that the students have high tendency to interact within their performance level group or close groups.

6.6 Conclusion and Future Work

Performance level prediction is important, because it can be used to adjust the learning materials to improve the learning experience of students. It becomes very crucial for helping at-risk students, especially in the first year of college or university. The result of performance level prediction can be used by a human instructor or by an intelligent agent in an intelligent tutoring system. In this chapter, a multi-channel decision fusion approach is proposed to determine the performance level as early as possible based on determining the performance levels in assignment categories such as homework and projects.

This approach consists of two phase of training and estimation. The advantages of the proposed method are:

- The student's performance level can be determined at early stages of a course, based on a few important assignment categories. In the case of our "Introduction to Computers and Programming" course, it can be determined up to 5 weeks after the beginning of the semester based on the homework and projects. In other words, evaluation via the assignments and the programming projects will accelerate the recognition of performance level.
- The system can determine which assignment category helps improve the quality of a student's performance. Consequently, a human instructor or an intelligent tutoring system can use this information to tailor the course for the best performance.

The importance of the proposed approach, compared to the other proposed approaches such as fuzzy skill level estimation, Bayesian networks, and Factorization methods, is in using assignment categories levels, rather than the overall performance levels to estimate the future performance levels.

It should be mentioned that since normal distribution is used to model the performance levels, at least 30 samples are needed for each level to correctly model the performance level. If lower number of samples is available, then Z or T distributions may be used.

The future work would focus on using neural networks for better learning the mapping between the performance level in assignment categories and the final performance level. Furthermore, we will study the use of fuzzy logic to better represent the fuzziness in the data. Also, as it was discussed in Sect. 6.5, the correlation between the assignment categories could be very important and it should be further investigated.

Although we planned to consider the learning style effects in the performance prediction, the learning style of the students at the engineering school might be limited to certain classes. Consequently, a wider study needed to be done to analyze the impact of the learning styles in performance prediction. Finally, the possibility of using the interaction of the user with system through non-assessment learning objects would be investigated.

Acknowledgements The authors like to thank Babak Araabi and Maryam Mirian for their feedback on this chapter. Furthermore, the authors like to thank the ITS group at the Advanced Robotics and Intelligent Systems Laboratory for their help throughout this research.

References

1. Singell, L.D., Waddell, G.R.: Modeling retention at a large public university: can at-risk students be identified early enough to treat? Res. High. Educ. **51**(6), 546–572 (2010)
2. Tinto, V.: From theory to action: exploring the institutional conditions for student retention. In: Smart, J.C. (ed.) Higher Education: Handbook of Theory and Research, vol. 25, pp. 51–89. Springer, Heidelberg (2010)
3. Donnelly, J.: Use of web-based academic alert system for identification of underachieving students at an urban research institution. Coll. Univ. **85**(4), 39–42 (2010)
4. Wass, R., Harland, T., Mercer, A.: Scaffolding critical thinking in the zone of proximal development. High. Educ. Res. Dev. **30**(1), 317–328 (2011)
5. Torabi, R., Moradi, P., Khantaimoori, A.R.: Predict student scores using bayesian networks. Procedia Soc. Behav. Sci. **46**, 4476–4480 (2012)
6. Ramaswami, M., Bhaskaran, R.: A CHAID based performance prediction model in educational data mining. Int. J. Comput. Sci. Issues **7**(1), 10–18 (2010)
7. Kotsiantis, S.B.: Use of machine learning techniques for educational purposes: a decision support system for forecasting students' grades. Artif. Intell. Rev. **37**(4), 331–344 (2012)
8. Baker, R., Pardos, Z., Gowda, S., Nooraei, B., Heffernan, N.: Ensembling Predictions of student knowledge within intelligent tutoring systems. In: Konstan, J., Conejo, R., Marzo, J., Oliver, N. (eds.) User Modeling, Adaption and Personalization. LNCS, vol. 6787, pp. 13–24. Springer, Heidelberg (2011)
9. Thai-Nghe, N., Drumond, L., Krohn-Grimberghe, A., Schmidt-Thieme, L.: Recommender system for predicting student performance. Procedia Comput. Sci. **1**(2), 2811–2819 (2010)
10. Ghazarian, A., Noorhosseini, S.M.: Automatic detection of users' skill levels using high-frequency user interface events. J. User Model. User-Adap. Inter. **20**(2), 109–146 (2010)
11. Thai-Nghe, N., Drumond, L., Horvath, T., Krohn-Grimberghe, A., Nanopoulos, A., Schmidt-Thieme., L.: Factorization techniques for predicting student performance. In: Santos O. C., Boticario J. G. (eds.) Educational Recommender Systems and Technologies: Practices and Challenges, pp. 129–153. IGI Global, Hershey (2012)
12. Chrysafiadi, K., Virvou, M.: Student modeling approaches: a literature review for the last decade. Expert Syst. Appl. **40**(11), 4715–4729 (2013)
13. Millán, E., Loboda, T., Pérez-de-la-Cruz, J.L.: Bayesian networks for student model engineering. Comput. Educ. **55**(4), 1663–1683 (2010)
14. Desmarais, M.C., d Baker, R.S.J.: A review of recent advances in learner and skill modeling in intelligent learning environments. User Model. User Adapt. Inter. **22**(1–2), 9–38 (2012)
15. Black, P.: Pedagogy in theory and in practice: formative and summative assessments in classrooms and in systems. In: Corrigan, D., Gunstone, R., Jones, A. (eds.) Valuing Assessment in Science Education: Pedagogy, Curriculum, Policy, pp. 207–229. Springer, Heidelberg (2013)
16. Romero, C., Ventura, S.: Educational data mining: a review of the state of the art. IEEE Trans. Syst. Man Cybern. Part C: Appl. Rev. **40**(6), 601–618 (2010)
17. Baker, R. S. J. d.: Data mining for education. In: McGaw, B., Peterson, P., Baker, E. (eds.) International Encyclopedia of Education (3rd edn.), vol. 7, pp. 112–118. Elsevier, Oxford (2010)
18. Chieu, V.M., Luengo, V., Vadcard, L., Tonetti, J.: Student modeling in orthopedic surgery training: exploiting symbiosis between temporal bayesian networks and fine-grained didactic. analysis. J. Artif. Intell. Educ. **20**(3), 269–301 (2010)

19. Theodoridis, S., Koutroumbas, K.: Pattern Recognition, 4th edn. Academic Press, Oxford (2008)
20. Han, J., Kamber, M., Pei, J.: Data Mining: Concepts and Techniques, 3rd edn. Elsevier, Oxford (2011)
21. Soman, K.P., Diwakar, S., Ajay, V.: Insight into Data Mining: Theory and Practice. Prentice Hall, India (2006)
22. Jia, B., Zhong, S., Zheng, T., Liu, Z.: The study and design of adaptive learning system based on fuzzy set theory. In: Cheok, Z.A.D., Müller, W., Zhang, X., Wong, K., (eds.) Transactions on Edutainment IV. LNCS, vol. 6250, pp. 1–11(2010)
23. Chrysafiadi, K., Virvou, M.: PeRSIVA: an empirical evaluation method of a student model of an intelligent e-learning environment for computer programming. Comput. Educ. **68**, 322–333 (2013)
24. Pardos, Z.A., Baker, R.S.J.d., Gowda, S.M., Heffernan, N.T.: The sum is greater than the parts: ensembling models of student knowledge in educational software. SIGKDD Explor. 13(2), 37–44 (2011)

Chapter 7
Predicting Student Performance from Combined Data Sources

Annika Wolff, Zdenek Zdrahal, Drahomira Herrmannova and Petr Knoth

Abstract This chapter will explore the use of predictive modeling methods for identifying students who will benefit most from tutor interventions. This is a growing area of research and is especially useful in distance learning where tutors and students do not meet face to face. The methods discussed will include decision-tree classification, support vector machine (SVM), general unary hypotheses automaton (GUHA), Bayesian networks, and linear and logistic regression. These methods have been trialed through building and testing predictive models using data from several Open University (OU) modules. The Open University offers a good test-bed for this work, as it is one of the largest distance learning institutions in Europe. The chapter will discuss how the predictive capacity of the different sources of data changes as the course progresses. It will also highlight the importance of understanding how a student's pattern of behavior changes during the course.

Keywords Predictive modeling · Education · Virtual learning environment · Student outcome

Abbreviations

ANOVA	Analysis of variance
CMS	Course management system

A. Wolff (✉) · Z. Zdrahal · D. Herrmannova · P. Knoth
Knowledge Media Institute, The Open University, Milton Keynes, MK7 6AA, UK
e-mail: a.l.wolff@open.ac.uk

Z. Zdrahal
e-mail: z.zdrahal@open.ac.uk

D. Herrmannova
e-mail: drahomira.herrmannova@open.ac.uk

P. Knoth
e-mail: p.knoth@open.ac.uk

A. Peña-Ayala (ed.), *Educational Data Mining*,
Studies in Computational Intelligence 524, DOI: 10.1007/978-3-319-02738-8_7,

CS	Course signals
GUHA	General unary hypotheses automaton
MOOC	Massive open online course
OU	Open university
SVM	Support vector machine
TMA	Tutor marked assessment
VLE	Virtual learning environment

7.1 Introduction

Predicting student performance, in time to make interventions for improving student performance and reducing drop out or failure, leads to benefits for both students and teaching institutions. In traditional classroom learning, tutors use a range of information sources to judge whom to help, including their personal interactions with the students. In distance education, where students interact with learning materials on a virtual learning environment (VLE), machine-learning methods can be applied to combined sources of student data to predict which students will benefit most from an intervention and allow tutors to better judge whom to offer their assistance to.

Whilst VLE's have been used to deliver course materials for quite some time, their use for really large-scale delivery is a recent phenomenon. Previous course statistics have focused largely on providing data for a whole course, after completion, using only demographic data and historical analysis. For example, Kabra and Bichkar [1] use decision trees to predict failing engineering students, using past performance as the main feature for building the tree, but less emphasis on demographic data such as age or background.

Baradwaj and Pal [2] also chose decision trees over other possible methods to predict student outcome using student performance on a course combined with other information such as attendance to lectures. Pandey and Sharma [3] did very similar work with decision trees, but included demographic data with their student performance data. This work focused on comparing different decision tree methods, these were J48, Simple CART, Reptree and NB tree. J48 produced the most accurate outcome for their dataset.

Real-time prediction, incorporating student activity data from a VLE is less explored, since VLE learning (especially on a large scale) is quite recent. This type of educational data mining is focused on real-time predictive analysis of large student data sets to produce immediately accessible information upon which students, tutors and faculties can act to effect learning outcomes [4]. The course signals (CS) system developed by Purdue [5, 6] is a well developed and tested system for mining data from the Purdue course management system (CMS) and presenting information to teachers using a colour-coded traffic light system for easy visualisation. The goal is early detection of struggling students and the improvement of student retention. The underlying predictive methods used are heavily weighted to take into account a students effort, which is defined by their interaction with the VLE compared to their peers. It also takes into account performance in the course to that point, prior academic history and student demographic data.

In this chapter we will propose how to incorporate the VLE data whilst still making use of the existing static data from demographic data sets, students past history and their assessment results. This type of predictive analytics from VLE data becomes ever more relevant with the emergence of Massive Open Online Courses (MOOC), such as FutureLearn and Coursera, and where even conventional teaching institutions offer online alternatives to traditional lecture-hall teaching.

First of all, the development of predictive models will be introduced. Next a case study will be introduced which will explore some of the issues encountered when developing such models in an institutional context. Finally, we will propose how the methods can extend to new context.

7.2 Defining the Problem

The work described in this chapter has been developed in the context of the OU. The OU is a purely distance learning institution that offers modules of study which can be undertaken in isolation or as part of a degree. Where the remainder of this chapter refers to *modules* these are equivalent to what are more commonly referred to as *courses* in other institutions. Extensive use is made of a VLE to deliver modules to large numbers of students. Tutors don't see students face to face and so predictive models can be very useful for identifying at-risk students.

In this chapter we are interested in developing models to predict students' performance within a single module. We do not attempt to answer the question as to what is the probability that the student will complete the degree or diploma, or that the selected curriculum will satisfy student's expectations etc.

The only contextual information that may be available is whether the student already has experience with another OU course. There are different tasks of predictive modeling depending on the stage of the module and the goal of modeling. The suitable modeling algorithms also depend on the available data.

Successful predictive modeling requires that models developed using the past presentation of the same module is valid for future presentations. This condition requires a certain level of module stability and similar demographic structure of cohorts across presentations. The satisfaction of the first condition can be guaranteed by selecting suitable modules. The second one must take into account changes of the socioeconomic conditions of students. Data used for constructing our predictive models have been selected so that both conditions are satisfied. The validity of models across different presentations has been included as a part of the task.

The module data for which we develop predictive models have the following structure. In early stages, no module-related data is available. The predictive model can be constructed only from demographic data and from the characteristics of the module.

As the module proceeds, additional student data become available as a result of a number of assessments and the data collected from student engagement with the VLE. Thus, predictive modeling can be more supported by these data types and rely less on the expected impact of demographic data.

In our scenario, each course has a well-defined criterion of success and failure, which is based on a student's performance during the course and on the final exam, which is required for most of the modules. The success criterion is not defined in terms of a student's engagement with the VLE, although this information is available for predictive modeling. The assessments and the final exam are marked on the scale from 0 to 100 and the values have associated weights in accordance with their importance for the module curriculum.

The student succeeds in the module if the weighted sum of scores of the required assessments and the final exam are above a predefined S_{pass}, typically $S_{pass} = 40$. The availability of data in time is schematically shown in Fig. 7.1.

A number of useful modeling tasks can be defined for different stages of the module presentation. We will first introduce four problem specifications that have both informed, and been refined as a result of the predictive modeling work described in the remainder of the chapter. Next we will introduce two further problem specifications that will form the basis of future work in this area.

7.2.1 Problem Specification 1

Input: Demographic data of registered students.

Goal: Calculate the probability that the student will submit the first assessment A_1 and that the score of this assessment will be higher than 40.

Problem specification 1 does not take into account students' behavior in the module. It may happen that the student registers but for some reason never gets engaged with the module. The solution to the problem (model) can be constructed prior to the start of the module to identify the most vulnerable part of the cohort. The problem can be extended if VLE information is available.

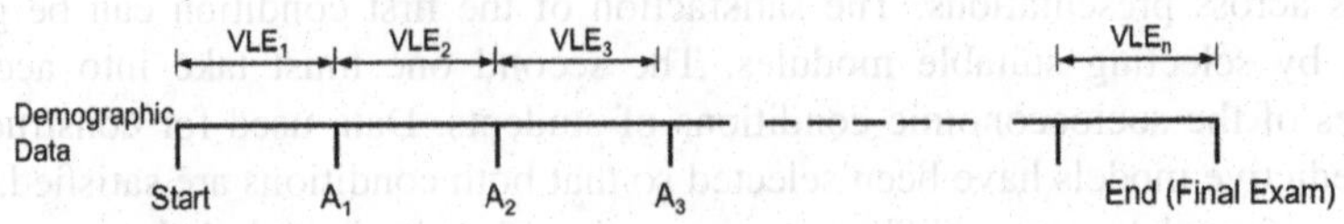

Fig. 7.1 Module data in time

7.2.2 Problem Specification 2

Input: Demographic data of registered students, scores of assessments A_1, …,A_k, and VLE activities up to the assessment A_{k+1}, i.e. Vle_1, …, Vle_k (Fig. 7.1).

Goal: Calculate the probability that the student will achieve a pass/fail at the final outcome.

This model gives a broad view of student outcome at different moments throughout the module. However, it does not address the problem of identifying students at the point of failure.

7.2.3 Problem Specification 3

Input: Demographic data of registered students and the data from student's activity prior to assessment A1, Vle_1 (in Fig. 7.1).

Goal: Calculate probability that the student submits the first assessment A1 and that the score of this assessment will be higher than 40.

The model can be built just before the deadline of A_1. The modeled value can be compared with the real score of A1 and the impact of the VLE for each individual student in the early stages can be evaluated. In the later stages of module presentation, more informative models can be constructed.

7.2.4 Problem Specification 4

Input: Demographic data of registered students, scores of assessments A1, …,A_k, and VLE activities up to the assessment A_{k+1}, i.e. Vle_1, …, Vle_k (in Fig. 7.1).

Goal: Calculate probability that the student scores more than a given value in assessment A_{k+1}.

7.3 Sources of Student Data

In order to develop a predictive model it is necessary to first understand the different data sources and what they can tell us about a student.

The models have been developed on a number of historic datasets from fairly typical OU modules, across different faculties, which meet the requirements laid out in the previous section. The modules are anonymous within this chapter. From this data, we have identified four main sources of student data that we will discuss within this chapter. We do not include communication data (such as email, telephone exchange, logs for service requests) as in our experience this type of data is hard to categorize accurately as often the content and/or purpose of the exchange is unknown. We will demonstrate how models are built and applied to the OU data, but with the goal of demonstrating more general predictive models that can apply in different contexts.

7.3.1 Student Activity Data from the Virtual Learning Environment

The OU is Europe's largest distance learning institution. As such, it has been a forerunner in the use of VLE to deliver module content to students. When a user interacts with a VLE their clicks are logged. The VLE can be configured to capture data at different levels of granularity. At the highest level, each click is counted, but not categorized. At the lowest level, each page access is identified with a time-stamp. In between, it is possible to categorize clicks according to different types of activity, such as access to learning materials, interactive quizzes, or forums.

7.3.2 Demographic Data

Demographic data usually includes the student's age, gender and postcode. Other demographic data might inform of a student's other employment, how their study is funded (e.g. if they are self-funded, receiving financial assistance, or being funded by their employer), if they have declared a disability, their marital status and if they have any dependents. The demographic make-up of a cohort can vary not only between institutions but also from module to module, but is not always easy to predict.

For example, it can be expected that the age group 65+ is more likely to enroll into an art course than into a course of computer hardware. However, a more detailed analysis of the OU data has revealed that in many cases these stereotypes are wrong. It is also possible to assume that members of an older age group will study out of personal interest rather than as a need for their job. This can lead to less importance being placed on taking an exam to achieve a formal qualification. This apparent 'failure' of the module is, in the mind of the student, actually a success.

These differences can be important to consider when developing predictive models. For this reason (and because this type of data can be considered sensitive) the exact nature of demographic findings will not be discussed here. Instead, methods applied using demographic data will be presented in fairly broad terms, but we suggest that when carrying out this type of work it is important to always repeat the process to find the best demographics for predicting outcome in a particular institutional and module context.

7.3.3 Past Study

Past study data indicates the student's past academic achievements. This can include both previous courses and past modules of the current course. For example, if the student has completed the first year of a course, then their outcome on this phase of learning can be included in data when predicting their performance on a 'current' second year of study.

7.3.4 Assessment Data

Most modules include some form of continuous assessment in the form of coursework or exams. Student performance on these assessments for a current module of study can be a good predictor of future performance.

7.4 Feature Selection and Data-Filtering

The different data sources can be combined to provide a set of features for building a classifier. This section describes a process for selecting which features to use in building models. It is important to use a number of features that is appropriate for the size of the datasets that will be used to build the predictive models, otherwise problems such as over-fitting can occur. There are a number of different methods that can be used for feature selection, especially for datasets with a large number of features (e.g. see [7] for a discussion). The methods used for features selection for the different data types available are now discussed.

Of the three types of data, only the demographic data is available for a student when they begin their studies. Analysis of variance (ANOVA) has been used to determine which demographic feature had the strongest effect on the prediction. The ANOVA was repeated for all of the modules. Some demographic information was found to be consistently predictive across modules whereas others, particularly age, varied. The ANOVA was mainly used to confirm knowledge about the predictive capacity of the various demographic features that was already known through past experience of running historical analysis on module data. Due to the relatively low number of demographic features available it was also possible to run correlations to eliminate features that were highly correlated.

The VLE activity data, in the form of clicks, becomes available as soon as the module starts. The VLE clicks were analyzed against success and failure. This analysis revealed that it is possible for students who do not click at all to still pass the course, whereas those who click a lot might fail. It is not possible to predict failure through counting clicks, but rather what is important is the amount of activity between assessment periods, and more crucially how this activity compares to the students own previous activity. Therefore, VLE data can usefully be represented in terms of an activity drop between one period of prediction and another. An additional strong trend that can be seen from analyzing VLE clicking patterns is that students often increase clicking immediately prior to submission of an assessment.

This can clearly be seen looking at data for two modules (known as modules A and C within this chapter), where the peaks of activity correspond with the period of time directly before submission of an assessment was due. Also to be seen is an overall trend of steadily decreasing activity over time (Figs. 7.2 and 7.3). Therefore, the period of prediction, when using VLE data, is most useful when it is conducted in time periods between assessments. This is in contrast to the Purdue Course Signals which, as mentioned previously, defines a measure of *effort* which is a student interaction with the VLE compared to their peers [5, 6].

Analysis of the different data categories within the VLE revealed that there was little benefit to filtering activity types from the data, since the most informative sources of data (the learning activities) were not fine-grained but were dominant. Other activities such as logging into change personal details or to do a blogging activity were too infrequent to impact on the accuracy of results. However, this was due to a large extent on the coarse-grained recording of the different activity types.

Students who did not engage at all with the VLE were filtered from the data set prior to building models, based on the previously mentioned findings that such students could equally pass or fail depending, presumably, on their (unknown) reason for not engaging with the VLE.

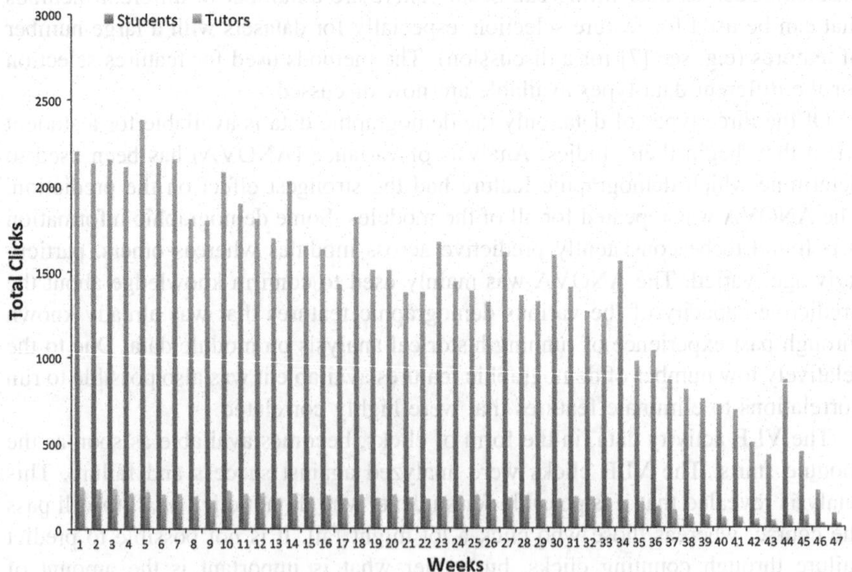

Fig. 7.2 Module A clicks

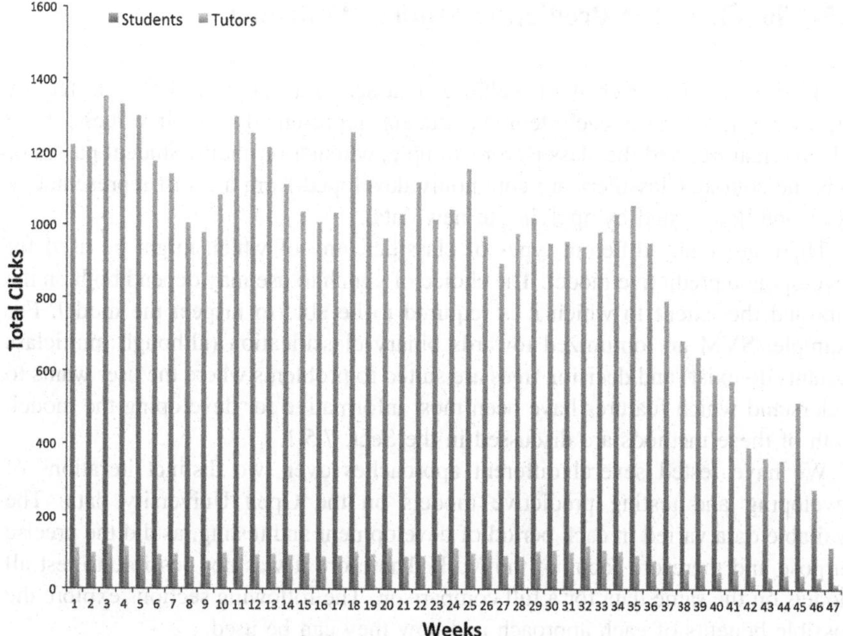

Fig. 7.3 Module C clicks

Assessment data can be either represented according to the assessment score or in terms of a defined threshold (e.g. the boundary between pass and fail). However, this presents a fairly naïve view of the importance of an individual assessment score in assessing a students progress.

This is due to the students having knowledge of how a module pass/fail is calculated, which allows them to apply strategies such as deliberately not submitting or failing assessments under circumstances where they can still achieve a pass on the module (e.g. at the end of the module their lowest assessment score will be omitted from the weighted average score). We later propose methods for addressing this issue. For the work conducted here, a simple model is used to find whether students are dropping below a pass level for an individual assessment at a given point in time.

Finally, once a student passes from one module to the next, there is the possibility to include information about their past study. This is the last data type to become available on the students educational pathway through the institution. This type of data can vary from a relatively simple flag to indicate whether the student is new or continuing (from some sort of previous study)—which in our case, was the level of data readily available. Also of interest might be to obtain assessment scores, final outcomes or even to analyze patterns of behavior during previous modules to determine if the student had used any strategies.

7.5 Classifiers for Predicting Student Outcome

A classifier assigns each item within a data set to a class, based on its feature values. So in this case, each item is a student, represented by their values for the selected features and the class is, for example, whether or not the student passes or fails the course. Classifiers are commonly developed from a set of representative cases and then tested by applying to new data.

There are many different types of classifier, any of which might be used for developing a predictive model. The choice of which to use may depend both on the data and the extent to which it is required to be able to inspect the model. For example, SVM are optimized towards binary classification (although multiclass variants do exist) and decision trees are suited to problems where the user wants to understand which features have been most informative for developing the model. Both of these methods are discussed in the Sect. 7.5.1.

We have tested several different approaches over two distinct iterations of developing and testing predictive models on the Open University data. The available data varied in each period of development and testing, as did the precise purpose and therefore focus of the work. Therefore it was not possible to test all models on the same data for a full comparison. The following sections explore the possible benefits of each approach and how they can be used.

7.5.1 Support Vector Machines and Decision Trees

In a first iteration of model building, SVM (the single binary classifier) and decision trees (specifically, C4.5 which is a version of Quinlan's ID3 [8]) were tested using as features either assessment scores only, VLE clicks only, or a combination of both (but without demographic data). The overall goal of this work was to establish whether it was possible to use the VLE data for predicting the student outcome and to identify a plausible model to use in this task. Models were developed and tested on historical data from three modules (which we will call modules A, B and C) using tenfold cross-validation. These models were implemented using WEKA [9].

Table 7.1 shows how the different modules compared in terms of the number of registered students and the total number of clicks within the VLE. The classes to predict were either *performance drop* (predicting that a student would pass below a threshold value in a future assessment) or *final outcome* (predicting pass/fail of the

Table 7.1 Profiles of the three selected modules for developing and testing models

Module ID	Number of students	Number of VLE clicks
A	4,397	1,570,402
B	1,292	2,750,432
C	2,012	1,218,327

module). These relate to problem specifications 2 and 3 respectively, albeit without including the demographic data. VLE clicks were counted within a time period between assessments. Different windows of time—periods between assessments—were investigated.

Overall, the SVM performed badly on this data set. This corresponds to previous findings such as in [2] and [3]. The decision-tree models, however, were demonstrated to be accurate at predicting both performance drop and final outcome. Figure 7.4 shows some results for predicting performance drop across the three modules.

It can be seen that both precision and recall were very high for Modules B and C. Precision (P) is a measure of the accuracy of the result (is the model accurately predicting failure, or did some of those students actually pass?). Recall (R) is a measure of how many relevant cases are found (were all failing students identified?).

The F-measure considers both precision and recall. Whilst recall was a little lower for Module A, a decent precision was still achieved, in other words, the algorithm was not incorrectly predicting when performance drop would occur, although some cases of performance drop were being missed for Module A. Interestingly, when predicting performance drop, the VLE clicking in the time immediately leading up to the assessment that was being predicted was the most informative, suggesting that a student who used to work with the VLE before but then stopped is likely to show a drop in their performance.

Figures 7.5, 7.6 and 7.7 compare the prediction of final outcome for Modules A, B and C at the time of the third assessment using VLE data, assessment data or both combined. These figures show that assessment data on its own is not very good for predicting final outcome.

VLE data on its own, or combined with assessment data, gave much better results. This depended to some extent on the module: prediction for Module B was marginally better when using only VLE data. The difference in the effect of VLE data on the prediction could be explained by differences in the way that VLE activities were used in different modules. Overall it looks as though the combined VLE and assessment data is best.

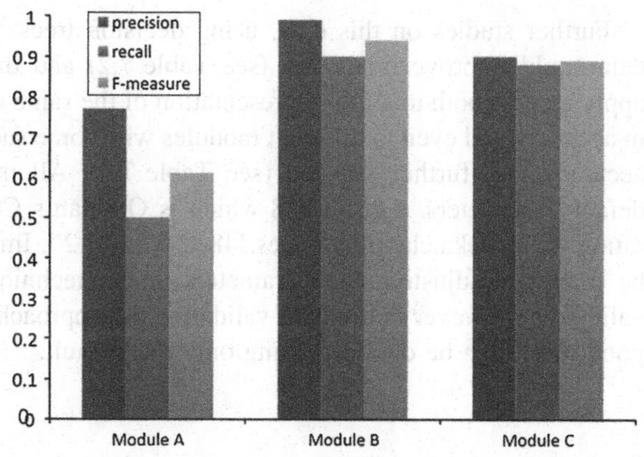

Fig. 7.4 Predicting performance drop, using VLE and TMA (Tutor Marked Assessment) data, with a window size $k = 3$

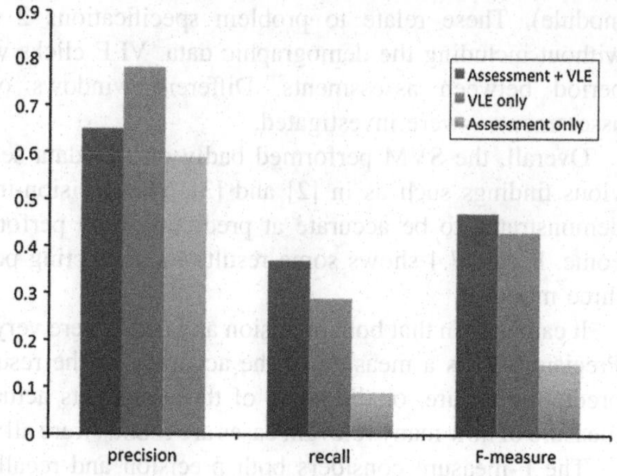

Fig. 7.5 Predicting final outcome at TMA 3 for Module A—comparing TMA, VLE and combined

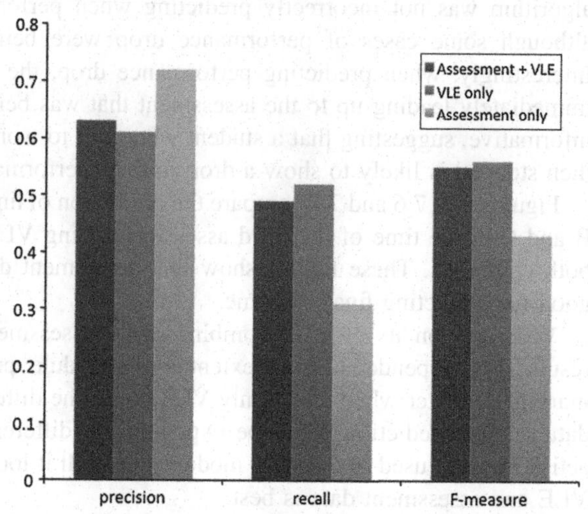

Fig. 7.6 Predicting final outcome at TMA 3 for Module B—comparing TMA, VLE and combined

Further studies on this data, using decision trees, revealed that demographic data could improve prediction (see Table 7.2) and that it was even possible to apply models both to a future presentation of the same module with little reduction in accuracy and even to different modules with some success, although in this case, accuracy was further reduced (see Table 7.3). All models in WEKA used the default parameters, e.g. for J48 which is Quinlan's C4.5 algorithm, the configuration was "weka.classifiers.trees.J48-C 0.25-M 2". Improvements could no doubt be made by adjusting the parameters, using techniques such as nested cross-validation, however in terms of validating the approach it was demonstrated that a good result can be obtained using only the default.

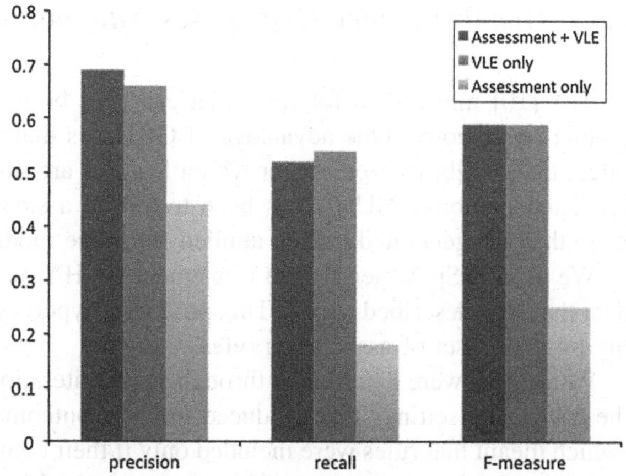

Fig. 7.7 Predicting final outcome at TMA 3 for Module C—comparing TMA, VLE and combined

Table 7.2 Comparing accuracy with and without demographic data

Assessment number	With demographics			Without demographics		
	p	R	F1	p	r	F1
2	0.62	0.23	0.34	0.73	0.21	0.32
3	0.7	0.37	0.49	0.65	0.37	0.47
4	0.7	0.35	0.47	0.74	0.33	0.46

Table 7.3 Applying model across modules

Assessment period		2			3			4		
Trained on	Applied to	P	r	F1	p	r	F1	p	r	F1
Module C	Module C (c-v)[a]	0.90	0.44	0.59	0.69	0.52	0.59	0.38	0.16	0.22
	Module C (all)[b]	0.91	0.44	0.60	0.75	0.59	0.66	0.94	0.29	0.45
	Module A	0.52	0.28	0.37	0.63	0.44	0.52	0.60	0.19	0.29
	Module B	0.84	0.42	0.56	0.72	0.40	0.51	0.95	0.26	0.41
Module A	Module A (c-v)	0.73	0.21	0.32	0.65	0.37	0.47	0.74	0.33	0.46
	Module A (all)	0.85	0.19	0.31	0.78	0.47	0.59	0.87	0.44	0.58
	Module C	0.98	0.17	0.29	0.63	0.63	0.63	0.38	0.65	0.48
	Module B	0.85	0.41	0.55	0.73	0.44	0.55	0.90	0.34	0.50
Module B	Module B (c-v)	0.79	0.54	0.64	0.63	0.49	0.55	0.65	0.49	0.56
	Module B (all)	0.82	0.55	0.66	0.96	0.61	0.74	0.96	0.36	0.52
	Module C	0.40	0.60	0.48	0.23	0.78	0.36	0.22	0.77	0.34
	Module A	0.26	0.86	0.39	0.25	0.43	0.31	0.14	0.91	0.24

[a] 'c-v' refers to tenfold cross-validation setting
[b] 'all' refers to experiments using the same complete data set for training and testing

7.5.2 General Unary Hypotheses Automaton

GUHA [10] mines data for rules that can then be used on unknown samples to predict an outcome. One advantage of GUHA is that, similar to decision trees, it offers the possibility to discover which features are implicated in determining the predicted outcome. GUHA has been tested on a small scale to see whether the rules that are generated can be applied from one module to the next.

We used LISp-Miner [11] to implement GUHA on the same historical module data that was described above. This produced hypotheses about failing students in the form of a set of association rules.

Parameters were determined through several iterations of building and applying the rules. The settings that produced the best outcome was 0.700 for confidence (which meant that rules were included only if their confidence value was a 70 % or higher) and 0.001 for support. The reason for the low value for support is due to the small number of cases to predict within the dataset, in other words there are not that many students within the dataset who will fail therefore requesting a high number of examples to fit a rule for failure will lead to no rules being returned. The maximum number of antecedents for any given rule was set to 5.

The GUHA method confirmed previous findings that click totals alone were not predictive. Therefore, the VLE feature selected for generating hypotheses was the percentage change in users owns clicking behavior, between one assessment and the next. Clicking categories were obtained by dividing VLE data into 8 categories of equal frequency. Assessments were also divided into categories. The model was trained to predict overall module failure.

It was demonstrated that GUHA rules that were learned from the test data of one module was accurate in predicting outcome using data from the same module in its next presentation. As an example, from a module presented first in 2010 (on which GUHA rules were learnt) and then tested on the same module in 2011, a fail in a particular assessment could predict failure with 88 % confidence in 2010 (537 cases) and 83 % confidence in 2011 (516 cases). GUHA produced a more specific version of this rule that included the change in VLE activity between two additional assessments that improved performance to 94 % confidence in both 2010 (472 cases) and 2011 (394 cases). Table 7.4 shows an example of these rules and results.

However, the drawbacks of GUHA were that it took a long time to generate the many plausible hypotheses from the data, and also the time it took to find the optimal parameter setup values and that it was not robust to change in these.

Table 7.4 Example of rules produced by GUHA and applied a cross module presentations

	2010		2011	
	Conf	Supp	Conf	Supp
Tma4(<0;40))	0.88	0.1696	0.83	0.15
Tma4(<0;40)) and Vle6_vle7(<−133;−30)… <0;1))	0.94	0.1578	0.94	0.13

7.5.3 Bayesian Networks and Regression

The next iteration of model building has investigated the use of Bayesian modeling for predicting student outcome, and compared it to other methods, namely linear regression, logistic regression and weighted average score. Of interest was the potential for including the demographic data that wasn't used for the previous models. This is particularly important since—as mentioned previously—at the start of the module this is all that is known about the student. Three historic data sets were used for development and testing.

One of these was Module A, as used in the previous examples. Two alternative modules were chosen for the other data sets. Data was available for 3 presentations of the module for modules A and D, and 2 presentations of module E. Table 7.5 shows enrollment figures for these modules (note that enrollment does not always equate to the number of students who end up registered on and starting the course).

Firstly, the impact of VLE data for modeling students performance is shown on an example of predicting the result of the first assessment, based only on their demographic data alone and then in combination with their interactions with the VLE system in the very first period of the module, i.e. prior to the deadline of the first assessment. The data was taken from Module A.

The example is simplified by considering only 3 demographic variables: gender *S* (female, male), previous experience *X* with the OU courses (new, continuing—n or c) and previous qualification *Q* divided into 5 disjunctive classes q1, q2, q3, q4 and q5. The interactions *V* with the VLE system were also quantified into 5 classes: 0, 1–20, 21–100, 101–800 and more than 801 interactions. These are denoted as v1, v2, …, v5 respectively.

The predicted hypothesis A1 states that the student submits the assessment and achieves a score above 40 points. The goal is to find out who are the students that do not satisfy the hypothesis, i.e. ~A1. Variables S, X, Q and V serve as evidence for proving/disproving the hypothesis.

This goal can be expressed as finding the value of conditional probability P(~A1|S, X, Q, V) for all values of S, X, Q and V. In this simplified problem the joint probability P (A1, S, X, Q, V) must be defined for 2 * 2 * 2 * 5 * 5 = 200 values of arguments which is computationally feasible. For conditional probability the Bayes formula holds, as in Eq. (7.1).

$$P(A1|S,X,Q,V) = \frac{P(S,X,Q,V|A1) * P(A1)}{P(S,X,Q,V)} \tag{7.1}$$

Table 7.5 Enrollment figures for Modules A, D and E

Module	Presentation 1	Presentation 2	Presentation 3	Total
Module A	4,568	3,149	4,253	12,060
Module D	4,100	2,939	3,244	10,283
Module E		555	544	1,099
				23,443

The available data set of a student cohort may include tens of multinomial demographic variables. Moreover, the VLE data can be further classified according to the activities in which the students are engaged. Such problems would not be practically solvable without providing additional domain knowledge. This is typically introduced by specifying which variables are mutually independent [12, 13]. In our simplified example we will use a strong assumption of conditional independence of evidence shown in Eq. (7.2) that would significantly reduce the computational complexity. Equation (7.1) can be rewritten to Eq. (7.3) where Z is a normalization factor.

$$P(S,X,Q,V|A1) = P(S|A1) * P(X|A1) * P(Q|A1) * P(V|A1) \tag{7.2}$$

$$P(A1|S,X,Q,V) = \frac{1}{Z} P(S|A1) * P(X|A1) * P(Q|A1) * P(V|A1) * P(A1) \tag{7.3}$$

The resulting algorithm is called naïve Bayes classifier, shown in Fig. 7.8. Since the assumption in Eq. (7.2) is usually not fully satisfied, the results produced by the naïve Bayes classifier are only an approximation of the Bayes classification according to Eq. (7.1).

As a part of attribute selection for this example, we have tested pairwise statistical dependencies of 10 different demographic variables including those selected in this example. In some cases, tests were parametrised by a third variable. In addition to the selected variables tests included age, address, motivation, employment and others. Out of the total of 45 tested pairs approximately half of them were dependent and this dependence could not be removed by different quantification and clustering of variables. The purpose of this experiment was twofold:

- Comparison of the results of *naïve Bayes classifier* of Eq. (7.3) with the full *Bayes classification* as defined by Eq. (7.1);
- Assess the impact of VLE data on the classification calculated only from demographic data. This evaluation makes it possible to judge to what extent the student results are affected by behavior during the course and whether their activities can "overwrite" their demographic predispositions.

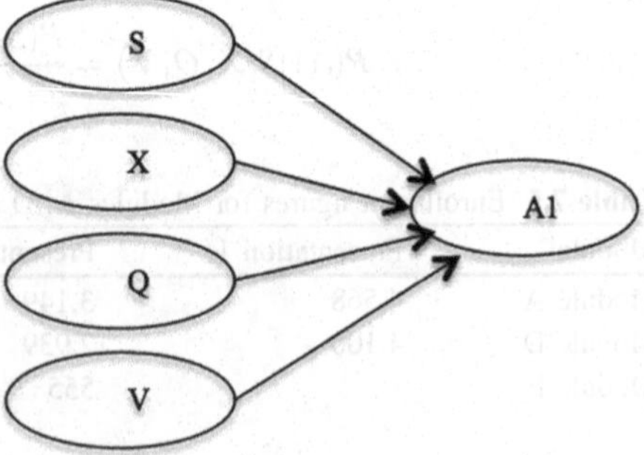

Fig. 7.8 Naïve Bayes classifier

Out of many studies, we describe four cases differentiated by their success probability based purely on demographic variables: one at the lower end, two in the middle of the scale and one at the high end of success scale. As explained earlier, we calculate the probability that the student does not submit the first assessment or submits and scores below 40 points. This obviously does not mean that the student will fail in the module, however it indicates his/her initial problems. The results can be found in Tables 7.6, 7.7, 7.8, 7.9, 7.10, 7.11, 7.12, 7.13.

Case 1 Highest probability of failing based on demographic data
Case 2 Medium high probability of failing based on demographic data
Case 3 Medium low probability of failing based on demographic data
Case 4 Lowest probability of failing based on demographic data

These cases illustrate that what the student does on the module quickly becomes more important to their chance of success than their demographic profile. Next, models were developed and tested for giving monthly predictions of a student's final outcome, using three types of data: demographic, assessment data and VLE activity.

Table 7.6 Only demographic data (case 1)

Demographic only	
Bayes classification P(~A1|S, X, Q)	0.23, i.e. 23 % chance of failing
Naïve Bayes classifier P(~A1|S, X, Q)	0.185, i.e. 18.5 % chance of failing

Table 7.7 Demographic data combined with VLE activities (case 1)

Clicks	Naïve (3) (%)	Full Bayes (1) (%)	Number of students
0	64	75	4
1–20	44	66	3
21–100	26	20	5
101–800	6.3	0	14

Table 7.8 Only demographic data (case 2)

Demographic only	
Bayes classification P(~A1|S, X, Q)	0.068, i.e. 6.8 % chance of failing
Naïve Bayes classifier P(~A1|S, X, Q)	0.077, i.e. 7.7 % chance of failing

Table 7.9 Demographic data combined with VLE activities (case 2)

Clicks	Naïve (3) (%)	Full Bayes (1) (%)	Number of students
0	39	34	4
1–20	22	19	3
21–100	11.2	11.8	5
101–800	2.4	1.5	14

Table 7.10 Only demographic data (case 3)

Demographic only	
Bayes classification P(~A1|S, X, Q)	0.061, i.e. 6.1 % chance of failing
Naïve Bayes classifier P(~A1|S, X, Q)	0.060, i.e. 6.0 % chance of failing

Table 7.11 Demographic data combined with VLE activities (case 3)

Clicks	Naïve (3) (%)	Full Bayes (1) (%)	Number of students
0	33	54	13
1–20	18	24	41
21–100	9	6	101
101–800	1.8	2.3	305

Table 7.12 Only demographic data (case 4)

Demographic only	
Bayes classification P(~A1|S, X, Q)	0.056, i.e. 5.6 % chance of failing
Naïve Bayes classifier P(~A1|S, X, Q)	0.045, i.e. 4.5 % chance of failing

Table 7.13 Demographic data combined with VLE activities (case 4)

Clicks	Naïve (3) (%)	Full Bayes (1) (%)	Number of students
0	26.4	20	10
1–20	14	13	27
21–100	6.5	14.3	56
101–800	1.3	2.6	189

A Bayesian model was built using Microsoft's Infer.NET [14]. A Bayesian network is a directed acyclic graph, with the nodes representing variables and the connections between them representing the conditional dependencies between them. Several versions of the model were developed and tested to determine the nodes and structure. An abstracted example of the final resulting model (masking demographic features) can be seen in Fig. 7.9.

Models for linear and logistic regression were also tested, using the same features that were used within Infer.NET, as well as a simple prediction based solely on grades of previous assessments in the module (the weighted average score). This weighted average score and the linear regression were run using the built in R functions. Logistic regression was implemented using WEKA.

7.6 Evaluation Framework

Results were evaluated by comparing errors in prediction. Figures 7.10 and 7.11 show a comparison of the tested methods and the drop in the error (the proportion of misclassified students) of the predictions over time for Module A and Module D. A benchmark (all unsuccessful) assumes failure.

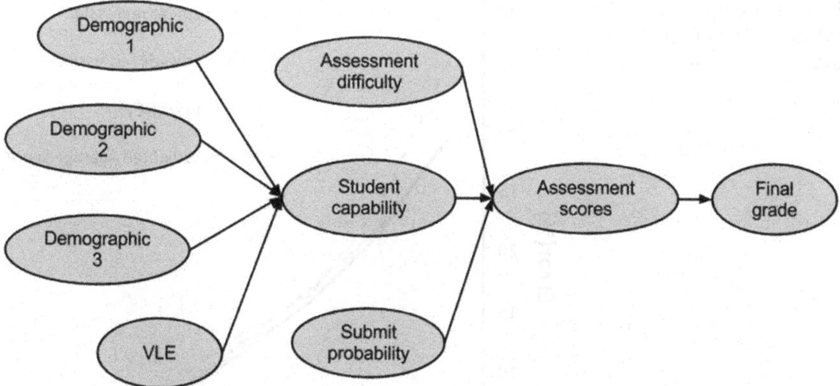

Fig. 7.9 An example of the Bayesian model in Infer.NET

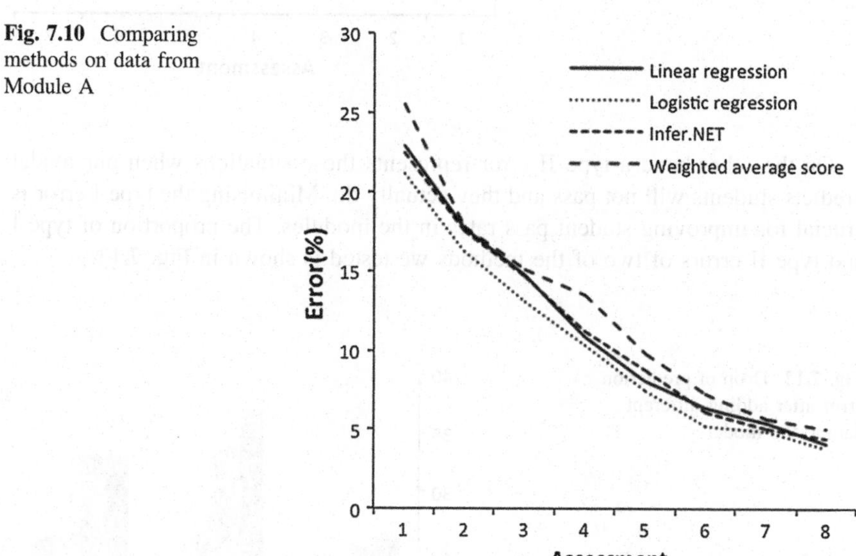

Fig. 7.10 Comparing methods on data from Module A

As these figures show, the accuracy of the predictions was always similar regardless of the method used. Also, accuracy increased with more data used, which suggests that selecting features is more critical for improving the quality of predictions rather than the actual model used. Figure 7.12 shows a comparison of a drop in error after adding demographic information, data from the VLE (from the first month since the start of the module) and scores from the first assignment to the model.

It is also possible to classify the prediction error into two groups of type I and type II errors. A type I error represents the situation when our model predicts that students will pass while they actually don't.

Fig. 7.11 Comparing methods on data from Module D

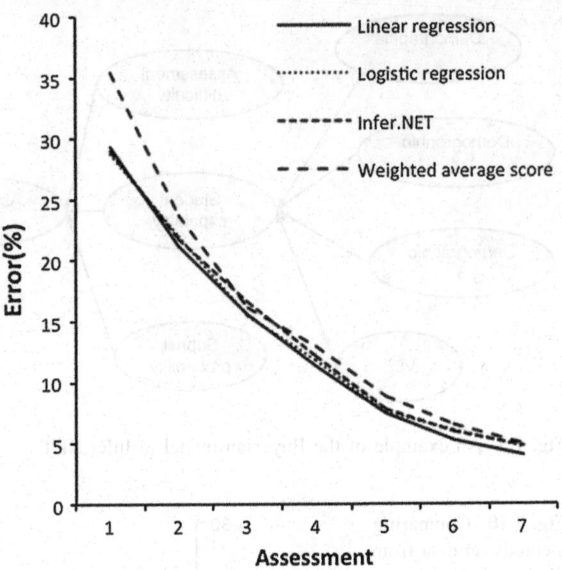

On the other hand a type II error represents those situations when our model predicts students will not pass and they actually do. Minimising the type I error is crucial for improving student pass rates in the modules. The proportion of type I and type II errors of two of the methods we tested is shown in Fig. 7.13.

Fig. 7.12 Drop of prediction error after adding different data to the model

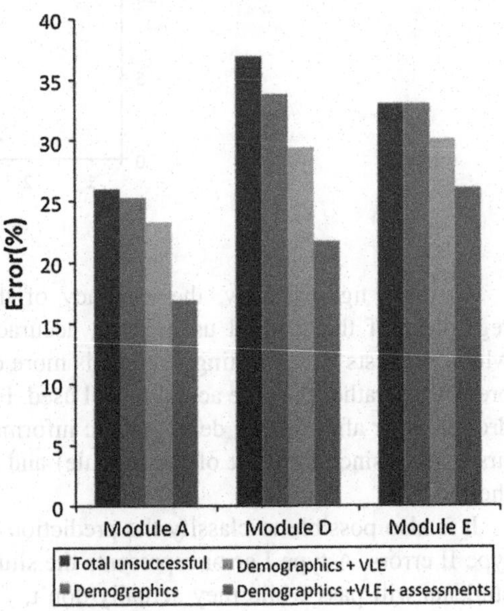

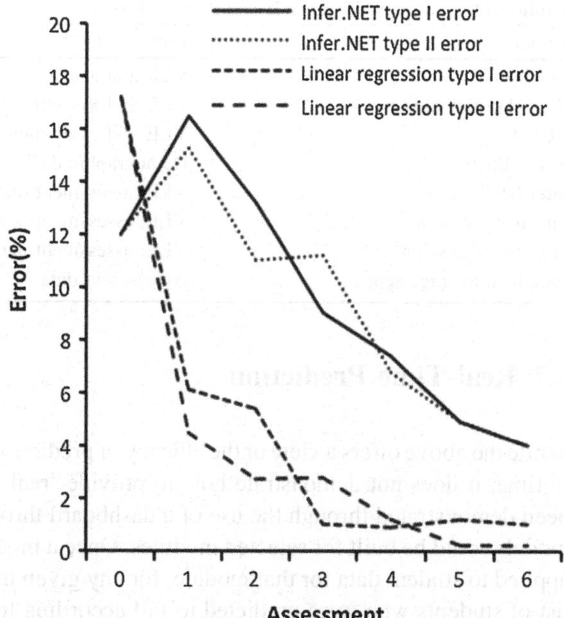

Fig. 7.13 Type I and type II prediction errors on Module D

As before, the models were tested on different presentations of the same module as well as on different modules, with good results. We have seen that subsequent presentations of the same module exhibit very similar patterns. As mentioned previously, our analysis has however shown that different modules exhibit somewhat different patterns and thus focusing on separate modules or group of modules during the design of the models could yield better results.

Overall, the outcome was fairly similar regardless of the method used, what is more important is the selection of features for building the model. Regression methods were much quicker to implement than the Bayes model, which required much more time invested to create the model and would require similar effort to adapt it for every new scenario, such as for different modules or if the existing module structure changed.

However, a benefit of the Bayes model is that the visual nature makes it useful for seeing how variables are related, and it also allows questions about different parts of the model, such as the difficulty of assessments or the ability of each student. Table 7.14 briefly summarises the methods used and the data they were applied to. The demographic data that was considered include education, financial assistance, occupation, ethnicity, continuation, gender, disability and postcode. However, not all of these demographic features were used.

Table 7.14 Summary of methods and data sources

Model	Data applied to
SVM	VLE and assessment data
C4.5 ID3 Decision trees	VLE and assessment data (separately and combined)
GUHA	VLE and assessment data
Naïve Bayes	Demographic data
Infer.NET	VLE, assessment and demographic data
Linear regression	VLE, assessment and demographic data
Logistic regression	VLE, assessment and demographic data
Weighted average score	Assessment data

7.7 Real-Time Prediction

While the above offers a view of the efficacy of prediction methods at different points in time, it does not demonstrate how to provide 'real time' information. This has been demonstrated through the use of a dashboard through which the decision-tree models could be built for selected modules. Once a model was built it could then be applied to student data for that module, for any given moment in time, to produce a list of students who were predicted to fail according to various 'risk' factors. One risk factor was the output of the predictive model. However, other risk factors were also included, such as whether or not they had engaged with the VLE, or if they had failed to submit, or received a below-pass mark score for an assessment.

None of these factors, in themselves, necessarily lead to failure, but when combined can add up to increased risk for an individual student. This information was presented through the dashboard using a traffic light system (Fig. 7.14).

Thus a lecturer could organize a list of students by their risk category (red, amber or green), then view which factors contributed to the risk assessment, as well as an overall view of how many students fell into each risk category. In Fig. 7.14, medium risk students have been selected and the risk factors are shown on the right hand side of the dashboard screenshot. Due to the need for a data warehouse to be fully in place to integrate the required features for building and running modules, the dashboard has not yet been used on 'live' data or used to offer real-time data to tutors from whom they can decide to offer appropriate support to at risk students.

7.8 Revisiting the Problem Specification in Light of Results

We previously described some problem specifications. We will now revisit and expand upon problem specification 4, and also introduce two new ones. These represent the proposed future work, based both on the findings from the described predictive modeling work and also based on information that came to light about differences in module structure, while the work was ongoing.

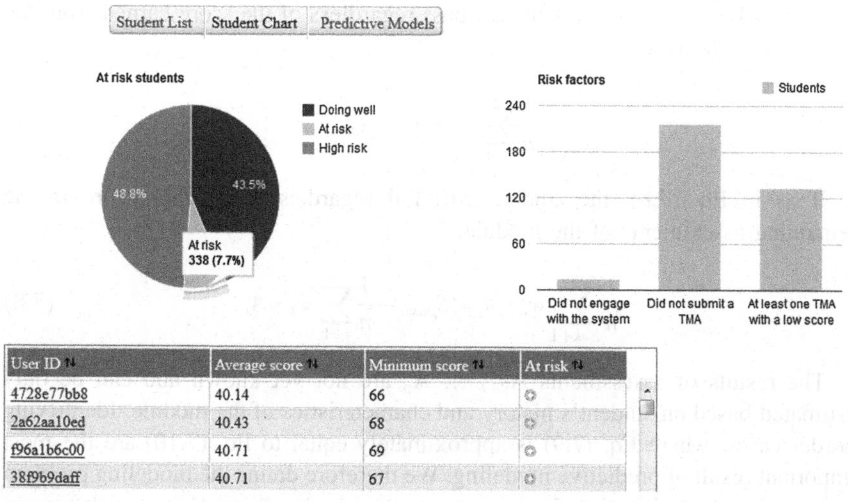

Fig. 7.14 Screenshot of the dashboard

7.8.1 Problem Specification 4 (Revised)

Problem specification 4 (in Sect. 7.2) was as follows:

Input: Demographic data of registered students, scores of assessments A1, …, A_k, and VLE activities up to the assessment A_{k+1}, i.e. Vle_1, …, Vle_k (in Fig. 7.1).

Goal: Calculate probability that the student scores more than a given value in assessment A_{k+1}.

Problem specification 4 has various modifications. For example, values A_i may not represent the real score but could be expressed on a coarse-grained scale. We may define 4 bands: 0–40, 40–60, 60–80, 80–100. If the overall objective is to identify students who are at risk of failing, getting to the band 0–40 indicates that the overall student's score moves towards the failing category. However, this may not always be the case. The pass/fail decision for the module depends on the overall weighted score; this is shown in Eq. (7.4)

$$S(1,n) = \frac{1}{n}\sum_{i=1}^{n} w_i * A_i \tag{7.4}$$

If the student is in the middle of the module, say after assessment A_k, his/her score earn so far is as in Eq. (7.5) and there is still Eq. (7.6) to be earned.

$$S(1,k) = \frac{1}{n}\sum_{i=1}^{k} w_i * A_i \tag{7.5}$$

$$S(k+1,n) = \frac{1}{n}\sum_{i=k+1}^{n} w_i * A_i \tag{7.6}$$

If as in Eq. (7.7), the student will pass regardless of the score earned from the next *n–k* assessments.

$$\frac{1}{n}\sum_{i=1}^{k} w_i * A_i \geq S_{pass} \tag{7.7}$$

If as in Eq. (7.8), the student will fail regardless of his/her score in the remaining assessments of the module.

$$\frac{1}{n}\sum_{i=k+1}^{n} w_i * A_i < S_{pass} - \frac{1}{n}\sum_{i=1}^{n} w_i * A_i \tag{7.8}$$

The results of assessments A_{k+1} … A_n are not yet known and can be only estimated based on student's history and characteristics of the module. Identifying border cases, where Eq. (7.9) is approximately equal to Eq. (7.10) are the most important result of predictive modeling. We therefore define the modeling problem as in problem specification 5.

$$\frac{1}{n}\sum_{i=k+1}^{n} w_i * A_i \tag{7.9}$$

$$S_{pass} - \frac{1}{n}\sum_{i=1}^{n} w_i * A_i \tag{7.10}$$

7.8.2 Problem Specification 5

Input: Demographic data of registered students, scores of assessments A1, …, A_k, weights of assessments w_1, …, w_n, and VLE activities up to the assessment A_{k+1}, i.e. Vle_1, …, Vle_k in Fig. 7.1.

Goal: Identify the subset of students whose score satisfy condition Eq. (7.11) where the score denoted with prime, A_i' is an estimate of scores A_{k+1} … A_n.

$$\frac{1}{n}\sum_{i=k+1}^{n} w_i * A_i' \approx S_{pass} - \frac{1}{n}\sum_{i=1}^{n} w_i * A_i \tag{7.11}$$

As previously discussed, after the k-th assessment there are students who based on their results are guaranteed to pass, those who already cannot satisfy the module conditions and who will fail and those on the border between the pass and the fail. The ultimate purpose of predictive modeling is to identify at risk students and provide them with additional support that would help them to pass.

The problem can be assessed on the bases of costs incurred and saved. Each student who fails incurs loss, expressed by cost C_i for the university (and for themselves). Also each intervention from the university has associated cost *C*.

For example in terms of university staff that intervenes, cost of identifying at risk students etc. This cost $C_{i\tau}$, is justified if the probability of useful and successful intervention, $P(\text{int})$ is high and $C_{\text{int}} < C_{loss} * P(_{i\tau}$. The probability $P(\text{int})$ after assessment A_k, depends on the difference Eq. (7.12) and on the estimate Eq. (7.13). The problem specification that takes into account costs can be defined as in problem specification 6.

$$\Delta_k = S_{pass} - \frac{1}{n}\sum_{i=1}^{n} w_i * A_i \tag{7.12}$$

$$\frac{1}{n}\sum_{i=k+1}^{n} w_i * A_i' \tag{7.13}$$

7.8.3 Problem Specification 6

Input: Demographic data of registered students, scores of assessments A1, …, A_k, weights of assessments $w_1, \ldots, w_n$, and VLE activities up to the assessment A_{k+1}, i.e. $Vle_1, \ldots, Vle_k$ in Fig. 7.1.

Goal: Identify the subset of students whose score from the past assessments and estimate score from the future assessments satisfy condition Eq. (7.14).

$$C_{int} < C_{loss} * P(int) \tag{7.14}$$

Note that for students who satisfy condition Eq. (7.15) probability *P*(*int*) because the intervention is not useful (students pass anyway). Similarly, *P*(*int*). for students who satisfy Eq. (7.16), because no intervention can help them.

$$\frac{1}{n}\sum_{i=1}^{k} w_i * A_i \geq S_{pass} \tag{7.15}$$

$$\frac{1}{n}\sum_{i=k+1}^{n} w_i * A_i < S_{pass} - \frac{1}{n}\sum_{i=1}^{n} w_i * A_i \tag{7.16}$$

7.9 Developing and Testing Models on Open University Data (A Case Study)

The above methods have been developed and tested on data from modules of the OU. Data at the OU was, as in many institutions, distributed across different departments and not held in a data warehouse. Similarly, the knowledge about how to interpret each data source with respect to students and modules was not centralized.

In order to fully understand what was meant by each feature name in each database, and the meaning of each value in the range of values for that feature, required a number of meetings with staff in different departments ranging from the IT department, the department already involved in statistical analysis of different types of student data (mainly historical) and also the teams responsible for the modules that had been chosen for development and testing. This was important since each module is structured differently, albeit within a reasonably well defined range of parameters.

These parameters include the exact criteria for passing the module, as well as the number of assessments that need to be passed and whether it is possible or not to substitute marks from one assessment for another. The process of obtaining and understanding data with respect to individual modules has been very informative for improving the models. This is important when extrapolating from the results obtained and applying them in other contexts, whether the new context is a different Open University module or a course from a different institution.

The OU results can inform some features to investigate when developing a predictive module and it can propose some general findings such as to focus on clicking behavior of different student types (heavy clickers compared to low clickers) but what is always needed is for each institution to combine their own data sources, develop their own understanding of how their data relates to the course that is on offer and also to investigate their own demographic data, since it is not certain that the demographics of OU students will be equivalent to other students, e.g. those doing a MOOC where the motivation, or age of students, might differ.

7.10 Beyond OU: Applying Models on Alternative Data Sources

The trialed methods are appropriate for predicting student performance from data in a distance learning environment, provided that appropriate feature selection and data filtering is carried out. Variations amongst how modules are structured and delivered, as well as differing student demographics for different types of learning, preclude simply taking a set of features from one scenario and plugging it into another.

However, taking into consideration the work described above in conjunction with other cited works for predicting student performance [1–6], it is fair to claim that the overall approach has been demonstrated to yield good results in different domains, when thinking in terms of different types of data to include in the model, instead of the individual features. The biggest difference between institutions is the source of the student activity data, which provides a measure of engagement.

In purely distance learning (including MOOC), student activity data is obtained from VLE clicking, as demonstrated throughout this chapter. In a blended learning environment, a student might spend some time in face to face teaching, such as a lecture or seminar. Students are sometimes even given a choice of which mode of learning to adopt at any time and can switch between one and the other. Therefore, it is not possible to rely purely on VLE data as a measure of engagement, but to also take into account attendance in lectures, etc., assuming that this data is available.

There is also the possibility to extend the models within an institutional context to take into account a students pathway across modules to a final outcome. Care must be taken in being sensitive to differing requirements of modules even within one specific program of study, especially if using VLE data or assessments, where the module structure dictates to a large extent how a student will behave.

7.11 Conclusions

This chapter has defined a set of problem specifications for predicting student performance. A series of models have been developed, tested and compared using data from Open University modules. The key findings are that VLE activity is a useful data source to include for predicting student outcome; however it should not be viewed as an absolute measure of engagement, but rather with reference to a student's own past behavior. Further, feature selection has a big impact on the reliability of a model generated from the data, regardless of which model type is chosen.

Even within the same institution, it has been demonstrated that different modules require different methods, since the module profiles differ in terms of what assessments a student must undertake and how they are required to interact with the VLE. Therefore, a suggested approach for developing predictive models is to first understand and model the different possible module parameters, then to classify modules according to a metadata profile. This would have the effect of simplifying the transferal of a predictive method developed for one module to another.

Acknowledgments We would like to acknowledge the help and support of JISC and the contribution from Microsoft Research.

References

1. Kabra, R.R., Bichkar, R.S.: Performance prediction of engineering students using decision trees. Int. J. Comput. Appl. **36**(11), 8–12 (2011)
2. Baradwaj, B., Pal, S.: Mining educational data to analyze student's performance. Int. J. Adv. Comput. Sci. Appl. **2**(6), 63–69 (2011)
3. Pandey, M., Sharma, V.K.: A decision tree algorithm pertaining to the student performance analysis and prediction. Int. J. Comput. Appl. **61**(13), 1–5 (2013)

4. Baepler, P., Murdoch, C.J.: Academic analytics and data mining in higher education. Int. J. Sch. Teach. Learn. **4**(2), 1–9 (2010)
5. Arnold, K.E., Pistilli, M.D.: Course signals at purdue: using learning analytics to increase student success. In: 2nd International Conference on Learning Analytics and Knowledge, pp. 267–270. ACM, New York (2012)
6. Pistilli, M.D., Arnold, K.E.: Purdue signals: mining real-time academic data to enhance student success. About Campus **15**(3), 22–24 (2010)
7. Peng, H., Long, F., Ding, C.: Feature selection based on mutual information criteria of max-dependency, max-relevance, and min-redundancy. IEEE Trans. Pattern Anal. Mach. Intell. **27**(8), 1226–1238 (2005)
8. Quinlan, J.R.: C4.5: Programs for Machine Learning. Morgan Kaufmann, San Francisco (1993)
9. Hall, M., Frank, E., Holmes, G., Pfahringer, B., Reutemann, P., Witten, I.H.: The WEKA data mining software: an update. SIGKDD Explor. **11**(1), 10–18 (2009)
10. Hájek, P., Holeňa, M., Rauch, J.: The GUHA method and its meaning for data mining. J. Comput. Syst. Sci. **76**(1), 34–48 (2010)
11. Rauch, J.: GUHA method and the LISp-miner system. In: Observational Calculi and Association Rules. Studies of Computational Intelligence, vol. 469, pp. 233–260. Springer, Heidelberg (2013)
12. Koller, D., Friedman, F.: Probabilistic Graphical Models. MIT Press, Cambridge (2009)
13. Bishop, C. M.: A new framework for machine learning. In: Zurada, J.M., Yen, G.G., Wang, J. (eds.) Computational Intelligence: Research Frontiers, IEEE World Congress on Computational Intelligence. LNCS, vol. 5050, pp. 1–24. Springer, Heidelberg (2008)
14. Minka, T., Winn, J., Guiver, J., Knowles, D.: Infer.NET 2.5, Microsoft Research, Cambridge (2012)

Chapter 8
Predicting Learner Answers Correctness Through Eye Movements with Random Forest

Alper Bayazit, Petek Askar and Erdal Cosgun

Abstract The aim of this research is to predict learners' achievement by using a data mining technique: Random Forest (RF). For this purpose, learners eye movements were recorded by an eye-tracker and their answers to questions were collected via an online assessment tool. Online tests were administered to the students and computer interface was divided into two equal parts, which includes web browser and image processing software. Questions were asked through the browser and participants pencil usage (mouse click counts) was recorded by graphic tablet via the software. Results showed that eye metrics and mouse click counts can be used to predict the answer correctness. While mouse click counts were found to be an important factor for predicting answers in questions that require quantitative operations, fixation count and visit duration metrics are found to be important in questions which include visual elements like graphics. Total fixation duration, number of mouse clicks, fixation count and visit duration were found being the most important eye metrics that predict answers in reasoning questions. Results also showed that changing the presentation modality of a question causes changes in relative importance of each eye metric.

Keywords Online assessment · Eye metrics · Random forest · Prediction · Learner answers

---

A. Bayazit (✉)
Department of Computer Education and Instructional Technologies, Hacettepe University, Hacettepe Universitesi, BOTE Bolumu 06800 Beytepe, Ankara, Turkiye
e-mail: alperbay@hacettepe.edu.tr

P. Askar
Faculty of Education, TED University, TED Universitesi, Egitim Fakultesi, Kolej 06420 Ankara, Turkiye
e-mail: petek.askar@tedu.edu.tr

E. Cosgun
Department of Biostatistics, Hacettepe University, Hacettepe Universitesi, Tıp Fakultesi, Biyoistatistik Bolumu, Sihhiye 06100 Ankara, Turkiye
e-mail: erdal.cosgun@hacettepe.edu.tr

A. Peña-Ayala (ed.), *Educational Data Mining*,
Studies in Computational Intelligence 524, DOI: 10.1007/978-3-319-02738-8_8,

Abbreviations

AOI	Area of interest
CART	Classification and regression trees
GA	Genetic algorithm
GRE	Graduate record examinations
LMS	Learning management systems
MCAS	Massachusetts comprehensive assessment system
mtry	Number of descriptors randomly sampled for potential splitting
RF	Random forest
SIS	Student information systems

8.1 Introduction

Assessments are used to determine the knowledge gained by students and to determine if adjustments need to be made to either the teaching or learning process. Online assessment is also used to assess cognitive and practical abilities. In an online assessment environment, students manage cognitive processes as she/he engages in question solving via computer screen.

As an instructor, if we knew the learners' cognitive processes during test taking and we could predict whether she/he answers a question incorrectly then we could give them more accurate feedback. Therefore eye-tracking technology can be used for that kind of information. Because eye movements reveal real-time information about learners' cognitive activities while processing multimedia information [1]. Additionally, fixated zones help researchers and instructors understand students' problem solving strategies [2].

In this study we aimed to predict learners' achievement by using a data mining technique: RF. Data mining can support recommendations that are appropriate for learners' situation, the time they have available, the devices they can access, their current role and their future goals [3]. The adoption of educational data mining by higher education as an analytical and decision making tool is offering new opportunities to exploit the untapped data generated by various student information systems (SIS) and learning management systems (LMS) [4].

Application of data mining algorithms usually has one of two purposes: description and prediction. Prediction attempts to discover relationships between variables, in order to predict the unknown or future values of similar variables [5]. Several researchers of the educational data mining community have used these variables to predict learner performance [6]. Thus, characteristics of successful and at-risk students can be identified [7].

The approach discussed in this chapter contributes to "prediction of users' behaviors and achievements" topic. Therefore, in terms of data mining, we created a predictive model which is based on eye movement metrics and evaluates learners' achievement. The application presented has classification task which uses Random Forest method and Classification and Regression Trees (CART) algorithm.

8.2 Background

8.2.1 Related Work

Pardos et al. [8] applied the ASSISTment online tutoring system to over 600 students. Each student used the system as part of their math classes 1–2 times a month, doing on average over 100+ state-test items, and getting tutored on the ones they got incorrect.

The ASSISTment system has 4 different skill models, each at different grain-size involving 1, 5, 39 or 106 skills. Their goal was to develop a model that will predict whether a student will get correct a given item.

They compared the performance of these models on their ability to predict a student state test score, after the state test was "tagged" with skills for the 4 models. The best fitting model was the 39 skill model, suggesting that using finer-grained skills models is useful to a point.

Each tutoring item, which they call an ASSISTment, is based upon a publicly released Massachusetts Comprehensive Assessment System (MCAS) item to which they added "tutoring." Students get this tutoring, referred to as scaffolding, when they answer an original question incorrectly.

They believe that the ASSISTment system has a better chance of showing the utility of fine grained skill modeling due to the fact that they can ask scaffolding questions that breaks the problem down in into parts. Allowing us to tell if the student answered incorrectly because she/he did not know one skill versus another. They have found good evidence that fine-grained models can produce better tracking of student performance as measured by ability to predict student performance on a state test.

Bidgoli et al. [9] presented an approach to classifying students in order to predict their final grade based on features extracted from logged data in an education web-based system. They design, implement, and evaluate a series of pattern classifiers and compare their performance on an online course dataset. A combination of multiple classifiers leads to a significant improvement in classification performance. Furthermore, by learning an appropriate weighting of the features used via a genetic algorithm (GA), they further improve prediction accuracy.

The GA is demonstrated to successfully improve the accuracy of combined classifier performance, about 10–12 % when comparing to non-GA classifier. This method may be of considerable usefulness in identifying students at risk early, especially in very large classes, and allow the instructor to provide appropriate advising in a timely manner. These studies especially focused on predicting students' performance by web based systems. In our research, eye movements were used for this purpose. In this view, our research will support the other studies and show different data can be used to predict students' performance.

8.2.2 Cognitive Processes

Attention. Studies in cognitive psychology show that, attention does not focus on more than one object at the same time. It is not possible for the cognitive system to process all of the rich inputs from sense organs altogether. Necessary information is selected by the attention and unnecessary one is eliminated in order to avoid the needles occupation of the system. Attention is mind's deliberately foregrounding one of the objects or events that appear simultaneously. Selective attention means focusing on the object which is thought to be more important and keeping other objects in the background when one looks at the spatial area [10].

Distinctive objects are recognized at 50 ms. This is called the "salience effect". According to Treisman, this process realizes in two steps. The first one is the pre-attention stage in which the medium is scanned. The second one is a stage in which the objects are discriminated according to their physical characteristics and coded on cognitive maps [10]. Which element is distinguished initially in the design of questions can be determined by the first fixation time and first fixation point.

According to theories that concern attention and selective attention, first fixation time and first fixation point may vary in different variations of the question. Distinctive features such as color, size, direction or movement should be considered when designing the question in order to highlight significant elements. However, using over-abundant or unnecessary elements may cause distraction. Drawing attention is achieved by underlining the negative auxiliary verb (not) in questions.

In general terms, attention is described as a set of mechanisms which selectively filters the inputs of a system, thereby reducing its work-load. The intense data input received by the eyes is filtered in many levels in the human brain, whereby constructing the visual attention system. The fact that data processing in the brain is limited in terms of physical resources (using oxygen or allocating neurons etc.) may be seen as the primary reason why such a system is needed. Since the selection process of the attention mechanism used for filtering the visual data shapes the information, it represents information more effectively, thus reducing the work-load of the higher cognitive functions [11]. Visual attention mechanism is involved in the following activities [12]:

- Selecting an area of interest.
- Selecting the values and characteristics of interest.
- Controlling the information in the neuron networks that form the visual system.
- Shifting to one selected area from another one in time.

Perception. In short terms, perception can be described as interpreting what we sense. Perceptions may be affected from pre-experiences, prejudices and our former hypotheses. Perceptual space is described as our short-term experiences. Studies on the perceptual space start with the eye because eye is easier to study when compared to all other senses [10]. In the design, perceptual space is the information that users obtain on the screen in an unit of time. When we consider that, reading practices for the perceptual space is limited to 4–5 words at a time; the font size on the screen has a significant effect on the distances between the objects and determining the sizes and positions of the graphical elements used for designing questions. Perception is particularly important for the spatial skill questions asked in GRE-like examinations. It is necessary to determine the effects of differences in the design of such questions on their perception, performance and responding time.

In his Generative Multimedia Learning Theory, Mayer [13] examined the presentation of a single material in different forms (auditory, visual, and motional) and clarified the presentations in learning environments utilizing various cognitive theories. According to Mayer's Spatial Contiguity Principle, when it is required to write explanations about visuals, it is more suitable to write the words on the emphasized places rather than presenting the visual and the text at the same place. Visuals and texts on the screen should be properly lined and associated to help reduce the eye fatigue and accelerate association on the mind. The content should be designed to be presented in full screen in order to avoid any distractions.

Some questions require interpreting table-graphics. Such questions may either contain a graphic, or the relevant values may be presented in tables. Considering the limited capacity of individuals, it is important to determine eye movements which help predict the correct answer in cases where questions are asked only as graphics, tables or texts. According to the Dual Coding theory, visual and auditory data is processed via separate data processing channels [14]. Each data processing channel has a limited capacity and processing data in separate channels is an active cognitive process which is designed to configure proper cognitive models [15].

8.2.3 Eye Movement Data

Tracking eye movements is a technique that examines where and how often an individual looks at and shifts his/her eyes from one point to another in any given moment. Thanks to tracking eye movements, researchers who work on Human–Computer interaction are able to study elements that affect visual data processing and disposability of the system interfaces. It also enables people with eye disorders to use their eye movements as an input device without using a mouse or a keyboard [16]. In this context, two main data, Fixation and Saccade, and their sub-derivatives come into prominence in gathering information about cognitive processes.

Fixation. Fixation point defines which data is processed at a given time. Fixation may be interpreted in different forms depending on the context in which the study is conducted. In an encoding task, dense fixations observed at a single point may indicate that, the relevant area might be important for completing the task. It may also be interpreted as; there might be a difficulty in encoding the relevant area. However, different meanings can be attributed to these data in a visual search. A large number of singular fixations or clusters of fixations indicate perplexities in recognition of the object [17]. According to Goldberg and Kotval [18] long lasting fixations are not as meaningful as fixations that last shorter.

Number of Fixation. It gives the number of fixations on objects which appear on the screen. Excessive number of fixations and fixations that have spread across the screen may be interpreted as; the user is "confused". Increased amount of fixations in a visual search means that, an ineffective search process has been recorded [18].

Fixation Duration. It gives in milliseconds (ms) for how long an object on the screen was processed. Increased fixation duration indicates a difficulty in processing the relevant object.

Number of Fixations Per Area of Interest. An area of interest is an area which has been pre-determined by the researcher. Usually menus, links, pictures and tables of importance are determined as areas of interest. Number of fixations on these areas indicates "the importance that users give" on these areas [16].

Repeat Fixations. Repeated fixation made on a certain object may mean that, the object "couldn't be identified" or the visual was not "seen" or it was "blurry" [18].

First Fixation Duration. It enables to determine the first distinctive element in the design. If a crucial element does not attract the primary attention, it may be interpreted as a problem with respect to utility.

Gaze Duration. It is the total fixation duration on a certain point. It may mean that, the user may have trouble with identifying that object or the object is located somewhere which is faulty in terms of context.

Saccade. Saccades are rapid ocular movements that are made by orienting the eyeballs in order to bring the desired section of the visual area over fovea. These movements can be both deliberate and reflex. Eye saccades may take up from 10 to 100 ms. Eye saccades are considered to be ballistic and stereotype. Stereotype is the repeated occurrence of certain movement templates. In a ballistic saccade, the target point of an eye saccade is predetermined.

The target needs to be determined before ballistic eye saccade begins; however, this target should be selected using low-clarity peripheral vision due to the fact that, the target tends to be out of the fovea [19].

Saccade is usually regarded as an indication to the fact that, focus of attention has been deliberately changed. No coding is made during saccades. Thus, no information is given about the recognition of an object. Nevertheless, repeated saccades and rollbacks may indicate a difficulty in recognizing the object [20]. Saccades that are observed during the reading process are quite small. They are usually made along two or three letters. Increased number of saccades may indicate difficulties in comprehending the text.

Number of Saccade. Increased number of saccades may indicate a poor design or an ineffective search process. Saccade Amplitude shows that, saccade occurs instantaneously onto a distant object that appeared afterwards and attracted attention all of a sudden. Regressive Saccades mean that, an object is slightly identified [18].

8.2.4 Random Forest

In last decades researchers prefer to investigate the non-linear relations between variables rather than descriptive statistics. Therefore we have used RF, one of the most popular and accurate method to predict the correctness of questions. It has been scientifically proved that RF has a highest accuracy among the tree-based methods (CART, CHAID, ID3) [21]. In this study, we prefer to use RF as a feature selector. The major reason for using the RF is that it gives the importance of variables.

This is very useful for instructors to understand the reasons of right, wrong, blank answers and have important information for the cognitive processes during question solving. Another reason is RF used bootstrap selection for variables and samples. This means: every step of iteration RF algorithm used all combinations of variables and samples. With this way our bias on predictive problems will be smaller than classical statistical methods. On the other hand, RF has invented as a Machine Learning method. It means researchers can use this method in small data sets for their future projects. They will create a base model and select important variables for their predictions. We have used "Random Forest" package from R.2.15.0 version.

Random Forests are a combination of tree predictors such that each tree depends on the values of a random vector sampled independently and with the same distribution for all trees in the forest. The generalization error for forests converges to a limit as the number of trees in the forest becomes large. The generalization error of a forest of tree classifiers depends on the strength of the individual trees in the forest and the correlation between them. The algorithm is shown as follows an illustrated in Fig. 8.1.

```
1: Draw a bootstrap sample from the data. Call those not in
   the bootstrap sample the ''out-of-bag'' data.
2: Grow a ''random'' tree, where at each node, the best split
   is chosen among m randomly selected variables. The tree is
   grown to maximum size and not pruned back.
3: Use the tree to predict out-of-bag data.
```

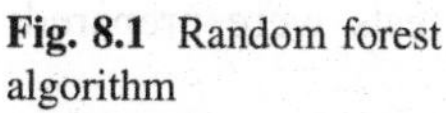

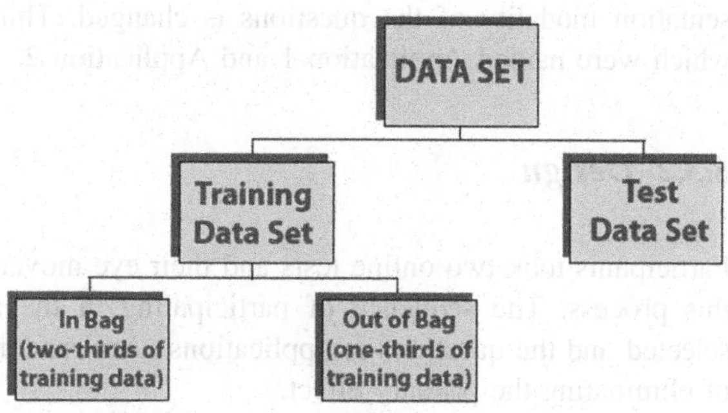

Fig. 8.1 Random forest algorithm

```
4: Use the predictions on out-of-bag data to form majority
   votes.
5: Repeat, N times and collected an ensemble of N trees.
   Prediction of test data is done by majority votes from
   predictions from the ensemble of trees.
```

[Random forest algorithm]

Variable Importance on RF Models. RF determines the relative importance of each sample, through various methods, such as calculation of the Gini Index, which assesses the importance of the variable and carries out accurate online assessment system's variable selection. In this study, variable importance is also used to investigate whether or not the changes in question presentation modality effects to the importance of each eye metric.

RF creates many decision trees. The most important advantage of creating high number of tree is finding an importance of all variables. RF uses two stage internal validation. First of all it divides data into training (80 %, 46 individuals in this work) and test set (20 %, 11 individuals in this work) then creates two data set from training set (inBag-2/3, out-of bag-1/3).

Therefore users do not have to do extra cross-validation analysis [21]. For each variable in the tree, permute the variables values and compute the out-of-bag error, compare to the original out-of-bag error, the increase is an indication of the variable importance. According to this indication variable importance means that average increase in out-of-bag error over all trees and assuming a normal distribution of the increase among the trees, determines an associated p value [21].

8.3 Method

8.3.1 The Purpose of the Study

The purpose of this study is to determine which eye movement metrics can be used for predicting the answers given to questions and the level of their contribution to predict the results. The study also aims to determine whether there are any changes in eye-metrics which are used for predicting the correct answer when the presentation modality of the questions is changed. Thus, two tests were prepared which were named Application-1 and Application-2.

8.3.2 Design

Participants took two online tests and their eye movements were recorded during this process. The sequence of participating in the applications was randomly selected and the questions in applications were randomly aligned for the purpose of eliminating the primary effect.

8.3.3 Pre-Application

Pre-application was conducted on 6 research assistants in order to specify problems that may arise during the application. Following arrangements were included as per the feedbacks obtained:

- Decreasing the number of questions for each session.
- Running tests on local network in case there are any glitches with the internet connection.
- Eliminating the time limitation in tests.
- Performing a pilot test with four questions before the exam in order to familiarize students with answering questions using graphic tablet and the online testing tool.

8.3.4 Application

Online tests were applied on 57 students between April and August 2011. Each test consisted of 6 Turkish GRE questions and a minimum of 4-week period was placed between the tests thereby eliminate recalling that may occur while answering similar questions.

All of the questions were asked to Turkish students in Turkish language. 28 participants took the Application-1 test and the remaining 29 took the Application-2 test which consisted of the same questions asked in a different presentation modality. Which participants would take the tests first was determined randomly. All of the participants were informed about the application and purpose of the study prior to the application.

Participants were familiarized with the graphic tablet and pen via a 4-question pilot test which was conducted before the application. User interface of the testing tool was also introduced. Participants were specifically mentioned that, there was no time limitation in the exam but they were supposed to answer the questions correctly as soon as possible. They were also instructed that, they could omit a question if they did not feel like answering it or get tired but they wouldn't be able to return to that question later on. Before the application, participants were asked to take any sitting position with which they felt comfortable and keep that position as long as possible in order to maintain calibration.

8.3.5 Study Group

A total of 57 postgraduate and undergraduate volunteers at Hacettepe University participated in the study. Some of them were seniors who were preparing for Turkish GRE and studying at the Department of Computer Education and Instructional Technology. Other students were studying at various other departments that accept students based on their quantitative and equal-weighted scores at the student placement examination. Participants were informed about the goal of the study and what they were expected to do during the data collection process; and they were accepted for the application by appointment.

8.3.6 Data Collection Instruments

Application–1 and Application–2 Tests. In order to reach the purpose of the study, 12 Turkish GRE questions which had been asked between the years 2004–2010 were selected. Research questions were also considered while selecting the aforementioned questions. In the end, a set of questions were determined which required equations, logical reasoning and spatial skills. The way these questions presented was altered and a parallel test was included.

Application–1, which consisted of the exact same questions that had been asked in the examinations, and Application–2, which consisted of the same questions asked in a different presentation modality, were transferred to online testing tool and they were presented via an internet browser. These questions are given in supplementary.

After altering the way the questions were presented, two experts who graduated in Mathematics Education were consulted about the equivalency of the questions. Required amendments were made on the questions based on the experts' opinions.

Online Assessment Tool. The online assessment tool was developed by the researcher. The tool was written in PHP programming language and it uses MySQL database. It also enables adding images to questions and options. The tool records answers given by each participant and reports the number of correct and incorrect answers at the end of the exam.

Time limitation feature of the testing tool was not used in order to gather more accurate data. Participants were not allowed to change their answers and they were instructed to pass to the next question after answering one. Correct answers were scored 1 point, wrong or omitted answers were scored 0 point.

Eye tracker. Tobii T70 series eye tracker device was used in the study in order to record eye movements that were made by participants while answering questions in the online testing tool. Eye metrics that were recorded by the eye tracker for the purpose of answering research questions are as follows [22]

- Total fixation duration
- First fixation duration
- Fixation count
- Visit duration
- Mouse click count
- Time from first fixation to next mouse click.

Total Fixation Duration–seconds. This metric measures the sum of the duration for all fixations within an Area of Interest (or within all AOIs belonging to an AOI group), thus the value used to calculate descriptive statistics is based on the number of recordings. If during the recording, the participant returns to the same media element then the new fixations on the Area of Interest (AOI) are included in the calculations of the metric. If at the end of the recording, the participant has not fixated on the AOI, the Total Fixation Duration is not computed and the recording will not be included in the descriptive statistics calculations.

First Fixation Duration–seconds. This metric measures the duration of the first fixation on an Area of Interest. When using AOI groups, the measured fixation corresponds to the first fixation on any of the AOIs belonging to the group. If at the end of the recording, the participant has not fixated on the AOI, the First Fixation Duration value is not be computed and that participant will thus not be included in the descriptive statistics calculations.

Fixation Count–count. This metric measures the number of times the participant fixates on an AOI or an AOI group. If during the recording the participant leaves and returns to the same media element, then the new fixations on the media will be included in the calculations of the metric. If at the end of the recording the participant has not fixated on the AOI, the Fixation Count value will not be computed and the recording will not be included in the descriptive statistics calculations.

Visit Duration–seconds. This metric measures the duration of each individual visit within an AOI (or AOI group). The value used to calculate descriptive statistics is based on the number of visits. A visit is defined as the interval of time between the first fixation on the AOI and the next fixation outside the AOI.

If during the recording the participant returns to the same media element then the new fixations on the AOI will be included in the calculations of the metric. If at the end of the recording the participant has not fixated on the AOI, the Visit Duration value will not be computed and thus that recording will not be included in the descriptive statistics calculations.

Mouse Click Count–count. This metric measures the number of times the participant left-clicks with the mouse on an AOI or an AOI group. If during the recording the participant leaves and returns to the same media element, then the new Mouse clicks on the media will be included in the calculations of the metric. If at the end of the recording the participant has not clicked on the AOI, the Mouse Click Count value will not be computed and the recording will not be included in the descriptive statistics calculations.

Time from First Fixation to Next Mouse Click–seconds. This metric measure how long it takes before a participant left-clicks with the mouse on an AOI or AOI group once he/she has fixated on it. The time measurement starts when the participant fixates on the AOI for the first time and stops when the participant clicks on the same AOI.

If during the recording the participant views a different media between the first fixation and the Mouse click, the time spent viewing the other media element is excluded from the calculations. If at the end of the recording the participant has not fixated and clicked on the AOI, the Time from First Fixation to First Mouse

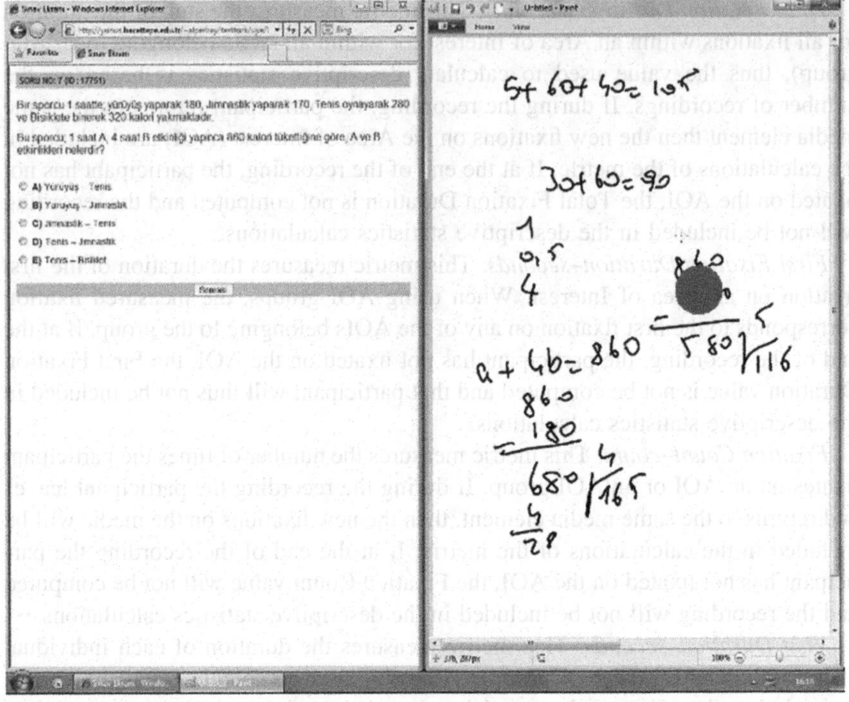

Fig. 8.2 Screen image for the question and solution area

Click value will not be computed and that recording will thus not be included in the descriptive statistics calculations.

Participants used graphic tablet and pen rather than paper-and-pencil for the purpose of monitoring their answering processes. A screen image for an exemplary question is shown below. As shown in Fig. 8.2, the computer screen was divided into 2 equal sections and on the left side of the screen, there is a browser window containing questions in the exam; and on the right side of the screen, there is an image processing software used for pencil usage.

8.4 Analyses of Results

We have used Random Forest method (RF) to predict the answer correctness. We have created 100 trees for this analysis. The one tuning parameter for RF is "*mtry*" which is the number of descriptors randomly sampled for potential splitting at each node during tree induction. This parameter can range from 1 to p (the number of predictors). We have used −p/3 = 2— as recommended for classification and regression problems [23].

In this study we have 2 class (Correct–Wrong answers) and 9 variables for RF classification model. While creating the RF model we used 6 variables. 3 questions were excluded because of having an unbalanced class distribution. They have only one or two correct class. This problem decreased our model performance. RF is not good at un-balanced databases. These questions exclude at the data pre-processing step [24].

Since our aim was to find the reason of correct-false answers, variable importance is the key point at this study. We applied RF to 6 variables and sort them according to their classification model. Variable importance has a relative scale. Most important variable gets "1.0", least important variable gets "0.0". "Time from First Fixation to Next Mouse Click" is the top variable for RF model.

Question-1 (Table 8.1) is the text version of the equation question and requires a pencil usage for the calculations. When random forest method was applied, time from first fixation to next mouse click, mouse click count and fixation count metrics were found to be the most important variables for predicting whether participants would answer this question correctly, incorrectly or whether they would omit it. Total fixation duration however, has the least significance in predicting the correct answer. Accuracy percentage for all the metrics in predicting the answer is 59 %.

Question-2 (Table 8.2) is the graphical version of Question-1. The presence of a graphic and the increased amount of visual materials in the question caused the visit duration and total fixation duration metrics to have higher importance in predicting the answer. Since the question also requires forming equations, mouse click count has 61 % significance in predicting the answer, which indicates a regression compared to text version of the question having 78 % significance.

"Time from first fixation to next mouse click" metric, which was the most important variable in the textual presentation modality, has been the least

Table 8.1 Question-1 analysis results for the random forest method

Eye-metrics	Variable importance
Time from first fixation to next mouse click	1.00
Mouse click count	0.78
Fixation count	0.78
Visit duration	0.59
First fixation duration	0.58
Total fixation duration	0.31

Table 8.2 Question-2 analysis results for the random forest method

Eye-metrics	Variable importance
Visit duration	1.00
Total fixation duration	0.91
Mouse click count	0.61
First fixation duration	0.53
Fixation count	0.42
Time from first fixation to next mouse click	0.38

important variable in the graphical presentation modality in predicting the accuracy of the question. Accuracy percentage for all the metrics in predicting participants' answer is 68 %.

Question-3 (Table 8.3), which contains a table also requires calculations and forming equations. Since it requires mathematical transactions, mouse click count metric is important in predicting the answer. Similarly, time from first fixation to next mouse click has a greater significance along with the visit count. First fixation duration and fixation count variables have primary significance in this table question in predicting, whether the answer is correct or incorrect. Accuracy percentage for all the metrics in predicting the answer is 58 %.

Question-4 (Table 8.4) represents the graphical modality of Question-3, which has a table modality. The values in the table are asked here graphically. It is seen that, mouse click count metric has a higher significance.

From this point of view, it can be considered that, participants did not need to note the values in the table, yet they did note the values in the graphic. Time from first fixation to next mouse click and fixation count variables is also important in predicting the accuracy of the question. However, visit count, first fixation duration and total fixation duration variables are not very important in predicting the accuracy of the question. Accuracy percentage for all the metrics in predicting the accuracy is 68 %.

Question-5 (Table 8.5) has table presentation modality. It requires participants to use the values in the table and form equations. It can be answered by using a pen, or by eye movements or by moving from the options. In this question, visit duration, first fixation duration and time from first fixation to next mouse click are the most important variables in predicting whether participants would answer it correctly, incorrectly or whether they would omit it. Mouse click count and

Table 8.3 Question-3 analysis results for the random forest method

Eye-metrics	Variable importance
Time from first fixation to next mouse click	1.00
Mouse click count	0.68
Visit duration	0.54
Total fixation duration	0.49
Fixation count	0.45
First fixation duration	0.26

Table 8.4 Question-4 analysis results for the random forest method

Eye-metrics	Variable importance
Mouse click count	1.00
Time from first fixation to next mouse click	0.66
Fixation count	0.50
Total fixation duration	0.37
First fixation duration	0.29
Visit duration	0.29

Table 8.5 Question-5 analysis results for the random forest method

Eye-metrics	Variable importance
Visit duration	1.00
First fixation duration	0.99
Time from first fixation to next mouse click	0.77
Total fixation duration	0.56
Fixation count	0.44
Mouse click count	0.32

fixation count are not as important as the other metrics. Accuracy percentage for all the metrics in predicting the accuracy of the answer is 57 %.

Question-6 (Table 8.6) has a graphical modality and it requires reading values in the graphic and calculating their percentages. Since it requires calculations, mouse click count is the most important variable in predicting the answer. Other important variables are total fixation duration, time from first fixation to next mouse click and visit duration. Fixation count metric is not very significant in this question in predicting the answer. Accuracy percentage for all the metrics in predicting the answer is 71 %.

Question-7 (Table 8.7) is a field measurement question. Whereas total fixation duration, fixation count and mouse click count are significant in this question in predicting the answer, visit duration and first fixation duration are not that significant. Given that it is impossible to make drawings on the question using a pen, some participants tried to answer the question by drawing it in the solution area. The ones who could not make drawings tried to calculate the area of the triangle by using eye movements. This indicates that, first the fixation duration and fixation count and then, the mouse click count should be important in predicting the

Table 8.6 Question-6 analysis results for the random forest method

Eye-metrics	Variable importance
Mouse click count	1.00
Total fixation duration	0.87
Time from first fixation to next mouse click	0.81
Visit duration	0.54
First fixation duration	0.47
Fixation count	0.32

Table 8.7 Question-7 analysis results for the random forest method

Eye-metrics	Variable importance
Total fixation duration	1.00
Fixation count	0.75
Mouse click count	0.63
Time from first fixation to next mouse click	0.61
First fixation duration	0.56
Visit duration	0.34

Table 8.8 Question-8 analysis results for the random forest method

Eye-metrics	Variable importance
Fixation count	1.00
Visit duration	0.67
Total fixation duration	0.62
Mouse click count	0.61
First fixation duration	0.52
Time from first fixation to next mouse click	0.52

Table 8.9 Question-9 analysis results for the random forest method

Eye-metrics	Variable importance
Mouse click count	1.00
Time from first fixation to next mouse click	0.91
Visit duration	0.76
Fixation count	0.70
Total fixation duration	0.58
First fixation duration	0.55

accuracy of the answer. Percentage for all the metrics in predicting the accuracy of the answer is 68 %.

Question-8 (Table 8.8) is a spatial skill question. It also requires mental rotation skills. A pen is not necessarily required to answer the question. It is possible to reach the answer using eye movements in order to determine how many times shape two is repeated in shape one. The analyses show that, the most important variables in predicting the answer are fixation count, visit duration, total fixation duration and mouse click count. First fixation duration and time from first fixation to next mouse click metrics are also important, but not as important as the others. Accuracy percentage for all the metrics in predicting the answer correctly is 66 %.

Question-9 (Table 8.9) is the same as Question-8 with inverted background and figure colors. It is a spatial skill question and the entire metrics are important in predicting the answer, just as in Question-8. Only this time, their order of importance is different. The most important variables in predicting the answer in this question are mouse click count, time from first fixation to next mouse click, visit duration and fixation count. The significance of the first fixation duration metric in predicting the answer is very similar with what it was Question-8. Accuracy percentage for all the metrics in predicting the answer is 61 %.

8.5 Conclusion and Discussion

The purpose of this study is to determine which eye movement metrics can be used for predicting the answers given to questions and whether these metrics change depending on the presentation modality. To that end, random forest method was used.

According to the obtained data, first fixation duration, total fixation duration, fixation count, visit duration, mouse click count and time from first fixation to mouse click are the metrics which could be used for predicting whether students' answers would be correct or incorrect.

Analyzed questions also include question pairs whose presentation modality has been changed. When these pairs were examined within themselves and when questions with the same presentation modality were examined, it was seen that metrics which were used for predictions changed from one question to another; and in some cases, similar metrics could be used for predictions.

Time from first fixation to mouse click, mouse click count and fixation count metrics predicted the answer in a textual question. On the other hand, total fixation duration and mouse click count metrics are the most important ones in a graphical question for predicting the answer. In a question with a table however, time from first fixation to mouse click, visit duration and total fixation duration metrics are the most significant ones for predicting whether the question will be answered correctly or incorrectly. When questions are examined in general, it can be claimed that, mouse click count metrics become prominent in questions requiring quantitative operations whereas fixation count or visit duration metrics come to the fore in questions containing visual elements like graphics.

Changing a question's presentation modality caused a change in metrics which were used for predicting the answer. For instance, whereas the time from first fixation to mouse click was the most important metric, it turned into the least important one when an equation question in textual modality was transformed into a graphical question. Likewise, whereas visit duration had little importance in predicting the answer, it became the most critical metric for predicting the answer in graphical modality. Nevertheless, there is no difference in metrics used for predicting the answer when a table question is transformed into graphical modality.

According to random forest analysis made on a table question and a parallel question which was transformed into graphical modality, the most important metrics in predicting the answer are the time from first fixation to mouse click and mouse click count. However, the significance of the visit duration has decreased while the significance of the fixation count has increased. In spatial skill questions, changing the presentation modality did not significantly affect the metrics used for predicting the correct answer. When the background and object colors were inverted in a spatial skill question, eye movement metrics had the same degree of significance in predicting the answer. In both presentation modalities, the most important metrics in predicting the answer are fixation count, visit duration and mouse click count. The least important metric is the first fixation duration.

Our application is very important for educational data mining field. Because in last decades researchers prefer to investigate the non-linear relations between variables rather than descriptive statistics. Therefore, we have used RF, one of the most popular and accurate method to predict the correctness of answers. The major reason for using the RF is that it gives the importance of variables. Another reason is RF used bootstrap selection for variables and samples. This means: every step of iteration RF algorithm used all combinations of variables and samples. With this way our bias on predictive problems will be smaller than classical statistical methods.

This approach extends to several educational systems (ES), such as: e-learning systems, intelligent tutoring systems and online quizzes. Students' eye movement behaviors can be recorded during the interaction with formative assessment part of given ES. The application of a data mining technique (Random Forest) assists to educational designers, teachers and students. A real-time feedback can be given to the students while they are interacting with the questions on the screen.

A notice can be given to educational designers and teachers by the system. So they can have information about students' cognitive processes and behaviors to make the questions more accurate by changing their presentation modalities.

8.6 Future Work

In this study we used random forest method to predict the correctness of answers. Other data mining techniques like support vector machine, naive Bayes and relevance vector machine can be used to analyze the eye metrics. We did not use the demographic variables like gender, age and educational level. Investigating their correlations with the eye-metrics by using different data mining techniques can help us to find the complex patterns and hidden interactions during the assessment process.

A.1 Appendix: Supplementary

A.1.1 Question-1

There are X, Y, Z, V sub tests in a 90-question test. Test X has 45, test Y has 19, test Z has 14 and test V has 12 questions. A student's score in this exam is calculated as follows:

- Each correct answer adds 2 points in test X, 1.5 points in test Y, 1 point in test Z and 0.5 point in test V.
- No point is reduced for incorrect or omitted questions.

According to data: A student who answers all of the questions in test V correctly, has 6 times more correct answers in test Y than in test X.

Now that this student has 38 correct answers in total and his score in the exam is 44, what is the number of correct answers he has in test Z?

A) 3
B) 4
C) 5
D) 6
E) 7

A.1.2 Question-2

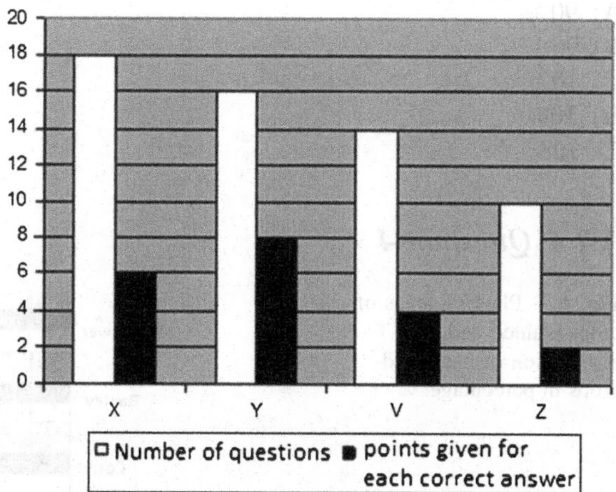

Fig. A.1 Number of questions in *X*, *Y*, *V*, *Z* tests and the points given for each correct answer

The Fig. A.1 shows the number of questions in 4 sub tests and the points given for each correct answer. No point is reduced for incorrect or omitted questions.

According to data: A student who answers all of the questions in test V correctly, has 3 times more correct answers in test Y than in test X.

Now that this student has 30 correct answers in total and his score in the exam is 132, what is the number of correct answers he has in test Z?

A) 2
B) 4
C) 6
D) 8
E) 10

A.1.3 Question-3

Table A.1 Planting areas of crops planted and the distribution of harvested crops in percentage

Crop	Planting area (%)	Harvested crop (%)
Wheat	60	50
Corn	25	30
Barley	10	8
Sunflower	5	12

The Table A.1 gives the planting areas of crops planted in a 1,200-decare land and the distribution of 600 tons of harvested crops in percentage.

According to this, how many tons greater is the harvested corn than the harvested sunflower?

A) 90
B) 96
C) 102
D) 104
E) 108

A.1.4 Question-4

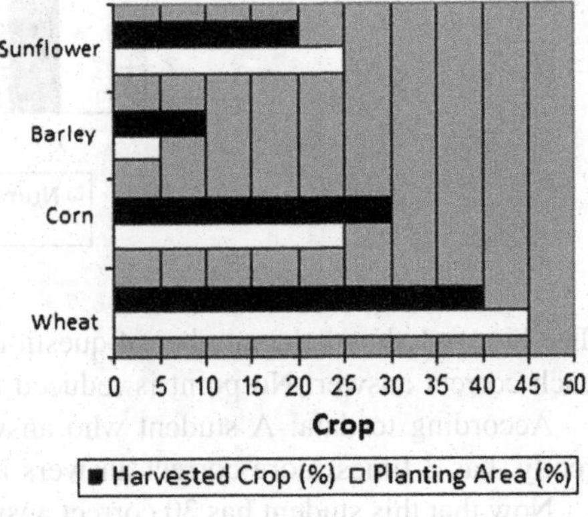

Fig. A.2 Planting areas of crops planted and the distribution of harvested crops in percentage

The Fig. A.2 gives the planting areas of crops planted in a 1,200-decare land and the distribution of 600 tons of harvested crops in percentage.

According to this, how many tons greater is the harvested corn than the harvested sunflower?

A) 15
B) 30
C) 45
D) 60
E) 75

A.1.5 Question-5

The Table A.2 shows some activities performed by a sportsman and the amount of calories he burned after an hour of performing these activities.

Table A.2 Activities and the amount of burned calories

Activity	Burned calorie
Walking	180
Gymnastics	170
Tennis	280
Biking	320

Now that this sportsman burns 1,130 calories by performing 1 h of activity A, 5 h of activity B; what are activities A and B?

A) Walking—Tennis
B) Walking—Gymnastics
C) Gymnastics—Tennis
D) Tennis—Gymnastics
E) Tennis—Biking

A.1.6 Question-6

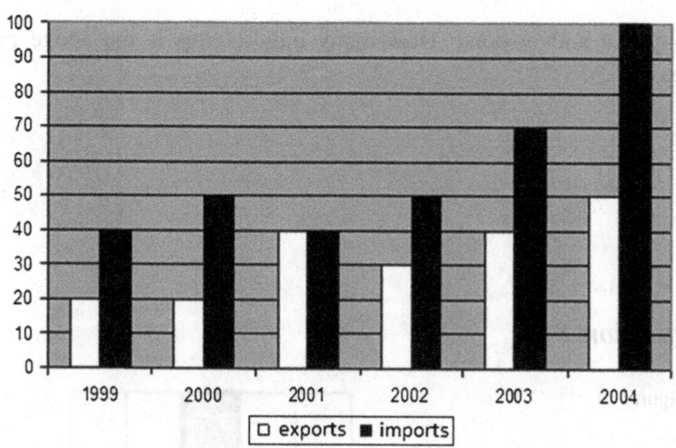

Fig. A.3 Amounts of a country's imports and exports from 1999 to 2004

The Fig. A.3 shows the amounts of a country's imports and exports from 1999 to 2004. How many percent is this country's exports in 2004 constitute of its total exports for six years?

A) 25
B) 28.5
C) 30.5
D) 35
E) 50

A.1.7 Question-7

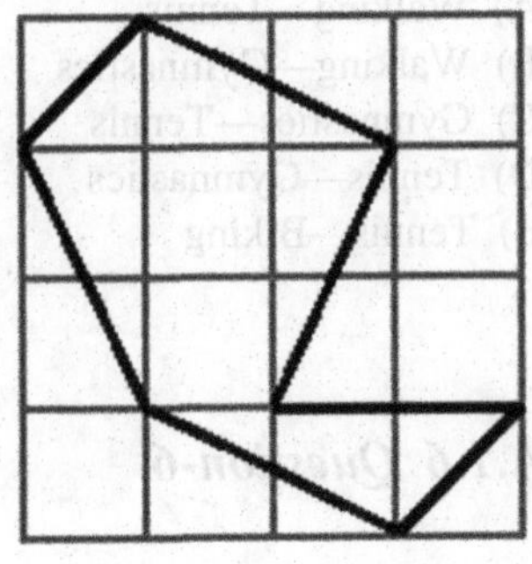

Fig. A.4 Polygonal zone in question-7

See the Fig. A.4 and respond: How many unit squares is the above polygonal zone's area?

A) 6
B) 6.5
C) 7
D) 7.5
E) 8

A.1.8 Question-8

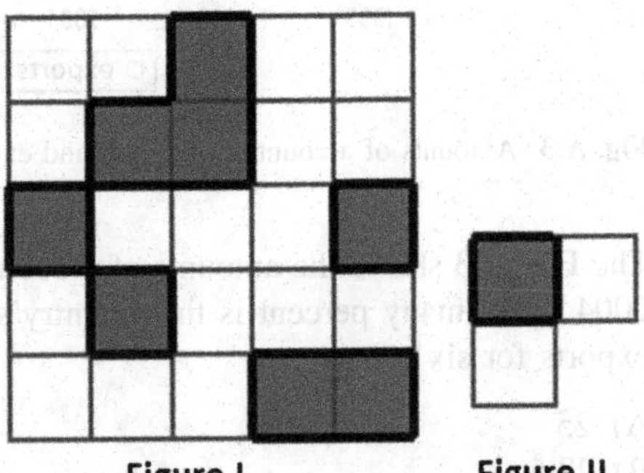

Fig. A.5 Figures in question-8

As for Fig. A.5, Mert wants to obtain parts in Fig. A.5II by slicing the square in Fig. A.5I.

How many parts can he obtain at most?

A) 4
B) 5
C) 6
D) 7
E) 8

A.1.9 Question-9

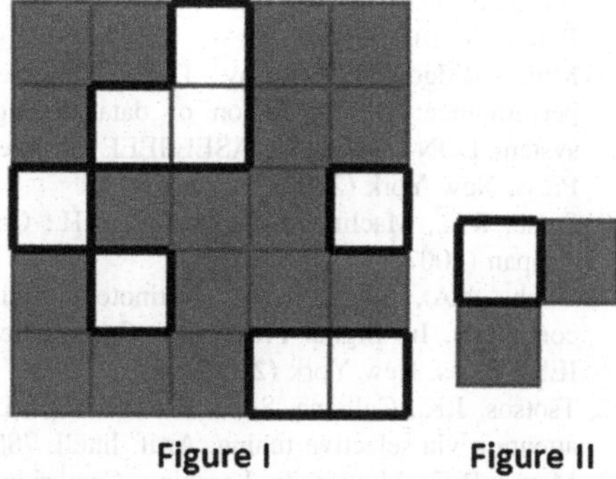

Fig. A.6 Figures in question-9

Regarding Fig. A.6, it is wanted to obtain parts in Fig. A.6II by slicing the square in Fig. A.6I.

How many parts can be obtained at most?

A) 4
B) 5
C) 6
D) 7
E) 8

References

1. Chuang, H.H., Liu, H.C.: Effects of different multimedia presentations on viewers' information-processing activities measured by eye-tracking technology. J. Sci. Educ. Technol. **21**(2), 276–286 (2012)
2. Tsai, M.J., Hou, H.T., Lai, M.L., Liu, W.Y., Yang, F.Y.: Visual attention for solving multiple-choice science problem: an eye-tracking analysis. Comput. Educ. **58**(1), 375–385 (2012)
3. Buckingham, S.S., Ferguson, R.: Social learning analytics. Educ. Technol. Soc. **15**(3), 3–26 (2012)
4. Abdous, M., He, W., Yen, C.J.: Using data mining for predicting relationships between online question theme and final grade. Educ. Technol. Soc. **15**(3), 77–88 (2012)
5. Xu, B., Recker, M.: Teaching analytics: a clustering and triangulation study of digital library user data. Educ. Technol. Soc. **15**(3), 103–115 (2012)
6. Verbert, K., Manouselis, N., Drachsler, H., Duval, E.: Dataset-driven research to support learning and knowledge analytics. Educ. Technol. Soc. **15**(3), 133–148 (2012)
7. Hung, J.L., Hsu, Y.C., Rice, K.: Integrating data mining in program evaluation of K-12 online education. Educ. Technol. Soc. **15**(3), 27–41 (2012)

8. Pardos, Z.A., Heffernan, N.T., Anderson, B.S., Heffernan, C.L.: Using fine-grained skill models to fit student performance with Bayesian networks. In: Romero, C., Ventura, S., Pechenizkiy, M., Baker, R.S.J.D. (eds.) Handbook of Educational Data Mining, Chapman & Hall/CRC Data Mining and Knowledge Discovery Series, pp. 417–426. CRC Press, Boca Raton (2010)
9. Minaei-Bidgoli, B., Kashy, D.A., Kortemeyer, G., Punch, W.F.: Predicting student performance: an application of data mining methods with the educational web-based system LON-CAPA. In: ASEE/IEEE Frontiers in Education Conference, pp. 1–6. IEEE Press, New York (2003)
10. Solso, R.L., Maclin, M.K., Maclin, O.H.: Cognitive Psychology. Pearson Education, Old Tappan (2007)
11. Salah, A.A., Saygin, A.P.: Retinotopy and selective visual attention in humans and computers. In: Signal Processing. Communication and Applications Conference, pp. 1–6. IEEE Press, New York (2008)
12. Tsotsos, J.K., Culhane, S.M., Wai, W.Y.K., Lai, Y., Davis, N., Nuflo, F.: Modeling visual attention via selective tuning. Artif. Intell. **78**(1–2), 507–545 (1995)
13. Mayer, R.E.: Multimedia Learning. Cambridge University Press, Cambridge (2001)
14. Anderson, J.R.: Cognitive Psychology and Its Implications. Worth Publishers, New York (2005)
15. Baddeley, A.D., Hitch, G.J.L.: Working memory. Psychol. Learn. Motiv.: Adv. Res. Theor. **8**, 47–89 (1974)
16. Poole, A., Ball, L.J., Phillips, P.: In search of salience: a response time and eye movement analysis of bookmark recognition. In: Fincher, S., Markopoulos, P., Moore, D., Ruddle, R. (eds.) People and Computers XVIII—Design for Life, pp. 363–378. Springer, London (2004)
17. Hyöna, J., Nurminen, A.: Do adult readers know how they read? Evidence from eye movement patterns and verbal reports. Br. J. Psychol. **97**(1), 31–50 (2006)
18. Goldberg, H.J., Kotval, X.P.: Computer interface evaluation using eye movements: methods and constructs. Int. J. Ind. Ergon. **24**(6), 631–645 (1999)
19. Duchowski, A.T.: Eye Tracking Methodology: Theory and Practice. Springer, Heidelberg (2003)
20. Rayner, K.: Eye movements in reading and information processing: 20 years of research. Psychol. Bull. **124**(3), 372–422 (1998)
21. Breiman, L.: Random forests. Mach. Learn. **45**(1), 5–32 (2001)
22. Tobii User Manuals http://www.tobii.com/Global/Analysis/Downloads/User_Manuals_and_Guides/Tobii_Studio2.2_UserManual.pdf
23. Svetnik, V., Liaw, A., Tong, C., Culberson, J.C., Sheridan, R.P., Feuston, B.P.: Random forest: a classification and regression tool for compound classification and QSAR modeling. J. Chem. Inf. Comput. Sci. **43**(6), 1947–1958 (2003)
24. Dahinden, C.: An improved random forests approach with application to the performance prediction challenge datasets. In: Guyon, I., Cawley, G., Dror, G., Saffari, A. (eds.) Hands-On Pattern Recognition: Challenges in Machine Learning, vol. 1, pp. 223–230. Microtome Publishing, Brookline (2011)

Part III
Assessment

Chapter 9
Mining Domain Knowledge for Coherence Assessment of Students Proposal Drafts

Samuel González López and Aurelio López-López

Abstract Often, academic programs require students to write a thesis or research proposal. The review of such texts is a heavy load, especially at initial stages. One feature evaluated by instructors is coherence, i.e. the interrelationship of the various elements of the text. We present a coherence analyzer, which employs latent semantic analysis (LSA) to mine existing corpora to further assess new drafts. We designed the analyzer as part of an Intelligent Tutoring System, considering seven common sections. After mining domain knowledge, experiments were done on graduate and undergraduate corpora to define a grading scale. Another experiment that involved human reviewers was set to validate the process. The technique allowed evaluating the coherence of the different sections, reaching an acceptable result and hinting that the level reached so far is adequate to support online review. An innovative exploration across sections was performed, uncovering a consistent interrelationship, according to methodology authors.

Keywords Coherence · Writing support · Latent semantic analysis · Intelligent tutoring system

Abbreviations

DM	Data mining
ITS	Intelligent tutor system
LSA	Latent semantic analysis
LSI	Latent semantic indexing
NMF	Non-negative matrix factorization
PLSA	Probabilistic latent semantic analysis

S. G. López · A. López-López (✉)
Instituto Nacional de Astrofísica, Óptica y Electrónica, Luis Enrique Erro # 1,
Santa María Tonantzintla, Puebla 72840, Mexico
e-mail: allopez@inaoep.mx

S. G. López
e-mail: sgonzalez@inaoep.mx

A. Peña-Ayala (ed.), *Educational Data Mining*,
Studies in Computational Intelligence 524, DOI: 10.1007/978-3-319-02738-8_9,

SPM Student progress module
SVD Singular values decomposition

9.1 Introduction

Academic programs or courses in educational institutions often conclude requiring students to elaborate a thesis or research proposal. A customary process followed by students is to write a first draft and then improve it after iterated reviews and recommendations of the instructor.

Some institutions provide guides that support students in structuring the proposal draft. However, this is insufficient in many cases, i.e. students often need help on how to structure and write every aspect of their draft. This demands that the academic advisor or instructor spends extra time on the reviewing process.

Data mining (DM), whose aim is to identify novel, potentially useful and understandable correlations or pattern from data, can adhere to one of two approaches: seek to build models or find patterns.

Educational data mining has similar aim and approaches but working on data obtained from educational settings [1]. An educational setting of interest is college education, where the heavy load of draft review can be ameliorated by the use of information technologies and methods.

One way to achieve this objective is to mine existing corpora of research proposals and theses to build models of different features (e.g. topics, language models, or argumentation) to analyze in new drafts of students. In particular, employing data mining, we can characterize the semantics of the domain of information technologies and computer science to assess coherence in drafts.

This chapter focuses on examining coherence in documents written in Spanish. Coherence is defined as the connection of all parts of a text into a whole [2]: the interrelationship of the various elements of the text. Therefore, coherence within proposal drafts is important because if a document does not have each of the elements related into a whole, or sections are not close to a topic, it would seem incoherent.

In this chapter, we present a global coherence analyzer, which employs Latent Semantic Analysis technique and tool to mine existing corpora of research proposals and theses to further assess proposal drafts of college students in information technologies and computer science. Its main aim is to help students to improve their coherence in drafts during the writing process, especially in the early stages. Furthermore, we intend that this analyzer, implemented in a system, indirectly helps the academic advisor by reducing the time dedicated to the draft review, enabling to focus on content.

We designed the analyzer considering seven common sections in drafts, and is in-tended as part of an intelligent tutor system (ITS), supporting students online. To assess global coherence after mining domain knowledge, experiments were done on graduate and undergraduate corpora to validate the process. Experiments involved human reviewers to compare the results of the analyzer with those of the reviewers, so we computed agreement measures. From mined domain knowledge, an exploration across sections was performed, as an additional validation procedure.

The results on coherence analysis reported here are parts of a larger project that may help students to evaluate their drafts early, and facilitate the reviewing process of the academic advisor.

The approach contributes to DM with a method to employ the results of latent semantic analysis (LSA) to grade and support students online to improve their writings, and a process for further exploration of mined knowledge.

This chapter is structured as follows. Next Section reviews previous related research. Section 9.3 describes the coherence analyzer. Section 9.4 details the data employed to mine and validate the experiments. Experiments validating the approach are presented in Sect. 9.5. Section 9.6 discusses results and their analysis. Section 9.7 includes an overview of the ITS. Finally, Sect. 9.8 details the conclusions and future work.

9.2 Background

Three themes are central to this research; coherence, data mining employing latent semantic analysis, and previous approaches for the mining of learners essays. We review concepts and related work in the following subsections.

9.2.1 Global Coherence

Coherence is classified based on its scope: global and local. Global coherence means that a document is related to a main topic, i.e. it is not consistent when its elements have no such main topic. Local coherence is defined within small textual units [3]. Recently, [4] reported a study of different factor conducing to cohesion and coherence in texts coming from student discussion forums.

An exploration of how foreign language learners express cohesion and coherence in their writings is reported in [5], employing topical structure analysis. An analysis of several methods for assessing coherence in the context of automated assessment of learners' responses is given in [6]. In [7], the authors define four aspects related to local and global coherence, one of which relates to the topic developed in the essay respect to the required topic by the teacher. Despite focusing on local coherence, [8] highlights specific areas of research for NLP in essay scoring. None of these studies of coherence is on proposal writings and they are predominately to grade essays already written, i.e. not to support directly the writing process.

9.2.2 Latent Semantic Analysis

LSA at first known as latent semantic indexing (LSI) [9], is an automatic indexing and retrieval technique, which was initially designed for improved detection of relevant documents on the basis of search queries. This is a dimensionality reduction technique based on statistical analysis that allows uncovering the implicit (latent) semantics (structure) in a collection of texts. Afterward, Landauer and Dumais developed the LSA technique [10].

They define the LSA as a theory and method for extracting and representing the contextual meaning of words in use, through statistical computation applied to a large corpus.

In [11], they evaluated the textual coherence using LSA technique. This paper shows the coherence prediction by analyzing statement by statement a set of four texts, with a 300-dimensional semantic space, which is constructed based on the first 2,000 characters of each of the 30,473 articles of the Encyclopedia of American Academic Groliers. After separation of the four individual sentences texts, the vector of each text was calculated as the sum of the weights (each term), subsequently being compared with the next vector, so the cosine of these two vectors showed the semantic relationship or coherence.

One of the discussions in this paper is whether the LSA technique is a model of text-level knowledge of an expert or novice. They state that it depends on the training that the LSA system has received in the application domain. This technique focuses on the latent semantic aspect, which is a relevant feature to our work.

Alternative techniques related to LSA are: Probabilistic Latent Semantic Analysis (PLSA) and Non-negative Matrix Factorization (NMF). PLSA [12] has a well-developed statistical foundation, defining a proper generative data model; and uses a generalization of Expectation Maximization algorithm for training, with some gains in performance.

NMF [13] applies a non-negativity constraint when factorizing a term-document matrix, leading to a more intuitive representation of documents as addition of topics. A comparison of four popular text mining methods is reported in [14], including LSA. An alternative way to mine regularities, and in consequence assess coherence, is proposed in [15].

9.2.3 Related Work

Several previous works have focused on evaluating educational aspects using the LSA technique. In the educational field, different kinds of documents are generated, such as documents written by teachers related to learning activities, student essays or textbooks [16]. Our work focuses on proposal drafts of undergraduate students, specifically in the Spanish language.

A different application of mining is presented in [17], where the aim is to reveal the processes involved during collaborative writing. In [18], they take a data mining perspective to do essay scoring, with LSA as one of the methods to consider content.

Given that the coherence analyzer is intended to help students when preparing their text, our work is also related to intelligent writing support systems and tools, such as Glosser [19] that supports students when writing essays by formulating questions and providing content clues to answer them, employing data mining. A more recent work [20] goes a step further by generating questions from student writings citations and content elements.

Despite in essence the coherence analyzer described in the chapter performs student text grading, this is done from text in process of improvement (i.e. prior to submission), not from static given text (post submission) as mainstream essay grading (e.g. [8, 18],). The approach also extends previous applications of LSA with a method to exploit its output to grade and support students online.

9.3 Analyzer Model of Global Coherence

Many text definitions include coherence as a necessary feature. Coherence in proposal drafts of students is important because if it is not present in each of the elements, the central idea loses all meaning. Different approaches have been proposed by researchers, some techniques have focused on the semantic aspect when seeking to achieve overall coherence evaluation, while other studies have worked the syntactic aspect, as a way to attack the local coherence. Both approaches were developed in different ways in [21, 22], but our work focuses in global coherence, as the first step to improve the proposal drafts of undergrad students.

Our model seeks to evaluate the global coherence in seven sections of a proposal draft: problem statement, justification, objective, research questions, hypothesis, methodology, and conclusions. The global coherence refers to the thematic similarity between the section subject to evaluation and the semantic space, mined from an existing corpus in the domain of computer science and information technologies.

For example, if the text under evaluation contains concepts thematically close to biology, their measure of coherence will be poor, since our corpus is of the

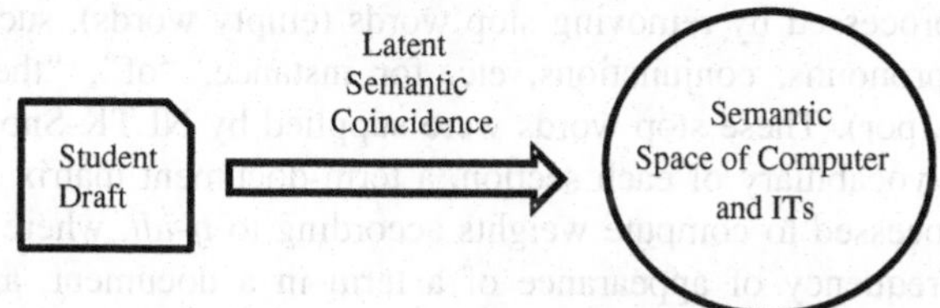

Fig. 9.1 Global coherence

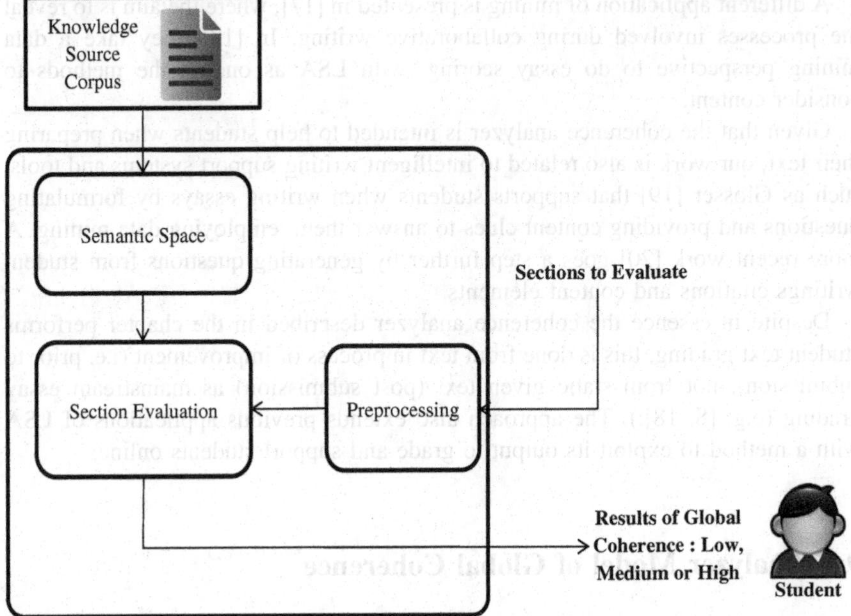

Fig. 9.2 Coherence evaluation model

computer and information technologies domain. Figure 9.1 depicts the concept of global coherence. Under this concept of global coherence, we designed our model as illustrated in Fig. 9.2.

Knowledge Source Corpus. The first step was to gather documents in Spanish, such as student theses and research proposals, previously reviewed and approved. Both kinds of documents were of under-graduate and graduate level. With this corpus, the semantic spaces were extracted for each section, i.e. there were seven corpora to mine. Corpus description is presented in detail in the following section.

Semantic Space. To extract the semantic space, terms of the input elements of a proposal draft were truncated (stemmed). Images, tables and figures were ignored. The goal of the stemming process is to reduce the variations of each word. For example the words "computer" and "computers" (in Spanish computadora, computadoras) are similar, so the process would produce a word stem "comput". We used the Freeling tool for stemming. In this way, many related terms are grouped, reducing the dimensionality of terms. Afterward, each corpus of the sections was processed by removing stop words (empty words), such as articles, prepositions, pronouns, conjunctions, etc. for instance, "of", "the", "by" (in Spanish de, la, por). These stop words were supplied by NLTK-Snowball.

Having the vocabulary of each section, a term-document matrix is built. This matrix was processed to compute weights according to *tf-idf*, where *tf* represents the absolute frequency of appearance of a term in a document, and *idf* is the inverse frequency of the term in the documents of the collection, i.e. the weight of a term in a document increases if this occurs frequently in such document and decreases if appears in many (most) of the documents.

LSA then reveals the (latent) meaning of words, discarding the words occasionally used in specific contexts and focusing on what is common in all contexts [23]. This is achieved by the core process in LSA, Singular Value Decomposition (SVD). So, after preprocessing the corpus, the algebraic SVD algorithm is used. SVD allows reducing the dimensionality of the original matrix to a more manageable number and also reduces noise or irrelevant information in the matrix. The SVD produces three matrices:

- Orthogonal Matrix U. Obtained by linear processing of number of columns in the original matrix A. This matrix represents terms as vectors in space of words.
- Transpose matrix V^T. Obtained by permuting the rows with columns, providing an orthogonal arrangement of the elements of the row. Through this transposition, documents are represented as vectors in space of words.
- Diagonal matrix Σ. Calculated by linear processing from number of rows, number of columns and the number of dimensions of the original matrix A. The diagonal matrix represents singular values of A. The singular value decomposition of the matrix is illustrated in Fig. 9.3

Once the three matrices are obtained, we can generate the matrix A, but depending on the singular values maintained, it would be a matrix close to matrix A, i.e. an approximation to A with the most relevant information.

Sections to Evaluate. These correspond to the sections that the student wants to evaluate, so they are analyzed one a time (i.e. there is no need to parse sections). Our analyzer allows evaluation of seven sections of a proposal: problem statement, justification, objective, research questions, hypothesis, methodology, and conclusions.

Preprocessing. This part of the model considers the stemming, stop word removal and computation of the *tf-idf* weights on the section to be evaluated. Once these processes have been applied, the text is ready to compare against the corresponding semantic space to measure similarity.

Section Evaluation. The section under evaluation is compared against the corresponding semantic space. For this purpose, the cosine similarity measure is applied to the input vectors obtained from the section and those vectors coding the semantic space. The expression for the computation is:

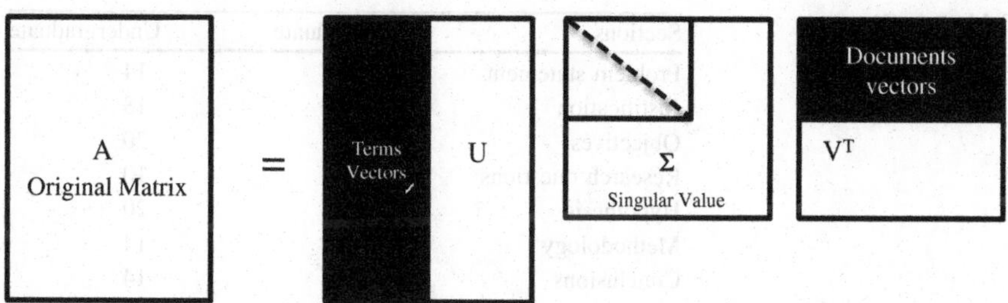

Fig. 9.3 SVD algorithm [24] representation

$$\cos(A, B) = \frac{A * B}{||A|| * ||B||}. \tag{9.1}$$

According to this expression, the similarity is one when the angle between the two vectors is 0°, that is, the vectors are pointing in the same direction and are parallel. This result expresses the highest coherence in the text. We get 0 when the vectors are orthogonal and correspond to no coherence at all.

Results of Global Coherence. Instead of reporting a numerical value as result of the coherence evaluation, the result is expressed in terms of three levels: High, Medium and Low. To achieve this qualitative scale of coherence, a process was applied, setting thresholds to determine each level.

This information was obtained taking as reference the graduate corpus, under the premise that the level of graduate students writing is better than those at undergrad level. Next, we describe the corpus used to extract the domain knowledge, as well as the threshold to define the discrete values for grading.

9.4 Data Description (Corpus)

We gathered a corpus of different elements in proposal documents in Spanish. We distinguished two kinds of student texts: graduate proposal documents, and undergraduate drafts. The first kind of texts includes documents reviewed and approved by faculty, so they are considered as reference or examples for knowledge mining. The second kind of documents is used as test examples. The whole corpus consists of a total of 410 collected samples as detailed in Table 9.1. The corpus domain is computing and information technologies. They were then preprocessed as detailed above.

9.5 Experiments

This experiment focused on evaluating the sections of student's proposal draft from the aspect of global coherence. We selected LSA because it captures the documents' latent semantic, something we want to mine from different sections in a proposal.

Table 9.1 Corpora

Sections	Graduate	Undergraduate
Problem statement	40	14
Justification	40	18
Objectives	60	20
Research questions	40	10
Hypothesis	40	20
Methodology	40	14
Conclusions	40	14
Total	300	110

9.5.1 Experimental Design

An experiment was set to validate the proposed online reviewing process, involving human reviewers to compare the results of the analyzer against their grades, calculating an agreement measure. In particular we computed agreement in terms of Fleiss or Cohen Kappa. From mined domain knowledge, an exploration across sections was performed, exploring their interrelationship.

All the collection was sent for evaluation to two or three instructors serving as reviewers, that have experience in advising students in the preparation of their drafts in the computing and information technologies. The reviewers did not know beforehand the level (graduate or undergraduate) of each sample. Each reviewer was requested to assign a level to each sample, using the scale: High, Medium and Low coherence, where the high level meant that the text has a strong coherence or relationship to the domain of computing and information technologies and the low level meant that the relationship is weak relative to the domain. Two examples of High and Low coherence in the objectives section are given next.

High Coherence. Analyze problems that arise in the system development of software architectures of Enterprise type.

We can observe that the word "systems" and "software" are very close to the domain, including the term "architecture" in the context of the previous terms fit within the domain of computing. Likewise, words with less thematic load such as "development" or "analyze" are often used in the domain.

Low Coherence. Identify the effect of feedback on the learning of the business leader, to allow being more effective.

Notice that even though terms like "learning" or "feedback" may have some proximity to the domain, the words or phrases "business", "leader" or "be more effective" are the central topic and are barely used in the domain of interest.

The assessments provided by our reviewers allowed to exclude those examples in our knowledge mining set considered low by at least two, or those where they did not agree, since they will bias the construction of the semantic spaces. For instance, if an objective was labeled as High by two or three of the human reviewers, this example will be part of our training corpus since, according to the reviewers; such objective is indeed highly coherent with the domain. In case that only one reviewer assigned High grade, this objective will not be part of the training corpus since there is a doubt about its coherence, and can introduce noise into the corresponding semantic space.

On the other hand, the assessments on the test set allow comparing the automatic evaluation of coherence, after extracting the semantic space and defining a grading scale. Once instructors evaluated the whole collection (training and test subsets) detailed in previous section, we then evaluate the level of agreement among them.

These human reviews allow getting the subset of examples subject to mine their knowledge, i.e. those contributing for the construction of the semantic space. Once we computed the semantic spaces, we can set the thresholds that define the scale after analyzing all samples that human reviewers assigned a high level of coherence (see Table 9.2).

The thresholds for levels High, Medium and Low in our system were established using as a basis the average obtained when evaluating the training corpus (elements labeled with a high level) with a cross-validation. It was a one-fold validation, i.e. the element was removed from the corpus and the semantic space was generated with the remaining examples.

Then, we calculated the standard deviation of the values obtained, and the high level is calculated as the average plus one sigma and low as the average as minus one sigma. Previously, we corroborate the normality of the data, with 95 % of confidence. With the use of one sigma for thresholds, we can ensure that the results will be in a close range to the average obtained with the best documents (labeled as high). In this case if the result is closest to the upper limit, it means that the text is closer to the domain of computing and shows strong evidence of global coherence.

Also having the semantic spaces for the different sections of our mining subset, then one can evaluate automatically the corresponding section in the test subset. Then, we have the elements to evaluate the level of agreement between the grade assigned by the system and by instructors.

9.5.2 Agreement Evaluation

Each section was tagged by human reviewers (two or three reviewers). For each section, Fleiss and Cohen Kappa measures [25] were computed, depending on the case to be presented, i.e. two or three evaluators. In addition, we calculated the Cohen Kappa to evaluate the level of agreement between the analyzer and human reviewers.

We proceed to describe the grades assigned by human reviewers and the level of agreement among them first, and then the result of agreement between the grade assigned by the coherence analyzer and humans. We also provide the values obtained from mined semantic spaces and used to define the thresholds determining the grading levels. These results are presented for each of the section under analysis at a time.

Table 9.2 Training and test corpus

Sections	Training	Test	Tagged as high level
Problem statement	40	14	23
Justification	40	18	20
Objectives	60	20	40
Research questions	40	10	36
Hypothesis	40	20	20
Methodology	40	14	27
Conclusions	40	14	24

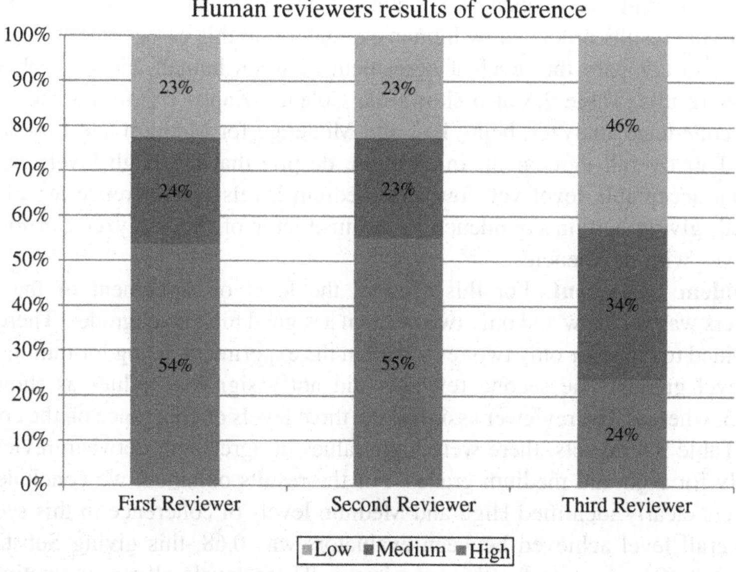

Fig. 9.4 Results of three human reviewers (objective)

Objective. Figure 9.4 shows the percentages of level of coherence assignment by each human reviewer. Note that the first and second human reviewer has similar percentages obtained in each of the levels. The third evaluator presented an inverse behavior to the first two reviewers; we assume that this rater was stricter than the other, when evaluating objectives.

The Fleiss Kappa coefficient of agreement was computed for the three reviewers considering the test corpus. Table 9.3 shows the Fleiss Kappa results for each level, for the objective section.

The reviewers had a Substantial agreement for the Low and High grades, and a Poor agreement in Medium grades. For the results obtained, we conclude that reviewers clearly identified High and Low levels but not those in the middle. The overall level achieved between evaluators was 0.54, this corresponds to a Moderate confidence of agreement for the experiment.

These levels allow automating the evaluation of the coherence analyzer. In particular, for the objective section, we got an average of 0.49 with a standard deviation of 0.17, resulting in the highest threshold of 0.64 and the lowest threshold at 0.28.

Table 9.3 Kappa for test corpus

Kappa	Reviewers (Fleiss)	Analyzer versus reviewers (Cohen)
High	0.6862	0.0000
Medium	−0.0378	0.2609
Low	0.7353	0.4218
Overall	0.5458	0.2237

Once the scale is defined, we evaluated the test samples with the aim to compare the results produced by human evaluators. In this case, Cohen's Kappa is pertinent to compare the level of agreement between human and our coherence analyzer results. Table 9.3 also shows the Cohen's Kappa results for the human versus coherence analyzer, being Fair and Moderate for Medium and Low levels, with a Fair overall agreement. In addition, despite that the High level does not reach an acceptable level yet, low and medium levels of coherence are already detected, giving certain confidence to the instructor of the analyzer can identify objectives with deficiencies.

Problem Statement. For this section, the level of agreement of the three reviewers was very low and only two of them assigned high level grades. Therefore, we decided to consider only two reviewers in the experiment, using for mining their high level grades. The second reviewer did not assign low values as shown in Fig. 9.5, whereas first reviewer assigned the three levels of coherence on the corpus.

As Table 9.4 depicts, there were high values of agreement between reviewers, but only for high and medium grades. For the results obtained, we conclude that reviewers clearly identified High and Medium levels of coherence in this section. The overall level achieved between evaluators was 0.68, this giving Substantial confidence of agreement for the experiment. These levels allow automating the evaluation of the coherence analyzer. For this section, after getting the semantic space, we obtained a low average of 0.127 with and standard deviation of 0.057, leading to setting the thresholds at 0.07 for Low and 0.18 for High.

As observed in the Kappa values between analyzer and reviewers, there is a Perfect level of agreement in High grades but a margin for improvement in the Medium grade since this is Fair as the overall agreement.

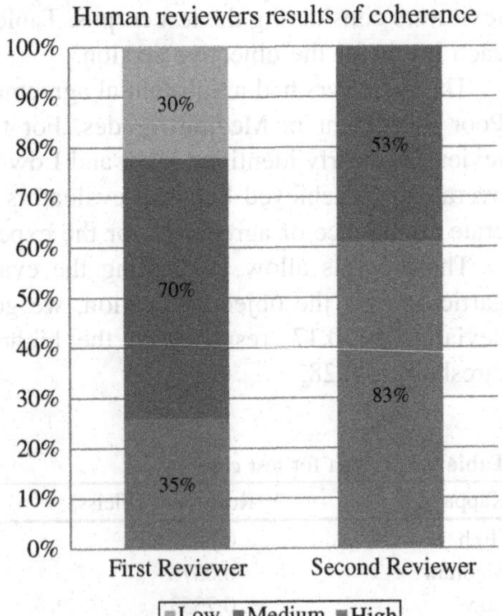

Fig. 9.5 Results of two human reviewers (Problem Statement)

Table 9.4 Kappa for test corpus

Cohen kappa	Reviewers	Coherence analyzer versus reviewers
High	1.000	1.0000
Medium	1.000	0.3300
Low	0.000	0.0000
Overall	0.680	0.4000

Since human reviewers did not agree on tagging problem statements with a low grade in the test set, we cannot expect any agreement with the analyzer. But, to find out if our approach can identify the low grades, we took examples labeled as low in graduate corpus. These examples were not included in the training set, but for exploration purpose, we evaluated the examples with the coherence analyzer and add them to previous results obtained with test set. With these results we computed the Cohen kappa between human reviewers and analyzer.

According to the results, the kappa showed an improvement for low and medium level. High level maintained the level of agreement, the medium and low level of Fair changed to Moderate, with 0.43 and 0.40 respectively. The overall agreement level was 0.49 which represents a Moderate level.

Hypothesis. Figure 9.6 shows the percentages of grades assigned by human reviewers, the first reviewer assigned the High grade more often, while the second reviewer had a more balanced performance. However, this is a normal behavior of human reviewers in the classroom. As in problem statement, we only used two of the human reviewers.

As Table 9.5 details, Kappa results between human reviewers were Acceptable with 0.301, similarly as the Kappa between our analyzer and human reviewers was Acceptable with 0.2558. However, it was lower than in the objective and problem statement sections. For the purpose of automating the evaluation of the coherence analyzer for the hypothesis section, we got an average of 0.636 with a standard deviation of 0.236, resulting in the high threshold of 0.87 and the low threshold at 0.4.

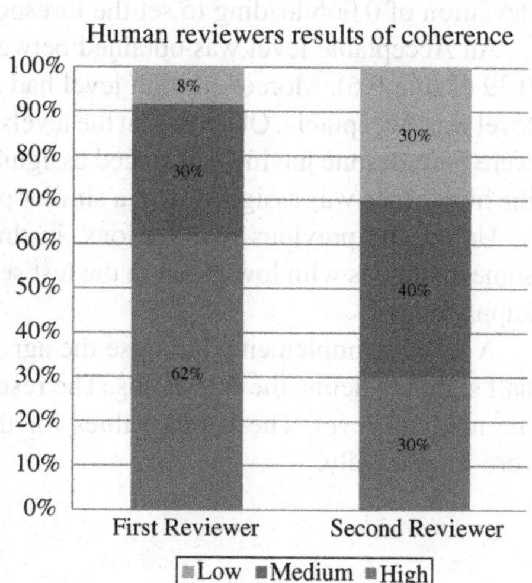

Fig. 9.6 Results of two human reviewers (Hypothesis)

Table 9.5 Kappa for test corpus

Cohen kappa	Reviewers	Coherence analyzer versus reviewers
High	0.3953	0.5294
Medium	0.2528	0.1428
Low	0.0000	0.0000
Overall	0.3010	0.2558

For particular levels, there is a Moderate value in the Kappa scale for High level, among human reviewers and our analyzer. The zero value of agreement among human reviewers affects the outcome of the analyzer to the low level. Although only examples with High level were used to mine, the human reviewers distribution was unbalanced. Low grades in Hypothesis presented a similar complication as the Problem Statement section, where reviewers did not agree tagging examples with low grade. Again, to find out whether our approach can identify low grades, we took the examples labeled as low in graduate corpus.

Then, we evaluated the examples with the coherence analyzer and add them to previous results obtained with test set. When executing Cohen kappa between human reviewers and the analyzer, the values high, medium and low were 0.6363, 0.111 and 0.333, respectively. It was observed that Kappa for High level is "Substantial". The medium level remains at "Slight" level and the Low moved from "Poor" to "Fair". The overall level of agreement was "Fair".

In this case, despite the medium grade did not reach an acceptable level, the low level reach an acceptable agreement. The analyzer can give certain confidence to the instructor that a hypothesis with deficiencies will be identify by our system, and can suggest students to improve their Hypothesis.

Justification. In this section, human reviewers had a more balanced distribution across the three coherence levels. The kappa values achieved were lower compared to the previous sections; even so the level is Acceptable or Fair. Figure 9.7 shows the percentages of levels assigned by the two reviewers. For the justification section, after computing the semantic space, we obtained an average of 0.137 with a standard deviation of 0.066 leading to set the thresholds at 0.07 for Low and 0.2 for High.

An Acceptable level was obtained between the reviewers and the analyzer with 0.39 (Table 9.6). Moreover, high level had a Moderate agreement and the medium level was Acceptable. Observe that the levels of agreement between human reviewers were Fair, despite having a balanced assignment of grades. The reason could be that the high grade was assigned with a similar percentage but not to the same samples.

Unlike the previous two sections, in this section the human reviewers tagged some examples with low grade in the test set, showing a Fair agreement in terms of kappa value.

A strategy implemented to raise the agreement results for low grades was using half sigma to define the thresholds. The results improved for low level, but affected the medium level. The kappa values for the High and Low level were 0.33 and zero respectively.

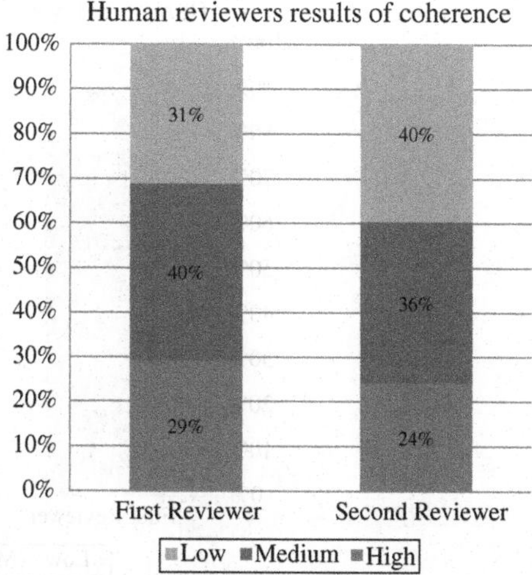

Fig. 9.7 Results of two human reviewers (Justification)

Table 9.6 Kappa for test corpus

Cohen kappa	Reviewers	Coherence analyzer versus reviewers
High	0.2200	0.5800
Medium	0.2075	0.3600
Low	0.2758	0.0000
Overall	0.2283	0.3900

Another attempted alternative to improve results was training a classifier (Naive Bayes), using as input vector the LSA value provided by the semantic space and the grade (class) assigned by the reviewers. As training examples, we used the set of graduate and undergraduate texts, evaluated as low and medium. After training, the classifier had a precision of 0.714 and recall of 0.5 for the low grade. The medium level reached a precision of 0.706 and recall of 0.857. The level of agreement was Acceptable in terms of kappa. These results indicate that the classifier is a promising alternative to predict medium and low grades for this section.

Conclusions. For this section the instructors identified the three grade levels at different rates (Fig. 9.8). The first reviewer was probably more rigorous than the second since assigned 35 % to high level, while the second reviewer duplicated the value, assigning a 65 % to high grade.

As expected from the percentages, the agreement results for this section were not satisfactory. The level of agreement between reviewers was 0.31, corresponding to the Acceptable level.

In this section, we got an average of 0.268 with a sigma of 0.247 allowing to set the thresholds for Low at 0 0.021 and for High level at 0.514. Also there was a 0.1666 level of agreement among human reviewers and the analyzer, this means a Slight level of agreement.

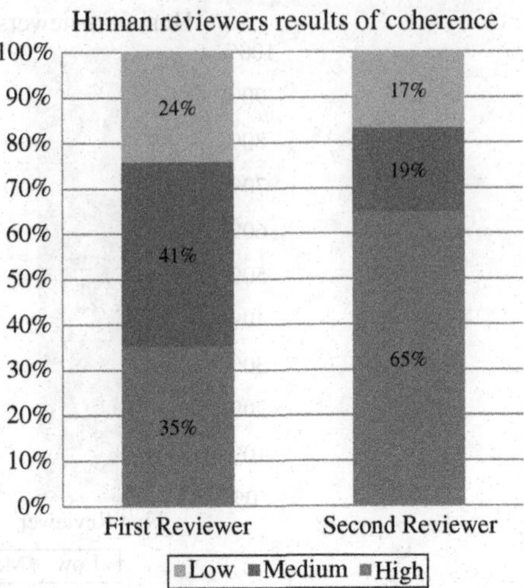

Fig. 9.8 Results of two human reviewers (Conclusions)

High and medium grades were Acceptable according to a kappa of 0.28. The value of agreement was zero for low grade. This was probably due to the low coincidence of examples labeled as high. As observed in results of previous sections, our analyzer results regarding human agreement levels are close, indicating that our analyzer is directly dependent on the level of agreement between humans.

In addition, the kappa level between human reviewers for low level was null, since none of the examples was graded as low (see Table 9.7). But to know whether our approach can identify low grades, we took examples labeled as low in the graduate corpus. The result again was unfavorable, since the values were low, according to previous values.

Subsequently, we decided to try a classifier (Naive Bayes) to improve results. For training, we used examples of the graduate corpus, tagged as medium and low. After training, we obtained the values of precision and recall.

The results were favorable, reaching a precision value of 1 and recall of 0.556 for the medium class, while for the low class reached a precision of 0.556 and recall of 1. Kappa value was of 0.447, higher than using thresholds.

These results indicate that for this section, the use of a classifier for predicting medium and low class seems more promising than using sigma to define the scale. The classifier was trained with medium and low classes, since our analyzer was built with the high class.

Table 9.7 Kappa for test corpus

Cohen kappa	Reviewers	Coherence analyzer versus reviewers
High	0.2857	0.2857
Medium	0.4000	0.2857
Low	0.0000	0.0000
Overall	0.3103	0.1666

Research Questions. Human reviewers had a similar assignment percentage of the low grade level. For medium and high percentages, they were unevenly (Fig. 9.9). This behavior was reflected in the values of Kappa. In the figure, we can notice that the first reviewer assigned a 30 % of medium grades, while the second reviewer assigned 55 %. This led to have an average of 0.432 with a sigma of 0.286, allowing to set the Low Level at 0 0.227 and the High level at 0.638, for this section.

We can observe in Table 9.8 that the Medium grade level had a zero percent agreement, which was expected since the level of agreement was very uneven between reviewers. For high grade, reviewers reached a value of 0.50 and for low grade reviewers obtained a kappa of 0.46, which corresponds to a Moderate agreement.

The agreement results between human reviewers and our analyzer were 0.33 for High and Low grades. This corresponds to an Acceptable level according to the range of kappa. We can notice clearly that the reviewers and our analyzer identified High and Low grades.

Methodology. Figure 9.10 shows that human reviewers had a significant difference between percentages of assignments of coherence level. This distribution was one of the reasons for low levels of agreement. For this section, one can notice quite unbalanced percentages of grades.

The Kappa for High level was 0.19 between reviewers, i.e. Slight agreement. An average = 0.315 with a standard deviation of 0.158 allowed to set the Low grade level at 0.156 and the High grade level at 0.474, for automating the grades of the analyzer for this section. Among human reviewers and our analyzer, a value of 0.12 of agreement was obtained for High grades. Both values are in poor performance based on Kappa. One possible cause is that the undergrad methodologies tend to have fewer steps and a simpler elaboration than graduate level methodologies.

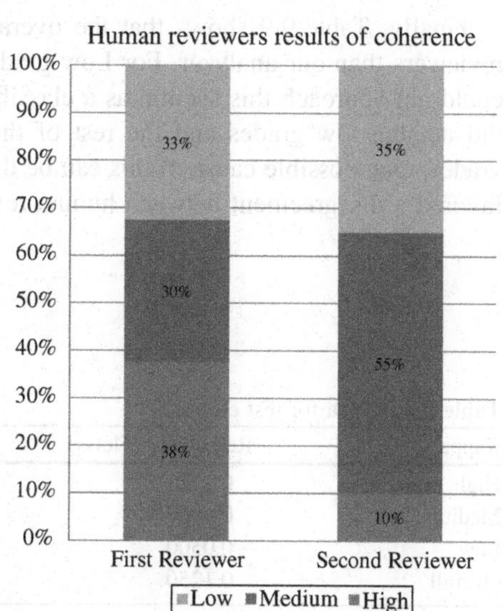

Fig. 9.9 Results of two human reviewers (Research Questions)

Table 9.8 Kappa for test corpus

Cohen kappa	Reviewers	Coherence analyzer versus reviewers
High	0.5000	0.3333
Medium	−0.0230	0.0000
Low	0.4666	0.3333
Overall	0.2727	0.2000

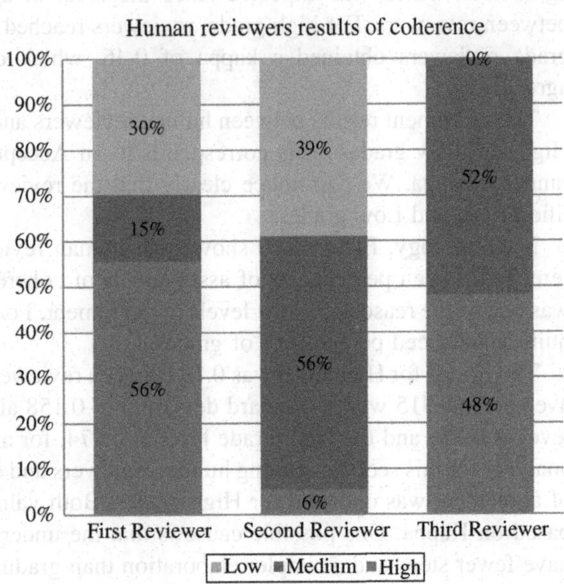

Fig. 9.10 Results of two human reviewers (Methodology)

Finally, Table 9.9 shows that the overall agreement values are lower among reviewers than our analyzer. For Low grade, the agreement amounts to zero. We could not approach this section as a classification task since one of the reviewers did not tag low grades and the rest of the reviewers did not coincide on their grades. One possible cause of this can be the variety in writing in this section that favored a disagreement between human reviewers.

Table 9.9 Kappa for test corpus

Kappa	Reviewers (Fleiss)	Analyzer versus reviewers (Cohen)
High	0.1923	0.1250
Medium	0.1900	0.2750
Low	−0.0500	0.0000
Overall	0.1250	0.1764

9.5.3 Across Section Exploration

Given that we mined the semantic spaces for the different sections, we were in the position of performing an analysis among sections. So, our second experiment allowed extracting and identifying a behavior pattern between the different sections evaluated. This exploration was motivated by the relationships that different authors state in research methodology. These relationships are suggested to students when they write their research proposal by their academic advisors.

The relationships found are from the perspective of global coherence, i.e. these are thematic relationships that allow identifying similar concepts. For example, from the corpus of research questions, ten items were taken randomly and were evaluated in the semantic space of the objectives section. The same was done with the remaining sections. It is noteworthy that these inter-sections coincide with what methodology authors suggest. These authors of methodology books suggest that once the objective is defined, this can suggest one or more research questions, which would lead the student to maintain coherence between these elements [26].

The diagram in Fig. 9.11 shows the relationships revealed among the different semantic spaces of sections. The intensity of the gray in the lines represents the strength or degree of relationship, darker color represents higher relation.

Inter-Relations. There is a high relationship among Objective, Research Questions and Hypotheses sections.

- The diagram shows a medium relation between Hypotheses and Research Questions.
- Another aspect shown in the diagram is the low relation between Objectives and Justification elements.
- Also Objective and Conclusions sections show a medium relation.
- Hypothesis, Research Questions and Conclusions showed a medium relation between semantic spaces and elements of their corresponding corpus.

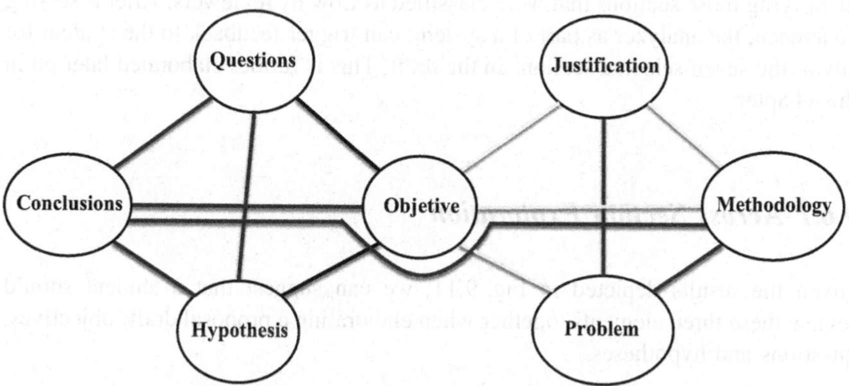

Fig. 9.11 Pattern of inter-relations among sections

These observed behaviors were revealed from corpora. Somehow, the recommendations that instructors provide to their students, lined up when crossing semantic spaces. This knowledge was supposed at the beginning of our experiments, but the detected behaviors reinforce the academic advisors recommendations (from the perspective of global coherence).

9.6 Analysis and Discussion of Results

We observed that the levels of agreement in the Low case is Moderate and Medium level is Fair, the overall level of agreement between humans and the analyzer was Fair. We conclude that the analyzer would have an acceptable support for the student and academic advisor in the process of preparing the proposal draft.

After comparing the statistical results in terms of the Kappa coefficient of agreement, we also performed a qualitative analysis between the results of coherence analyzer and the process of reviewing a proposal draft, i.e. the advisor would expect that the analyzer was a first filter so that when the drafts reach him, at least have a Medium or High Level.

Under this premise, the results of our analyzer match the concept of a strict filtering reviewer, because it provided low and medium values in most test sections. We can observe that if our system does not achieve at this time a higher level of agreement in the High grade level, this is not a problem since the analyzer is being strict to assign the high level.

In the experiment, the analyzer evaluated as Medium the few highest levels assigned by the reviewers. If the analyzer behaves more flexible and allows high level to sections that have to be of a medium or low level, this could cause a burden to the academic advisor, failing to support in review.

Finally we note that between the coherence analyzer and human evaluators, the agreement is Moderate for low levels, bringing confidence that the analyzer is identifying those sections that were classified as Low by reviewers. After assessing coherence, the analyzer as part of a system, can trigger feedback to the student for any of the seven selected sections in the draft. This is further elaborated later on in this chapter.

9.6.1 Across Section Exploration

Given the results depicted in Fig. 9.11, we can suggest that a student should review these three elements together when elaborating a proposal draft: objectives, questions and hypotheses.

The diagram also showed a medium relation between hypotheses and research questions sections, suggesting that the questions should be considered when drafting the hypothesis.

Another detail shown in the diagram is the low relation between Objectives and Justification sections. This relationship can be caused by the varied nature of justifications, since these can be economic, efficiency, capacity, to response to a need for the project, and so on, and could not be related to the stated objectives.

Hence, a student can have some freedom to write independently these two section when writing the proposal. This does not mean that both sections do not have to agree with the Problem Statement.

9.7 System Overview

Despite in essence the coherence analyzer described in the chapter performs student text grading, this is intended for text in process of improvement (i.e. prior to submission). In consequence, our approach final aim is to support students online. So, the coherence analyzer is embedded in an ITS, to enable students to improve their first draft, working on each section at a time, either by typing or pasting for analysis. In addition, students receive feedback so they can improve the document in progress.

9.7.1 Intelligent Tutoring System

The intelligent tutor is illustrated in Fig. 9.12 where the Domain Module includes information (material) concerning the definition of global coherence and what is expected to contain the different sections, regarding the concept of coherence. Also we present material about the structure that should contain a proposal draft.

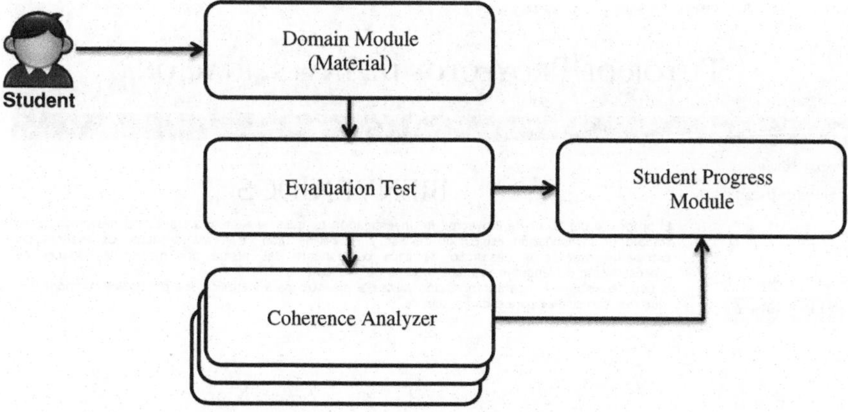

Fig. 9.12 Model of intelligent tutoring system

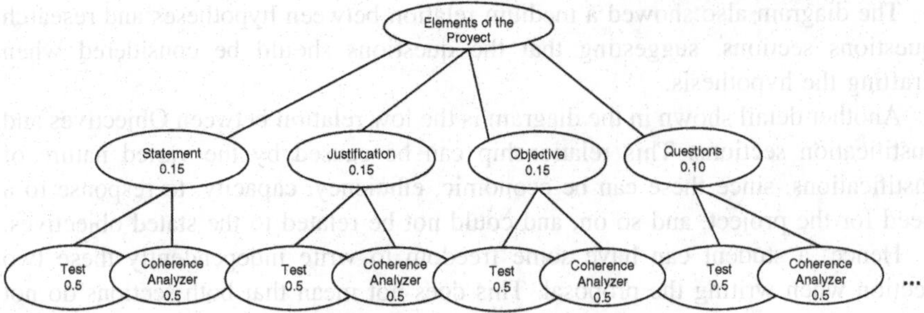

Fig. 9.13 Network used in student progress model of tutor

A test is applied to validate the reading of materials and then practical exercises are suggested and applied using the Coherence Analyzer to achieve a high level of coherence in the student text productions. The results of the test and coherence analysis are sent to the student progress module (SPM) to update the knowledge state of the student, represented in a network. The SPM records the student's progress in the network representation, which is partially depicted in Fig. 9.13 (only four of the seven sections are illustrated to avoid clutter the diagram). When the student completes the test, the value of the test node element is updated and the SPM calculates the student's progress for the parent node, considering the weights assigned to each question in the test.

Similarly, when doing the exercises with the Coherence Analyzer, the corresponding node in the network is updated and the SPM estimates the student's progress for the parent node using the weights assigned. Figure 9.13 illustrates the weights assigned to each node according to the experience of instructor.

For instance, in the Test node of the Problem Statement, a weight of 50 % of the parent node (Statement) is assigned, which includes five questions to verify that the student has read the pertinent information.

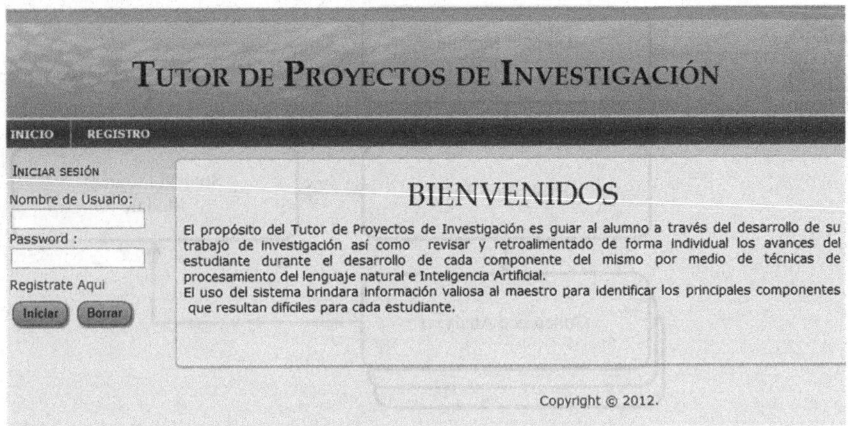

Fig. 9.14 Coherence analyzer (in Spanish)

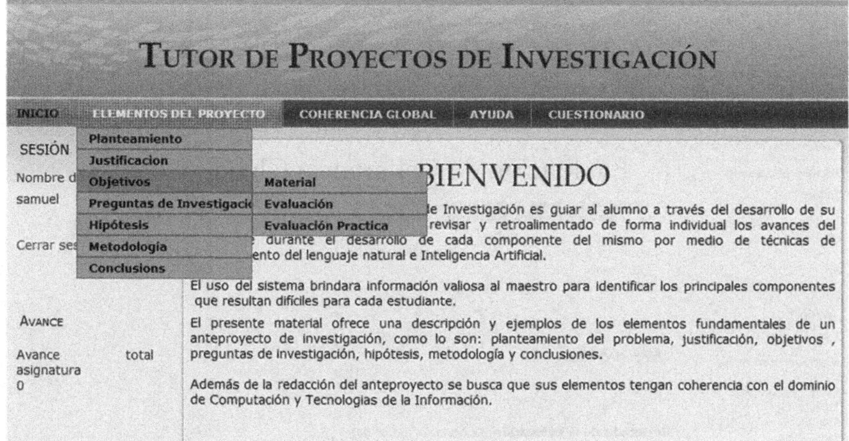

Fig. 9.15 Main menu of ITS (in Spanish)

9.7.2 Web Interface

The ITS is developed in PHP for easy access via web and the network structure is stored in a MySQL database, the coherence analyzer is developed in Python given the easy access to processing tools for natural language.

Figure 9.14 shows the graphical interface (in Spanish) of the tutoring system in which we can observe the login section to the left.

Figure 9.15 depicts the menu on the top to access the elements (sections) of the writing project (in Spanish Elementos del proyecto). For each element, there are three sections: material, test, and practical evaluation.

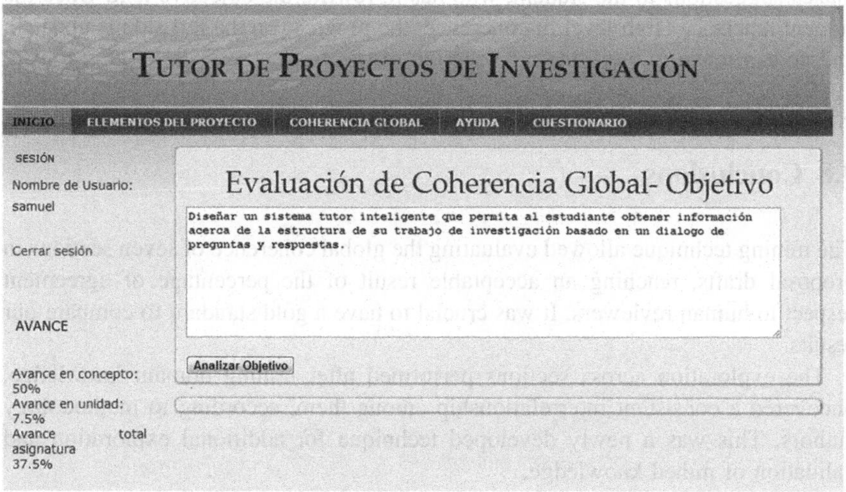

Fig. 9.16 Coherence analyzer (in Spanish)

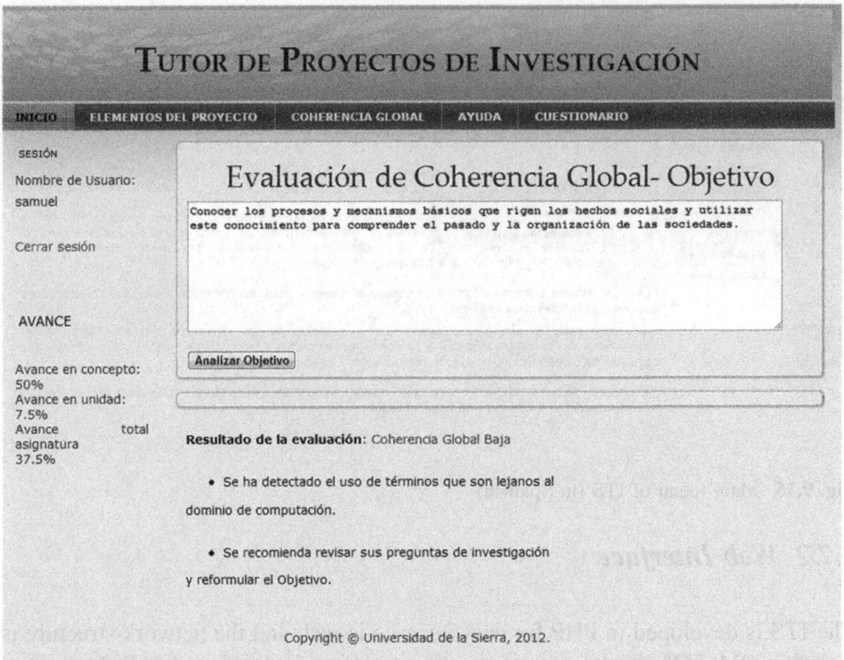

Fig. 9.17 Results of coherence analyzer (in Spanish)

Figure 9.16 below shows the coherence analyzer and the report of student progress (in Spanish *Avance*) in percentages in the bottom left part of the screen. This screen snapshot also illustrates an objective text ready for coherence analysis.

The report generated by the coherence analyzer can be seen in Fig. 9.17. In this case the level of coherence found in the text is Low. In consequence, the tutor makes suggestions to the student, who has to rewrite the objective text. Once the student reaches a High level in coherence, the progress on the left side is updated, and he can move to work the next section of the draft.

9.8 Conclusions

The mining technique allowed evaluating the global coherence of seven sections in proposal drafts, reaching an acceptable result of the percentage of agreement respect to human reviewers. It was crucial to have a gold standard to compare our results.

The exploration across sections performed after mining domain knowledge, uncovered a consistent interrelationship among them, according to methodology authors. This was a newly developed technique for additional exploration and validation of mined knowledge.

We will continue increasing the size of the corpus, so that the analyzer has a wider coverage, since the computing and information technologies domain is quiet extensive and constantly growing. We also need additional good examples for certain sections (e.g. conclusions or justification) to mine and improve their assessment.

In these initial experiments, the evaluation of coherence analysis was important to identify the student level, but could be improved by evaluating additional aspects in texts such as lexical richness [27] or local coherence. This will help students to improve their writing, and academic adviser would have more time to review the contents of the proposal documents.

We expect that this computational tool generates in students a motivation to develop their proposal drafts and this analyzer will contribute to the advance in their writings. We currently have a web interface for the student to evaluate the draft in coherence analysis. Bringing our model to a different domain does not seem too challenging, neither moving it to a different language, assuming similar language processing resources and corpus are available.

The approach discussed in this chapter contributes to the following topics: (a) web mining of educational sources; (b) mining of assessment produced by the learner educational system interactions; (c) DM applied to the personalization of educational content and services; and (d) information repositories oriented to the educational field.

As far as we are aware of related work, this coherence analysis is the first to mine existing resources by proposal sections and specific for computer science and information technologies. Besides coherence, we also plan to mine language models to guide in the formulation of specific sections in proposal texts. Also we are in the process of developing a method to identify answers to methodological questions within the elements and objective justification of a proposal draft. In addition, it has the potential to be extended to other engineering domains (e.g. electrical, electronics, control, mechanical, etc.).

We foresee an experiment that includes a pilot test with a control and experimental group of students.

Acknowledgments We thank the reviewers: Rene Castro M., Claudia I. Esquivel L., J. Miguel García G., Ramón Cárdenas G., Israel Chávez G., Orlando Madrid M., and Raúl Beltran Q. This research was supported by CONACYT, México, through the scholarship 1124002 for the first author. The second author was partially supported by SNI, México.

References

1. Luan, J.: Data mining and its applications in higher education. New Dir. Inst. Res. **2002**(113), 17–36 (2002)
2. Vilarnovo, A.: Coherencia textual: ¿Coherencia Interna o Coherencia Externa? Estudios de Lingüística **6**, 229–240 (1990)

3. Louwerse, M.M.: A concise model of cohesion in text and coherence in comprehension. Revista Signos **37**(56), 41–58 (2004)
4. Skogs, J.: Subject line preferences and other factors contributing to coherence and interaction in student discussion forums. Comput. Educ. **60**(1), 172–183 (2013)
5. Medve, V.B., Takac, V.P.: The influence of cohesion and coherence on text quality: a cross-linguistic study of foreign language learners written production. In: Piechurska-Kuciel, E., Szymańska-Czaplak, E. (eds.) Language in cognition and affect. Second language learning and teaching, pp. 111–131. Springer, Heidelberg (2013)
6. Yannakoudakis, H., Briscoe, T.: Modeling coherence in ESOL learner texts. In: 7th Workshop on the Innovative Use of NLP for Building Educational Applications, pp. 33–43. Association for Computational Linguistics, Stroudsburg (2012)
7. Higgins, D., Burstein, J., Marcu, D., Gentile, C.: Evaluating multiple aspects of coherence in student essays. In: Human language technology conference/North American chapter of the Association for Computational Linguistics, pp. 185–192. Association for Computational Linguistics, Boston (2004)
8. Miltsakaki, E., Kukich, K.: Evaluation of text coherence for electronic essay scoring systems. Nat. Lang. Eng. **10**(1), 25–55 (2004)
9. Deerwester, S., Dumais, S.T., Furnas, G.W., Landauer, T.K., Harshman, R.: Indexing by latent semantic analysis. J. Am. Soc. Inf. Sci. **41**, 391–407 (1990)
10. Landauer, T., Dumais, S.: A solution to Plato's problem: the latent semantic analysis theory of acquisition, induction, and representation of knowledge. Psychol. Rev. **104**, 211–240 (1997)
11. Foltz, P., Kintsch, W., Launder, T.: Textual coherence using latent semantic analysis. Discourse Process. **25**, 285–307 (1998)
12. Hofmann, T.: Probabilistic latent semantic indexing. In: 22nd international conference on research and development in information retrieval, pp. 50–57. ACM, NY (1999)
13. Lee, D.D., Seung, H.S.: Learning the parts of objects by non-negative matrix factorization. Nature **401**(6755), 788–791 (1999)
14. Lee, S., Baker, J., Song, J., Wetherbe, J.C.: An empirical comparison of four text mining methods. In: 43rd Hawaii international conference on system sciences, pp. 1–10. IEEE Computer Society, Washington (2010)
15. Zhang, M., Yang, H., Ji, D., Teng, C., Wu, H.: Discourse coherence: lexical chain, complex network and semantic field. In: Ji, D., Xiao, G. (eds.) Chinese Lexical Semantics. LNCS, vol. 7717, pp. 756–765. Springer, Heidelberg (2013)
16. Dessus, P.: An overview of LSA-based systems for supporting learning and teaching. In: Dimitrova, V., Mizoguchi, R., du Boulay, B., Graesser A. (eds.) Conference on Artificial Intelligence in Education: Building Learning Systems that Care: From Knowledge Representation to Affective Modeling, pp. 157–164. IOS Press, Amsterdam (2009)
17. Southavilay, V., Yacef, K., Calvo, R.A.: Process mining to support students' collaborative writing. In: Baker, R.S.J.D., Merceron, A., Pavlik Jr., P.I. (eds.) 3rd International Conference on Educational Data Mining, pp. 257–266. International Educational Data Mining Society, Pittsburgh (2010)
18. Jiang, H., Huang, G., Liu, J.: The research on CET automated essay scoring based on data mining. In: Zhou, M., Tan H. (eds.) Advances in Computer Science and Education Applications. Communications in Computer and Information Science, vol. 202, pp. 100–105. Springer, Heidelberg (2011)
19. Villalón, J., Kearney, P., Calvo, R.A., Reimann, P.: Glosser: enhanced feedback for student writing tasks. In: International conference on advanced learning technologies, pp. 454–458. IEEE Computer Society, Washington (2008)
20. Liu, M., Calvo, R.A., Rus, V.: Automatic question generation for literature review writing support. In: Aleven, V., Kay, J., Mostow, J. (eds.) Intelligent Tutoring Systems. LNCS, vol. 6094, pp. 45–54. Springer, Heidelberg (2010)

21. Higgins, D., Burstein, J.: Sentence similarity measures for essay coherence. In: 7th international workshop on computational semantics, pp. 77–88. Tilburg University, Tilburg (2007)
22. Vasile, R., Nobal, N.: Automated detection of local coherence in short argumentative essays based on centering theory. In: Gelbukh, A. (ed.) Computational Linguistics and Intelligent Text Processing. LNCS, vol. 7181, pp. 450–461. Springer, Heidelberg (2012)
23. Kintsch, W.: On the notions of theme and topic in psychological process models of text comprehension. In: Louwerse, M., Van Peer, W. (eds.) Thematics, Interdisciplinary Studies, pp. 157–170. John Benjamins Publishing, Amsterdam (2002)
24. Berry, M.W., Dumais, S.T., O'Brien, G.W.: Using linear algebra for intelligent information retrieval. Soc. Ind. Appl. Math. **4**, 573–595 (1995)
25. Landis, J.R., Koch, G.G.: The measurement of observer agreement for categorical data. Biometrics **33**(1), 159–174 (1977)
26. Hernández, R.: Metodología de la Investigación. Mc Graw Hill, México (2006)
27. García-Gorrostieta, J.M., González-López, S., López-López, A., Carrillo, M.: An intelligent tutoring system to evaluate and advise on lexical richness in students writings. In: Hernández-Leo, D., Ley T., Klamma, R., Harrer, A. (eds.) EC-TEL 2013. LNCS, vol. 8095, pp. 548–551. Springer, Heidelberg (2013)

Chapter 10
Adaptive Testing in Programming Courses Based on Educational Data Mining Techniques

Vladimir Ivančević, Marko Knežević, Bojan Pušić and Ivan Luković

Abstract Designers of student tests, often teachers, primarily rely on their experience and subjective perception of students when selecting test items, while devoting little time to analyse factual data about both students and test items. As a practical solution to this common issue, we propose an approach to automatic test generation that acknowledges required areas of competence and matches the overall competence level of target students. The proposed approach, which is tailored to the testing practice in an introductory university course on programming, is based on the use of educational data mining. Data about students and test items are first evaluated using the predictive techniques of regression and classification, respectively, and then used to guide the test creation process. Besides a genetic algorithm that selects a test most suitable to the aforementioned criteria, we present a concept map of programming competencies and a method of estimating the test item difficulty.

Keywords Programming competencies · Concept maps · Test creation · Classification of test items · Genetic algorithms

Abbreviations

ARC Area coverage
C1 Criterion1

V. Ivančević (✉) · M. Knežević · B. Pušić · I. Luković
University of Novi Sad, Faculty of Technical Sciences, Trg Dositeja Obradovića 6, 21 000 Novi Sad, Serbia
e-mail: dragoman@uns.ac.rs

M. Knežević
e-mail: marko.knezevic@uns.ac.rs

B. Pušić
e-mail: bpusic@uns.ac.rs

I. Luković
e-mail: ivan@uns.ac.rs

A. Peña-Ayala (ed.), *Educational Data Mining*,
Studies in Computational Intelligence 524, DOI: 10.1007/978-3-319-02738-8_10,

C2	Criterion2
CAT	Computerized adaptive testing
CBA	Computer-based assessment
CR	Correct ratio
DF	Difficulty
DM	Data mining
EDM	Educational data mining
FIT	Fitness
FTS	Faculty of Technical Sciences
GA	Genetic algorithm
GC	Generation count
IC	Item count
IRT	Item response theory
M	Mean
MAX	Maximum
MDF	Mean difficulty
MF	Mean fitness
MGC	Max generation count
MH	Math
MIN	Minimum
MT	Mean completion time
NDF	Natural difficulty
NDFC	Natural difficulty category
OWL	Web ontology language
PAS	Past assignment
PLADS	Programming languages and data structures
PS	Population size
PTS	Past test
RA	Random approach
RDF	Resource description framework
SC	Student capacity
SCR	Student capacity rank
SD	Standard deviation
SDF	Standard deviation for fitness
SGC	Student group capacity
SK	Skewness
SPDF	Specified difficulty
SVM	Support vector machine
TPS	Test pool size
TS	Test
TSR	Test ratio
WRST	Wilcoxon rank sum test

10.1 Introduction

Computerized testing represents an area that has emerged together with the rising popularity of personal computers and their increased availability. With their introduction into schools and universities, a large number of students could be swiftly evaluated and graded using tests administered in computer classrooms. Furthermore, the switch to digital tests provides numerous benefits to both teachers and students.

The teacher's burden of grading each individual test using a same key is greatly reduced because the solution and grading process need to be specified only once while a computer may execute it as many times as necessary. Moreover, computer tests exist only in digital form, which eliminates the need for official storage space, as well as the time and costs associated with the copying of test forms. Additional advantage is that students may retrieve test results immediately after the completion of the test, thus obtaining timely feedback on their performance. Testing based on computers is suited primarily to tests with precisely defined structure and solutions. Multiple choice tests, which are in widespread use across different levels of education and research, fit this format very well.

Although this rigidity may appear to exclude many other forms of educational assessment and impose severe restrictions on the testing process, there have been many studies devoted to "intelligent" approaches to computerized testing that allow for very complex evaluation of student knowledge. The most important activity in the testing process, irrespective of the testing format, is the creation of a test. A test designer is responsible for forming a set of test items that encompass all relevant areas at the prescribed level of knowledge. This activity is sensitive to changes, as seemingly minor test modifications may cause noticeable differences in student performance. The knowledge gap between teachers and students, together with other distinctions between these two groups, may introduce additional difficulties into test creation. A teacher may assume that students have acquired necessary competencies, although the teaching process may not have been completely successful. For these reasons, we devise an approach that could overcome some of the aforementioned problems.

Our goal is to create software infrastructure for the computerized testing of programming knowledge that supports adaptation of tests to the competence of a target student group in a university course. We focus on the automatic creation of static tests for introductory university courses on structured programming, particularly the C language, and data structures. Tests created in this manner could be used as exercises tailored to a student group or even as finely tuned assessments determining the final grade. They should provide good separation of students into different classes according to their level of programming knowledge and be concise yet thorough in the examination. In this manner, generated tests could overcome the following problems:

- The bias of an examiner during the construction of tests.
- Long testing times, which is especially important in settings with limited technical facilities and many students.
- Unsatisfactory correspondence between the test results, and actual student understanding and competencies, which is usually caused by the unadjusted difficulty of the administered test.

The proposed approach is not necessarily restricted to computerized testing, but it may be best applied in such setting, since data required for test generation should be in electronic format. The identified problem is suitable for the application of educational data mining (EDM) [1] because the testing process may yield large data sets and the tuning of the test creation algorithm depends on the analysis of historical data concerning students and test items.

The foundation for the administration of such tests would be the Otisak testing system [2], which is a web-based software solution extensively used at the Faculty of Technical Sciences (FTS) in Novi Sad, Serbia. For this reason, the implementation of the proposed approach is tailored to the testing practice employed in an introductory programming course and information available at FTS. In the Otisak system, assessment of student knowledge is primarily conducted using multiple choice and short answer tests on programming. We intend to utilize assessment logs generated by this system that include student scores and copies of individual tests with recorded student answers. In the analysis of these logs, we may rank and evaluate test items [3] by utilizing classification algorithms.

Moreover, by mapping test items to programming language concepts, we form a concept-based foundation for the automatic creation of comprehensive programming tests. Information about previous student performance on related tests may be used in the additional refinement of tests, i.e., we mine available records with the goal of creating a student model [4].

This would allow for explicit acknowledgement of differences between specific groups of students during the construction of tests. In other words, testing may be considered adaptive because generated tests are suited to the actual competencies of the target groups of students. Therefore, the two key requirements that drive test generation are adequate coverage of the specified areas of competence and specified difficulty, which may be automatically calculated for a group of students.

In order to acknowledge both requirements, we create a genetic algorithm (GA) [5] that is designed to discover the best test with respect to the two criteria. Moreover, we also evaluate the efficacy of the generated tests [6, 7], by comparing them with those created using a random method generator. The research activities that lead to the implementation of the approach include:

- Creation of a formal model of programming concepts and competences that is suitable for a university course on programming (primarily for the C language).
- Mapping of previously used test items to programming concepts and competencies.
- Classification of test items according to their difficulty.
- Formal definition of the structure of a student profile.

- Estimation of current student competency.
- Creation of an algorithm for test generation that acknowledges student competencies (or explicitly specified difficulty), test item classes, and test comprehensiveness.
- Evaluation of tests created in this manner.

The rest of this chapter is organized in the following manner: in Sect. 10.2, an overview of the related work on computerized testing is given; in Sect. 10.3, background information about the testing method and the programming course is provided; in Sect. 10.4, a model of programming concepts and competencies is presented; in Sect. 10.5, we demonstrate how item difficulty and student competence may be predicted; in Sect. 10.6, the genetic algorithm for test creation is presented; in Sect. 10.7, an application of the proposed approach is illustrated; and in Sect. 10.8, the chapter is closed with concluding remarks and ideas for future research.

10.2 Related Work

Automated preparation and administration of tests in programming has become a necessity because competencies related to the computer and programming skills have become the norm in various disciplines and many professionals need to be rapidly trained. However, this field also has a relatively long history, as many custom solutions have been built over the past few decades.

As discussed in [8], computer-based assessment (CBA) brings many benefits to higher education. Contrary to the popular opinion, CBA allows various types of assessment, which are beneficial for students. The author illustrates how multiple choice tests, which are sometimes regarded as suitable for the evaluation of factual knowledge and too simple for advanced assessment, may be used to give complex problems to students. Their integration with randomization techniques, despite high initial costs to set up, could save significant time, especially in environments that are stable. In the context of the course from which the assessment data are retrieved for the study presented herein, multiple choice items have also proved to be a valuable tool when assessing more advanced competencies in programming. Thus, the proposed approach to generation of multiple choice tests could be viewed as a beneficial method of test creation, both in terms of the shortened design time and the reduced possibility of error, but under the condition that the individual test items are carefully designed.

Another group of valuable programming assessments are laboratory exams, where students have to write a working program in the computer laboratory, while being observed by the invigilators [9]. This also represents one of the applied methods in the course that we analyze. However, the authors also employ a special web system that is used in the whole assessment process, from the presentation of program specification, to program testing and storing of the final result.

More information about the rich history of automated assessment systems and techniques in computer sciences may be found in [10]. The authors also identify some of the problems associated with automated assessment and recognize the need for a human assessor, while the assessment systems should act as a support in the education process. A study on more recent advances in automatic assessment for programming exercises is presented in [11]. According to the authors, one of the big problems in this area is the lack of open solutions that may be freely applied by others. This leads to the proliferation of in-house solutions and the need to implement standard system features from scratch. The creation of a student test, together with the selection of adequate test items, is a delicate activity that is not so rarely based on intangible factors guiding the process. An important part of designing a test typically relies on the experience of a teacher/designer to estimate the difficulty of items.

Item Response Theory (IRT) [12] provides a powerful framework for test construction and tuning, in which items have a central role. The probability of a correct response to an item is modelled by a set of item parameters and depends on the examinee's ability. As a result, such approach allows for shorter assessment times, as well as adaptive testing, which represents assessment tailored to the individual ability. It has also boosted the development of computerized adaptive testing (CAT). However, despite the relatively long history and considerable research effort behind this theory, its strong requirements, difficult interpretation, and formal background have most likely been the reasons for its slow adoption among teachers. Owing to the static nature of tests and the lack of a dynamic assessment system in the analyzed course, we could not fully utilize the main benefits provided by IRT. This has partially motivated us to adopt a somewhat different approach to test adaptation that could be more acceptable for some educational settings.

Moreover, as reported in [3], Proportion Correct, a simple item difficulty estimate based on the proportion of correct answers, outperformed other more advanced estimates. By relying on the proportion of correct answers to estimate item difficulty, we defined an easily understandable set of item difficulty classes and a process in which items could be categorized. We also supported addition of new items, which is typical of the analyzed course, and allowed for immediate use of items without having to conduct experimental difficulty estimation. For this purpose, among several variables, we also utilized the expert estimate of item difficulty in the categorization of new items. During the creation of an item difficulty classifier, we corroborated the finding reported in [3] that the expert estimate is not always one of the best methods, since we found that the number of different concepts associated with an item correlates better with the proportion of correct answers as opposed to the expert estimate.

Formalizing the representation of knowledge in some area is another important element in the test design that allows for an approach which is less subjective and, most probably, less error-prone. Ontologies are typically used to represent concepts and their relationships in a domain. An educational ontology for the programming in the C language is presented in [13]. The authors also provide a number of guidelines on creating a "beautiful ontology", i.e., one that should be clear, symmetrical, and well-organized. The actual ontology is publicly available in [14].

However, ontologies were originally devised to specify concepts that should be shared on the Semantic Web and automatically processed by computers. Such environment implies that a single ontology for some domain should be created, published, and shared by participants that are not necessarily known in advance. Unlike the aforementioned knowledge representations, our model of the programming knowledge is a particular solution, a concept map specifically created to match the requirements in a university course on programming and data structures. The omission of the required concepts would render our map incomplete, while the inclusion of concepts not discussed in the course would further complicate the map without any actual benefit for the people involved. Moreover, its primary users, teachers involved with the course, may not be that familiar with ontologies in general. Another benefit is that a concept map may be more readily comprehended and modified, if needed. More information about the ontologies and concept maps is given in the introductory part of Sect. 10.4.

10.3 Background

The creation of estimations of students and test items requires mining of logs [15] from the student testing system, which have to be parsed and imported into a specially designed database. Since the complexity of student models and item difficulty estimates depends on the richness of the extracted information, the proposed approach is primarily tailored to the quality of the available data. In order to implement and evaluate the approach, we relied on the testing logs collected during the organization of a university course on programming. More information about the environment from which the data originate, as well as about the data set, may be found in the following two subsections.

10.3.1 Environment

The data used in this study represent records associated with student tests that were organized as part of the Programming Languages and Data Structures (PLADS) course. This course is held at FTS (University of Novi Sad, Serbia), as a first year introductory course on programming for students of Computing and Control.

All the programming is taught using the C programming language, with the special attention devoted to the data structures. More information about the structure of the course may be found in [16].

The final grade in the course is determined by the score in pre-exam assessments, which are conducted in practical classes during the semester; and the score in the written theory exam, which is organized at the end of the semester. However, we restrict our analysis to the tests conducted during practical classes, which are held in a computer laboratory. There are two programming assignments, which require writing a C program, and typically three to four tests on programming.

The laboratory was specially designed for courses in computer science [17]. It features a computer-based student testing system named Otisak, which supports multiple choice, multiple response, and fill-in-the-blank test items.

Student activity during testing, together with test items and student answers, is recorded in the form of electronic logs and database entries. These records represent the main source of data to implement and test the approach proposed in this chapter.

10.3.2 Data Set

The data produced by the Otisak system is imported to a separate database that is used solely in the analysis of data about previous tests, student scores, and individual items. The database schema is shown in Fig. 10.1.

The stored data cover the period from 2008 to 2013. They are related to several computer science courses held at FTS, including the PLADS course.

There is basic information about each conducted test (table *Test*), assessed student (table *Students*), test items (table *Question*), test item options (table *QuestionAnswer*), and the corresponding course (table *Course*).

For each test, there are records describing which students took the test (table *StudentTakenTest*) and which test items (questions) were used (table *QuestionTest*). For each test item that a student was answering, there are records about the system events (table *AnswerLog*).

For test items, there are records about the covered knowledge areas (table *QuestionInfo*) and item difficulty (table *QuestionDifficulty*) with respect to some course.

There is information about 1,055 tests (including all courses, and many tests for the debugging and preparatory purposes), while 124 tests are related to the analysed introductory course on programming.

10.4 Modeling Programming Knowledge

In order to generate comprehensive tests about programming, basic information about a test item (e.g., for multiple choices items a designer provides a stem and a set of options) should be extended with the specification of important areas and competencies covered by that particular item.

The simplest solution would be to create an unordered list of areas and competencies, and to assign to each test item one or more list elements. Although the

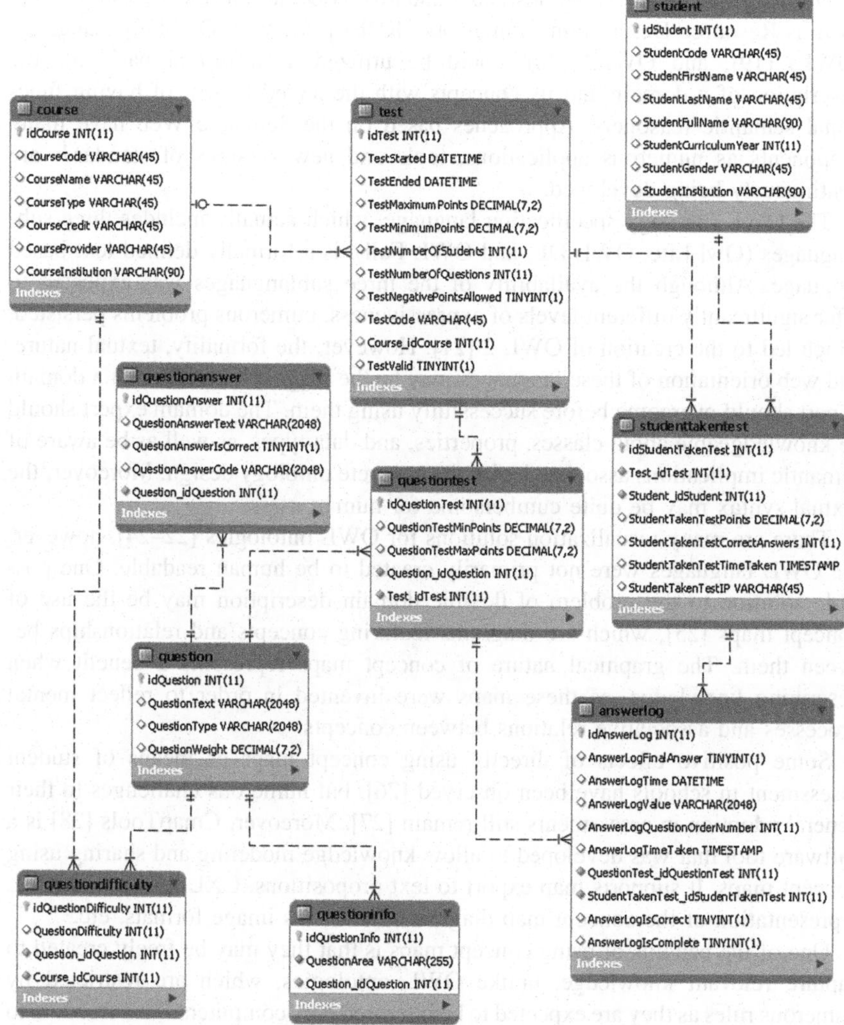

Fig. 10.1 The test log database schema

creation of such list would not require a special set of skills except possessing domain expertise, the final product would be largely impractical as any non-trivial domain or subdomain typically features at least several hundred concepts.

As a result, navigating a large unordered list in search for an adequate concept would be very time-consuming and tiring for a domain expert. Furthermore, such representation of the domain may not adequately transfer domain knowledge to non-experts, namely students who could only benefit from the access to the domain model.

On the other hand, ontologies and standards associated to the Semantic Web, such as Resource Description Framework (RDF) [18], Web Ontology Language (OWL) [19], and OWL2 [20], could be utilized as a formal basis for the description of a domain and its concepts with the added benefit of having functional semantic reasoners. Approaches based on the Semantic Web have many proponents as numerous applications, tools and new versions of standards are continuously being developed.

The OWL ontology specification language, which actually includes three sublanguages (Owl Lite, OWL DL, and OWL Full), is a formally defined text-based language. Although the availability of the three sublanguages was expected to offer significantly different levels of expressiveness, numerous problems persisted, which led to the creation of OWL 2 [21]. However, the formality, textual nature, and web orientation of these languages may be the biggest obstacles that a domain expert should overcome before successfully using them. The domain expert should be knowledgeable about classes, properties, and data types, as well as be aware of semantic implications associated with the concrete ontology design. Moreover, the textual syntax may be quite cumbersome for human users.

There are many visualization solutions for OWL ontologies [22–24], however, the OWL languages were not primarily created to be human readable. One possible solution to the problem of flexible domain description may be the use of concept maps [25], which are diagrams featuring concepts and relationships between them. The graphical nature of concept maps represents a benefit when describing knowledge, as these maps were invented in order to reflect mental processes and associative relations between concepts.

Some positive effects of directly using concept maps as means of student assessment in schools have been observed [26], but numerous challenges to their general adoption in assessments still remain [27]. Moreover, CmapTools [28] is a software tool that was developed to allow knowledge modeling and sharing using concept maps. It supports map export to text propositions, CXL format (an XML representation of the concept map diagram), numerous image formats, etc.

One of the benefits of using concept maps is that they may be freely created to capture relevant knowledge, unlike OWL ontologies, which are restricted by numerous rules as they are expected to be interpreted by computers. This freedom to create arbitrary concepts and relationships without the overhead caused by many formalisms also facilitates communication between domain experts who are creating a concept map, as well as between students who are exploring the modeled domain.

The structure of the map diagram is not limited to tree structures but allows for any graph, i.e., several connections may point to a single concept. In case the map needs to be programmatically processed, its CXL representation may be parsed. For these reasons, we have opted for a concept map to express the programming knowledge. We primarily require that the knowledge representation may be relatively easily created and understood by a layperson, as well as programmatically accessed and analysed when estimating individual test items. In the remainder of this section, we propose a model of programming competencies and concepts, which was created using the CmapTools software.

10.4.1 Programming Knowledge Overview

In order to simplify the concept map and its parsing, a tree structure of concepts was created by a teaching assistant from the analyzed course, together with a set of auxiliary links that are not parent–child links, which are known as cross-links. In the resulting knowledge model, excluding the cross-links, each concept may have zero or more child concepts. On the other hand, each concept has exactly one parent concept, save for the root concept, which does not have a parent. A link between concepts (typically marked with a noun) includes a linking phrase (marked with a verb) and a set of connections between the concepts and the linking phrase.

The resulting concept map was created to match the knowledge outcomes of the PLADS course organized at FTS. It includes 309 concepts, 93 linking phrases, and 401 connections. It has a single root concept labeled *Programming in C*. All of the relevant concepts are divided into two groups: one containing concepts related to general competencies associated with structured programming and the other containing specificities of the C language. The root concept of the first group is *Programming Competencies*, while the second group starts from the concept *C Language Elements*. Between the two subtrees starting from the two abovementioned concepts, there are many crosslinks matching programming competencies with concrete constructs and keywords of the C language. The first two levels of the concept map are presented in Fig. 10.2. The concept map represents a solution that has been created for a concrete university course, as well as the test generation problem which we are attempting to solve.

This is mainly evident from the structure of the *Programming Competencies* subtree, which includes some common programming principles embodied in *Algorithmic Thinking* and *Code Styling*. Although these principles are universal, the exact structure and level of detail with which they are presented to students may differ between institutions. Moreover, teachers are expected to design a course according to the restrictions imposed by a study program, such as class duration and frequency.

For these reasons, some teachers may find the concept map too detailed or even limited in scope, depending on the objectives of the programming courses that they teach. On the other hand, the *C Language Elements* subtree contains the technical

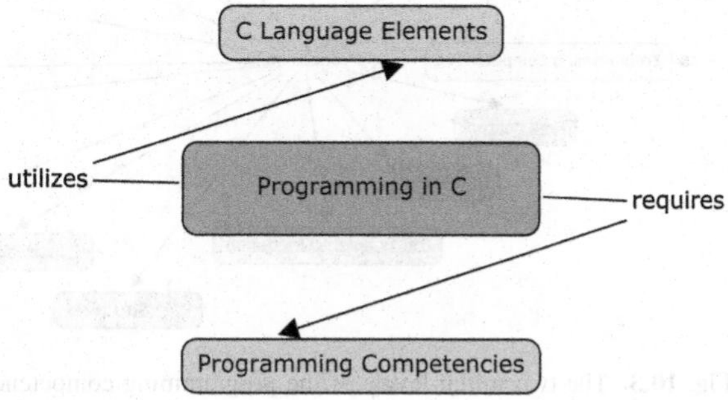

Fig. 10.2 The two initial levels of the programming knowledge model

terms from C that are standardized, which makes this portion of the map reusable in different institutional and educational contexts.

In the following subsection, some of the general concepts from the two subtrees are presented. In the model, each concept that denotes a subtree corresponding to a module graded in the practical portion of the programming course has an underlined label—there are 19 such concepts in total. In order not to clutter the diagrams with numerous connections, cross-links are excluded from the provided map excerpts.

10.4.2 Modeling Programming Competencies

Programming competencies are organized by areas, such as *Variable Manipulation*, *String Manipulation*, *Flow Control*, *Command Line Arguments*, *Numeral Systems*, and *Data Structures* (see Fig. 10.3). They typically correspond to the mastery of a basic command (or a set of them) that have similar outcomes in different programming languages. The presented portion of the concept map includes skills that are typical of the structured programming paradigm and mostly transferable between various languages, e.g., C, C++, and Java. The listed concepts are further decomposed. For example, the *Data Structures* concept encompasses 29 other concepts.

The majority of these competencies are graded (13 in total), while student knowledge in areas *Algorithmic Thinking* and *Code Styling*, which are taught and encouraged throughout the course, is not explicitly assessed during computerized testing.

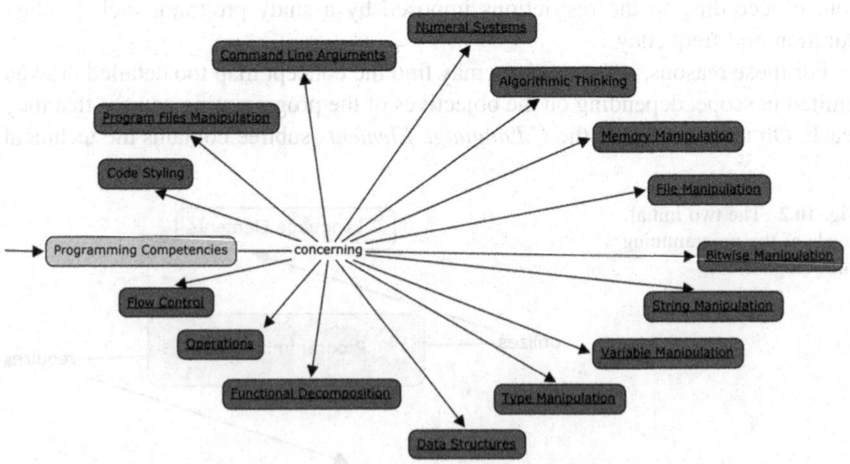

Fig. 10.3 The two initial levels of the programming competencies subtree

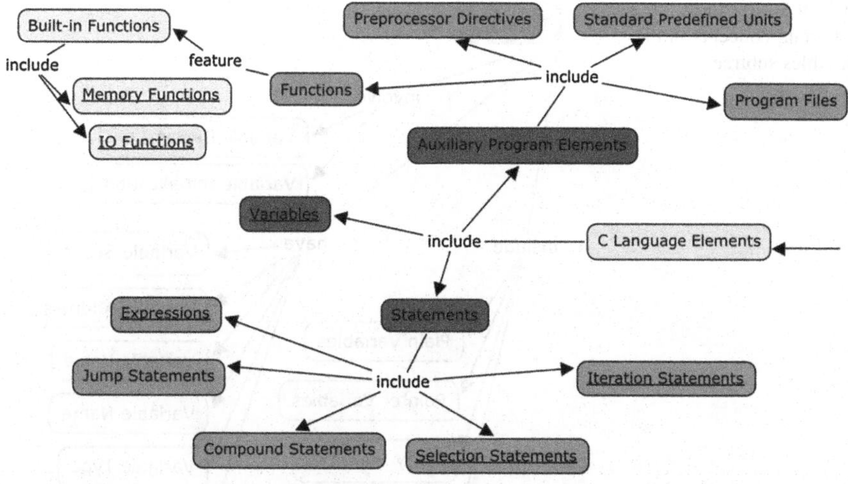

Fig. 10.4 The most important concepts in the C Language elements subtree

10.4.3 Modeling Programming Concepts of the C Language

We organized the key constructs and capabilities of the C language into three broad categories: Variables, Statements, and Auxiliary Program Elements. Some of the most prominent members of these categories are presented in Fig. 10.4.

There are in total 6 key areas belonging to the *C Language Elements* sub tree of the knowledge model (underlined in Fig. 10.4), which are thoroughly covered in student assessments: *Variables*, *Expressions*, *Selection Statements*, *Iteration Statements*, *IO Functions* and *Memory Functions*.

The suggested "taxonomy" should not be viewed as a strict scientific overview of a generic programming language, but as a model of the features of the C language, as presented in the educational context of an introductory course on programming. Many of the directly linked concepts do not conform to the "is-a" relationship. They rather represent arbitrary associations between concepts, in the way they may be mentally formed by students during classes and individual study. As a result, a concept may be further linked to other concepts that designate subclasses, properties, or behaviour of their parent concept. An example of these relationships may be observed for the *Variable* concept (see Fig. 10.5).

For instance, program variables may be classified as plain, pointer, array, or matrix variables. Each variable has several properties: scope, address, value, name, and type. A variable may be declared or initialized. However, not all information about variables in C is covered by the *Variable* subtree of the model.

There may be a cross-link to the *Statements* subtree, namely the *Expressions* concept, whose subtree is the most populous in the model. *Expressions* in C may involve variables, thus specifying additional operations applicable to variables. Due to the complexity of many concepts, information about a single concept

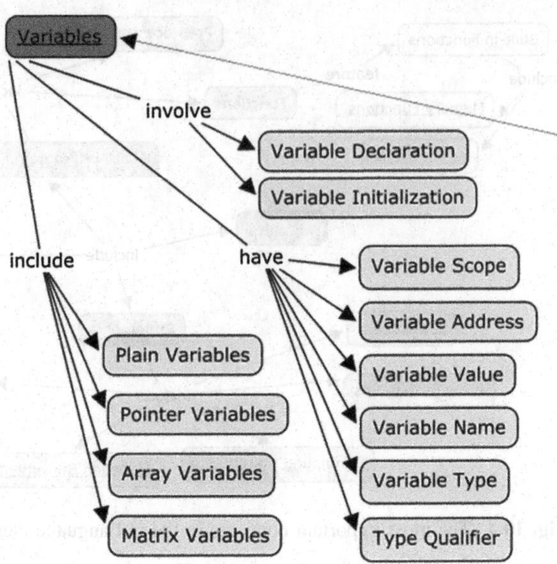

Fig. 10.5 The most important concepts in the variables subtree

cannot be contained within a single subtree. In order to create an intuitively understandable model, the model designer has to choose global demarcation lines between concept groups, which may be connected using cross-links when needed.

10.5 Estimating Test Difficulty

Student performance in an assessment is primarily influenced by the requirements level of the assessment (test difficulty) and the competence of the assessed students. When creating a new test, both factors should be taken into account. The test should cover all topics relevant for the particular assessment and have an adequate level of difficulty, as determined by the difficulty of individual test items. However, these two requirements are not easy to fulfill in practice. The testing time is often limited by the available time during regular classes, which restricts the number of items in the test and, consequently, the test comprehensiveness.

Furthermore, teachers may not always correctly estimate the difficulty of a single item. They may misjudge the knowledge of students, thus creating a test in which students perform very bad or very well. An algorithm for automatic test creation, which attempts to overcome these problems, utilizes available data about items, students and previous tests, in order to generate a test satisfying the aforementioned conditions. By performing educational data mining, we estimate the difficulty of items and future performance of a group of students who need to be assessed.

As a result of the process, the predicted values may be passed to the test generation algorithm, which further uses them as guidance during the automatic selection of items for a test. Once the selection of items is finished, the resulting test may be administered to the target students. In this manner, we obtain a test with the required number of fixed items that is tailored to the estimated ability of a student group as a whole and not to the individuals. In the following two subsections, we demonstrate how item difficulty and student performance may be represented and predicted in the context of the PLADS course at FTS.

10.5.1 Estimating Test Item Difficulty

Tests used in the practical assessment of students in the analyzed course primarily include multiple choice questions as items. In order to build a test of the required difficulty, for each possible item there should be a numerical estimate of its difficulty. With this information, the test difficulty could be calculated as the arithmetic mean of the difficulty values matching the pertaining items.

The individual item difficulty is neither explicitly modeled nor evaluated differently for each student, because, in the analyzed course, a test is always administered to a student group and each student receives the same set of items that cannot be modified once the test has started. The simplest solution would be to calculate item difficulty using the percentage of correct answers for an item in the past tests. However, there are two problems with such solution.

The first problem is that the proposed solution is possible only when there are sufficient records in the assessment log about each item. For the analyzed course, we observed that the percentage of correct answers for an item may increase if that item is often repeated in different assessments, most probably because assessed students readily share information about the completed tests with their peers. For this reason, new items are constantly being added to the item pool. Nonetheless, these items do not have their own percentage of correct answers and, hence, their difficulty cannot be estimated. In case that the item's difficulty is unknown, there is a non-negligible risk that a new test item might be too difficult (or easy) for students, which may lead to an unjustified change in scores and overall student performance. The second problem is related to the meaning of the difficulty estimate. The percentage of correct answers for an item is an exact value, but for teachers this value alone may not be sufficient to understand the difference between items with respect to their difficulty or know exactly for which percentage of correct answers the item becomes difficult. In other words, there is no suitable interpretation of these values for the purpose of test creation.

In order to remedy the identified problems, in this subsection, we present how the difficulty of test items may be represented to offer a more manageable interpretation for teachers. Furthermore, we also demonstrate how the newly defined difficulty may be estimated for an item in two scenarios: when the item has been extensively used in past assessments and when the item is previously unknown or rarely used. For the purpose of illustrating the estimation process, as well as the identification of important variables related to item difficulty, we have selected a sample of 172 items from the test records database.

All of the items were annotated with the matching programming concepts defined in the model from Sect. 10.4. As demonstrated in this section, concepts mapped to an item also provide valuable information about the item's difficulty. The individual items were chosen so that they reflect different type of programming questions which typically appear in student assessments. For each item, there is also the percentage of correct answers (*CorrectRatio*, *CR*) recorded during past assessments, which is the key variable in the estimation of item difficulty.

In order to add the meaning to the difficulty of an item that has been extensively used, we decided to first simplify the estimation problem, by discretizing the percentage of correct answers. The percentage range for the analysed sample is automatically divided into three intervals using the Jenks natural breaks classification method [29], whose implementation is available in R environment for statistics and data analysis [30]. As a result, we obtain the categories for the difficulty of test items (*NaturalDifficultyCategory*, *NDFC*). The generated partition, including the meaning for each category, is shown in Table 10.1, while the histogram for *CR* is presented in Fig. 10.6.

As our aim was to provide a limited number of difficulty levels that are sufficiently separated but enclose a comparable number of items, according to the distribution of *CR* values, we defined exactly three readily interpretable categories, where easy items are denoted by 1, items of moderate difficulty by 2, and difficult items by 3.

In the automatic calculations involving item difficulty, we are using more precise (and more informative) values in addition to the three integers. For an item *it* with a known value of *CR*, the exact value of *NaturalDifficulty* (*NDF*), which lies inside the interval [0.5, 3.5], may be calculated using the formula (10.1):

$$NDF(it) = ndc - 0.5 + (U_{ndc} - CR(it))/(U_{ndc} - L_{ndc}), \tag{10.1}$$

where *ndc* is the natural difficulty category of the item *it*, L_{ndc} the lower bound of the item's *CR* for the category *ndc*, and U_{ndc} the upper bound of the item's *CR* for the category *ndc*. In this manner, if the *CR* value of an item is the midpoint of the interval defined by L_{ndc} and U_{ndc}, the corresponding *NDF* is equal to the item's *NDFC*, while the *NDF* for the other *CR* values typically falls somewhere between the integer values matching the three categories.

Table 10.1 Natural segmentation of the percentage of correct answers

Natural difficulty category	Correct ratio range		Number of items in the category
	Lower bound	Upper bound	
1 (Many correct answers)	0.76 (exclusive)	1 (inclusive)	50
2 (Majority of correct answers)	0.4688 (exclusive)	0.76 (inclusive)	65
3 (Minority of correct answers)	0.0323 (inclusive)	0.4688 (inclusive)	57

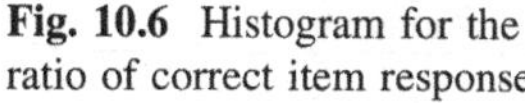

Fig. 10.6 Histogram for the ratio of correct item response

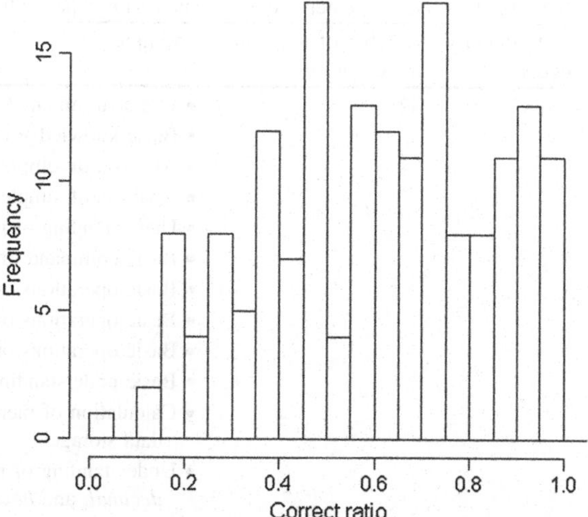

In the estimation of the difficulty of a new or rarely used item, i.e., when the item's *CR* value is unknown or not representative of the item's difficulty, we predict *NDFC* of an item using other item-related variables. In this scenario, the exact difficulty (*NDF*) is not predicted but only the corresponding category (class) due to the predictive quality of the available variables.

The simplest solution would be to have an expert give a difficulty estimate for each new test item (*ExpertDifficulty*). For this purpose, a teaching assistant from the analysed course rated all available test items on an integer scale from 1 to 3, where 1 denotes an easy item while 3 denotes a difficult item.

The principal criterion for assigning labels was the complexity of the problem presented in a test item. An item is considered easy (label 1) if it requires basic reproduction of some piece of information presented in the course, or an analysis of a simple statement written in C. A moderate item (label 2) is the one that typically combines from two to four language constructs (or a combination of practical and theoretical ideas) and requires an analysis of their interaction or relationship.

A difficult item (label 3) requires a careful analysis of a program code featuring a combination of advanced concepts (typically including pointers) with complex flow controls and strong dependencies between presented code sections.

An overview of the expert classification of test items is given in Table 10.2. For each difficulty category, we present the label, number of items from the sample that are assigned to that particular label, and examples of what is typically being evaluated in that category.

However, the expert estimate of the difficulty of an item has a weak-to-moderate correlation with *CR*, which is closely related to *NDFC*, and the items are less evenly distributed between the categories as opposed to *NDFC*. Therefore, we included other item-related variables in the estimation, for which we hypothesized

Table 10.2 Overview of categories associated with expert difficulty estimation

Expert difficulty category	Number of items in the category	Examples
1—Easy	42	• Use of terminal • Basic knowledge of types • Analysis of simple selection statements • Analysis of simple iteration statements • Understanding of memory addresses • Basic command of memory functions • Basic operations on arrays • Basic operations on strings • Basic operations on files • Basic understanding of C structures • Calculation of memory requirements for concrete data storage • Understanding of numeral systems (*binary*, *octal*, *decimal*, and *hexadecimal*)
2—Moderate	99	• Use of pointers • Advanced use of expressions (in selection and iteration statements) • Advanced string operations • Bitwise operations (operators and masks) • Functions (declaration, definition, calling) • Advanced knowledge on types (duality of certain types and type conversion) • Use of command line arguments • Use of memory functions for custom data structures • Analysis of moderately complex code
3—Difficulty	31	• Analysis of very complex code featuring competencies from all categories

that there should be a positive relation to item difficulty. For each item in the analyzed sample, we calculated three additional values:

- AreaCoverage$_{\text{all}}$ (ARC$_{\text{all}}$). A number of *all* 19 relevant areas from the knowledge model (underlined concepts in the model from Sect. 10.4) that are at least partially covered by the item, i.e., the item is associated to a concept within one of the 19 relevant subtrees.
- StemLength. A number of characters in the stem, where an occurrence of multiple consecutive whitespace characters counts as a single character.
- KeywordCount. A number of the reserved words in the C language that appear in the stem.

In Table 10.3, for the three aforementioned variables and the expert difficulty estimate, we present mean (*M*), standard deviation (*SD*), minimum (*MIN*), maximum (*MAX*), skewness (*SK*), and Pearson's correlation coefficients with respect to *CR*. With respect to correlation with *CR*, the best variable is AreaCoverage$_{all}$,

Table 10.3 Summary statistics and correlation for predictor variables

Variable	M	SD	MIN	MAX	SK	Correlation to CR
ExpertDifficulty	1.936	0.65	1	3	0.063	−0.276
$\text{AreaCoverage}_{\text{all}}$	0.142	0.069	0.1	0.26	0.238	−0.392
StemLength	175.878	55.32	74	325	0.464	0.032
KeywordCount	4.395	2.306	0	12	−0.105	−0.150

which surpasses even the expert estimate (*ExpertDifficulty*), the second best variable.

On the other hand, *KeywordCount* exhibits weak correlation, while *StemLength* appears to be linearly unrelated to CR. As a result, the estimation of the difficulty of an item without its *CR* value is performed using a classifier that is created including the following independent variables:

- ExpertDifficulty.
- $\text{AreaCoverage}_{\text{all}}$.
- StemLength.
- KeywordCount.

The only dependent variable is:

- NDFC.

Although the preliminary analysis of variables *StemLength* and *KeywordCount* does not indicate that they have significant predictive quality, they are included in the classifier because their presence offered small improvements in the prediction rate, as discovered in the initial experiments with the classifier.

For illustrative purpose, after the data preparation phase on the set of the 172 test items, we tested four different types of classifiers, which are available as part of the R environment. The classifier performance was evaluated with respect to the training error, cross-validation error (for 10 folds and 100 experiments), and Fleiss' kappa.

The results are presented in Table 10.4. The support vector machine (SVM) classification using the Crammer-Singer native multi-class method [31] has the lowest errors overall and the top kappa value. The best results for this algorithm are obtained using the radial basis function (Gaussian) kernel with parameters $C = 80$ (the parameter in the cost function) and $sigma = 80$ (the kernel function

Table 10.4 Classifier performance on the testing set

Classifier	Training error	Cross-validation error	Kappa
Support vector machine	0.081	0.490	0.877
K-nearest neighbor	0.081	0.515	0.877
Decision tree	0.267	0.531	0.597
Naive Bayes	0.467	0.506	0.277

parameter), as indicated by the results of a two-step grid search for good parameter values.

The first error value, which most probably is lower than it could be expected for a set of new items, illustrates how much the model is misclassifying the training data. The second error value, which is calculated when performing the k-fold cross validation, generally provides a more realistic perspective on the performance of the classifier on new data. However, this error describes a classifier trained without using one kth of the potentially valuable data. The kappa value indicates the agreement with the used data over the one expected by chance.

Given all the aforementioned information, the difficulty estimate (*Difficulty*, *DF*) for any available item may be made in the following manner (10.2):

$$DF(it) = \begin{cases} 3, & valid\ CR(it) \wedge NDF(it) > 3 \\ NDF(it), & valid\ CR(it) \wedge 1 \leq NDF(it) \leq 3 \\ 1, & valid\ CR(it) \wedge NDF(it) < 1 \\ NDFC(it), & not\ valid\ CR(it) \end{cases} . \tag{10.2}$$

When there is a representative *CR* value for an item, we may calculate a more precise estimate, while, for all the other cases, the trained SVM classifier is utilized to estimate the item difficulty by predicting the item's *NDFC*.

By using the difficulty estimates of available items, we may evaluate the student capacity to do well in assessments with respect to item difficulty, as well as parameterize the test generation algorithm to construct a test matching the capacity of a group of students.

10.5.2 Estimating Student Capacity

In the proposed approach, the term student capacity denotes the ability of a student to perform well in a given course, i.e., achieve good scores in the assessments conducted by course teachers. There are three primary requirements when estimating student capacity for the purpose of test creation. First, there should be a predictive model for student scores in programming. Second, there should be a measure of student capacity that is related to student scores and has meaning for teachers, similarly to the case of item difficulty from the previous subsection. Third, there should be a clearly defined relationship between the measure of student capacity and measure of item difficulty because such relationship would allow direct comparison between the capacity of a student group and the difficulty of a test with its pertaining items. In this manner, the quality of an automatically generated test could be evaluated with respect to the capacity of the target student group. In the remainder of this subsection, we present our solutions to the three issues.

It is difficult to predict the exact performance of students in tests that are held during the PLADS course because the course is organized in the winter semester (the first semester) of the first year of study, when there are still almost no data about the academic achievement of students. For the initial portion of the semester, the only available information includes student scores from the university entrance exam. Given the fact that the entrance exam is used to evaluate the proficiency in mathematics, the resulting student score is not directly related to the score in a programming test. However, there is moderate positive correlation between the score in the entrance exam on mathematics and score in the programming tests in the analyzed course.

The value of the Pearson's correlation coefficient for the two variables in the academic year 2011–2012 is 0.477 (0.475 for the Spearman's correlation coefficient). Moreover, once the initial programming assessments are completed, it is possible to improve capacity prediction by utilizing student scores from the already completed programming tests and assignments.

Owing to the preliminary findings, we opted for the multiple linear regression as means for estimating student score in some of the n_a programming tests that are conducted during a semester. The independent variables include the entrance exam score in mathematics (*Math*, *MH*), the total score in the first i programming assignments (*PastAssignment*, *PAS*), and the total score in the first i programming tests (*PastTest*, *PTS*), while the dependent variable is the total score in the $n_a - i$ remaining tests in the programming course (*Test*, *TS*). The i value is an integer from $[0, n_a - 1]$. For a student s, the regression model is of the following form (10.3):

$$TS(s,i) = \beta_3(i) \cdot PTS(s,i) + \beta_2(i) \cdot PAS(s,i) + \beta_1(i) \cdot MH(s) + \beta_0(i), \quad (10.3)$$

where β_3, β_2, and β_1 are the regression coefficients for the predictors *PTS*, *PAS*, and *MH* respectively, while β_0 is the intercept. The regression coefficients and intercept depend on the number of already completed tests in a semester, which is marked by i. Because each test marks a milestone during a semester, a semester part is defined as a period between two consecutive tests, including the special case of a period before the first test.

Therefore, an i value also denotes the $(i + 1)$th part of the semester. As students complete programming tests, i.e., progress from one semester part to the next one, there is more information about students' programming knowledge. This information, which is present in the cumulative test scores, may be used to predict performance in the remaining tests. However, it also changes throughout the semester.

As a result, for each part of the semester, we utilize a different regression formula with its own set of values of the coefficients and intercept. For illustrative purpose, in Table 10.5, we present information about all regression formulae for 144 students enrolled in the winter semester of the academic year 2011–2012, when there were three programming tests ($n_a = 3$).

The values of the regression coefficients, together with the allowed ranges for *TS*, *PTS*, *PAS*, and *MH*, indicate that, as a semester progresses (i increases), the *PTS* variable becomes more important and the variables *PAS* and *MH* less

Table 10.5 Details about regression formulae for student programming performance

i	Allowed range				β_3	β_2	β_1	β_0	R^2	p-val
	TS	PTS	PAS	MH						
0	0–40	/	/	0–60	/	/	0.230	21.253	0.198	$<10^{-7}$
1	0–30	0–10	0–15	0–60	0.387	0.342	0.087	13.507	0.271	$<10^{-7}$
2	0–10	0–30	0–15	0–60	0.198	0.090	0.022	−0.147	0.228	$<10^{-7}$

important in the prediction. This is evident primarily from the more rapid decrease in the values of β_2 and β_1 as opposed to β_3.

Other regression models were also evaluated with respect to Eq. (10.3): models excluding the *MH* predictor, models with the added squared term for *PTS* and/or *PAS*, and models with the added interaction between *PTS* and *PAS*. Nonetheless, when compared to the original model in terms of the residual standard error, the other models performed worse or, for certain *i* values, equally well but with the added burden of unnecessary terms.

In order to link the predicted *TS* value for a student to the matching difficulty of test items, we introduce auxiliary variables. For (10.4) a non-empty set of n_{items} test items *Items*, let:

$$rank : Items \rightarrow \{1, \ldots, n_{items}\}, \tag{10.4}$$

be a ranking function that assigns to each item a different integer, so that an item *it* has a lower rank value when compared to all the other items with a higher *CR* value, while the ranks of items with equal *CR* may be ordered according to items' identifiers within the database presented in Sect. 10.3.2. The formula (10.5) for calculating *TestRatio* (*TSR*) is:

$$TSR(s,i) = TS(s,i)/TS_{\max}(i), \tag{10.5}$$

where $TS_{max}(i)$ is the maximum allowed value for *TS*(*s*, *i*), which is given in Table 10.5 for different *i* values.

The formula (10.6) for *StudentCapacityRank* (*SCR*) is

$$SCR(s,i) = \begin{cases} ceil(n_{items} \cdot (1 - TSR(s,i))), & TSR(s,i) \neq 1 \\ 1, & TSR(s,i) = 1 \end{cases}. \tag{10.6}$$

For each student, we may estimate the DF value corresponding to the student's capacity by using the Eq. (10.2) in the Eq. (10.7) for *StudentCapacity* (*SC*):

$$SC(s,i) = DF(rank^{-1}(SCR(s,i))). \tag{10.7}$$

For a non-empty set of n_s students *Students*, the *StudentGroupCapacity* (*SGC*), which is the difficulty of a matching test for that student group (10.8), is within the [1, 3] range and calculated as the arithmetic mean of individual values of *SC*:

$$SGC(Students, i) = (1/n_s) \cdot \sum_{s \in Students} SC(s, i). \tag{10.8}$$

Equations (10.4–10.8) formally describe a process responsible for matching student capacity to item difficulty. The predicted student performance is expressed as a ratio (*TSR*) between the predicted score and the maximum score in Eq. (10.5). For a given student performance, there is a matching rank *SCR* from 1 (best) to n_{items} (worst). The process of transforming the student performance ratio to its rank is started by sorting existing test items in the decreasing level of difficulty, as expressed by *CR*, using the rank function from Eq. (10.4). Next, the bottom *TSR* percentage of ranked items is removed and the rank of the least difficult item remaining (or the most difficult item removed) becomes the rank matching the student capacity, which is expressed in Eq. (10.6).

Finally, the student capacity is mapped to the difficulty of the item with the same rank in Eq. (10.7). For example, if a student's predicted score ratio in a test is 0.9 (90 %), then the student is expected to give a correct answer for 90 % of items, on the average. For this scenario, the student's capacity equals the difficulty of the item that separates the top 10 % of items from the rest. In this manner, for each student, the corresponding capacity is defined as the difficulty of the most difficult item for which that student is expected to give a correct answer.

10.6 Test Generation Algorithm

The test generation algorithm is designed as a genetic algorithm that searches for a combination of test items, which, as a group, should cover as many specified areas as possible, but at the same time be as close to the specified mean difficulty as possible. As the stated problem belongs to the field of multi-criteria optimization, in this case exhibited by the need to attain the specified coverage and difficulty, genetic algorithms are chosen as the method of test construction. Therefore, if a set of available test items, together with a set of arguments, is passed to the algorithm, the output is a combination of test items matching the aforementioned criteria and argument values.

Once the test generation process is finished, we may conduct an assessment, in which all target students have to give answers for the same items within the generated test. As with any genetic algorithm, there are several common steps. First, a random group (*population*) of solutions (*individuals*) is generated and set as the current population. Next, the population evolves through the specified number of iterations (*generations*), with the possibility of terminating the process early if the population becomes homogenous (becomes entirely composed of similar or same solutions) or a solution of a desired quality is found.

In each generation, some individuals are selected (*selection*) according to their *Fitness* (*FIT*), which represents the quality of an individual and is calculated using a custom fitness function. The selected individuals enter the phases of *crossover* (two or more individuals are combined to generate a new individual) and then *mutation* (selected new individuals are randomly modified), thus producing new individuals, which, as a group, are generally expected to be better than the current population. The new individuals comprise a new population, which replaces the current population and enters the next iteration. The concrete elements that have to be specified in this generic procedure include:

- Algorithm arguments.
- Structure of a solution.
- Selection process.
- Crossover process.
- Mutation process.
- Fitness function for the solution.

The following algorithm arguments are required in the proposed approach:

- Items. The set of potential test items, where each item has an identifier, difficulty, which is calculated using Eq. (10.2) from Sect. 10.5.1, and list of knowledge areas that it covers.
- ItemCount (IC). The exact number of chromosomes (test items) in individuals, i.e., the desired number of items in the generated test.
- SpecifiedDifficulty (SPDF). The student group capacity of students who would take the generated tests, which may be manually set to a value from [1, 3] or calculated using Eq. (10.8) from Sect. 10.5.2.
- ConceptMap. The concept model containing all knowledge areas (presented in Sect. 10.4).
- Required. The set containing knowledge areas (presented in Sect. 10.4) that needs to be covered by the generated test.
- PopulationSize (PS). The size of the population, i.e., the exact number of individuals within the population (preferably an even number).
- MaxGenerationCount (MGC). The maximum number of generations, after which the algorithm should terminate.
- FitnessThreshold. The minimum fitness measure that leads to the termination if observed in an individual (the observed individual is considered the best solution).
- Convergence. The maximum deviation in mean population fitness over a specified number of latest generations that leads to the termination.
- CrossoverCount. The number of chromosomes that will be exchanged between individuals during the crossover phase.
- MutationChance. The chance that a mutation will occur in an individual.
- Elitism. Indicator about whether to allow elitism, which is a process when best fitted individuals (elites) skip the crossover phase and directly enter the new population.
- ElitismMutation. The indicator about whether to allow elitism mutations, which is a case when an elite individual, who directly passes to the new population, also undergoes the mutation phase.

Each individual contains an array of different test item identifiers. The length of the array is equal to the chromosome count specified before algorithm execution. Before the crossover phase, pairs of individuals are randomly formed, where the number of pairs is equal to one half of the population size. However, since the proposed algorithm employs roulette wheel selection, fit individuals, which have a high fitness value, are more likely to transit to the crossover phase and, consequently, propagate their chromosomes (a set of test items) to the next generation.

During the crossover, the specified number of chromosomes (item identifiers) is swapped between the individuals who were coupled in the selection process. In the mutation phase, given the initially specified mutation chance, a randomly selected value corresponding to a valid test item identifier is set as a new value of a single chromosome. The target chromosome is either selected randomly or it represents an item whose difficulty varies the most from *SPDF* within the individual. The fitness function is one of the key elements in the algorithm. For a set of test items encompassed by the individual, the fitness function provides a numerical indicator of the quality of the test that would contain these items. It takes as input an individual *ind* and calculates to which extent the two criteria are satisfied:

- Criterion1 (C1). The required knowledge areas (*Required*) are covered by the items within the individual *ind*.
- Criterion2 (C2). The mean difficulty of the items within the individual *ind* matches the specified student group capacity (*SPDF*).

The first criterion is expressed by the following formula (10.9):

$$C1(ind) = ARC_{required}(ind)/n_{required}, \tag{10.9}$$

where *ind* is an individual (a set of items), $ARC_{required}$ the number of the required areas from the *Required* set (an argument passed to the algorithm) that are covered by the individual *ind*, and $n_{required}$ the total number of the required areas (the cardinality of the *Required* set). The second criterion is expressed by the following formula (10.10):

$$C2(ind) = (2 - |MDF(ind) - SPDF|)/2. \tag{10.10}$$

where *MeanDifficulty* (*MDF*) is the mean difficulty for the test items enclosed within the individual *ind*, and *SPDF* the desired difficulty of the generated test (an argument passed to the algorithm). For both criteria, the allowed range is [0, 1], where 0 denotes the worst fitness and 1 the best fitness of an individual. The fitness function is of the following form (10.11):

$$FIT(ind) = 0.5 \cdot C1(ind) + 0.5 \cdot C2(ind). \tag{10.11}$$

For some typical use scenarios in practice, with the presented fitness formula, area coverage may have greater influence on the fitness measure of an individual than the distance between the obtained and specified test difficulty. However, this may be a case more acceptable than the opposite situation because good knowledge coverage of the test is one of the primary goals in the analyzed course. In case the opposite criterion may need to be encouraged, the constant factors in the two addends from the fitness formula may be modified.

Moreover, as evidenced in the Sect. 10.5.1, area coverage of a test item is positively correlated with the difficulty of the items. If a great coverage of all possible areas is required together with a less demanding test featuring just a few items, which may be one of the typical scenarios in practice, the proposed approach is generally expected to discover a solution matching both criteria only to some extent. However, this tradeoff may be somewhat avoided by increasing the number of required items or extending the pool of potential items with those that individually satisfy such requirements.

10.7 Application and Results

For teachers who are primarily interested in obtaining a test for student assessment, there are three especially important algorithm arguments: the exact number of test items (*IC*), student group capacity of target students (*SPDF*), and knowledge areas that need to be covered by the test (*Required*). For the purpose of illustrating the use and performance of the algorithm, we formulate three distinct assessment scenarios:

- S1. The creation of a 5-item test of low difficulty ($IC = 5$, $SPDF = 1.5$), which is an example of a short assessment that may be frequently administered.
- S2. The creation of a more difficult test with 10 items ($IC = 10$, $SPDF = 2.5$), which is an example of an assessment that requires greater concentration and competence from students.
- S3. The creation of a very difficult test with 20 items ($IC = 20$, $SPDF = 3$), which is an example of an assessment useful for discerning between the best students.

In all three cases, the target test was created from a set of 172 items and expected to cover all 19 knowledge areas from the analyzed course. The proposed approach is compared to the random approach incorporating a random generator that randomly chooses items to create a specified number of tests (*TestPoolSize*, *TPS*) and then selects the best one as the solution. The random approach represents a benchmark, as its variant is currently used to generate tests in the analyzed programming course.

The tests from the two groups were evaluated using the metrics built within the fitness function of the genetic algorithm. For each scenario, an experiment was conducted in $N = 10$ iterations using the two approaches. In each iteration, two tests were automatically generated, one using the random approach (RA), the other using the proposed genetic algorithm approach (GA). In all three scenarios, the most important settings for the GA approach were:

- FitnessThreshold = 1
- CrossoverCount = 0.4 × *IC*
- MutationChance = 20 %
- Elitism = true
- ElitismMutation = false

In order to facilitate the comparison of the two approaches, we chose such a *TPS* value for the RA approach so that it leads to the mean completion time which approximately matches the one of the GA approach. The results of the evaluation are presented in Table 10.6. For both approaches, there are the mean fitness of the solution (*MF*), standard deviation for the fitness of the solution (*SDF*), and mean completion time (*MT*) in seconds. For the GA approach, there are also the argument *PS*, and mean generation count (*GC*). For the RA approach, there is also *TPS*. The comparison of solution fitness between the two approaches is done using the Wilcoxon Rank Sum test (*WRST*). The obtained results indicate that, for each of the three scenarios, the GA approach significantly outperforms the RA approach for similar completion times, as evidenced by the differences in *MF* values and the *p*-values for the significance test. Moreover, in each scenario, the GA approach always yielded solutions with the same fitness value, i.e., the GA approach produced consistently good solutions.

The GA convergence, as represented by the change of mean population fitness across generations, is illustrated in Fig. 10.7. The scenario *S1*, which required the

Table 10.6 Comparison of the GA and RA approach for N = 10 iterations

S	GA					RA				WRST (N = 10)
	PS	GC	MF	SDF	MT	TPS (k)	MF	SDF	MT	
S1	100	149.7	0.843	0	4.43	39	0.82	0.008	4.45	W = 95 $p \ll 0.01$
S2	400	22.9	1	0	9.58	48	0.895	0.012	9.83	W = 100 $p \ll 0.01$
S3	500	73.7	0.95	0	93.5	281	0.842	0.012	93.88	W = 100 $p \ll 0.01$

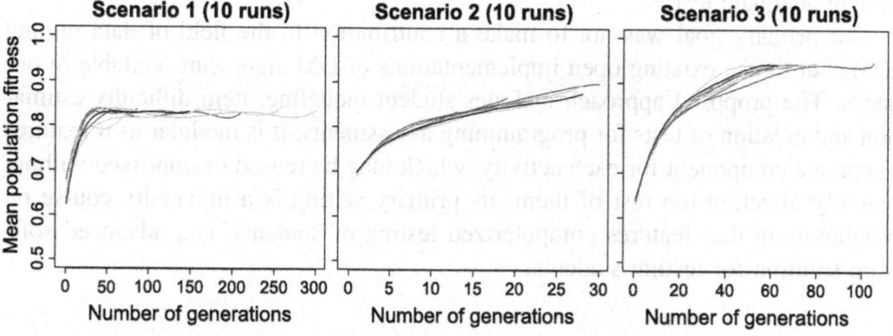

Fig. 10.7 The genetic algorithm convergence for three scenarios

most generations and resulted in the suboptimal solution, was the most demanding most likely because of the insufficient number of test items for the posed requirements. On the other hand, with the increased number of items in *S2*, the GA search ended prematurely after fewer generations due to the discovery of the perfect solution, which had maximum fitness. However, when the further increased number of items was coupled with the need for the extreme difficulty, as in the case of *S3*, the search could not yield the perfect solution but the population managed to converge.

These findings suggest that the proposed GA approach may provide solutions for different assessment scenarios and generate tests in reasonable time. As a result, it should be considered for use in practice in the analyzed course.

10.8 Conclusion and Future Work

In the application of the proposed approach to generation of computer based tests on programming, we utilize the EDM techniques to estimate the difficulty of a test. The difficulty estimate is one of the two important pieces of information that is used to guide the search for a good test, which is conducted using a genetic algorithm. Four variables are used to train a multi-class SVM classifier for the estimation of the difficulty of a new test item, while a regression model is used to estimate the competency of students that should take the test. With this information, the proposed algorithm for test creation attempts to find a test with the minimum difference between the test difficulty and student competency.

The other important piece of information is the test coverage of important programming areas. For the purpose of automatic calculation of coverage, an extensive concept map of the programming competencies and C language elements was designed. Furthermore, each test item was annotated with the matching knowledge areas specified in the concept map. As a result, the algorithm also favors tests that cover more of the required knowledge areas. The benefits of the proposed approach are demonstrated in an evaluation where the presented algorithm is compared to a solution that randomly searches within the item space to find an adequate test.

Our primary goal was not to make a contribution to the field of data mining (DM) but to use existing open implementations of DM algorithms suitable to our needs. The proposed approach includes student modeling, item difficulty estimation and creation of tests for programming assessments. It is modular as it features a separate component for each activity, which may be reused or improved without severely affecting the rest of them. Its primary setting is a university course on programming that features computerized testing of students, i.e., advanced software solution for testing students.

However, the approach could also be applied in any environment where automatic construction of programming tests may provide benefits, including assessments within distance learning systems, as well as within the primary and secondary education. On the other hand, the performance of the approach may be significantly influenced by the quality of the assessment data set. It is required that previous records feature a comprehensive set of student data and items corresponding to various levels of competence and difficulty, respectively. The proposed predictive models should be trained and used on such representative data sets.

There are several possible directions for the future research on the presented issue. The quality of estimates of item difficulty and student competence is also tightly related to the structure of available data. For the purpose of improving these predictions, we may deploy additional mechanisms that would record additional variables related to student performance in programming tests. By merging the accessibility of the proposed approach with the formality and power of IRT, we could further enhance the precision in the assessment process and reduce testing times.

Moreover, we may also create an ontology matching the presented concept map, as a way of providing additional means to its use in different environments. In some scenarios, it may be needed to generate a similar test for different groups of students, where each test should have as few common items with other tests as possible. For this purpose, the algorithm may be extended to generate a set of non-overlapping tests, while the fitness function would have to be modified to include this additional criterion.

Acknowledgments The research was supported by the Ministry of Education, Science and Technological Development of the Republic of Serbia, Grant III-44010, Title: Intelligent Systems for Software Product Development and Business Support based on Models. The authors are very grateful to their colleagues from the Chair of Applied Computer Science at the Faculty of Technical Sciences (University of Novi Sad, Serbia), who have contributed to the Otisak testing system and participated in the organization of the Programming Language and Data Structures course, thus making the presented study possible.

References

1. Romero, C., Ventura, S.: Educational data mining: a review of the state of the art. IEEE Trans. Syst. Man Cybern. Part C Appl. Rev. **40**(6), 601–618 (2010)
2. Živanov, Ž., Rakić, P., Stričević, L., Pušić, B., Suvajdžin, Z., Hajduković, M.: Computer aided student examination. Info M **7**(25), 45–53 (2008)
3. Wauters, K., Desmet, P., Noortgate, W.V.D.: Acquiring item difficulty estimates: a collaborative effort of data and judgment. In: Pechenizkiy, M., Calders, T., Conati, C., Ventura, S., Romero, C., Stamper, J. (eds.) 4th International Conference on Educational Data Mining, pp. 121–127. International Educational Data Mining Society, Eindhoven (2011)
4. Peña-Ayala, A., Sossa-Azuela, H., Cervantes-Pérez, F.: Predictive student model supported by fuzzy-causal knowledge and inference. Expert Syst. Appl. **39**, 4690–4709 (2012)
5. Holland, J.H.: Adaptation in Natural and Artificial Systems. MIT Press, Cambrigde (1992)
6. Barla, M., Bieliková, M., Ezzeddinne, A.B., Kramár, T., Šimko, M., Vozár, O.: On the impact of adaptive test question selection for learning efficiency. Comput. Educ. **55**(2), 846–857 (2010)

7. Feng, M., Heffernan, N.: Can we get better assessment from a tutoring system compared to traditional paper testing? Can we have our cake (Better Assessment) and eat it too (Student Learning During the Test)? In: Alven, V., Kay, J., Mostow, J. (eds.) Intelligent Tutoring Systems. LNCS, vol. 6095, pp. 309–311. Springer, Heidelberg (2010)
8. Thelwall, M.: Computer-based assessment: a versatile educational tool. Comput. Educ. **34**(1), 37–49 (2000)
9. Daly, C., Waldron, J.: Assessing the assessment of programming ability. ACM SIGCSE Bull. **36**(1), 210–213 (2004)
10. Douce, C., Livingstone, D., Orwell, J.: Automatic test-based assessment of programing: a review. J Educ. Resour. Comput. **5**(3), 4 (2005)
11. Ihantola, P., Ahoniemi, T., Karavirta, V., Seppälä, O.: Review of recent systems for automatic assessment of programming assignments. In: 10th Koli Calling International Conference on Computing Education Research, pp. 86–93. ACM, New York (2010)
12. Baker, F.B.: The Basics of Item Response Theory. ERIC Clearinghouse on Assessment and Evaluation, Washington (2001)
13. Sosnovsky, S., Gavrilova, T.: Development of educational ontology for C-programming. Int. J. Inf. Theor. Appl. **13**, 303–308 (2006)
14. Sosnovsky, S.: C Programming Language Ontology, http://www.sis.pitt.edu/~paws/ont/c_programming.rdfs
15. Zhou, M., Xu, Y., Nesbit, J.C., Winne, P.H.: Sequential pattern analysis of learning logs: methodology and applications. In: Romero, C., Ventura, S., Pechenizkiy, M., Baker, R.S.J.d. (eds.). Handbook of Educational Data Mining, Chapman & Hall/CRC Data Mining and Knowledge Discovery Series, pp. 107–121. CRC Press, Boca Raton (2010)
16. Faculty of Technical Sciences in Novi Sad, Accreditation of the Study Programme; Computing and Control Engineering, http://www.ftn.uns.ac.rs/_data/planovi/2012/engleski/osnovne/ftn_e2.pdf
17. Rakić, P., Stričević, L., Živanov, Ž., Suvajdžin, Z., Hajduković, M.: Computer classroom: deployment and exploitation. Info M **6**(21), 9–13 (2007)
18. Klyne, G., Carroll, J.J., McBride, B.: Resource description framework (RDF): concepts and abstract syntax. W3C Recommendation **10** (2004)
19. McGuinness, D.L., Van Harmelen, F.: OWL web ontology language overview. W3C Recommendation **10** (2004)
20. Motik, B., Patel-Schneider, P.F., Parsia, B., Bock, C., Fokoue, A., Haase, P., Smith, M.: OWL 2 Web ontology language: structural specification and functional-style syntax. W3C Recommendation **27,** 17 (2009)
21. Grau, B.C., Horrocks, I., Motik, B., Parsia, B., Patel-Schneider, P., Sattler, U.: OWL 2: the next step for OWL. Web Semant.: Sci., Serv. Agents World Wide Web **6**(4), 309–322 (2008)
22. Falconer, S.: OntoGraf, http://protegewiki.stanford.edu/wiki/OntoGraf (2010)
23. TopBraid Composer, http://www.topquadrant.com/products/TB_Composer.html
24. Krivov, S., Williams, R., Villa, F.: GrOWL: a tool for visualization and editing of OWL ontologies. Web Semant.: Sci., Serv. Agents World Wide Web **5**(2), 54–57 (2007)
25. Novak, J.D., Cañas, A.J.: The theory underlying concept maps and how to construct and use them. Technical report, Florida Institute for Human and Machine Cognition (2008)
26. Novak, J.D.: Learning, Creating, and Using Knowledge: Concept Maps as Facilitative Tools in Schools and Corporations. Taylor & Francis, New York (2010)
27. Ruiz-Primo, M.A., Shavelson, R.J.: Problems and issues in the use of concept maps in science assessment. J. Res. Sci. Teach. **33**(6), 569–600 (1996)
28. Cañas, A.J., Hill, G., Carff, R., Suri, N., Lott, J., Eskridge, T., Carvajal, R.: CmapTools: a knowledge modeling and sharing environment. In: Concept maps: Theory, Methodology, Technology 1st International Conference on Concept Mapping, vol. 1, pp. 125–133. Universidad Pública de Navarra, Pamplona (2004)
29. Bivand, R.: ClassInt: Choose Univariate Class Intervals. R package version 0.1-19 (2012), http://CRAN.R-project.org/package=classInt

30. R Core Team.: R: A Language and Environment for Statistical Computing, Manual. R Foundation for Statistical Computing (2013)
31. Karatzoglou, A., Smola, A., Hornik, K.: Achim Zeileis, A.: Kernlab—an S4 package for kernel methods in R. J. Stat. Softw. **11**(9), 1–20 (2004)

Chapter 11
Plan Recognition and Visualization in Exploratory Learning Environments

Ofra Amir, Kobi Gal, David Yaron, Michael Karabinos and Robert Belford

Abstract Exploratory Learning Environments (ELEs) are open-ended software in which students build scientific models and examine properties of the models by running them and analyzing the results (Amershi and Conati, Intelligent tutoring systems. LNCS. Springer, Heidelberg, 463–472, 2006); Chen (Instr Sci, 23(1–3):183–220, 1995); (Cocea et al., 2008). ELEs are generally used in classes too large for teachers to monitor all students and provide assistance when needed (Gal et al., 2008). They are also becoming increasingly prevalent in developing countries where access to teachers and other educational resources is limited (Pawar et al., 2007). Thus, there is a need to develop tools of support for teachers' understanding of students' activities. This chapter presents methods for addressing these needs. It presents an efficient algorithm for intelligently recognizing students' activities, and novel visualization methods for presenting these activities to teachers. Our empirical analysis is based on an ELE for teaching chemistry that is

O. Amir (✉)
Harvard University, School of Engineering and Applied Sciences, 29 Oxford Street, Cambridge, MA 02138, USA
e-mail: oamir@seas.harvard.edu

K. Gal
Department of Information Systems Engineering Ben-Gurion Boulevard, Ben-Gurion University, Building 93, Beer Sheva 84105, Israel
e-mail: kobig@bgu.ac.il

D. Yaron · M. Karabinos
Department of Chemistry, Carnegie-Melon University, 4400 Fifth Avenue, Pittsburgh, PA 15213, USA
e-mail: yaron@cmu.edu

M. Karabinos
e-mail: mk7@cmu.edu

R. Belford
Department of Chemistry and Biochemistry, University of Arkansas, A 119 Chemistry Building, Fayetteville, AR 72701, USA
e-mail: rebelford@ualr.edu

A. Peña-Ayala (ed.), *Educational Data Mining*, Studies in Computational Intelligence 524, DOI: 10.1007/978-3-319-02738-8_11,

used by thousands of students in colleges and high schools in several countries (Yaron et al., Science, 328(5978), 584–585, 2010).

Keywords Plan recognition · Visualization · Exploratory learning environments · Recognition algorithm · Virtual labs

Abbreviations

AI	Artificial intelligence
CCD	Create correct device action
ELEs	Exploratory learning environments
ITS	Intelligent tutoring systems
MS	Mix solution
MSC	Mixing solution components
MSI	Mixing the solution using an intermediate flask
SDP	Solving the dilution problem

11.1 Introduction

There are several aspects to students' interactions that make plan recognition in ELEs particularly challenging. First, students can engage in exploratory activities involving trial-and-error, such as searching for the right pair of chemicals to combine in order to achieve a desired reaction. Second, students can repeat activities indefinitely in pursuit of a goal or sub-goal, such as adding varying amounts of an active compound to a solution until a desired molarity is achieved.

Third, students can interleave between activities, such as preparing a solution for a new experiment while waiting for the results of a current experiment. Explicitly representing all possible combinations of these activities is computationally infeasible.

The recognition algorithm presented in this paper addresses these challenges by using a recursive grammar to generate plan fragments for describing key chemical processes in the lab. The algorithm receives as input students' complete interaction sequence with the software, as well as a grammar describing possible activities. It expands activities from the grammar using a heuristic that chooses (possibly non-contiguous) actions from students' interaction and outputs a hierarchical plan that explains how the software was used by the student. The algorithm was evaluated using real data obtained from students using the ELE to solve six representative problems from introductory chemistry courses. Despite its incompleteness, the algorithm was able to correctly infer students' plans in all of the instances given that appropriate grammar rules were available. It was able to identify partial solutions in cases where students failed to solve the complete problem, as well as capture interleaving plans.

We used two novel visualization methods to present students' activities to teachers. One of the methods visualized the plans that were inferred by the recognition algorithm. The second method visualized students' actions over a timeline. A user study with chemistry teachers was conducted that compared these visualization methods with a baseline technique consisting of movies showing the students' application window during their work. The results showed that teachers preferred the temporal- and plan-based methods over the movie visualization, despite the fact that the movie was easier to learn. Both the plan- and temporal-based visualization methods were found useful by teachers, and improved teachers' understanding of student performance. These visualization methods will be incorporated into a separate application that will be available for use by teacher and student users of Virtual Labs.

These results demonstrate the efficacy of combining computational methods for recognizing users' interactions with intelligent interfaces that visualize how they use flexible, open-ended software. It is a first step in creating systems that provide the right machine-generated support for their users. For teachers, this support consists of presenting students' performance both after and during class. For students, this support will guide their problem-solving in a way that maximizes their learning experience while minimizing interruption.

This chapter integrates and extends a past study for recognizing students' activities in ELEs Amir and Gal [7] in several aspects. First, it introduces novel visualization methods of students' work with exploratory learning environments, one of which is informed by the recognition algorithm. Second, it demonstrates the efficacy of these visualization methods in the real world by showing they support teachers in the analysis of student performance in ELEs. Lastly, it evaluates the recognition algorithm on a significantly larger scale.

The rest of this chatper is organized as follows. Section 11.2 presents related work in two different areas: plan recognition and student assessment. Section 11.3 presents the ELE domain which is the focus of our empirical methodology. Section 11.4 presents the plan recognition algorithm and demonstrates its performance on student data.

Section 11.5 describes a user study for comparing different visualization methods of students' activities to teachers. Section 11.6 concludes this work and discusses its significance for the goal of creating collaborative systems in exploratory domains.

11.2 Related Work

The work reported in this book chapter relates to two different areas of prior work and a range of approaches within each: plan recognition and assessment of students' activities with software. The subsections below discuss related work in these two areas respectively.

11.2.1 Plan Recognition

Plan recognition is a cornerstone problem in artificial intelligence (AI) which aims to infer an agent's goals and plans given observations of its actions. Applications of plan recognition can be found in a wide range of fields, such as natural language dialog Carberry [8]; Grosz and Sidner [9], software help systems Baueret al. [10]; Mayfield [11], story understanding Wilensky [12]; Charniak and Goldman [13] and human–computer collaboration Lesh et al. [14].

Past works have used plan recognition to infer students' plans from their interactions an ELE for teaching statistics Gal et al. [4]; Reddy et al. [15]; Gal et al. [16]. Specifically, Reddy et al. [15] proposed a complete algorithm which modeled the plan recognition task as a Constraint Satisfaction Problem (CSP). Gal et al. [4] devised a heuristic algorithm that matched actions from students' logs with the recipes for the given problem. These approaches do not support recursive grammars, which are essential for capturing the type of exploratory activities that characterize the ELE in our setting, such as indefinite repetition. We further extend these works by visualizing students' activities to teachers.

Other works have implemented plan recognition techniques to model students' activities in Intelligent Tutoring Systems (ITS) VanLehn et al. [17]; Conati et al. [18, 19]; Anderson et al. [20]; Corbett et al. [21]; Vee et al. [22]. In these systems, the tutor takes an active role in students' interactions, providing feedback and hints. Plan recognition has also been used to recognize users' activities when programming in UNIX Blaylock and Allen [23], or interacting with medical diagnosis and email notification systems Bauer [24]; Horvitz [25]; Lesh [26]. All of the above settings are significantly more constrained than ELEs, severely limiting the amount of exploration that students can perform. Thus these approaches are not suitable for recognizing students' activities in ELEs. Our work also extends the plan recognition literature more generally. Traditional approaches to plan recognition Kautz [27]; Lochbaum [28] did not consider incomplete information of the agent, mistakes, extraneous actions, interleaving and multiple plans, which are endemic feature of ELEs.

More recently, Geib and Goldman [29] proposed a probabilistic model of plan recognition that recognized interleaving actions and output a disjunction of plans—rather than a single hierarchy—to explain an action sequence.

It also accounted for missing observations (e.g., not seeing an expected action in a candidate plan makes another candidate plan more likely). Our work is distinct from this approach in several ways. First, the settings studied by Geib and Goldman do not account for agents' extraneous actions, which are common to students' interactions in ELEs. Second, we show the efficacy of our approach on real-world data obtained from students using pedagogical software, whilst Geib and Goldman use synthetic data.

11.2.2 Assessment of Students' Activities

The visualization methods in this paper relate to several strands of research for analyzing and assessing students' interactions with pedagogical software. Some systems work on-line, visualizing predefined features of students' interactions to teachers. The following describe notable examples. The student tracking tool Pearce-Lazard et al. [30]; Gutierrez-Santos et al. [31] is part of the MiGen project for improving 11–14 year-old students' learning of algebraic generalization. This tool monitors students' activities during their sessions with an ELE for teaching algebra. The tracking tool visualizes "land-marks" which occur when the system detects specific actions or repetitive patterns carried out by the student.

The FORMID-Observer Gueraud et al. [32] monitors students' activities in simulation-based work sessions with the FORMID-Learner. A teacher can specify specific situations to be monitored representing certain system states, possible student mistakes, and tests that can be triggered by the student. These activities are visualized in the teacher interface which shows the situations and results of validation requests of each student, using a coloring scheme of green for correct activities and red for incorrect activities.

Other systems work post hoc, and generate reports to teachers based on students' complete interaction histories. These systems do not display the students' activities but rather summarize performance measures such as the number of hints requested and success rates in problems. Relevant examples include the ASSISTment system Feng and Heffernan [33] and Student Inspector Scheuer and Zinn [34].

Lastly, data mining techniques have been used to analyze students' performance with pedagogical software. The DataShop Koedinger et al. [35] system generates learning curves reports for students, error reports and general performance reports of students. The Tool for Advanced Data Analysis in Education (TADA-Ed) Merceron and Yacef [36] system discovers correlations between students' mistakes in different problems. Sao Pedro et al. [37] and Montalvo et al. [38] trained decision tree detectors to identify two types of students' planning approaches in microworlds, a simulation based educational software. Their modelling is based on features such as action frequencies and latency between actions. Amershi and Conati [39] have used data mining techniques to cluster and classify students' interaction behaviors in ELEs as either effective or ineffective for learning. Kardan and Conati [40] extended this work to extract association rules of each cluster and use these rules for both online classification of new learners as well as for post analysis of the behaviors that were effective for learning.

Our work differs from these data mining approaches in that it provides an individual analysis of students' problem-solving behavior. For example, while the approach described in Kardan and Conati [40] will classify a student as belonging to either a high-learning gain group or a low-learning gain group, our approach provides a temporal and hierarchical visualization of the student's interaction.

11.3 The Virtual Labs Domain

In this section we describe the ELE that provides the empirical setting for this paper. Virtual Labs simulates a real chemistry lab and used in the instruction of college and high school chemistry courses worldwide. It allows students to design and carry out experiments which connect theoretical chemistry concepts to real world applications Yaron et al. [6]. We will use the "dilution problem", posed to students that use VirtualLabs in an introductory chemistry course, as a running example to demonstrate our approach.

Your objective is to prepare a solution containing 800 milliliters (ml) or more of HNO_3 with a desired concentration of 7 M[1] You are allowed a maximal deviation of 0.005 M in either direction.

To solve this problem in VirtualLabs, students are required to pour the correct amounts of HNO_3 and H_2O to an empty flask which will contain the diluted solution. Despite the simplicity of this problem, students solve it in different ways. A possible solution for this problem is to repeatedly mix varying quantities of HNO_3 with H_2O until achieving the required concentration. We describe a student's interaction adapted from our empirical analysis which follows this paradigm. The student began by pouring 100 ml of an HNO_3 solution with a concentration of 15.4 M to a 100 ml intermediate flask, and transferred the content of the intermediate flask to an empty destination flask.[2]

This activity was repeated four times, resulting in 400 ml of HNO_3 in the destination flask. The student proceeded to dilute this solution by mixing it with 510 ml of H_2O. This activity was carried out in two steps, one adding 10 ml of H_2O (using an intermediate flask of 10 ml) and another adding 500 ml of H_2O (using an intermediate flask of 500 ml). At this point the molarity of HNO_3 in the destination flask was too low (6.666 M), indicating that too much H_2O had been poured. To raise the concentration to the desired level, the student began to pour small amounts of HNO_3 to the destination flask using an intermediate 10 ml flask, while checking the concentration level of the resulting compound. The student first poured 10 ml of HNO_3, then poured another 10 ml of HNO_3, and finally added 5 ml of HNO_3 to the destination flask, which achieved the desired concentration of 7 M.

Figure 11.1 shows a snapshot of Virtual Labs taken right after the student added 510 ml of H_2O to the destination flask. The panel on the left shows a stockroom of chemicals which can be customized for different activities.

One of the flasks, labeled "15.4 M HNO_3" (outlined in red in the figure) contains an HNO_3 solution with a concentration of 15.4 M. The middle panel shows the "workbench" of the student, used to carry out activities in the

[1] In chemistry, 'M' denotes the measure of Molar concentration of a substance.

[2] Intermediate flasks are commonly used in Virtual Labs to help measure solutions accurately, as in a physical laboratory.

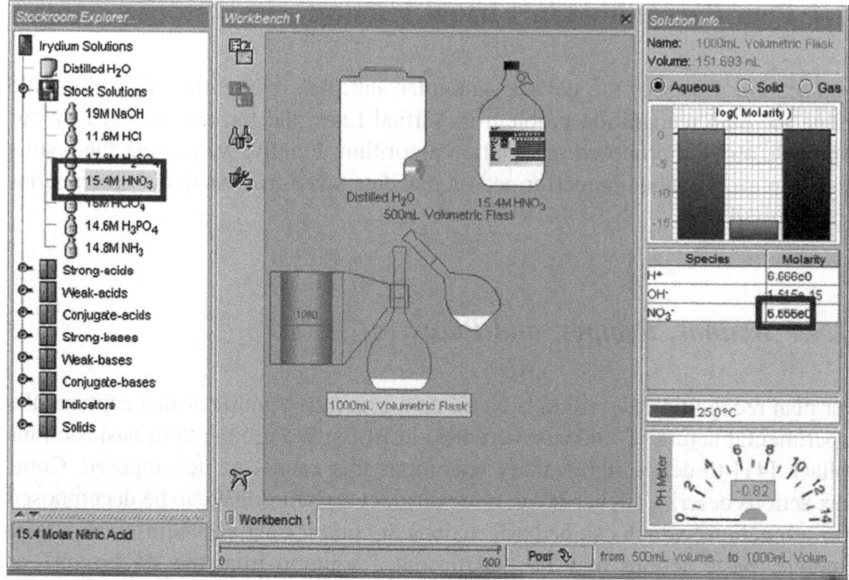

Fig. 11.1 Snapshot of interaction in virtual labs

laboratory. This panel shows the flask containing HNO_3 with a concentration of 15.4 M, the H_2O flask, and the destination flask (a 1,000 ml volumetric flask). It also shows one of the intermediate flasks used by the student (a 500 ml volumetric flask). The "Solution Information" panel on the right shows the volume and concentration of selected compounds. It shows that the concentration level of HNO_3 in the destination flask is 6.77 M (outlined in red in the figure).

The student's interaction described above highlights several aspects endemic to scientific inquiry that are supported by the Virtual Labs software. First, the concept of titration, which repeatedly adding a measured compound to a solution until a desired result, is achieved. This is apparent in the student repeatedly adding small quantities of HNO_3 to the destination flask. Second, the interleaving of actions that relate to different activities. This is apparent in the student beginning to pour HNO_3 to the destination flask, then switching to pour H_2O, and then returning to pour more HNO_3. Lastly, performing exploratory actions and mistakes. This is apparent in the student adding too much H_2O to the destination flask, and proceeding to increase the concentration of the compound by adding more HNO_3.

Whereas the student in the example interaction described above used a trial-and-error approach to solve the dilution problem, there are other possible solution strategies. For example, students can pre-calculate the exact amounts of H_2O and HNO_3 that should be mixed to achieve the desired molarity. After calculating these quantities students can proceed to immediately mix them in Virtual Labs and achieve the desired concentration.

11.4 Plan Recognition in Virtual Laboratories

This section describes the devised grammar and plan recognition algorithm. We define the plan recognition problem in Virtual Labs, the formalisms used by our approach, and the proposed recognition algorithm. Finally, we present the results of an empirical evaluation performed on real data taken from students' interactions with VirtualLabs.

11.4.1 Actions, Recipes, and Plans

Our plan recognition algorithm is based on a generative grammar that captures the experimental nature of students' activities in ELEs. We use the term basic actions Pollack [41] to define rudimentary operations that cannot be decomposed. Complex actions describe higher-level, more abstract activities that can be decomposed into sub-actions, which can be basic actions or complex actions themselves. In our example, basic actions may be "taking out a solution from the stockroom" or "pouring 10 ml of H_2O to an intermediate flask", while complex actions may consist of "solving the dilution problem", or "mixing together H_2O and HNO_3 four times".

A recipe for a complex action specifies the sequence of operations required for fulfilling the complex action, called sub-actions. Formally, a recipe is a set of sub-actions and constraints such that performing those sub-actions under those constraints constitutes completing the action. The set of constraints is used to (1) specify required values for action parameters; (2) enforce relationships among parameters of (sub-) actions, such as chronological order; and (3) bind the

Fig. 11.2 Recipes for **a** solving the dilution problem; **b** repetition of activities; **c** using intermediate flasks

(a)

$\underline{SDP}[s\_id_1, vol_1, s\_id_2, vol_2, d\_id_1] \rightarrow$
$\underline{MSC}[s\_id_1, d\_id_1, sc_1 = H_2O, vol_1]$,
$\underline{MSC}[s\_id_2, d\_id_2, sc_2 = HNO_3, vol_2]$

$d\_id_1 = d\_id_2$

(b)

$\underline{MSC}[s\_id_1, d\_id_1, sc_1, vol = vol_1 + vol_2] \rightarrow$
$\underline{MSC}[s\_id_1, d\_id_1, sc_1, vol_1]$,
$\underline{MSC}[s\_id_2, d\_id_2, sc_2, vol_2]$

$s\_id_1 = s\_id_2, d\_id_1 = d\_id_2, sc_1 = sc_2$

$\underline{MSC}[s\_id, d\_id, sc, vol] \rightarrow MS[s\_id, d\_id, sc, vol]$

(c)

$\underline{MSC}[s\_id, d\_id, sc, vol] \rightarrow \underline{MSI}[s\_id, d\_id, sc, vol]$

parameter values of a complex action to the value of the parameters in its constituent sub-actions.

Figure 11.2a presents a recipe for the complex action of Solving the Dilution Problem (SDP) composed of two complex sub-actions for Mixing Solution Components (MSC), namely H_2O and HNO_3. In our notation, complex actions are underlined, while basic actions are not. Actions in VirtualLabs are associated with identifiers that bind to recipe parameters. For example, the parameters of the action MSC[s_id_1; d_id_1; $sc_1 = H_2O$; vol_1] of pouring $= H_2O$ in Fig. 11.2a identify the source flask (*s_id*) from which a source chemical (*sc*) is poured, the destination flask (*d_id*), and the volume of the solution that was poured (*vol*). The constraints for this recipe require that the destination flask identifier for both MSC actions is the same ($d\_id_1 = d\_id_2$) in addition to specifying the type of chemicals in the mix ($sc_1 = H_2O$ and $sc_2 = HNO_3$).

Recipes may be recursive, capturing activities that can repeat indefinitely, as in titration. This is exemplified in the recipe shown in Fig. 11.2b for the complex action (MSC) of adding a solution component of volume *vol* from flask s_id_1 to flask d_id_1. The constituent actions of this recipe decompose the MSC action into two separate MSC actions for adding vol_1 and vol_2 of the solution using the same source and destination flask. This recipe effectively clusters together repetitive activities. Also shown is the "base-case" recipe for MSC that includes a Mix Solution (MS) basic action.

Figure 11.2c presents another recipe for an MSC complex action which decomposes into a constituent sub-action for Mixing the Solution using an Intermediate flask (MSI).[3] We say that a recipe for a complex action is *fulfilled* by a set of sub-actions if there is a one-to-one correspondence from each of the sub-actions to one of the recipe's constituents that meets the recipe constraints. For example, in the student's interaction described in Sect. 11.3, the complex sub-actions for mixing H_2O with HNO_3 fulfill the recipe for the complex action SDP of solving the dilution problem. These actions are labeled "1, 2" and "14" in Fig. 11.3a.

A *plan* is a set of complex and basic actions such that each complex action is decomposed into sub-actions that fulfill a recipe for the complex action. A hierarchical presentation of a (partial) plan used by the student to solve the dilution problem is shown in Fig. 11.3a. Time is represented in the Figure from top to bottom, thus crossing edges signify interleaving between actions.

The hierarchy emanating from the root node SDP (the action labeled "1") shows that the student was able to solve the dilution problem by mixing together 425 ml of HNO_3 from flask ID 1 (the action labeled "2") with 510 ml of H_2O from flask ID 4 (the action labeled "14") in destination flask ID 2. These actions further decompose to their respective constituent actions. For example, the path in bold, from left to right, shows part of the plan for the complex action of pouring

[3] For brevity, we omit the recipes for the MSI action. The complete recipe library for the dilution problem can be found in Sect. 11.8.

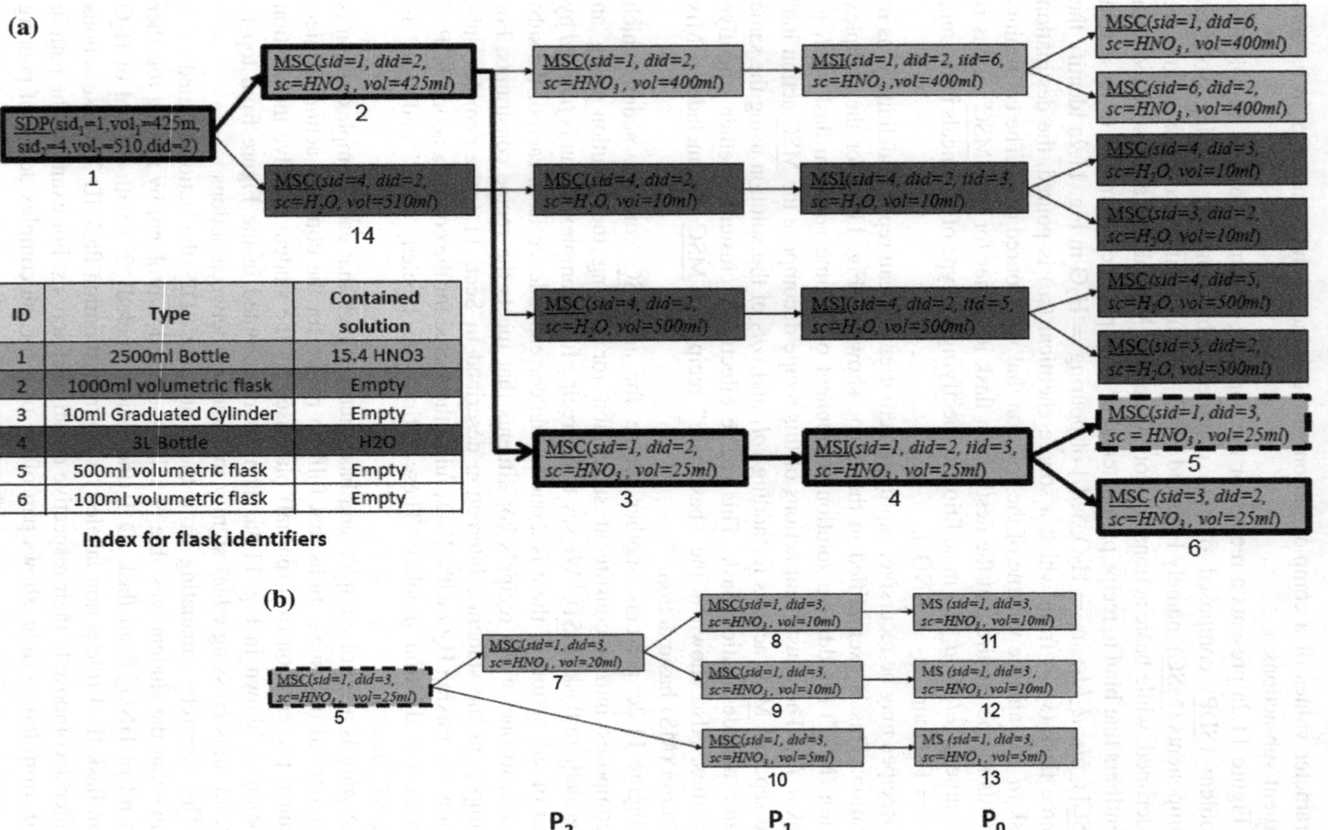

Fig. 11.3 **a** A partial plan for the dilution problem corresponding to the student's interaction described in Sect. 3; **b** a plan for the MSC complex

425 ml of HNO_3 from flask ID 1 to flask ID 2 (the action labeled "2"). Here, the student poured 25 ml of HNO_3 from flask ID 1 to flask ID 2 (the action labeled "3") using intermediate flask ID 3 (the action labeled "4"). The action labeled "4" is decomposed to the two subactions for pouring the solution from flask ID 1 to intermediate flask ID 3, and pouring from flask ID 3 to the destination flask ID 2 (actions labeled "5"and "6"). For brevity, we do not expand the complex actions in Fig. 11.3a down to the leaves.

Figure 11.3b describes the student's use of titration. This plan expands the action of pouring 25 ml from flask ID 1 to flask ID 3 (action labeled "5") down to the basic-level actions corresponding to the student's interaction with the software (the three MS actions at the leaves). The constituents of this action consisted of two separate pours from flask ID 1 to flask ID 3, one pouring 20 ml (action labeled "7") and the other pouring 5 ml (action labeled "10"). The action labeled "10" was further decomposed to the basic action of adding 5 ml of HNO_3 to flask ID 3 (action labeled "13").

11.4.2 The Plan Recognition Algorithm

As described in Sect. 11.3, students take diverse approaches to solving the dilution problem. They perform an indefinite number of mixing actions, choose whether to use intermediate flasks and interleave activities. For example, Fig. 11.3a shows the constituent sub-actions of the action labeled "14" occurred in between the constituent sub-actions of the action labeled "2". This reflects that the student interleaved the actions for adding HNO_3 and H_2O. A brute-force approach involves non-deterministically finding all ways in which a complex action may be implemented in students' interaction sequences. Such an approach was used by Reddy et al. [15] in an ELE for teaching statistic. Due to the exploratory and repetitive nature of students' actions in Virtual Labs, naively considering each of these possibilities is not possible.

The proposed algorithm shown in the program code for Bottom-up plan recognition method, incrementally builds a plan which describes students' activities with Virtual Labs. The algorithm BUILDPLAN(R,X) receives as input a finite action sequence representing a student's interaction, denoted X, and the set of recipes for the given problem, denoted R. At each step t, the algorithm maintains an ordered sequence of actions, denoted P_t and an open list OL. The action sequence P_0 is initialized with the original action sequence, X. During each step, the algorithm attempts to replace subsets of actions from P_t with the complex actions they represent. Each of the complex actions in P_t is a partial plan that explains some activity in the user's interaction. The algorithm iterates over the recipes in R (step 3) according to the following (partial) ordering criteria: if the complex action $\underline{C_2}$ is a constituent sub-action for a recipe for a complex action $\underline{C_1}$, then recipes for action $\underline{C_2}$ are considered before the recipes for action $\underline{C_1}$.

Note that the recipe language allows for cycles, but in practice recipes cannot be applied indefinitely in Virtual Labs, because interaction sequences are finite. An ordering over recipes can always be created (possibly by duplicating or renaming actions), such that it meets the sorting constraint. The algorithm repeatedly searches for a match for each recipe R_C for action C in the open list by calling the function FindMatch(R_C,OL) (step 5), which is described later in this section. FindMatch(R_C,OL) returns a set of actions $M_C \in$ OL such that M_C *fulfills* R_C.

For each match M_C that fulfills R_C, BUILDPLAN performs the following: First, the values of the parameters in C are set based on the values of the parameters of the actions in M_C and the restrictions specified in the recipe R_C (step 7). This incorporates into C the effects arising from carrying out the constituent actions in R_C. Second, the action C is added to the action sequences in P_{t+1} and OL, in the position held by the latest action in M_C (step 8). This is done to preserve the temporal ordering of the actions in the open list, which facilitates checking temporal constraints when matching recipes to actions in the open list. Adding the action to OL supports recursive recipes, in that it allows the action C itself to be part of the action set that fulfills R_C in the next iteration. Third, the action C in P_{t+1} is made a parent of all of the actions in R_C in P_t (step 10). This creates the hierarchy between a complex action in P_{t+1} and its constituent actions in P_t. Finally, the actions in M_C are removed from both the open list OL and P_{t+1} (step 11). Removing the actions in M_C from the open list prevents the same actions from fulfilling more than one recipe. Once no more matches for R_C can be found, (i.e., FindMatch(R_C, OL) returns Ø), the BUILDPLAN algorithm proceeds to consider a new recipe, and terminates once all recipes have been considered.

FINDMATCH, shown in the program code for the algorithm for finding a match using depth-first search, iterates over the actions in the open list OL performing a complete depth-first search for actions that together fulfill the complex action C, as defined by the recipe. The algorithm maintains an action set denoted M_C, which at each step of the algorithm contains a subset of actions from the open list that match the sub-actions in the recipe. At each step, the algorithm removes the next action a_P from the open list (step 8), and attempts to add it to the current match M_C. The procedure makes use of the EXTENDS function, a Boolean function that takes as input an action a_P, a partial match M_C, and recipe R_C (step 9). The function EXTENDS returns true if a_P can be added to M_C, such that (1) a_P corresponds to one of the constituent sub-actions of R_C and is not already in M_C and (2) the addition of a_P to M_C will not violate any of the recipe constraints in R_C. For example, given M_C = Ø, the action MSC[*sid* : 1; *did* : 3, *sc* : H2O; *vol_1* : 100] extends the recipe for SDP shown in Fig. 11.2a. If the action a_P extends the recipe, it is added to the match M_C, and a recursive call to FINDMATCH is performed, with the updated open list and match.

Each time FINDMATCH is called, it performs a call to the Boolean function FULFILLS(M_C, R_C) (step 12), which returns true if M_C is a complete match for the recipe R_C. We then say that M_C *fulfills* R_C. For example, the actions MSC[*sid* : 1, *did* : 3, *sc* : H_2O, *vol_1* : 100] and MSC[*sid* : 2, *did* : 3, *sc* : HNO_3, *vol_1* : 200] fulfill the recipe for SDP shown in Fig. 11.2a. Note that M_C can include both basic and complex actions.

```
procedure BuildPlan (R,X).
P0 ← X
for RC∈ SortRecipes (R) do
Pt+1, OL ← Pt
MC = FindMatch (RC,OL)
while MC ≠ ∅ do
BindParams (C, MC, RC)
Add C to OL and Pt+1 positioned after last a ∈ MC
for all a ∈ MC do
Create a branch from C in Pt+1 to a in Pt
Remove MC from OL and Pt+1
MC = FindMatch (RC, OL)
```

[Bottom-up plan recognition method]

The algorithm backtracks when it does not succeed in finding a match, by removing a_P from M_C and searching for another action to add to the match. It is therefore complete and guaranties to find a match for R_C, given that there is a subset of actions in the open list which fulfill the given recipe. Note that a match can contain non-continuous actions, as long as the constraints defined in the recipe hold, thus allowing for interleaving plans to be found.

We demonstrate this process using the plan in Fig. 11.3b describing the student's use of titration. At step P_1, the MS basic action (labeled "11") was chosen to match the recipe for the complex MSC action (labeled "8") using the second recipe in Fig. 11.2(b). At step P_2, the MSC actions labeled "8, 9" were chosen to match the recipe for the MSC action labeled "7".

```
procedure FindMatch(RC,OL) ▷RC: a recipe, OL: open list
return FindMatch (RC, OL, null)
procedure FindMatch (RC, OL, MC) ▷MC: a partial match
if FulFills (MC, RC) then
return (MC, OL)
OL′ ← OL
for aP ∈OL do ▷ aP: an action
remove aP from OL′
if Extends (aP, MC, RC) then
Add aP to MC
(MC,OL) = FindMatch (RC, OL′, MC)
if FulFills (MC; RC) then
return (MC, OL)
remove aP from MC
return (null, OL)
```

[Algorithm for finding a match using depth-first search]

We note that BUILDPLAN is capable of inferring multiple hierarchies, representing students' failed attempts to solve a problem, or exploratory activities that are exogenous to the actual solution path. Such behavior occurred in our empirical evaluation that is described in the next section.

Although FINDMATCH is complete, BUILDPLAN is a greedy algorithm. Once an action set M_C matches a recipe R_C, it does not backtrack and consider any of the actions in M_C for alternative recipes. Thus, it may fail to recognize a student's plan.

The complexity of BUILDPLAN is dominated by the complexity of the FINDMATCH algorithm, denoted C_{FM}. Let $|R|$ and $|X|$ be the number of recipes in R and the number of actions in the action sequence X, respectively. Then, BUILDPLAN calls FINDMATCH at most $|X|$ times per recipe, yielding an overall complexity of $O(|R| \cdot |X| \cdot C_{FM})$. Since FINDMATCH was implemented as a depth first search, its complexity is exponential in the size of the action sequence X, which dominates the complexity of the overall approach.

11.4.3 Empirical Methodology

We evaluated the algorithm on real data consisting of 20 students' interactions with VirtualLabs. These interactions were sampled from a depository of log files describing homework assignments of over 100 students from an R1 private university in a second semester general chemistry course (the sessions with VirtualLabs were not controlled in any way). The sampled interactions included students' solutions to six problems intended to teach different types of experimental and analytical techniques in chemistry, taken from the curriculum of introductory chemistry course using VirtualLabs (students were not repeated across problems). One of these was the dilution problem that was described in Sect. 11.3. A detailed description of all of the problems is given in Sect. 11.7. For diversity, the chosen students' logs varied greatly in size, ranging from 20 actions to 187 actions.

The recipes were created by transforming written descriptions of students' possible solution processes for each problem. These written descriptions were obtained from a domain expert who is a chemistry researcher and one of the developers of VirtualLabs. In addition, we also randomly sampled 5–6 of the students' logs for each problem from the depository of homework assignments described above and added recipes if they were not already given by the domain expert. The log files used in process of creating recipes were not used in the evaluation of the algorithm.

We ran the algorithm on each of the 20 log files using the recipe library of the corresponding VirtualLabs problem. The outputted plans ranged in depth from 3 to 21 levels. The algorithm was evaluated by the domain expert. For each problem instance, the domain expert was given the plan(s) outputted by BUILDPLAN, as well as the student's log. We consider the inferred plan(s) to be "correct" if the

Table 11.1 Performance measures for the recognition algorithm

	N	Log size	Plan size	Plan depth	Run-time (s)
Coffee	4	33.25	41.75	12	0.28
Oracle	4	92.75	57.75	6.25	1.06
Dilution	4	63	39.75	8	0.54
Unknown acid	4	54.25	56.25	12	0.8
Camping	2	76	31.5	5	0.4
Coffee 2	2	67.5	62	12	1.0
Overall	20	63	48.45	9.35	0.68

domain expert agrees with the complex and basic actions at each level of the plan hierarchy that is outputted by the algorithm. If the student was able to complete the problem, the outputted plan(s) represent the student's solution process. Otherwise, the outputted plan(s) represent the students' failed attempts to solve the problem.

The results revealed that BUILDPLAN correctly inferred students' plans for 19 out of the 20 problem instances. Specifically, the algorithm was able to capture trial-and-error approaches as well as explorations and mistakes. For instance, one of the students performed three separate attempts to solve the dilution problem. The first two attempts resulted in a wrong molarity of the solution, and after each of these unsuccessful attempts the student started over using different flasks. The algorithm represented each of these three attempts in a separate plan hierarchy. This is an important capability, as it allows teachers to gain important insights regarding students' problem solving processes by reviewing their plans. We demonstrate this capability in the user study described in Sect. 11.5.

The reason for the sole incorrect plan was revealed to be a recipe that was lacking a temporal constraint for enforcing an ordering between its constituent actions. It is important to note that this incorrect inference was not caused by the greediness of the BUILDPLAN algorithm, but by an incomplete recipe data base. This does not impede on the algorithms correctness, as it was always able to infer students' plans given that recipes were available.

Table 11.1 summarizes the performance of the algorithm according to several measures: *N*, representing the number of instances for each problem; log size, representing the size of the interaction history that serves as input to the algorithm; plan size, representing the number of nodes in the plan(s) outputted by the algorithm; plan depth, representing the length of the longest path in the inferred plan(s); run time of the algorithm (in seconds) on a commodity quad-core computer. All of the reported results were averaged over the different instances in each problem. As shown in the table, the overall average time for inferring students' plans was 0.68 s, with a relatively high variance (std. 0.79), due to the diversity of the students' interactions and the experimental processes required to solve each of the problems. The longest time to infer students' plans occurred for interactions relating to the "oracle" problem (1.06 s.) and "coffee 2" problem (1.0), which also resulted in the largest plans (57.75 and 62 nodes respectively). The key determinant of the algorithm's runtime was the size of the log that described the student's interaction. These results show the feasibility of using the proposed algorithm in practice, as students' interactions are finite and limited.

11.4.4 Complete Algorithms

In this section we present two plan recognition algorithms that are complete. Both algorithms work by converting the plan recognition problem into one or more constraint satisfaction problems and using standard techniques for their solution. A limitation of this approach is that it is constrained to non-recursive grammars, in which actions cannot be repeated indefinitely. To this end we employed a different exploratory learning environment called TinkerPlots, used world-wide to teach students in grades 4–8 about statistics and probability Konold and Miller [42].

TinkerPlots is an educational software system used world-wide to teach students in grades 4 through 8 about statistics and mathematics [42]. It provides students with a toolkit to actively model stochastic events, and to create and investigate a large number of statistical models [43]. As such, it is an extremely flexible application, allowing for data to be modeled, generated, and analyzed in many ways using an open-ended interface.

To demonstrate our approach towards recognizing activities in TinkerPlots we will use the following running example, called: The probability of rain on any given day is 75 %. Use TinkerPlots to compute the probability that it will rain on each of the next four consecutive days. This problem is a simple example drawn from a set of problems posed to students using TinkerPlots in schools and to subjects during our data collection procedure.

One of the possible approaches towards modeling this problem in Tinker Plots are shown in Fig. 11.4. The top part of the figure shows a sampler object containing "spinner" devices used to model distributions. The spinner device in the left-hand model contains two possible events, "rain" and "sun". The likelihood of "rain" is three times that of "sun", as determined by the surface area of these events within the spinner. Each draw of this sampler will sample the weather for a given day. The number of draws is set to four, making the sampler a stochastic model of the weather on four consecutive days.

The basis of the complete approaches make use of a structure called a plan tree for representing and reasoning about recipes in the database, essentially a search tree for capturing the set of possible plans consistent with the recipe database. A plan tree has two types of nodes: AND nodes, whose children represent actions that must be carried out to complete a recipe, and or nodes, whose children represent a choice of recipes for completing an action. The root, action C, is an OR node. For each recipe for C, a child AND node is added to the root and labeled with the sub-actions of that recipe. The children of this AND node are the plan trees of each sub-action. A branch terminates when a basic action is reached, as a basic action has no recipe by definition.

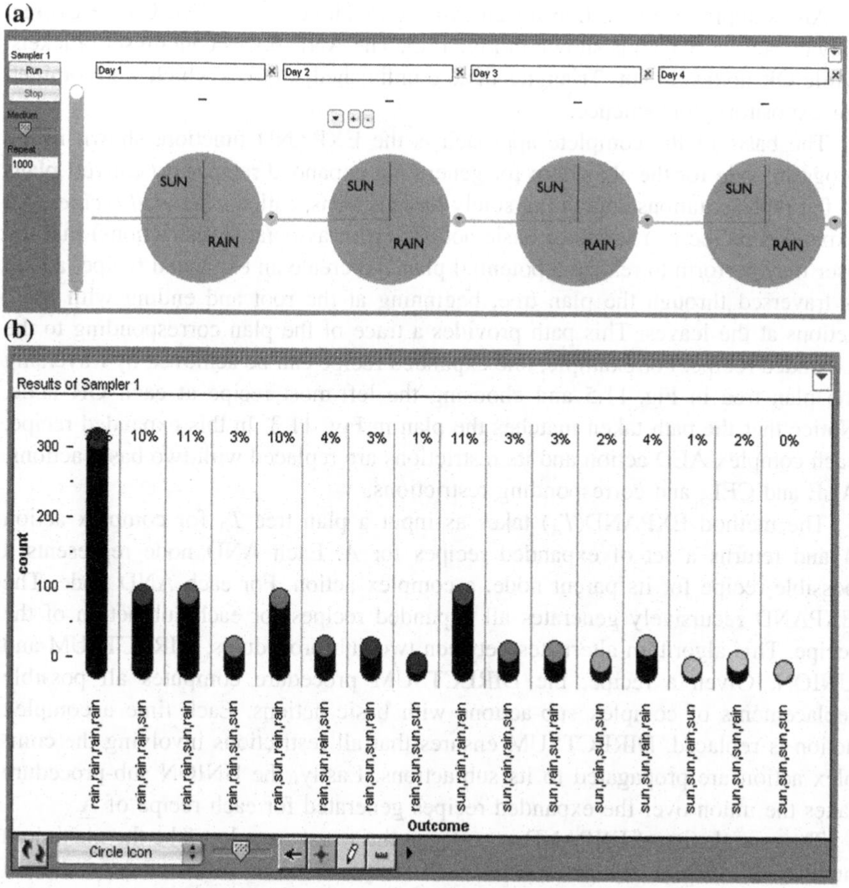

(a)

(b)

Fig. 11.4 Snapshots of tinkerplots interaction when solving the problem. **a** Using four spinners. **b** Plotted results

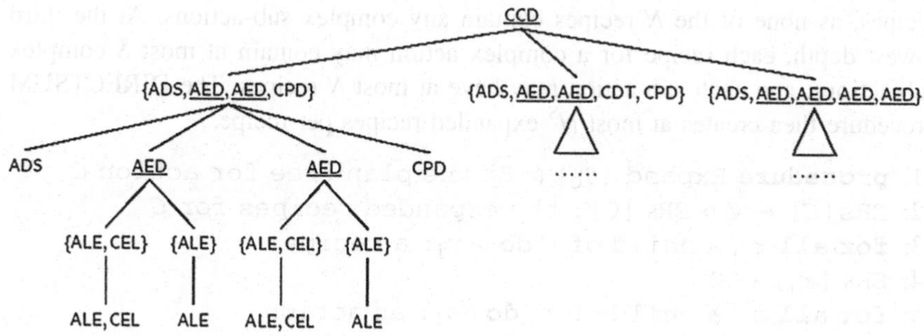

Fig. 11.5 A partial plan tree for the CCD complex action

An example of a plan tree for an activity in TinkerPlots called Create Correct Device Action (CCD) is shown in Fig. 11.5. The AND nodes contain set brackets, while OR nodes do not. Triangles denote unfinished subtrees which were omitted for expository convenience.

The basis of the complete approach is the EXPAND function, shown in the program code for the algorithm for generating expanded recipes, to convert plans to flat representations containing solely basic actions, called *expanded recipes*. An expanded recipe is a series of basic actions (with associated restrictions) that the user may perform to realize a potential plan. To create an expanded recipe, a path is traversed through the plan tree, beginning at the root and ending with basic actions at the leaves. This path provides a trace of the plan corresponding to the expanded recipe. For example, one expanded recipe can be achieved by traversing the plan tree in Fig. 11.5 and choosing the left-most recipe at each OR node. Notice that the path taken matches the plan in Fig. 11.3. In this expanded recipe, each complex AED action and its restrictions are replaced with two basic actions, ALE and CEL, and corresponding restrictions.

The method EXPAND(T_A) takes as input a plan tree T_A for complex action A and returns a set of expanded recipes for A. Each AND node represents a possible recipe for its parent node, a complex action. For each AND node, The EXPAND recursively generates all expanded recipes for each sub-action of the recipe. This algorithm alternates between two sub-procedures, DIRECTSUM and UNION. Given a recipe, the DIRECTSUM procedure computes all possible replacements of complex sub-actions with basic actions. Each time a complex action is replaced, DIRECTSUM ensures that all restrictions involving the complex action are propagated to its sub-actions. Lastly, the UNION sub-procedure takes the union over the expanded recipes generated for each recipe of A.

The complexity of EXPAND is costly in the worst case. Let S be the maximum number of *complex* sub-actions for each recipe, N be the maximum number of recipes for a *single* complex action, and C be the number of distinct complex actions. A plan tree has depth of at most $C + 1$, as we do not allow for recursive recipes. At the lowest depth of the plan tree, all actions are basic and do not have recipes. At the second lowest depth, complex actions have at most N expanded recipes, as none of the N recipes contain any complex sub-actions. At the third lowest depth, each recipe for a complex action may contain at most S complex sub-actions, and each sub-action may have at most N recipes. The DIRECTSUM procedure then creates at most N^S expanded recipes per recipe.

```
procedure Expand (T_C) ▷ T_C: the plan tree for action C
ERs[C] ← ∅ ▷ ERs [C] : the expanded recipes for C
for all r_j, a child of C do ▷ r_j: a recipe
ERs [r_j] ← ∅
for all a_i, a child of r_j do ▷a_i: an action
ERs [r_j] ← DirectSum (Expand (T_ai), ERs [r_j])
ERs [a] ← Union(ERs [a], ERs [r_j])
if ERs [a] = ∅ then
```

```
ERs [a] ← {a}
return ERs [a]
```

[Algorithm for generating expanded recipes]

The UNION procedure collects the expanded recipes resulting from each recipe for that action, resulting in a maximum of $N(N)^{S}$, or N^{S+1}, recipes. At the fourth lowest depth, each complex action can again have at most N recipes with at most S complex sub-actions in each. Each of these S sub-actions can contain at most $N(N)^{S}$ expanded recipes. So, the DIRECTSUM and UNION procedures create at most $N(N(N)^{S})^{S}$, or N^{S^2+S+1}, expanded recipes per recipe. Continuing this reasoning, the top level action can have at most (11.1) recipes, yielding an overall complexity of $N^{0(S^C)}$.

$$N^{\sum_{i=0}^{C-1} S^i} \tag{11.1}$$

Constraint Satisfaction Algorithm. In this subsection we explain how to combine an expanded recipe and action sequence to create a constraint satisfaction problem (CSP). A solution to the resulting CSP is the plan representing the users' activities. Formally, a CSP is a triple (X, Dom, C), where $X = \{x_1,\ldots, x_n\}$ is a finite set of variables with respective domains Dom $= \{D_1,.., D_n\}$, each a set of possible values for the corresponding variable, $D_i = \{v_1^i, v_k^i\}$, and a set of constraints $C = \{c_1,\ldots,c_m\}$ that limit the values that can be assigned to any set of variables.

The algorithm CONVERTTOCSP, shown in the program code for converting an expanded recipe and action sequence to a CSP, receives as input an expanded recipe E_A and an action sequence **X** and returns a CSP. If a solution exists for this CSP, a subset of the actions in **X** realize the expanded recipe E_A. We first show how to create variables in the CSP, and we use as a reference Fig. 11.6, which provides a graphical representation of the CSP resulting from some action sequence and expanded recipe. We used a graphical layout suggested by Dechter [44]. Note that parameters belonging to actions are not pictured unless they participate in some constraint.

Let $S = \{s_1,\ldots, s_n\}$ and R be the set of sub-actions and restrictions in the expanded recipe, respectively. Each action in S becomes a unique variable in the CSP by calling the subroutine ADDVARIABLEANDDOMAIN(s, **X**). Based on

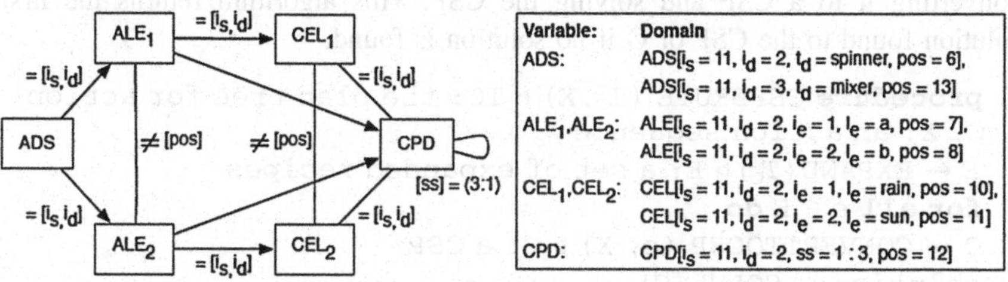

Fig. 11.6 CSP resulting from an action sequence and an expanded recipe

the expanded recipe, six variables are added at this time: ADS, ALE_1, CEL_1, ALE_2, CEL_2, and CPD. These variables appear, outlined, in the graph of Fig. 11.6.

```
procedure ConvertTOCSP(E_A = (S,R), X) ▷ EA: an expanded recipe S and restrictions R for complex action A, X: an action sequence
for all s ∈ S do ▷ S: a set of sub-actions
AddVariableAndDomain (s, X)
for all r ∈ R do ▷ R: a set of restrictions
AddRestrictionConstraint (r)
for all s ∈S do
AddRedundancyConstraint (S)
```

[Converting an expanded recipe and action sequence to a CSP]

Each variable's domain is then derived from the actions in the action sequence. For each occurrence of action *s* in the action sequence, a value is added to the domain of *s* in the CSP. The right-hand box of Fig. 11.6 gives the resulting domain for each variable based on the action sequence.

Lastly, we add restrictions to our CSP. For each restriction *r* in *R* over actions $\{s_1,\ldots, s_m\}$ in *S*, a constraint over the corresponding CSP variables is added to the CSP using the ADDRESTRICTIONCONSTRAINT(*r*) subroutine. Directed edges in the Fig. 11.6 represent temporal constraints between two variables. Undirected edges represent other parametric constraints. The edge from ADS to ALE_1 expresses the constraint ADS $\prec$ ALE_1 as well as the constraint ADS$[i_s, i_d]$ = $ALE_1[i_s, i_d]$.

For variables corresponding to the same action, additional redundancy constraints are added using the ADDREDUNDANCYCONSTRAINT subroutine. These constraints ensure that such variables are assigned distinct values, as these variables share the same domain. An example is the constraint connecting the ALE_1 and ALE_2 variables, which requires that these variable assignments have distinct pos parameters.

A solution for a CSP provides a match between an expanded recipe and an action sequence. In this section we present two algorithms that use CSPs to output a plan from an action sequined X for a desired complex action *C* given a set of recipes *R*.

The algorithm shown in the program code for brute force algorithm takes a brute force approach, calling EXPAND to generate each expanded recipe for *C*, converting it to a CSP and solving the CSP. This algorithm returns the first solution found to the CSP or Ø if no solution is found.

```
procedure CSPBRUTE (T_C,X) ▷ TC: the plan tree for action C, X: an action sequence
E ← EXPAND(T_C) ▷ E: a set of expanded recipes
for all e ∈ E do
C ← CONVERTTOCSP (e, X) ▷ C: a CSP
solution ← SOLVE(C)
```

```
if solution ≠ ∅ then
return solution
return ∅
```

[Brute force algorithm]

The complexity of CSPBRUTE can be analyzed in terms of the FINDMATCH2 and EXPAND procedures. Recall that calling EXPAND results in at most $N^{O(S^C)}$ expanded recipes, where N is the maximum number of recipes for a single complex action. In the worst case, all expanded recipes are considered, and for each expanded recipe a CSP solver must be run. The complexity of this CSP solver can be bounded by the complexity of a complete backtracking search, which we have seen to be $|X|!/S!$. So, an overall worst-case complexity of CSPBRUTE is (11.2).

$$N^{O(S^C)} O\left(\frac{|X|!}{S!}\right) \tag{11.2}$$

To evaluate the complete approach, we collected interaction sequences of people's interaction with TinkerPlots. Each subject received an identical 30 min tutorial about TinkerPlots and was then asked to complete four problems in succession; these problems are detailed in Sect. 11.7. TinkerPlots is equipped with a logging facility that records the basic actions that make up users' action sequences. As in the VirtualLabs domain, we noted whether each problem was solved, and we constructed the (possibly multiple) plans used to solve the problem. The analyzed user logs range in length from 14 to 80 actions. The average length of an interaction sequence for problems collected from adult subjects was 35 actions. Adults solved the assigned problems 70 % of the time. In contrast, the average length of an interaction sequence for problems collected from students was 68 actions. Students solved the assigned problems 60 % of the time. Also, people engaged in exploratory behavior using the software. For example, there were on average 15 exogenous actions in each problem that was obtained from adults. As expected, the complete approaches were able to achieve perfect performance on all of the logs. They also took reasonable time, measuring from 2 to 4 s on the logs.

11.5 Visualizing Students' Activities

This Section presents visualization methods that were designed for the purpose of presenting students' activities to teachers. It then describes a user study that evaluated these different methods with chemistry teachers.

[4] We used the Prefuse package to implement this application Heer et al. [45].

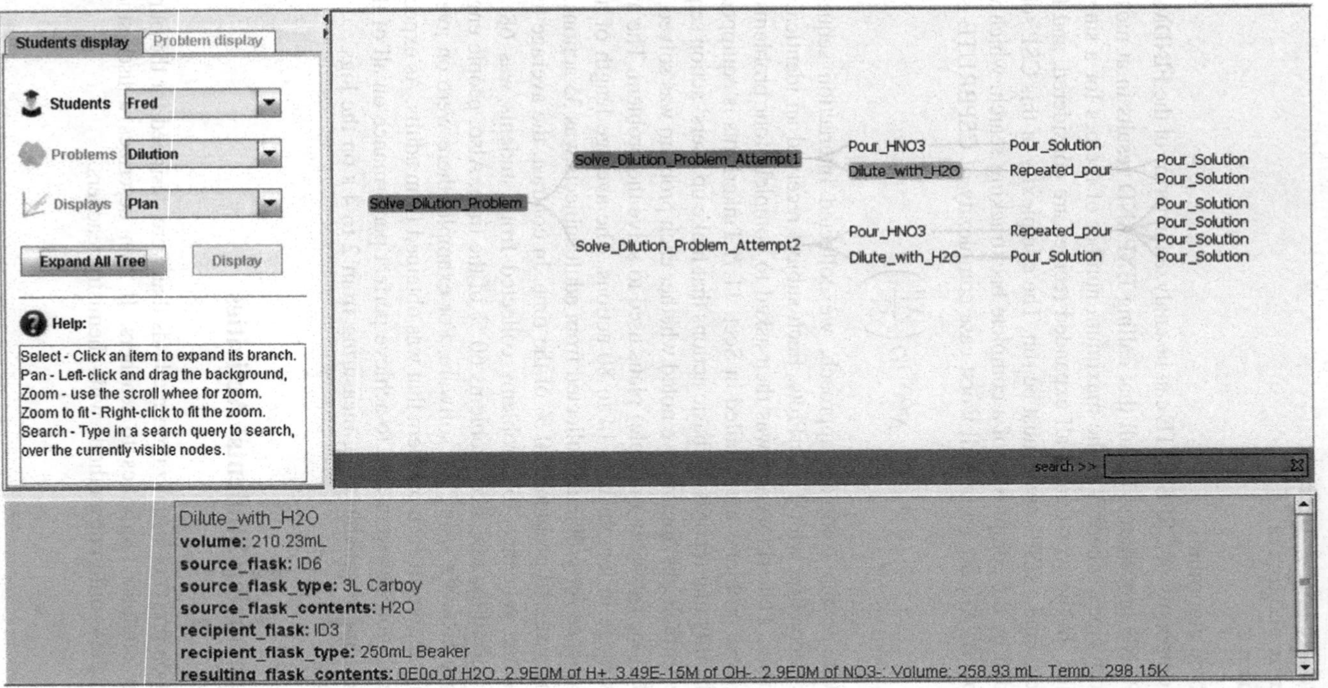

Fig. 11.7 A temporal visualization of a student's solution to the dilution problem

11.5.1 Visualization Methods

We hypothesized that showing students' plans to teachers would facilitate their understanding of students' work. In addition, we wished to evaluate an alternative visualization method that emphasizes the temporal aspects of students' interactions, which is lacking in the plan visualization. We therefore used the following three visualization methods that differ in the type of data they present as well as the way in which this data is presented.

The *plan visualization method* presents students' plans as they are inferred by the recognition algorithm. The plan is presented using an interactive interface that enables to explore the plan tree.[4] An example of this visualization on a student's plan for solving the dilution problem is shown in Fig. 11.7. The plan is presented as a tree. Each of the nodes in the tree represents a student's activity. The leaves of the plan represent the basic actions of the student that constitute students' interactions with VirtualLabs. The other nodes represent higher level activities that were inferred by the algorithm. As shown by the nodes "solve dilution problem attempt 1" and "solve dilution problem attempt 2", the student made two attempts at solving the dilution problem. The descendants of these nodes decompose the activities that constitute each of the attempts. When clicking on a node in the plan, the parameters of the action that corresponds to this node are displayed in the information panel shown at the bottom of the Fig. 11.7. As can be seen, the complex action "dilute with H_2O" consisted of pouring a total of 210.23 ml of H_2O to dilute the acid. The "resulting flask contents" shows the solution consistency in the flask after this dilution activity. The child node of the "dilute with H_2O" action is "repeated pour". Clicking on this node will show the two separate pours from the H_2O bottle that comprises the dilution activity.

The *Temporal visualization* presents students' interactions over a time line. The vertical axis displays the objects used by the student, while the horizontal axis displays students' actions in the order in which they were created. An example of a temporal visualization of a student's interaction with VirtualLabs when solving the dilution problem is shown in Fig. 11.8. This student's interaction consisted of mixing solutions in flasks, and each arrow in the figure represents one of these mixing actions. The base of the arrow represents the source flask, while the head of the arrow indicates the recipient flask. Thicker arrows correspond to larger volumes of solution being mixed. The information panel at the bottom of the figure describes the parameters of the mixing action represented by the boxed arrow in Fig. 11.8. It shows that the student poured 743.8 mL of H_2O, to a 1,000 ml Volumetric Flask. Also shown is the resulting consistency of the solution in the recipient flask.

The *Movie visualization* describes students' actions exactly as they occurred during their interactions with VirtualLabs, and is analogous to a teacher that is looking over the shoulder of a student. This is the only type of support that is currently available to teachers. This visualization replays the actions from the log in the order they were created by the student, but does not reflect the actual

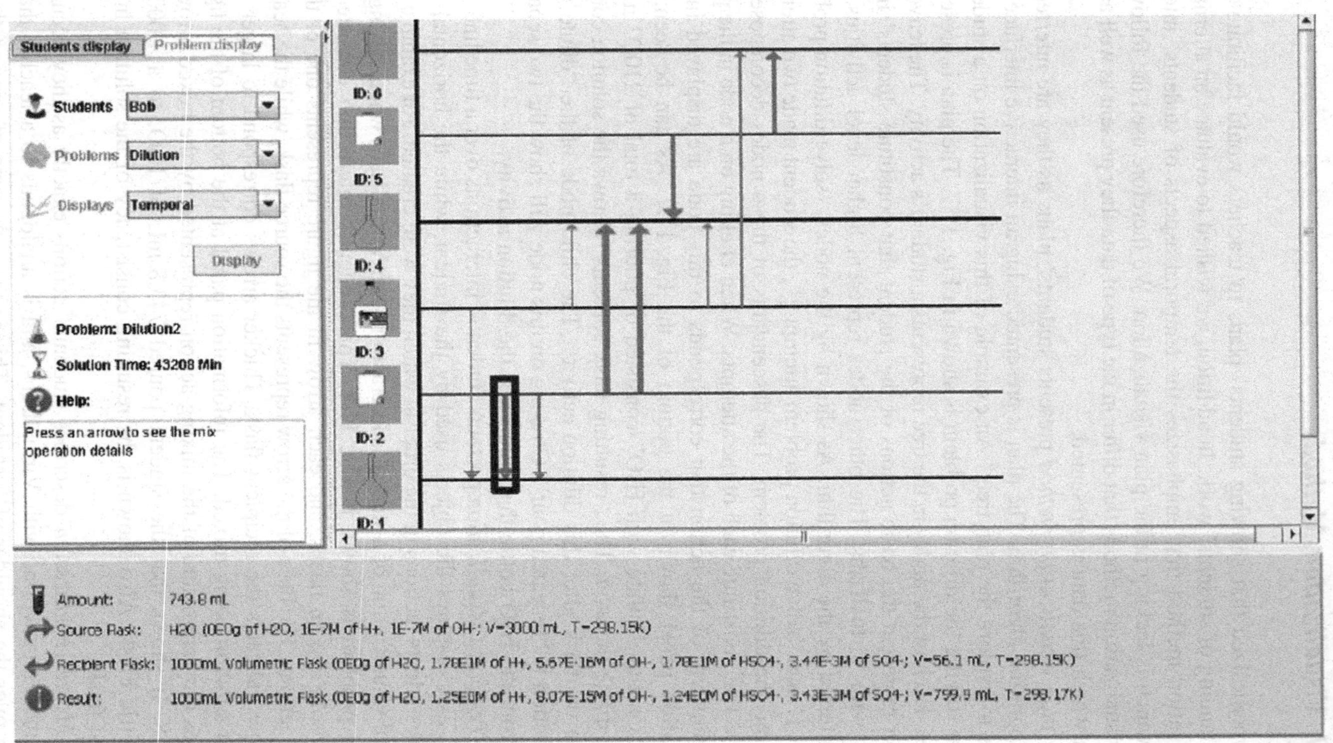

Fig. 11.8 A plan visualization of a student's solution to the dilution problem

passage of time between students' actions. The movie can be stopped, rewound and fast-forwarded to focus on the students' display at particular points in their interaction. A snapshot of this visualization for one of the students solving the dilution problem is shown in Fig. 11.9. In the snapshot the student is pouring NH_3 to a 500 ml Erlenmeyer flask. On the right side of the figure, the current contents of the selected flask are shown (in the "Solution Info" panel).

These three visualizations differ widely in the way they present information to teachers. First, both the movie and the temporal visualization methods render students' activities directly from the log. The plan visualization supersedes these visualizations in that it also visualizes higher level activities as inferred by the recognition algorithm. Second, the movie presents snapshots of the user's application window, while the temporal and the plan visualizations present a more expansive account of the student's work-flow. In particular, the temporal and plan visualization specify the amount of solution being poured from flask to flask, while this information is not directly shown in the movie.

To illustrate these differences, we describe how teachers and researchers may use each visualization method to identify that a student made several attempts to solve the dilution problem. Using the movie visualization, teachers need to keep track of which flasks the student used to mix acid with H_2O, and pause the movie after each mixing action to observe the resulting concentration of the solution in the flask in the "Solution Info" panel. Because the movie visualization presents a single action at each time-frame, it can be difficult to distinguish whether a mixing action using a new flask represents the commencement of a new attempt to solve the problem or an exploratory action (or a mistake). Using the temporal visualization, teachers can observe the set of flasks used by the student to dilute the acid, and the pouring actions that are associated with each flask.

To characterize the activities making up each of the student's attempts, teachers need to identify the relevant actions over the time line, starting from the action that poured acid to a new flask and terminating in the pouring action that resulted in the diluted solution. The temporal aspect of this presentation makes it easy to identify such sets of pouring actions when they occur close together in time. This is illustrated in Fig. 11.8, in which the three contiguous actions pouring solutions into Flask ID 1 and the 3 contiguous actions pouring solutions into Flask ID 3 represent two distinct attempts (and the next 4 pours represent additional two distinct attempts). However, this procedure may be difficult to do when students' interactions are long, or when students interleave activities, as any two adjacent actions may belong to different attempts.

Lastly, the plan visualization separates each of the students' dilution attempts into a separate branch, and the nodes in each branch comprise those pouring actions that characterize each attempt. This is illustrated in Fig. 11.7, in which each attempt aimed at solving the problem is a sub-plan that emanates from the "Solve_Dilution_Problem_Attempt1" and "Solve_Dilution_Problem_Attempt2" nodes. However, the plan does not order students' actions along a time line, and thus it is difficult to recognize the order in which actions were performed.

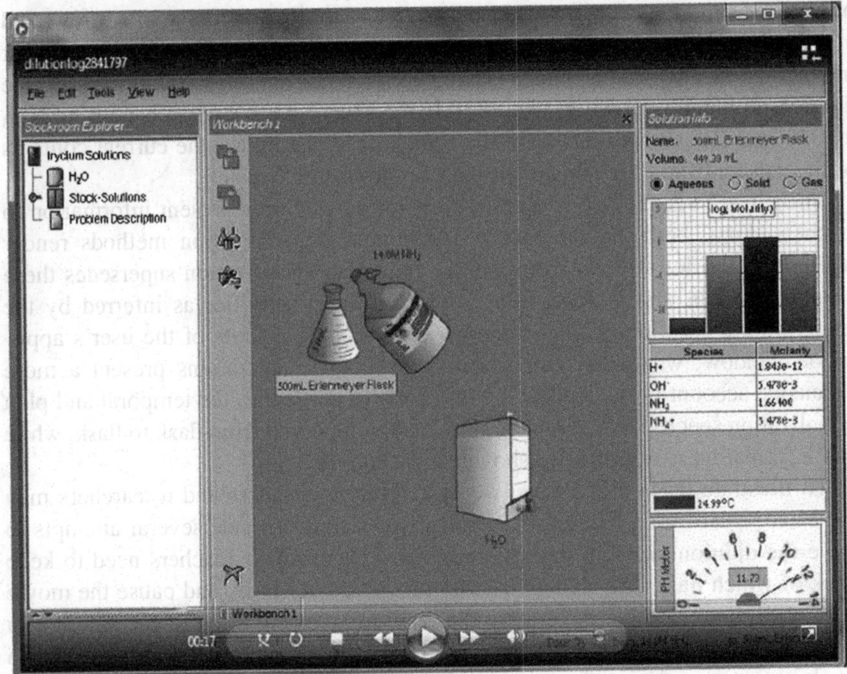

Fig. 11.9 A movie visualization of a student's solution to the dilution problem

11.5.2 Empirical Methodology

A user study with chemistry educators was conducted to evaluate these three visualization methods. The goals of this study were to determine how each of these visualizations contributes to teachers' analysis of students' work in VirtualLabs and which visualization methods teachers found helpful.

The interactions in the study were taken from the log files of students solving two problems (out of the six problems for which we collected data). These log files were also used in the evaluation of the plan recognition algorithm, such that the plans were validated as correct by a domain expert prior to the user study.

One of the two problems was the dilution problem described in Sect. 11.3. The other problem (called "coffee") required students to add the right amount of milk to cool a cup of coffee down to a desired temperature. These problems differed in the type of reasoning they demanded from students. The dilution problem was characterized with longer, more complex student solutions. For example, students solving the dilution problem used more intermediate flasks and more attempts to solve the problem. To illustrate, the average log size of solutions to the dilution problem was 51 actions, whereas the average log size of solutions to the coffee problem was 29.67 actions. Thus, we were able to evaluate the visualization methods on two problems with significantly different solution processes.

Seventeen participants took part in the study. Fifteen of the participants were graduate students of chemistry and chemistry engineering serving as teaching assistants (14 students from Ben-Gurion University, and one student from the Weizmann Institute). Two of the participants were a professor of chemistry from the University of British Columbia who uses VirtualLabs in the classroom and a professor of education and technology at Haifa University with a master degree in chemistry.

All participants received an online survey that included all of the materials used in the study (tutorials of VirtualLabs and the visualization methods, and questionnaires for the evaluation of the visualizations). The participants first watched an identical video tutorial of VirtualLabs and were asked to perform several tasks using the software to demonstrate their understanding. Participants were also provided with an identical tutorial about each of the three visualization methods. Each subject was presented with three student interactions solving the same problem. Each of these interactions was shown using one of the three visualization methods, and the order in which the visualizations were presented varied across participants. To avoid biasing the participants, each interaction that was visualized was chosen from a different student. For each problem, participants were presented with interactions that were similar in length and complexity of the student's solution.

Each participant was asked to comment on the visualization methods by answering the following questions: (1) Based on the presentation can you tell whether the student solved the problem?; (2) Based on the presentation can you tell how the student solved the problem?; (3) Assuming you were using VirtualLabs in your class, would you be likely to use this presentation style to understand students' work after a classroom session?[5] After seeing all of the visualizations, participants were asked to quantitatively compare between the different methods according to the same set of criteria, and were also asked to compare how easy it was to learn how to use the different methods. For this comparison participants were requested to rate each visualization method using a Likert scale of 1–7 (where 1 stands for "strongly disagree", 7 stands for "strongly agree", and 4 being a neutral answer of "neither"). Finally, participants were asked whether they preferred one visualization to the others, and how they would combine some or all of the visualizations. One of the researchers was present throughout each of the sessions, answered any questions participants had about the visualization methods, and validated that teachers' conclusions about students' work was correct not.[6] That is, when participants reported that they could infer whether or how a student solved the problem, the researcher validated that their

[5] We also asked participants to explain each of their answers. The full questionnaire can be found in Sect. 11.9.

[6] The researcher was physically present in the laboratory with all of the graduate students from Ben-Gurion University, and used VoIP technology (Skype) to connect with the other three participants.

inference was correct. In all cases participants that reported to have understood students' solutions, did so correctly.

11.5.3 Results

In this section we present the analysis of the responses we received from the participants in the user study. First, we describe a qualitative analysis of participants' responses with regards to the visualizations, followed by a quantitative analysis of their comparisons of the different visualization methods.

Qualitative Analysis of the Visualization Methods. We first describe participants' responses with regards to the *movie visualization*. When asked if the movie visualization demonstrated *whether* the student had solved the problem, participants' responses depended on the type of problem they were shown. Most of the participants (7 out of 9) who viewed solutions to the coffee problem claimed that they were able to tell whether the student had solved the problem, while the other two participants reported that this information was not apparent to them. Those participants that inferred how the students solved the problem did so by observing the final contents of the flasks used by the student. A typical response was "Yes, by following the temperature and volume meters on the right side and watching the actions the student took."

Half of the participants (4 out of 8) who viewed solutions to the dilution problem could not infer whether the student had solved the problem. These participants reported that the movie was too fast and difficult to follow. A typical response was "No, I can not. The added amounts are not clear and the appearance and disappearance of elements on the screen are confusing." Only 2 of 8 participants reported that the movie clearly demonstrated whether the problem was solved by the student, in contrast to 7 out of 9 participants who could infer this information from the coffee problem. A possible explanation for this discrepancy is the length of students' interactions. The average length of students' interactions for the dilution problem was significantly longer than the average length of students' interactions for the coffee problem. This made it more difficult for participants to keep track of students' actions using the movie visualization.

The participants expressed a more homogeneous opinion when asked whether they understood *how* the student had solved the problem, and these responses were not dependent on the problem shown. Ten of the participants reported that the movie enabled them to determine how the student solved the problem. A typical response explained, "It is easy to see the steps the student used to solve the problem." Three of the participants stated that they were not able to determine how the student solved the problem. One of them stated "The presentation [visualization] created a confusion. Irrelevant steps, such as moving flasks were shown, which created a confusion and made it difficult for me to distinguish important actions." Four participants claimed they could determine how the problem was solved in general terms, but were missing the exact quantities mixed by the students. Lastly, only four of the participants reported they would be likely to use the movie visualization in their class. The other participants found it to be too slow to be useful. A typical response was "This method seems to be much slower that could be a problem when checking 30 students or so…".

When evaluating the *temporal visualization*, all participants but one answered that this visualization method clearly demonstrated *whether* and *how* the student had solved the problem, and would be likely to use this method in their class. However, two participants were concerned that if a problem solution would require many steps, the method may not be useful. One of them explained: "[…] If an exercise requires moving many solutions from beaker to beaker, and lots of mixing, this method might not be as clear and become a bit messy."

In their evaluation of the *plan visualization*, all but one of the participants reported that the visualization demonstrated *whether* the student had solved the problem, and 14 out of the participants found that the plan visualization demonstrated *how* the students solved problems. Several of the participants specifically commented on the higher level activities that were represented in this visualization: "The presentation [visualization] focuses on the important actions and summarizes the student's activities.", or "I can read the final concentration in an easy way at each stage and the [different] attempts of the student are very clear."

Three of the participants commented that they have found the plan visualization difficult to understand. One of them explained: "This presentation [visualization] was more complicated. I had to click the nodes and observe the volumes and students actions at each step." In all, 14 out of 17 participants reported they would be likely to use the plan visualization in their classroom.

After observing and evaluating all of the visualization methods, we asked participants which visualization method they preferred. Six participants expressed a strict preference for the plan visualization and six participants preferred the temporal visualization. Only one participant strictly preferred the movie visualization over the other two proposed visualizations. Another participant stated an equal preference for the movie and temporal visualizations, while one participant claimed that he would prefer using the temporal visualization for simple problems, and the plan visualization for complex problems. Table 11.2 summarizes the qualitative responses described in this section. The table only includes the number of participants who expressed a non-ambiguous positive answer to each of the questions discussed above.

Finally, we were interested to see whether teachers would want to use a combination of some or all of the visualizations, given their distinct differences. Seven participants suggested combining the temporal visualization with the plan visualization. Three participants suggested combining the temporal visualization with the movie visualizations, while two participants suggested combining the plan visualization with the movie, and one participant said he would want to combine all visualizations. Several participants indicated in their response that they believe teachers may have different preferences, and therefore suggested to provide all visualizations and let the teacher choose which of them to use. They further envisioned using different visualizations for different purposes, for example

Table 11.2 Summary of qualitative responses

	Demonstrates whether the student solved the problem	Demonstrates how the student solved the problem	Likely to be used	Strictly preferred
Movie	9/17	10/17	4/17	1/17
Temporal	16/17	16/17	16/17	6/17
Plan	16/17	14/17	14/17	6/17

For each visualization the table shows the number of participants who expressed a clear positive response to each of the questions

using the plan visualization to get a one-image quick view of the solution structure, and then use the temporal visualization for a more in depth exploration of solutions they found more interesting.

Analysis of Quantitative Responses. After observing and separately evaluating each of the visualizations, participants were asked to make a quantitative comparison of the different methods. Fig. 11.10 shows the average score given by participants to each of the visualization methods when using a Likert scale of 1–7. $N = 17$ for each of the questions as all participants responded to all questions. As shown in the Fig. 11.10, the movie had a higher average score than the other methods with respect to ease of learning (Mean = 6.23, STD. = 1.35). The plan visualization was the hardest to learn (Mean = 4.17, STD. = 2.19), and also exhibited the highest variance in scores.

The plan visualization scored highest with respect to demonstrating whether the student solved the problem (Mean = 5.94, STD. = 1.89), closely followed by the temporal visualization (Mean = 5.53, STD. = 1.94). The movie was ranked last (Mean = 4.18, STD. = 2.27). The temporal and plan visualization methods scored highest with respect to demonstrating how the student solved the problem (Mean = 5.88, STD. = 1.65), followed by the plan visualization (Mean = 5.53, STD. = 1.56). Again, the movie visualization was ranked last (Mean = 4.65, STD. = 2.06). The plan (Mean = 5.88, STD. = 1.54) and temporal (Mean = 5.88, STD. = 1.54) visualizations scored highest with respect to being used in the classroom. The movie was ranked last (Mean = 3.41, STD. = 2.2).

We used the Friedman non-parametric test for analysis of variance to distinguish between participants' quantitative responses and found a significant effect of visualization type in all of the questions ($X^2 > 6.3$, $P < 0.05$). Post-hoc analysis with Wilcoxon Signed-Rank Test was conducted with a Bonferroni correction applied.

Median scores for ease of learning were 7 (3 to 7), 6 (1 to 7) and 4 (1 to 7) for the movie, temporal and plan visualizations respectfully. Both the movie and temporal visualizations were significantly easier to learn than the plan visualization ($Z < -2.55$, $P < 0.011$). The plan, with median score 7 (1 to 7), was significantly more helpful than the movie, with median score 5 (1 to 7), when inferring whether the student solved the problems ($Z = -2.61$, $P = 0.003$).

No significant difference was found between the temporal visualization with median 6 (1 to 7) and the other visualizations ($Z > -2.1$, $P > 0.031$). The temporal

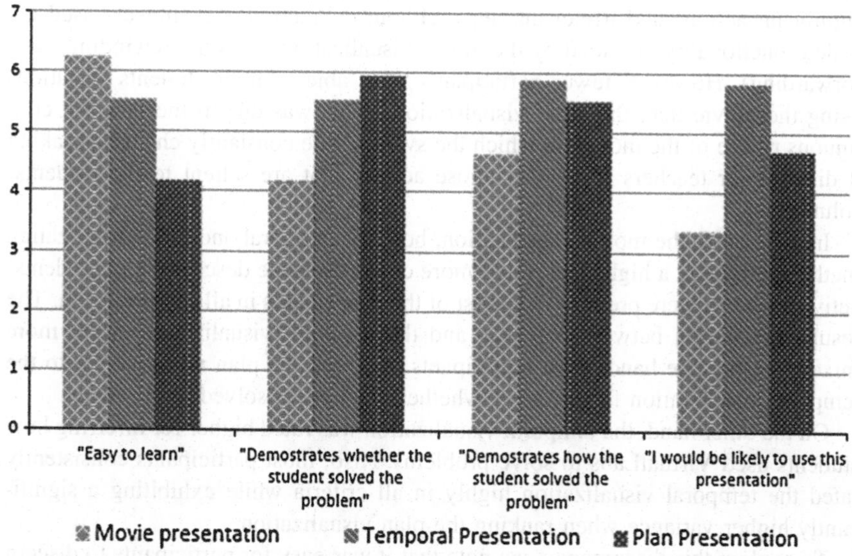

Fig. 11.10 Average scores for the quantitative questions

visualization was found to be significantly more helpful than the movie when inferring how the student solved the problem ($Z = -2.23$, $P = 0.011$) with medians 6 (1 to 7) and 5 (1 to 7) respectfully. There was no significant difference between the plan with median 6 (1 to 7) and the other visualizations ($Z > -1.51$, $-P > 0.075$). Lastly, the temporal visualization with median 6 (1 to 7) was found significantly more likely to be used than the movie visualization with median 3 (1 to 7) ($Z = -3.05$, $P = 0.001$). The differences between the plan visualization with median 5 (1 to 7) and both temporal and movie visualizations were not significant ($Z > -1.96$, $P > 0.027$).

11.5.4 Discussion

A challenge to performing this user study was the relatively large overhead involved in teaching participants about the three visualization methods, and the requirement that participants have prior teaching experience in chemistry. The fact that many of our conclusions reported above were found to be statistically significant is striking given these limitations.

The study revealed, unsurprisingly, that the movie was the most intuitive visualization style and the easiest to learn. However, it was also ranked the least useful for understanding students' work, as can be attested by one of the participants: "Even though the movie style is the easiest to learn it is the hardest to use." The movie was a "playback" of students' work in the lab, and presented both significant actions and irrelevant steps without distinction. All subjects used the added functionality provided by the movie visualization (pausing, rewinding, fast-forwarding). However, fewer participants were able to infer students' solutions using the movie than the other visualizations. This was due to the inherent continuous nature of the movie, in which the system state constantly changes, making it difficult for teachers to identify those actions that are salient to the students' solution.

In contrast to the movie visualization, both the temporal and plan visualization methods provided a higher level and more comprehensive description of students' activities. They were preferred by most of the participants in all of the criteria. The results comparing between the plan and the temporal visualizations were more mixed. On the one hand, most participants preferred the plan visualization to the temporal visualization for inferring whether the student solved the problem.

On the other hand, the temporal visualization was rated higher for inferring how students used VirtualLabs to solve problems. Also, most participants consistently rated the temporal visualization highly in all criteria while exhibiting a significantly higher variance when ranking the plan visualization.

To explain this discrepancy, we note that it was easy for participants to discern whether the student solved the problem by looking at the root of the plan hierarchy, while this information was not explicitly represented in the temporal visualization. We hypothesize that the hierarchical nature of the plan visualization was harder for participants to learn than the temporal visualization. This may explain why they preferred the temporal visualization to the plan visualization when inferring how students solve problems.

We found a 0.7 correlation between the likelihood of using the plan visualization and its ease of learning, and a 0.56 correlation between determining how the student solved the problem and its ease of learning. There was limited time in our lab study for participants to practice the plan visualization method, which was harder to understand than the other methods. However, this result suggests that teachers who understand the plan visualization are likely to adopt it, and that the plan visualization may be very useful to teachers in practice.

Lastly, the diversity of participants' suggestions for combining the various visualizations methods and the possible uses of the visualizations emphasizes the need to adjust to different educators preferences. There is no "silver bullet" visualization that is most useful in all cases and to all users. This was supported by participants' responses which suggested different uses of the visualization methods, and their suggestions for combining the different methods.

11.6 Conclusion and Future Work

This chapter presented novel methods and algorithms for augmenting existing pedagogical software for science education. It addressed two main problems: automatic recognition of students' activities in open-ended pedagogical software and the visualization of these activities to teachers in a way that supports their analysis of students' interactions with such software. To address the first problem, the chapter presented a general plan recognition algorithm for exploratory learning environments. The algorithm was successfully able to recognize students' plans when solving six separate problems in VirtualLabs, as verified by a domain expert. To address the second problem, the paper presented novel methods for visualizing students' interactions with VirtualLabs. Both of these methods were preferred by participants in a user study to a movie of students' interactions with the software.

Our long term goal is the design of collaborative systems for supporting the interaction of students and teachers in a variety of pedagogical domains. These tools embody the principals of collaborative decision-making, in that the system provides the best possible support for its users while minimizing the amount of intervention.

Our future work will extend the methods and algorithms proposed in this chapter in order to build such collaborative systems. To do so we intend to extend our work on both the plan recognition and visualization methods. One limitation of the plan recognition approach is the reliance on domain experts to construct appropriate recipes in a formal way.

In future work we will design novel methods for automatically extracting recipes and allowing teachers to design recipes in a straightforward way. We also intend to design new plan recognition algorithms that recognize students' activities in real time, during their interaction with the software. We will construct computer agents that use these recognition algorithms to generate interventions with the student while minimizing the amount of intrusion.

We will extend the work on visualization methods to study how other types of state-based visualizations affect teachers' understanding of students' activities, such as showing selected snapshots of students' interactions. Also we intend to develop aggregate visualization methods for describing groups of students.

Although our techniques were demonstrated on one software system their applicability has been shown to other open-ended pedagogical software Gal et al. [16]. We also plan to apply our approach to other types of domains in which users engage in exploration, such as Integrated Development Environments (IDEs).

11.7 Experimental Problems

We detail the six VirtualLabs problems used in our empirical evaluation.

DILUTION: You are a work study for the chemistry department. Your supervisor has just asked you to prepare 500 ml of 3 M HNO_3 for tomorrow's undergraduate experiment. In the stockroom explorer, you will find a cabinet called "Stock Solutions". Open this cabinet to find a 2.5 L bottle labeled "11.6 M HNO_3". The concentration of the HNO_3 is 15.4 M. Please prepare a flask containing 500 ml of a 3 M ($\pm$0.005 M) solution and relabel it with its precise molarity. Note that you must use realistic transfer mode, a buret, and a volumetric flask for this problem. Please do any relevant calculations on the paper supplied. As a reminder, to calculate the volume needed to make a solution of a given molarity, you may use the following formula: $C_1V_1 = C_2V_2$

ORACLE: Given four substances A, B, C, and D that are known to react in some weird and mysterious way (an oracle relayed this information to you within a dream), design and perform virtual lab experiments to determine the reaction between these substances, including the stoichiometric coefficients. You will find 1.00 M solutions of each of these chemical reagents in the stockroom.

COFFEE: During the summer after your first year at Carnegie Mellon, you are lucky enough to get a job making coffee at Starbucks, but you tell your parents and friends that you have secured a lucrative position as a "Java engineer". An eccentric chemistry professor (not mentioning any names) stops in every day and orders 250 ml of house coffee at precisely 95 °C. He then adds enough milk at 10 °C to drop the temperature of the coffee to 90 °C. (a) Calculate the amount of milk (in ml) the professor must add to reach this temperature. Show all your work, and circle the answer. (b) Use the Virtual Lab to make the coffee/milk solution and verify the answer you calculated in (a). Hint: the coffee is in an insulated travel mug, so no heat escapes. To insulate a piece of glassware in Virtual Lab, Mac-users should hold down the command key while clicking on the beaker or flask; Windows users should right click on the beaker or flask. From the menu that appears choose "Thermal Properties". Check the box labeled "insulated from surroundings". The temperature of the solution in that beaker or flask will remain constant.

COFFEE 2: During the summer after your first year at Carnegie Mellon, you are lucky enough to get a job making coffee at Starbucks, but you tell your parents and friends that you have secured a lucrative position as a "java engineer." An eccentric chemistry professor (not mentioning any names) stops in every day and orders 250 ml of Sumatran coffee. The coffee, initially at 85 °C. is way to hot for the professor, who prefers his coffee served at a more reasonable 65.0 °C. You need to add enough milk at 5.00 °C, to drop the temperature of the coffee.

How much milk do you add? Calculate the amount of milk (in ml) you must add to reach this temperature. In the previous part of the problem, you solved it assuming that both coffee and milk have the same specific heat capacities and densities as water. Since milk is a mixture of water, fat and proteins, its specific heat capacity is likely to be different than the one assumed. Solve again the same problem determining the specific heat of milk and considering it in your calculations. Assume the density is 1.000 g/ml for milk and coffee and the specific heat capacity is 4.184 J/(g °C) for coffee.

CAMPING: You and a friend are hiking the Appalachian Trail when a storm comes through. You stop to eat, but find that all available firewood is too wet to start a fire. From your Chem 106 class you remember that heat is given off by some chemical reactions; if you could mix two solutions together to produce an exothermic reaction, you might be able to cook the food you brought along for the hike. Luckily, being the dedicated chemist that you are, you never go anywhere without taking along a couple chemical solutions, just for times like this. The

Virtual Lab contains aqueous solutions of compounds *X* and *Y* of various concentrations. These compounds react to produces a new compound, *Z*, according to the reaction: $x + y \rightarrow z$. The following activities will guide you in using this reaction to produce the heat needed to warm up your food. Use the virtual lab to measure the enthalpy of the reaction shown above.

UNKNOWN ACID: The "Homework Solutions" cabinet contains a solution labeled "Unknown Acid", which is a weak mono-protic acid with an unknown *Ka* and with an unknown concentration. Your job is to determine the concentration and *Ka* to two significant figures.

11.8 The Recipe Library for the Dilution Problem

This section lists the complete recipe library for the dilution problem. Table 11.3 provides a key to the action abbreviations used in the recipes.

11.8.1 Dilution Problem Recipes

1. MSC[*sc*, *dt*; *sid*, *did*, *vol*, *scd*, *dcd*, *rcd*] → MS[*sc*, *dt*; *sid*, *did*, *vol*, *scd*, *dcd*, *rcd*]
2. MSC[*sc*, *dt*, *sid*, *did*, $vol = vol_1 + vol_2$, scd_2, dcd_2, rcd_2] → MSC[*sc*, *dt*, *sid*, *did*, vol_1, scd_1, dcd_1, rcd_1], MSC[*sc*, *dt*, *sid*, *did*, vol_2, scd_2, dcd_2, rcd_2] $sid_1 = sid_2$, $did_1 = did_2$, $scd_1 = scd_2$
3. MSI[*sc*, *dt*, *sid*, *did*, *vol*, *scd*, *dcd*, *rcd*] → MSC[*sc* : H_2O, *dt*, *sid*; *did*; *vol*; *scd*, *dcd*, *rcd*]
4. MSI[*sc*, *dt*, *sid*, *did*, *vol*, *scd*, *dcd*, *rcd*] → MSC[*sc* : 15:4 M HNO_3, *dt*, *sid*, *did*, *vol*, *scd*, *dcd*, *rcd*]
5. MSI[sc_1, dt_2, *sid*1, did_2, vol_1] → MSI[sc_1 : H_2O, dt_1, sid_1, did_1, vol_1, scd_1, dcd_1, rcd_1], MSC[sc_2, dt_2, sid_2, did_2, vol_2, scd_2, dcd_2, rcd_2][0] $did_1 = sid_2$, $rcd_1 = scd_2$

Table 11.3 Abbreviation key for complex actions used in recipes

Abbreviation	Action	Meaning
MS	Mixing solution	Basic solution mix operation as observed directly from log files
MSC	Mixing solution component 2	A complex action representing repeated mixing of a solution to the same destination
MSI	Mixing solution through intermediate flask	A complex action representing the use of intermediate flasks when mixing solution
SDP	Solve dilution problem	The root of the plan(s), composed of mixing H_2O and the solution to be diluted

6. $\underline{\text{MSI}}[sc_1, dt_2, sid_1, did_2, vol_1] \rightarrow \underline{\text{MSI}}[sc_1: 15{:}4\ \text{M HNO}_3, dt_1, sid_1, did_1, vol_1, scd_1, dcd_1, rcd_1], \underline{\text{MSC}}[sc_2, dt_2, sid_2, did_2, vol_2, scd_2, dcd_2, rcd_2][0]\ did_1 = sid_2, rcd_1 = scd_2$
7. $\underline{\text{MSC}}[sc, dt, sid, did, vol, scd, dcd, rcd] \rightarrow \underline{\text{MSI}}[sc, dt, sid, did, vol, scd, dcd, rcd]$
8. $\underline{\text{MSC}}[sc, dt, sid, did, vol = vol_1 + vol_2, scd_2, dcd_2, rcd_2] \rightarrow \underline{\text{MSC}}[sc, dt, sid, did, vol_1, scd_1, dcd_1, rcd_1], \underline{\text{MSC}}[sc, dt, sid, did, vol_2, scd_2, dcd_2, rcd_2]\ sid_1 = sid_2, did_1 = did_2, scd_1 = scd_2$
9. $\underline{\text{SDP}}[sc, dt, sid, did, vol = vol_1 + vol_2, scd_2, dcd_2, rcd_2] \rightarrow \underline{\text{MSC}}[sc: H_2O, dt, vol_1, did_1], \underline{\text{MSC}}[\text{sc}: 15{:}4\ \text{M HNO}_3, dt, vol_2, did_2]$

11.8.2 Recipes Explanation

Recipes Explanation: Recipes 1 and 2 capture repeated pouring activities, where users pour the same solution from the same source flask to the same destination flask (1 is the base of the recursion). Recipes 3 and 4 capture the activity of using an intermediate flask when pouring H_2O (i.e. pouring from flask 1 to flask 2 and then from flask 2 to flask 3). Recipes 5 and 6 are the same as 3 and 4, only for HNO_3. Recipes 7 and 8 are the same as 1 and 2, only now they can capture higher level activities which served the same overall goal (for example pouring from flask 1 to flask 2 through intermediate flask 3, and pouring from flask 1 to flask 2 through intermediate flask 4, both serve the same goal of pouring from flask 1 to flask 2). Recipe 9 forms the root of a plan, as it is composed of the pouring actions that involved H_2O and those of pouring HNO_3.

11.9 User Study Questionnaire

After observing each of the visualization methods, the participants responded to the following questions:

- Based on the presentation, can you tell WHETHER the student solved the problem? Please describe how you can tell whether the student solved the problem, or why you can't.
- Based on the presentation, can you tell HOW the student solved the problem? Please describe how the presentation helps you understand the student solution, or what information is missing.
- Assuming you were using VirtualLabs in your class, would you be likely to use this presentation style to understand a student's work after a classroom VirtualLabs session? Why?
- Additional comments. For example: What are the problems of this presentation style? How would you improve it? What information did you find helpful? What information was missing?

In the second part of the questionnaire participants stated their level of agreement (on a scale of 1–7) with the following statements with regards to each of the visualization methods:

- This presentation style was easy for me to learn.
- This presentation style demonstrates WHETHER the student solved then problem.
- This presentation style demonstrates HOW the student solved the problem.
- Assuming I would be using VirtualLabs in my class, I am likely to use this presentation style to understand a student's work after a classroom VirtualLabs session.

There was also space for additional comments after each of these statements. Finally, participants responded to the following two open questions:

- Did you prefer one style to all of the others? If so, which? Would you use one or some of the styles rather than the other/s to visualize students' work?
- Would you combine some or all of these presentation styles together? If so, can you list, for each presentation style, which aspects of a students' interaction are best visualized by that style?

References

1. Amershi, S., Conati, C.: Automatic Recognition of Learner Groups in Exploratory Learning Environments. In: Ikeda, M., Ashley, K.D., Chan, T.W. (eds.) Intelligent Tutoring Systems. LNCS, vol. 4053, pp. 463–472. Springer, Heidelberg (2006)
2. Chen, M.: A methodology for characterizing computer-based learning environments. Instr. Sci. **23**(1–3), 183–220 (1995)
3. Cocea, M., Gutierrez-Santos, S., Magoulas, G.D.: The Challenge of Intelligent Support in Exploratory Learning Environments: A Study of the Scenarios. In: Gutierrez-Santos, S., Mavrikis, M. (eds.) 1st International Workshop in Intelligent Support for Exploratory Environments (ISEE-2008), vol. 381, CEUR-WS, Maastricht (2008)
4. Gal, Y., Yamangil, E., Rubin, A., Shieber, S.M., Grosz, B.J.: Towards Collaborative Intelligent Tutors: Automated Recognition of Users' Strategies. In: Woolf, B.P., Aïmeur, E., Nkambou, R., Lajoie, S. (eds.) Ninth International Conference on Intelligent Tutoring Systems (ITS 2008), LNCS, vol. 5091, pp. 162–172. Springer, Heidelberg (2008)
5. Pawar, U. S., Pal, J., Toyama, K.: Multiple mice for computers in education in developing countries. In: Conference on Information and Communication Technologies and Development, pp. 64–71. University of California, Berkeley (2007)
6. Yaron, D., Karabinos, M., Lange, D., Greeno, J.G., Leinhardt, G.: The chemcollective-virtual labs for introductory chemistry courses. Science **328**(5978), 584–585 (2010)
7. Amir, O., Gal, Y.: Plan Recognition in Virtual Laboratories. In: Walsh, T. (ed.) 22nd International Joint Conference on Artificial Intelligence (IJCAI), pp. 2392–2397. AAAI Press, Menlo Park (2011)
8. Carberry, S.: Plan Recognition in Natural Language Dialogue. MIT Press, Cambridge (1990)
9. Grosz, B.J., Sidner, C.L.: Plans for Discourse. In: Morgan, J.L., Pollack, M.E., Cohen, P.R. (eds.) Intentions in Communication, pp. 417–444. The MIT Press, Cambridge (1990)

10. Bauer, M., Biundo, S., Dengler, D., Koehler, J., Paul, G.: PHI—Logic-based Tool for Intelligent Help Systems. In: Bajcsi, R. (ed.) 13th International Joint Conference on Artificial Intelligence (IJCAI), pp. 460–466. Morgan Kaufmann, San Francisco (1993)
11. Mayfield, J.: Controlling inference in plan recognition. User Model. User-Adap. Inter. **2**(1), 55–82 (1992)
12. Wilensky, R.: Why John married Mary: understanding stories involving recurring goals. Cogn. Sci. **2**(3), 235–266 (1978)
13. Charniak, E., Goldman, R.P.: A Bayesian model of plan recognition. Artif. Intell. **64**(1), 53–79 (1993)
14. Lesh, N., Rich, C., Sidner, C.L.: Using Plan Recognition in Human-Computer Collaboration. In: Kay, J. (ed.) Seventh International Conference on User Modeling, pp. 23–32. Springer, New York (1999)
15. Reddy, S., Gal, Y., Shieber, S.M.: Recognition of Users' Activities Using Constraint Satisfaction. In: Houben, G.J., McCalla, G., Pianesi, F., Zancanaro, M. (eds.) User Modeling, Adaptation, and Personalization, LNCS, vol. 5535, pp. 415–421. Springer, Heidelberg (2009)
16. Gal, Y., Reddy, S., Shieber, S., Rubin, A., Grosz, B.: Plan recognition in exploratory domains. Artif. Intell. **176**(1), 2270–2290 (2012)
17. VanLehn, K., Lynch, C., Schulze, K., Shapiro, J.A., Shelby, R.H., Taylor, L., Treacy, D.J.: Weinstein, A, Wintersgill, M.C.: The Andes physics tutoring system: lessons learned. Int. J. Artif. Intell. Educ. **15**(3), 147–204 (2005)
18. Conati, C., Gertner, A.S., VanLehn, K., Druzdzel, M.J.: On-line Student Modeling for Coached Problem Solving Using Bayesian Networks. In: Jameson, A., Paris, C., Tasso, C. (eds.) Sixth International Conference on User Modeling, pp. 231–242. Springer Wien, New York (1997)
19. Conati, C., Gertner, A.S., VanLehn, K.: Using Bayesian networks to manage uncertainty in student modeling. User Model. User-Adap. Inter. **12**(4), 371–417 (2002)
20. Anderson, J.R., Corbett, A.T., Koedinger, K.R., Pelletier, R.: Cognitive tutors: lessons learned. J. Learn. Sci. **4**(2), 167–207 (1995)
21. Corbett, A., McLaughlin, M., Scarpinatto, K.C.: Modeling student knowledge: cognitive tutors in high school and college. User Model. User-Adap. Inter. **10**(2–3), 81–108 (2000)
22. Vee, M.H.N.C., Meyer, B., Mannock, K.L.: Understanding novice errors and error paths in object-oriented programming through log analysis. In: Workshop on Educational Data Mining at the 8th International Conference on Intelligent Tutoring Systems (ITS 2006), pp. 13–20. Jhongli (2006)
23. Blaylock, N., Allen, J.: Recognizing Instantiated Goals Using Statistical Methods. In: Kaminka (ed.) Workshop on Modeling Others from Observations, pp. 79–86, Edinburgh (2005)
24. Bauer, M.: Acquisition of user preferences for plan recognition. In: Fifth International Conference on User Modeling, pp. 105–112. User Modeling Incorporated, Kailua-Kona (1996)
25. Horvitz, E.: Principles of mixed-initiative user interfaces. In: ACM SIGCHI Conference on Human Factors in Computing Systems, pp. 159–166. ACM, New York (1999)
26. Lesh, N.: Adaptive goal recognition. In: 15th International Joint Conference on Artificial Intelligence (IJCAI), pp. 1208–1214. Morgan Kaufmann, San Francisco (1997)
27. Kautz, H. A.: A formal theory of plan recognition. Ph. D Thesis, University of Rochester (1987)
28. Lochbaum, K.E.: A collaborative planning model of intentional structure. J. Comput. Linguist. **24**(4), 525–572 (1998)
29. Geib, C.W., Goldman, R.P.: A probabilistic plan recognition algorithm based on plan tree grammars. Artif. Intell. **173**(11), 1101–1132 (2009)
30. Pearce-Lazard, D., Poulovassilis, A., Geraniou, E.: The Design of Teacher Assistance Tools in an Exploratory Learning Environment for Mathematics Generalisation. In: Wolpers, M., Kirschner, P.A., Scheffel, M., Lindstaedt, S., Dimitrova, V. (eds.) Sustaining TEL: From

Innovation to Learning and Practice, LNCS, vol. 6383, pp. 260–275. Springer, Heidelberg (2010)
31. Gutierrez-Santos, S., Geraniou, E., Pearce-Lazard, D., Poulovassilis, A.: Design of teacher assistance tools in an exploratory learning environment for algebraic generalisation. IEEE Trans. Learn. Technol. **5**(4), 366–376 (2012)
32. Gueraud, V., Adam, J.M., Lejeune, A., Dubois, M., Mandran, N.: Teachers need support too: Formid-observer, a flexible environment for supervising simulation-based learning situations. In: 2nd International Workshop on Intelligent Support for Exploratory Environments, pp. 19–28. Brighton (2009)
33. Feng, M., Heffernan, N.T.: Towards live informing and automatic analyzing of student learning: reporting in assistment system. J. Interact. Learn. Res. **18**(2), 207–230 (2007)
34. Scheuer, O., Zinn, C.: How did the e-learning session go? The student inspector. In: 2007 Conference on Artificial Intelligence in Education, pp. 487–494. IOS Press, Amsterdam (2007)
35. Koedinger, K.R., Baker, R.S.J.D., Cunningham, K., Skogsholm, A., Leber, B., Stamper, J.: A Data Repository for the EDM community: The PSLC DataShop. In: Romero, C., Ventura, S., Pechenizkiy, M., Baker, R.S.J.D. (eds.). Handbook of Educational Data Mining, Chapman and Hall/CRC Data Mining and Knowledge Discovery Series Boca Raton, pp. 43–55. CRC Press, Boca Raton (2010)
36. Merceron, A., Yacef, K.: Tada-ed for educational data mining. Interact. Multimedia Electron. J. Comput. Enhanced Learn. **7**(1), 267–287 (2005)
37. Sao Pedro, M.A., Baker, R.S.J., Montalvo, O., Nakama, A., Gubert, J.D.: Using Text Replay Tagging to Produce Detectors of Systematic Experimentation Behavior Patterns. In: Baker, R.S.J.D., Merceron, A., Pavlik Jr., P.I. (eds.) 3rd International Conference on Educational Data Mining, pp. 181–190. International Educational Data Mining Society, Pittsburgh (2010)
38. Montalvo, O., Baker, R.S.J., Sao Pedro, M.A., Nakama, A., Gobert, J.D.: Identifying Students Inquiry Planning Using Machine Learning. In: Baker, R.S.J.D., Merceron, A., Pavlik Jr., P.I. (eds.) 3rd International Conference on Educational Data Mining, pp. 141–150. International Educational Data Mining Society, Pittsburgh (2010)
39. Amershi, S., Conati, C.: Combining unsupervised and supervised classification to build user models for exploratory learning environments. J. Educ. Data Min. **1**(1), 18–71 (2009)
40. Kardan, S., Conati, C.: A Framework for Capturing Distinguishing User Interaction Behaviours in Novel Interfaces. In: Pechenizkiy, M., Calders, T., Conati, C., Ventura, S., Romero, C., Stamper, J. (eds.) 4th International Conference on Educational Data Mining, pp. 159–168. International Educational Data Mining Society, Eindhoven (2011)
41. Pollack, M.E.: Plans as complex Mental Attitudes. In: Morgan, J.L., Pollack, M.E., Cohen, P.R. (eds.) Intentions in Communication, pp. 77–103. The MIT Press, Cambridge (1990)
42. Konold, C., Miller, C.: TinkerPlots Dynamic Data Exploration 1.0. Key Curriculum Press. URL http://www.keypress.com/x5715.xml (2004)
43. Hammerman, J.K., Rubin, A.: Strategies for managing statistical complexity with new software tools. Stat. Educ. Res. J. **3**(2), 17–41 (2004)
44. Dechter, R.: Constraint Processing. Morgan Kaufmann, San Francisco (2003)
45. Heer, J., Card, S.K., Landay, J.A.: Prefuse: A toolkit for interactive information visualization. In: SIGCHI conference on human factors in computing systems, pp. 421–430. ACM, New York (2005)

Chapter 12
Finding Dependency of Test Items from Students' Response Data

Xiaoxun Sun

Abstract In this chapter, we propose a new approach to find the most dependent test items in students' response data by adopting the concept of entropy from information theory. We define a distance metric to measures the amount of mutual independency between two items, and it is used to quantify how independent two items are in a test. Based on the proposed measurement, we present a simple yet efficient dependency tree searching algorithm to find the best dependency tree from the students' response data, which shows the hierarchical relationship between test items. The extensive experimental study has been performed on synthetic datasets, and results show that the proposed algorithm for finding the best dependency tree is fast and scalable, and the comparison with item correlations has been made to confirm the effectiveness of the approach. Finally, we discuss the possible extension of the method to find dependent item sets and to determine dimensions and sub-dimensions from the data.

Keywords Entropy · Independency · Response data · Correlation

Abbreviations

MI(A, B)	Mutual information measure
ISA	International school's assessment
PISA	OECD's programme for international student assessment

X. Sun (✉)
Australian Council for Educational Research, 19 Prospect Hill Road, Camberwell, VIC 3146, Australia
e-mail: xiaoxun.sun@acer.edu.au

A. Peña-Ayala (ed.), *Educational Data Mining*, Studies in Computational Intelligence 524, DOI: 10.1007/978-3-319-02738-8_12,

12.1 Introduction

Data mining is the analysis step of the knowledge discovery in databases process, and it is the process of discovering novel and potentially useful information and patterns from large data sets. There are different data mining technologies lying at the intersection of artificial intelligence, machine learning, statistics and database systems.

The goal of data mining is to extract useful and previously unknown information out of large complex data collections. Data mining techniques have been applied to many other fields. In the context of educational research, educational data mining refers to developing methods for exploring the unique types of data that come from educational settings, and using existing data mining or developing new methods to better diagnose students' performance and design tests that better suits students.

Students' response data contain the responses of students to a set of test questions. It can be used to determine the knowledge of a student has learned, and it can also be used to discover the relationship between the test items latent or underlying attributes.

Such relationship may take the form of attempting to find out which variables are most strongly associated with a single variable of particular interest, or may take the form of attempting to discover relationships between any two variables are strongest.

Students' response data are beneficial to both test developers and course instructors. Students' response data contains valuable information that can be used to improve the effectiveness of test items, and for course instructors, students' performance on the test is important to instructors for the guidance and improvement of teaching.

In this chapter, we apply the concept of entropy to propose a distance metric to evaluate the amount of mutual information among records in students' response data, and propose a method of constructing best dependency tree from the data. Finally, the experimental studies on synthetic data sets show the effectiveness and efficiency of the proposed method. We discuss the possible extension of the method to handle the dependency structure among item sets and to determine dimensions and sub-dimensions.

12.2 Related Work

The work is related to the application of dependency tree of information theory in data mining and databases. The dependency tree has been used in finding dependency structure in the features which improve the classification accuracy of the Bayes network classifiers [1, 2] used the dependency tree to represent a set of frequent patterns, which can be used to summarize patterns into few profiles.

Kovács and Szántai [3] presented large node dependency tree, in which the nodes are subsets of variables of dataset. The large node dependency tree is applied to density estimation and classification. As far as its application to educational database, fewer results are obtained. In this chapter, we apply the concept of entropy to capture the amount of mutual information among records in students' response data.

There is some existing research applied to students' response data, [4] reported an approach to classify students in order to predict their final year grade based on the features extracted from logged data in an educational web-based system, [5] used clustering algorithms on students' response data to characterize similar behaviour groups in unstructured collaboration, [6] developed a hybrid system of association rule mining algorithm with fuzzy set theory and inductive learning algorithm to find embedded information that could be fed back to teachers for refining or reorganizing the teaching materials and test, [7] applied genetic algorithms together with data mining classification process to improve the prediction accuracy.

The focus of the existing research is given to students' performance extracted the students' response data rather than discovering the relationship among test items based on the students' response data. This work is loosely related to the association rule mining. In the context of educational data mining, association rule mining technique has been used to perform crucial analysis in the educational environment.

Romero et al. [8] used association rule mining technique to discover patterns from student online course usage and [9] introduced a Test Result Feedback model that analyzes the relationships between students' learning time and the corresponding test results.

12.3 Mutual Independency Measure

The work presented in this chapter is based on information theory, and is related to the application of dependency tree of information theory in students' response data. In this section, we first introduce the concept of entropy, and the mutual information measure, which captures the mutual independency between test items.

12.3.1 Preliminaries

In the information theory, the main concept is entropy. It is defined to measure the expected uncertainty or the amount of information provided by a certain event. We feel more surprised when an unlikely event happens than a likely one occurs. One useful measure of the extent of surprise of an event is to use the logistic function.

Suppose the probability of an event happening is p then the extent of surprise of such event can be defined as $-log_k p$, in which k refers to the base of the logistic function. From this definition, it can be seen that the less the probability is, the higher the amount of information the event would provide. Given the example of students' response data, items that have been answered correctly by a small portion of students contains much more useful information for course instructors than the items that have been answered fully correct.

The choice of the base of the logarithm is not important as long as it is used consistently, as the change of the base only results in a rescaling. For the simplicity, base 2 logarithm will be used in this chapter. Formally, given a random variable X, the entropy of X is defined by:

$$H(x) = -\sum_x P(X = x) \log_2 P(X = x) \tag{12.1}$$

with $0\log_2 0 = 0$ by convention. It can be shown that $0 \leq H(X) \leq \log_2 |X|$. If X is uniformly distributed, $H(X) = \log_2 |X|$. This is because for uniform distribution, $P(X = x) = 1/|X|$ for all $x \in X|$.

For the students' response data, X could be the item of the test, x is the student's score on that item. If the item X is dichotomously scored (either correct or incorrect), $P(X = x)$ could be the proportion of students who answered correctly or incorrectly. In the rest of the chapter, we consider the dichotomously scored items.

12.3.2 Mutual Information Measure

In order to study the interaction between items, we use the concept of conditional entropy. The conditional entropy measures the uncertainty about X when Y is known. The conditional entropy $H(Y|X)|$ of a random variable Y given X is then defined as:

$$H(Y|X) = -\sum_{x,y} p(x,y) \log_2 p(y|x)| \tag{12.2}$$

where $p(x,y)$ is the joint distribution of variables X and Y. The conditional entropy has the following properties.

Proposition 1 *Let $H(Y|X)$ be the conditional entropy for Y given X, then,*

$$0 \leq H(Y|X) \leq H(Y) \tag{12.3}$$

$$H(X,Y) = H(X) + H(Y|X) = H(Y) + H(X|Y) \tag{12.4}$$

$$H(X,Y) \leq H(X) + H(Y) \tag{12.5}$$

The proof of **Proposition 1** is given in [10]. According to the proposition, the conditional entropy $H(Y|X)$ can be rewritten as: $H(Y|X) = H(X,Y) - H(X)$, which provides an alternative and easy way to compute the conditional entropy $H(Y|X)$. We adopt the conditional entropy to measure the mutual information, which is a distance metric.

Definition 1 (*Mutual Information Measure*) The mutual information measure with regard to two random variables A and B is defined as:

$$MI(A,B) = H(A|B) + H(B|A) \tag{12.6}$$

Mutual information measure is a measure of how independent is the two random variables when the value of each random variable is known. Two events A and B are independent if and only if their mutual information measure achieves the maximum $H(A) + H(B)$. Therefore, the less the value of the mutual information measure is, the more dependent the two random variables are. According to this measure, A is said to be more dependent on B than C, if $MI(A,B) \leq MI(A,C)$.

Theorem 1 *The mutual information measure* $MI(A,B)$ *satisfies the following properties:*

$$\text{(non-negativity) } MI(A,B) \geq 0 \tag{12.7}$$

$$\text{(symmetry) } MI(A,B) = MI(B,A) \tag{12.8}$$

$$\text{(triangle inequality) } MI(A,B) + MI(A,C) \geq MI(A,C) \tag{12.9}$$

It is easy to verify that $MI(A,B) = 0$ if and only if there is a one-to-one function mapping between A and B. Since when $H(B|A) = 0$ B is a function of A, then when $MI(A,B) = 0$ if and only if $H(B|A) = 0$ and $H(A|B) = 0$, i.e., there is a one-to-one function mapping between A and B. In this sense, the mutual information measure $MI(A,B)$ we defined is a distance metric.

12.3.3 Finding the Best Dependency Tree

Dependency tree was introduced by Chow and Liu [10], in which they introduced an algorithm for fitting a multivariate distribution with a tree (i.e., a density model that assumes that there are only pair wise dependency between variables). In the maximum likelihood sense, the dependency tree is the best tree to fit the dataset, and it uses mutual information measure to estimate the dependency of two random variables.

The dependency tree has been used in finding dependency structure in the features which improve the classification accuracy of the Bayes network classifiers [1]. Ma et al. 2] used the dependency tree to represent a set of frequent patterns, which can be used to summarize patterns into few profiles. Kovács and Szántai [3] presented a large node dependency tree, in which the nodes are subsets of variables of datasets.

The large node dependency tree is applied to density estimation and classification.

Definition 2 (*Dependency Matrix*) Given students' response data T with n test participants $\{r_1, r_2, \ldots r_n\}$, where each participant responds to m test items $\{A_1, A_2, \ldots A_m\}$, the dependency matrix D_T is defined as:

$$D\_T = (MI(i,j))_{mxm} \tag{12.10}$$

where $MI(i,j)$ is the mutual information measure, $i,j \in \{A_1, A_2, \ldots, A_m\}$. With the dependency matrix, we could construct a fully connected weighted graph $G = (V, E, \omega)$ where $V = \{v_1, v_2, \ldots, v_m\}$ is the set of vertices, which corresponds to the test items in T, and for each pair of vertices (v_i, v_j) there is an edge e_{ij} connecting them, and $\omega(e_{ij})$ refers to the weight of each e_{ij} between v_i and v_j, which can be obtained from the dependency matrix.

Definition 3 (*Subgraph*) Let $G = (V, E)$ be a graph, where V is the set of vertices and E is the set of edges. We say $G' = (V', E')$ is a subgraph of G if $V' \in V$ and $E' \in E$.

Definition 4 (*Dependency Tree*)

Let $G = (V, E, \omega)$ be a weighted graph. A dependency tree T_G of G is a subgraph that is a tree and connects all the vertices together. A best dependency tree is a dependency tree with weight less than or equal to the weight of every other dependency trees.

The following theorem (Theorem 2) proves the subgraph property of the best dependency tree, which is that any sub-structure of the best dependency tree is also the best dependency tree of that sub-structure.

Theorem 2 (Subgraph Property) *Let T_G be the best dependency tree of G with the set of vertices G. Let G' be a subgraph of G with the set of vertices V', if G is a tree, then G is also the best dependency tree in G.*

We observe that $\omega(e_{ij})$ represents to what extent vertex v_i (or attribute A_i) is dependent on v_j (or A_j). Although, in the worst case, any pair of attributes can be dependent, however, as stated in [10].

We could simplify by using an approximation which ignores the conditions on multiple attributes, and retaining only dependency in at most a single attribute at a time, which results in a tree-like structure. It is easy to see that in the fully connected weighted graph G, there are a large number of trees, each of which represents a unique approximation dependency structure.

Here, in order to reduce the uncertainty in the dataset and maximize the mutual information among the attributes simultaneously, we find a best dependency tree from the fully connected graph G with the least total weight based on our proposed mutual information measure.

Theorem 2 provides a theoretical foundation for a greedy algorithm for finding the best dependency tree. Because of the subgraph property, we can start constructing the best dependency tree with the following steps. First, the candidate edges are sorted in increasing order of their weights (i.e. mutual information measure).

Then, starting with an empty set E_0, the algorithm examines one edge at a time (in the order resulting from the sort operation), checks if it forms a cycle with the edges already in E_0 and, if not, adds it to E_0. The algorithm ends when $m - 1$ edges have been added to E_0, where m refers to the number of vertices in G. The algorithm for finding the best dependency tree is included in Algorithm 1.

12.3.4 An Example

For better understanding, in this section, a simple example will be included to show the concepts that have been introduced before. Table 12.1 shows a sample student response data which contains 12 students and their responses to 6 questions. All questions are dichotomous items, and the students' responses have been scored as 0 (wrong) and 1 (right).

As discussed, for each question, we defined the probability $P(Q_i = x)$ as the fraction of rows whose projection onto Q_i is equal to x, where $x \in \{0, 1\}$. In this example, it is equivalent to say that $P(Q_i = 1(0))$ is the proportion of students who have answered Q_i correctly (incorrectly).

For instance $P(Q_1 = 1) = 1/3$, $P(Q_3 = 0) = 1/6$ and $P(Q_1 = 1, Q_2 = 0) = 1/12$. The entropy can be calculated as $H(Q_1) = 0.9183$, similarly, $H(Q_2) = 0.9183$ and $H(Q_1, Q_2) = 1.5546$. By using Proposition 1(2), the conditional entropy can be calculated. In out example, $H(Q_1|Q_2) = 0.6363$, and $H(Q_2|Q_1) = 0.7433$.

Table 12.1 Sample data

Student	Q1	Q2	Q3	Q4	Q5	Q6
Student 1	0	0	0	1	1	1
Student 2	0	1	1	0	1	0
Student 3	1	1	0	1	0	0
Student 4	0	0	1	1	1	1
Student 5	0	1	1	1	0	0
Student 6	0	0	1	0	0	1
Student 7	1	1	1	0	0	1
Student 8	0	1	1	0	0	0
Student 9	1	1	1	0	1	1
Student 10	0	1	1	1	0	1
Student 11	0	1	1	1	0	0
Student 12	1	1	1	1	1	1

1. Compute the mutual information measure between every item pair in T and construct the dependency matrix D_T. There are $C_m^2 = m(m-1)/2$ weight need to be calculated, if T has m items.
2. Construct a fully connected graph, where the nodes correspond to items in T. The weight of each edge refers to their mutual information measure.
3. Order the edges in an increasing order of its weights.
4. Starting from empty set E_0.
5. Choose the edge with minimum weight, and if it forms cycle with the edges already in E_0, then skip to next.
6. Otherwise add the edge into E_0 Repeat Step 5 until there are $m-1$ edges selected.

(Algorithm 1. Finding best dependency tree)

By applying Algorithm 1, the best dependency tree of Table 12.1 is shown in Fig. 12.1. We can summarize patterns from the best dependency tree. For summarizing the item set into k patterns, we delete the $k-1$ largest weight edges from the dependency tree. For instance, from Fig. 12.1, three patterns can be observed $\{\{Q_2, Q_3, Q_6\}, Q_1, Q_5\}$ in T_G. According to Theorem 2, the best dependency tree of questions Q_2, Q_3, Q_6 is also a best dependency tree that could be constructed from the fully connected graph with vertices Q_2, Q_3, Q_6.

From the structure of the best dependency tree, it might be used to determine all the dimensions and sub-dimensions of the data. In Fig. 12.1, it seems that 6 questions can be divided into 3 dimensions, where questions $\{Q_2, Q_3, Q_6\}$ belong to dimension 1, and Q_1 and Q_5 are loaded in second and third dimension. From the definition of dependency tree, the most dependent items are organized in a tree structure, and this method can be used to discover or confirm the dimensionality of the data.

For example, in this scenario, $\{Q_2, Q_3, Q_6\}$ might be the math questions, Q_1 might be the literacy question, and Q_5 might be the science question. If all the questions are maths questions, this classification may suggest three sub-dimensions in the maths questions that have been observed from the data.

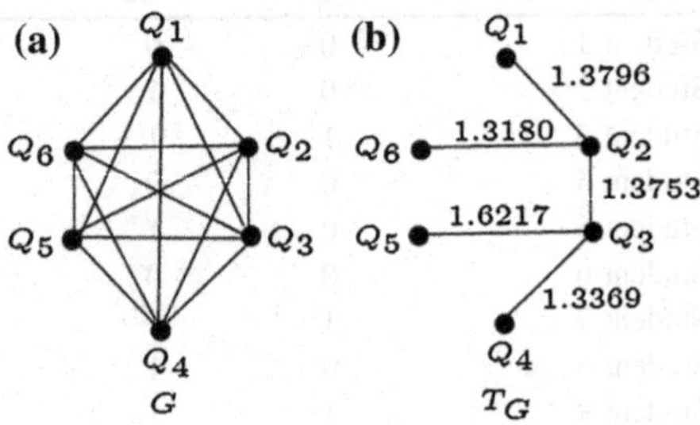

Fig. 12.1 **a** Fully connected graph G, **b** the best dependency tree of G

12.3.5 Extensions

In this chapter, we apply the entropy-based distance metric to measure the mutual information between two test items, and the method described in the chapter can be easily extended to be capable of handling two item sets, each of which consists of different items. For example, suppose 6 items in Table 12.1 can be organized into 3 groups G_1, G_2, and G_3 where $G_1 = \{Q_1, Q_2)$, $G_2 = \{Q_3, Q_4)$ and $G_3 = \{Q_5, Q_6)$.

The best dependency tree can be constructed from 3 item sets rather than 6 items. This extension is useful in the sense that it could provide the dependency relationship in a higher level. Still using Table 12.1 as an example, G_1, G_2, and G_3 may refer to different sub-strands in a test, and the extended method is able to generate the dependency between different sub-strands, which makes entropy-based dependency method superior to the traditional correlation method.

As mentioned in Sect. 12.3.4, the generated best dependency tree could be used to determine dimensions and sub-dimensions of the data. This can be done by summarizing patterns from the best dependency tree, and k patterns could be found by deleting the first $k - 1$ largest edges from the dependency tree.

12.4 Proof-of-Concept Experiments

The aim of the experimental studies is twofold. First, we compare the method of calculating mutual dependency with the method of calculating items' correlation, and this is to make sure that the method of constructing dependency tree indeed finds most correlated items. Second, we compare the scalability of the proposed algorithm for finding best dependency tree.

12.4.1 Data

In this experiment, we used the three synthetic data sets generated from normal distributions $N(0, 1)$. They represent the students' responses from short, medium and long tests respectively. The data generation specifications are shown in Table 12.2.

In addition to the synthetic data, we also used the International School's Assessment (ISA). The ISA assessment program (reading, mathematical literacy

Table 12.2 Data specification

Dataset	Number of items	Number of cases
Data1	5	50
Data2	20	200
Data3	50	500

and writing) is designed especially for students in international schools in Grades 3–10. It is based on the internationally endorsed reading, mathematical literacy and scientific literacy frameworks of the OECD's Programme for International Student Assessment (PISA). The data used for this chapter is Mathematics Grade 4 in 2012 October's sitting. The test contains 32 questions, and there are both multiple choice and open response questions, and all the questions are scored as 0 or 1.

12.4.2 Results on Synthetic Data Sets

First, we evaluate the running time of the dependency tree algorithm. Figure 12.2 shows the execution time of calculating the dependency among items and constructing the dependency tree. Figure 12.2a plots the time overhead comparison for simulated datasets data1, data2 and data3, which contains 5, 20 and 50 items respectively. From the graph, it can be seen that the time overhead is increasing as the number of items increases. This is expected, as the number of items increases, the number of combinations of 2-item sets is growing rapidly as well, and it takes

Fig. 12.2 Comparisons of execution time when constructing dependency tree, **a** comparison or running time, **b** comparison or running time for 50 item

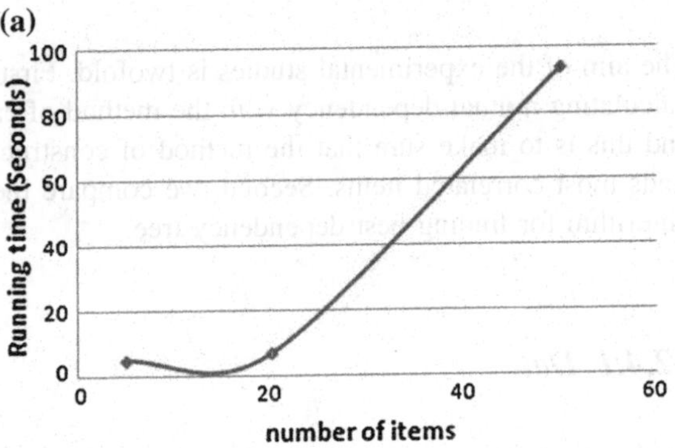

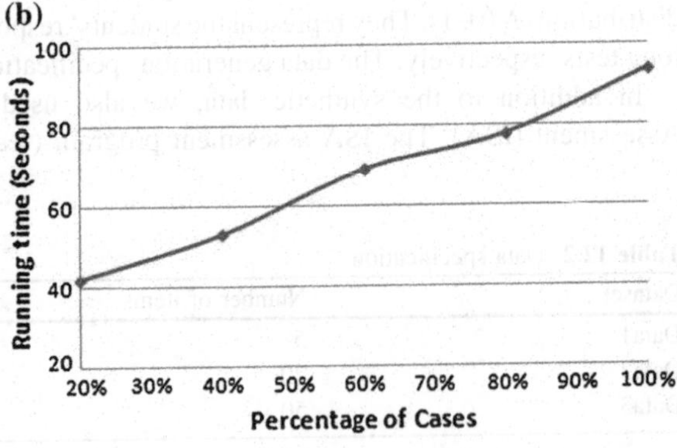

more time to compute the mutual dependency and construct the dependency matrix.

Figure 12.2b presents the comparison of the running time by using simulated dataset data3 by varying the case percentage from 20 to 100 %. As seen from the graph, the running time is increasing as the number of cases increases, and the plot is almost a line, which proves the scalability of the algorithm when applying to large datasets.

Next, we compare the mutual independency measure with the correlation coefficients. The correlation is a way to measure how associated or related two variables are. Since the mutual independency measures how independent two variables are. The value of the correlation ranges from −1 to +1.

When the value of correlation is greater than 0, it is called positive correlation. In a positive correlation, as the values of one of the variables increase, the values of the second variable also increase. Likewise, as the value of one of the variables decreases, the value of the other variable also decreases.

When the value of correlation is less than 0, it is called negative correlation, and in a negative correlation, as the values of one of the variables increase, the values of the second variable decrease.Likewise, as the value of one of the variables decreases, the value of the other variable increases.

No matter whether the correlation is positive or negative, the absolute value of the correlation reflects the strength of the correlation. As the correlation score gets closer to 1, it is getting stronger. In this study, we compare the strength of the correlation against the mutual information measure between two variables. In general, the more independent two variables are, the more related two variables should be. In this sets of experiments, we graph the relationship between mutual independency measure and the correlation of two test items.

Figure 12.3 shows the relationship between the mutual independency measure and the correlations. Figure 12.3a plots the relationship among items in a short test, Fig. 12.3b plots the relationship among items in a medium test, and Fig. 12.3c plots the relationship among items in a long test.

From Fig. 12.3a, since 5 items are included in the simulated data, the number of 2-item combinations is $C_5^2 = 10$, for Fig. 12.3b, 20 items will produce $C_{20}^2 = 190$ different combinations of 2-item sets and for Fig. 12.3c, there are $C_{50}^2 = 1225$ combinations for 50 items.

From all figures, it can be seen that the slopes of the regression line are negative, which confirms the fact that the more the mutual independency between two variables, the less correlated they are. R^2 is the indication of how strong the linear relationship is. In all cases, the values of R^2 are greater than 0.65, and in Fig. 12.3a, R^2 indicates a strong linear relationship, while that relationship is stronger in Figs. 12.3b, c.

Figure 12.4 displays the best dependency tree structure calculated from the simulated data1 and data2. There are two patterns$\{Q_1, Q_3, Q_4\}$, $\{Q_2, Q_5,\}$ observed from Fig. 12.4a and in Fig. 12.4b, all the questions are highly dependent on Q_{13}.

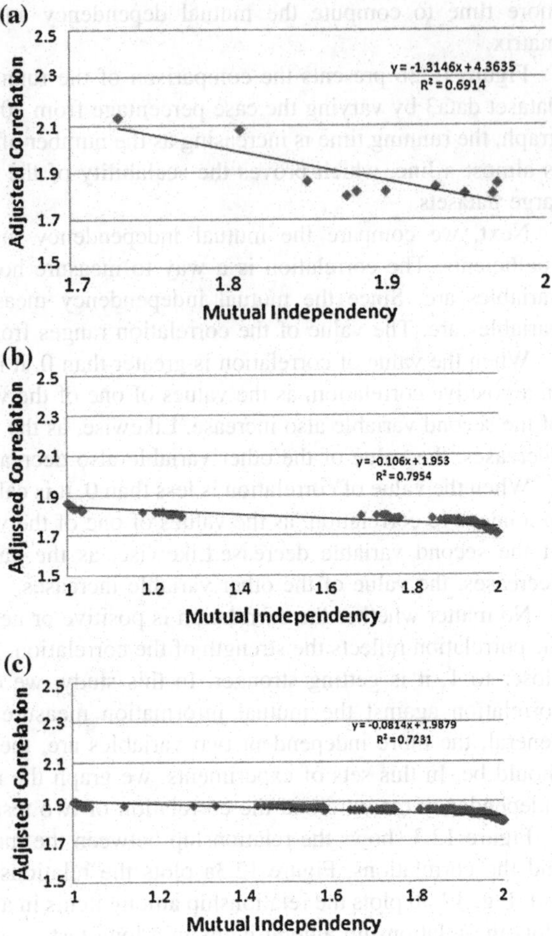

Fig. 12.3 Comparisons between mutual independency measure and correlations **a** comparison between correlation and mutual independency for 5 items **b** comparison between correlation and mutual independency for 20 items. **c** Comparison between correlation and mutual independency for 50 items

12.4.3 Results on Real Data

As discussed in the previous section, the entropy-based distance metric can be used to identify the substrands of each test. The aim of this subsection is to validate the accuracy of substrand identification by the new proposed method. In the study of real data, for example, questions 13 and 21 are showing high mutual information, and from the test description, these two questions belong to substrand "Quantity" in mathematics test. On the contrary, questions 5 and 28 are showing very low mutual information, and these two questions belong to substrands "Space and Shape" and "Quantity" respectively.

From the best dependency tree, the mutual information between two items reflect the best dependency, and we calculate the agreement between the entropy-based method and the item descriptions, and in the ISA Grade 4 Mathematics data,

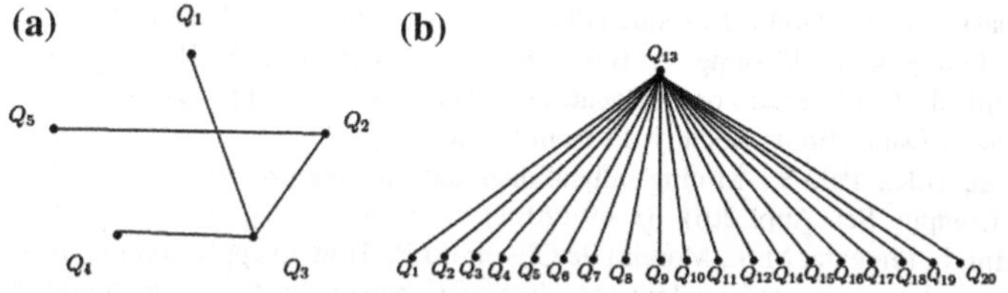

Fig. 12.4 The depedency tree sturcture of two data sets **a** best dependency tree constructed from data1 **b** best dependency tree constructed from data2

78 % of substrands confirms with what have been defined the item descriptions. Such a high agreement ensures that the entropy-based distance metric enables accurate sub-dimension detection.

12.5 Conclusions and Future Work

In this chapter, we apply the concept of entropy to propose a distance metric to evaluate the amount of mutual information among records in students' response data, and propose a method of constructing dependency tree from the data. The experimental results confirm the effectiveness and efficiency of the proposed method.

There is some potential work on the research agenda. First, the information theory based method presented in this chapter finds the dependent item pairs, and it can be extended to calculate the dependency between item sets. Second, it would be interesting to apply the association rules mining method to see whether there are any pattern with the rules and the mutual information measure.

Acknowledgment I would like to thank anonymous reviewers for their useful comments on this chapter. This research is supported by Australian Research Council (ARC) ARC-SRI: Science of Learning Research Centre (SLRC).

References

1. Rodrigo, M.M.T., Baker, RSJd: Comparing learners' affect while using an intelligent tutor and an educational game. Res. Pract. Technol. Enhanced Learn **6**(1), 43–66 (2011)
2. Ma, O., Sun, A., Yuan, Q., Cong, G.: Topic-driven reader comments summarization. In: 21st ACM International Conference on Information and Knowledge Management, pp. 265–274, ACM, New York (2012)
3. Kovács, E., Szántai, T.: On the approximation of a discrete multivariate probability distribution using the new concept of t-cherry junction tree. In: Marti, K., Ermoliev, Y., Makowski. M. (eds.) Coping with Uncertainty. LNEMS, vol. 633, pp. 39–56. Springer, Heidelberg (2010)

4. Falakmasir, M. H., Habibi, J.: Using educational data mining methods to study the impact of virtual classroom in e-learning. In: Baker, R. S. J. d., Merceron, A., Pavlik Jr., P. I. (eds.) 3rd International Conference on Educational Data Mining, pp. 241–248. International Educational Data Mining Society, Pittsburgh (2010)
5. Baradwaj, B.K., Pal, S.: Mining educational data to analyze students' performance. Int. J.Adv. Comput. Sci. Appl **2**(6), 63–69 (2011)
6. Gasparini, I., Pimenta, M.S., Moreira de Oliveira, J.P.: How to apply context-awareness in an adaptive e-learning environment to improve personalization capabilities? In: 30th International Conference of the Chilean Computer Science Society, pp. 161–170. IEEE Computer Society, Washington (2011)
7. Vizhi, J.M., Bhuvaneswari, T.: Data quality measurement on categorical data using genetic algorithm. Int. J. Data Min. Knowl. Manage. Process **2**(1), 33–42 (2012)
8. Romero, C., Romero, J. R., Luna, J. M, Ventura, S.: Mining rare association rules from e-learning data. International conference on educational data mining. In: Baker, R. S. J. d., Merceron, A., Pavlik Jr., P.I. (eds.) 3rd International Conference on Educational Data Mining, pp. 171–180. International Educational Data Mining Society, Pittsburgh (2010)
9. Yen, N.Y., Shih, T.K., Jin, Q., Hsu, H.H., Chao, L.R.: Trend of e-learning: the service mashup. Int.J. Distance Educ. Technol **8**(1), 69–88 (2010)
10. Chow, C., Liu, C.: Approximating discrete probability distributions with dependence trees. IEEE Trans. Inf. Theory **14**(3), 462–467 (1968)

Part IV
Trends

Chapter 13
Mining Texts, Learner Productions and Strategies with *ReaderBench*

Mihai Dascalu, Philippe Dessus, Maryse Bianco, Stefan Trausan-Matu and Aurélie Nardy

Abstract The chapter introduces *ReaderBench*, a multi-lingual and flexible environment that integrates text mining technologies for assessing a wide range of learners' productions and for supporting teachers in several ways. *ReaderBench* offers three main functionalities in terms of text analysis: cohesion-based assessment, reading strategies identification and textual complexity evaluation. All of these have been subject to empirical validations. *ReaderBench* may be used throughout an entire educational scenario, starting from the initial complexity assessment of the reading materials, the assignment of texts to learners, the detection of reading strategies reflected in one's self-explanations, and comprehension evaluation fostering learner's self-regulation process.

Keywords Cohesion-based discourse analysis · Topics extraction · Reading strategies · Textual complexity

---

M. Dascalu (✉) · S. Trausan-Matu
University Politehnica of Bucharest, 313 Splaiul Independentei,
060042 Bucharest, Romania
e-mail: mihai.dascalu@cs.pub.ro

S. Trausan-Matu
e-mail: stefan.trausan@cs.pub.ro

P. Dessus · M. Bianco
Laboratoire des Sciences de l'Education, University Grenoble Alpes,
BP 47, 38040 Grenoble, France
e-mail: philippe.dessus@upmf-grenoble.fr

M. Bianco
e-mail: maryse.bianco@upmf-grenoble.fr

A. Nardy
Laboratoire de linguistique et didactique des langues étrangères et maternelles,
University Grenoble Alpes, BP 25, 38040 Grenoble, France
e-mail: aurelie.nardy@upmf-grenoble.fr

A. Peña-Ayala (ed.), *Educational Data Mining*,
Studies in Computational Intelligence 524, DOI: 10.1007/978-3-319-02738-8_13,

Abbreviations

AA Adjacent agreement
CAF Complexity, accuracy and fluency
CSCL Computer supported collaborative learning
DRP Degree of reading power
EA Exact agreement
FFL French as foreign language
ICC Intra-class correlations
LDA Latent Dirichlet allocation
LMS Learning management system
LSA Latent semantic analysis
NLP Natural language processing
POS Part of speech
SVM Support vector machine
TASA Touchstone Applied Science Associates, Inc
Tf-Idf Term frequency – inverse document frequency
WOLF WordNet Libre du Français

13.1 Introduction

Text mining techniques based on advanced Natural Language Processing (NLP) and Machine Learning algorithms, as well as the ever-growing computer power, enable the design and implementation of new systems that automatically deliver to learners summative and formative assessments using multiple sets of data (e.g., textual materials, behavior tracks, meta-cognitive explanations). New automatic evaluation processes allow teachers and learners to have immediate information on the learning or understanding processes. Furthermore, computer-based systems can be integrated into pedagogical scenarios, providing activity flows that foster learning. ReaderBench is a fully functional framework based on text mining technologies [1].

It may be seen as a cohesion-based integrated approach that addresses multiple dimensions of learner comprehension, including the identification of reading strategies, textual complexity assessment and even Computer Supported Collaborative Learning (CSCL). In the later context, special emphasis is given to considering participant involvement and collaboration [2]. However, this facility will not be introduced in this chapter for readability's sake.

In addition to a fully functional NLP processing pipeline [3], in terms of Educational Data Mining, *ReaderBench* encompasses a wide variety of techniques: Latent Semantic Analysis (LSA) [4], Latent Dirichlet Allocation (LDA) [5] and specific internal processes addressing topics extraction, extractive summarization, identification of reading strategies, as well as textual complexity assessment, all deduced from a cohesion-based underlying discourse structure. In this context, *ReaderBench* provides teachers and learners information on their reading/writing activities: initial textual complexity assessment, assignment of texts to learners, capture of self-explanations reflected in pupil's textual verbalizations and reading strategies assessment [2].

The remainder of this chapter is as follows. The next section introduces a general perspective in terms of educational applications. The third section is an overview on how learner comprehension can be modeled and predicted while introducing an educational scenario that makes use of a wide variety of educational activities covered by *ReaderBench*. The fourth section is centered on cohesion, the core text feature from which almost all *ReaderBench* measures are computed. The next four sections present the main functionalities of our system: topic extraction, cohesion analysis, reading strategies analysis and textual complexity assessment. The ninth section focuses on the validation of the provided facilities, while the latter section compares *ReaderBench* to other systems, highlighting pros and cons for each approach.

13.2 Data and Text Mining for Educational Applications

Learning analytics aims at measuring, collecting, analyzing and "reporting data about learners and their contexts, for purposes of understanding and optimizing learning and the environments in which it occurs" (Society for Learning Analytics Research, http://www.solaresearch.org/). While the main focus of this approach is to gather data about learners (e.g., their behavior, opinions, affects, social interactions), very few researches are performed to infer what learners actually understand and the learning contexts are rarely taken into account (e.g., which learning material is used, to what pedagogical intent, within which educational scenario) [6].

Educational data analyzed from computer-based approaches typically comes from two wide categories: *knowledge* (e.g., textual material from course information) and *behavior* (e.g., learners' behavior in Learning Management Systems from log analysis). Whereas a substantial amount of research is centered on behavioral data [7], relatively few researches encompass the analysis of textual materials, presented beforehand to learners. Raw data is ideally and easily computable with data mining techniques [8], but the inferences necessary to uncover learners' cognitive processes are far more complex and involve comparisons to human experts judgments.

Our approach stems from a very broad idea. Cohesion, seen as the relatedness between different parts of texts, is a major determinant of text coherence and has been shown to be an important predictor of reading comprehension [9]. In turn, cohesion analyses can be applied on a wide range of data analyses in educational contexts: text readability and difficulty, knowledge relatedness, chat or forum group replies. The next section addresses learner comprehension, its relationships with textual complexity, and how the comprehension level can be inferred from learner's self-explanations.

13.2.1 Predicting Learner Comprehension

Learner's comprehension of textual materials during reading depends both on text properties and on the learner's reading skills. It has long been recognized that the comprehension performance differs according to lexical and syntactical complexity, as well as to the thematic content and to how information is structured [10, 11]. Of particular importance are the cohesion and coherence properties of texts that can support or impair [12] understanding and, moreover, interact with reader's personal characteristics [11, 13]. On the reader's side, his/her background knowledge and the reading strategies (s)he is able to use to process the provided information are also strong predictors of reading comprehension, in addition to personal word recognition abilities and semantic skills [14–16].

Therefore, our aim consists of designing an integrated system capable of supporting a wide range of educational activities, enabling in the end three kinds of work-loops (see Fig. 13.1) in which teacher/learners can be freely involved, thus triggering self-regulated learning [17]. It is worth noting that these three loops do not generate behavioral data per se, to be analyzed in turn in the automatic system.

The first loop addresses *reading*: learners read some material (e.g., course text, narrative) and can, at any moment, get information about its textual organization. The second one is a *gist selection loop*, which is a bit more interactive than the previous. Learners produce keywords or select main sentences of the read texts and submit their selection to either the teacher or the automatic evaluation system, which prompts feedback. The third is a *writing loop* that gives learners the opportunity to develop at length what they understood from the text (e.g., summaries) or the way they understood (strategies self-explanation). Besides these three loops, the teacher can use automatic tools to select appropriate textual materials according to learners' level. Each of the main activities presented in in Fig. 13.1, either tutor or learner centered, are presented from a computational point of view in subsequent sections. The remainder of this section elaborates more on the two main factors of text understanding: *textual features* (through textual complexity) and *readers' abilities* (through the identification of reading strategies).

13.3 Textual Complexity Assessment for Comprehension Prediction

Teachers usually need valid and reliable measures of textual complexity for their day-to-day instruction in order to select materials appropriate to learners' reading level. This proves to be a challenging and cumbersome activity since different types of texts (narrative, argumentative or expository) place different demands on different reading skills [18, 19]. For example, McNamara and her colleagues [19] found that narrative texts contain more familiar words than scientific texts, but that they have more complex syntactic sentences, as well. Narratives were also found

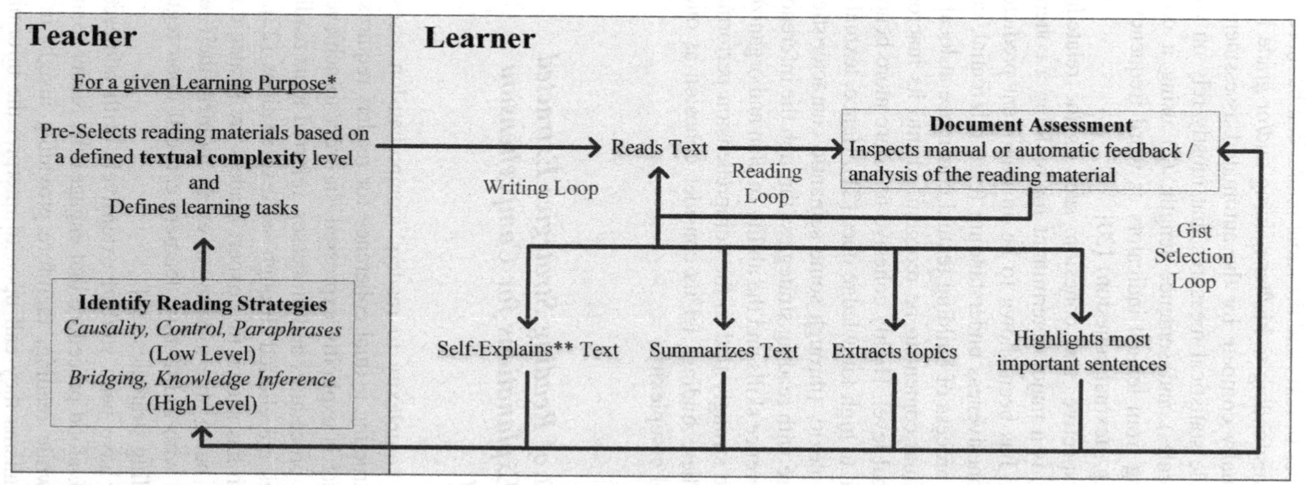

Information set
* The text may be: a narrative, a course, a case study, etc.
** The Learner's production may be: a self-explanation, a summary, a brainstorming, a critics

Fig. 13.1 General educational scenario envisioned for *ReaderBench*

to be less cohesive than science expository texts, the latter more strongly requiring background knowledge. In conclusion, different skills must be involved in comprehending different types of texts and the same reader can be more or less able to comprehend a text corresponding to his/her reading and/or grade level.

Two approaches usually compete for the automated assessment of text complexity: (1) using simple statistical measures that mostly rely on word difficulty (from already-made scales) and sentence length; (2) using a combination of multiple factors ranging from lexical indicators as word frequency, to syntactic and semantic levels (e.g., textual cohesion) [20].

As an in-depth perspective, text cohesion, seen as the relatedness between different parts of texts, is a major determinant for building a coherent representation of discourse and has been shown to be an important predictor of reading comprehension [9]. Cohesiveness understanding (e.g., referential, causal or temporal) is central to the process of building textual coherence at local level, which, in turn, allows the textual content to be reorganized into its macrostructure and situation model at global level. Highly cohesive texts are more beneficial to low-knowledge readers than to high-knowledge ones [21]. Hence, textual cohesion is a feature of textual complexity (through some semantic characteristics of the read text) that might interfere with reading strategies (through the inferences made by a reader). Moreover, inference skills and the ability to plan and organize information have been shown to be strongly tied to the comprehension performance of more complex texts [18]. These findings let us consider cohesion as one of the core determinants of textual complexity.

13.3.1 The Impact of Reading Strategies Extracted from Self-Explanations for Comprehension Assessment

Moving from textual complexity to readers' comprehension assessment is not straightforward. Constructing textual coherence for readers requires that they are able to go beyond what is explicitly expressed. In order to achieve this, readers make use of cognitive procedures and processes, referred to as reading strategies, when those procedures are elicited through self-explanations [22]. Research on reading comprehension has shown that expert readers are strategic readers. They monitor their reading, being able to know at every moment their level of understanding. When faced with a difficulty, learners can call upon regulation procedures, also called reading strategies [23].

Reading strategies have been studied extensively with adolescent and adult readers using the think-aloud procedure that engages the reader to auto-explain at specific breakpoints while reading, therefore providing insight in terms of the comprehension mechanisms they call upon to interpret the information they are reading. In other words, reading strategies are defined here as "the mental processes that are implicated during reading, as the reader attempts to make sense of the printed words" [24, p. 40].

Four types of reading strategies are mainly used by expert readers [25]. *Paraphrasing* allows the reader to express what she understood from the explicit content of the text, and can be considered the first and essential step in the process of coherence building. *Text-based inferences*, for example causal and bridging strategies build explicit relationships between two or more pieces of information in texts. On the other hand, *knowledge-based inferences* build relationships between the information in text and the reader's own knowledge and are essential to the situation model building process. *Control strategies* refer to the actual monitoring process when the reader is explicitly expressing what she has or has not understood. The diversity and richness of the strategies a reader carries out depend on many factors, either personal (proficiency, level of knowledge, motivation), or external (textual complexity).

We performed an experiment [26] to extend the assessment of reading strategies with children ranging from 3rd to 5th grade (8–11 years old). Children read aloud two stories and were asked at predefined moments to self-explain their impressions and thoughts about the reading material. An adapted annotation methodology was devised starting from McNamara's [25] coding scheme, covering the strategy items: paraphrases, textual inferences, knowledge inferences, self-evaluations, and "other". The "other" category is very close to the "irrelevant" category [25] as it aggregates irrelevant, as well as unintelligible statements. Two dominant strategies were identified: paraphrases and text-based inferences; text-based inferences frequency increases from grade 3 to 5, while erroneous paraphrases frequency decreases; knowledge-based inferences remain rare, but their frequency doubled from grade 3 to 5, amounting from 4 to 8 % of the identified reading strategies within the appropriate verbalizations.

Three results are noteworthy. Firstly, self-explanations are a useful tool to access the reading strategies of young children (8–11 years old) who already dispose of all the strategies older children carry out. Secondly, we found a relation between the ability to paraphrase and to use text-based inferences, on one hand, and comprehension and extraction of internal text coherence traits, on the other. A better comprehension in this age range is tied to less false paraphrases and more text-based inferences ($R^2 = 0.18$ for paraphrases and $R^2 = 0.16$ for text-based inferences). Thirdly, mediation models [27] showed that verbal ability partially mediates the effect of text-based inferences and that age moderates this mediating effect. The effect of text-based inferences on reading comprehension is mediated by verbal ability for the younger students while it becomes a direct effect for older students. Starting from the previous experiments and literature findings, one of the goals of *ReaderBench* is to enable the usage of new texts with little or no human intervention, providing both textual complexity assessments on these texts, and a fully automatic identification of reading strategies as a support for teachers. The textual complexity assessment aims at calibrating texts before providing them to learners.

13.4 Cohesion-Based Discourse Analysis: Building the Cohesion Graph

Text cohesion, viewed as lexical, grammatical and semantic relationships that link together textual units, is defined within our implemented model in terms of: (1) the *inverse normalized distance between textual elements* expressed in terms of the number of textual analysis elements in-between; (2) *lexical proximity* that is easily identifiable through identical lemmas and semantic distances within ontologies [28]; (3) *semantic similarity* measured through LSA [4] and LDA [5].

Additionally, specific natural language processing techniques [3] are applied to reduce noise and improve the system's accuracy: spell-checking (optional) [29, 30], tokenizing, splitting, part of speech tagging [31, 32], parsing [33, 34], stop words elimination, dictionary-only words selection, stemming [35], lemmatizing [36], named entity recognition [37] and co-reference resolution [38, 39].

In order to provide a multi-lingual analysis platform with support for both English and French, *ReaderBench* integrates both *WordNet* [40] and a serialized version of *Wordnet Libre du Français* (*WOLF*) [41]. Due to the intrinsic limitations of *WOLF*, in which concepts are translated from English while their corresponding glosses are only partially translated, making a mixture of French and English definitions, only three frequently used semantic distances were applicable to both ontologies: path length, Wu–Palmer [42] and Leacock–Chodorow's normalized path length [43].

Afterwards, LSA and LDA semantic models were trained using three specific corpora: *TextEnfants* [44] (approx. 4.2 M words), *Le Monde* (French newspaper, approx. 24 M words) for French, and *Touchstone Applied Science Associates* (TASA) corpus (approx. 13 M words) for English.

Moreover, improvements have been enforced on the initial models: the reduction of inflected forms to their lemmas, the annotation of each word with its corresponding part of speech through a NLP processing pipe (only for English as for French it was unfeasible to apply to the entire training corpus due to the limitations of the Stanford Core NLP in parsing French) [45–47], the normalization of occurrences through the use of term frequency-inverse document frequency (*Tf-Idf*) [3] and distributed computing for increasing speedup [48, 49].

LSA and LDA models extract semantic closeness relations from underlying word co-occurrences and are based on the bag-of-words hypothesis. Our experiments have proven that LSA and LDA models can be used to complement one other, in the sense that underlying semantic relationships are more likely to be identified, if both approaches are combined after normalization.

Therefore, LSA semantic spaces are generated after projecting the arrays obtained from the reduced-rank Singular Value Decomposition of the initial term-doc array and can be used to determine the proximity of words through cosine similarity [4]. From a different viewpoint, LDA topic models provide an inference mechanism of underlying topic structures through a generative probabilistic process [5]. In this context, similarity between concepts can be seen as the opposite of the Jensen-Shannon dissimilarity [3] between their corresponding posterior topic distributions.

From a computational perspective, the LSA semantic spaces were trained using a Tagged LSA engine [45] that preprocesses all training corpora (stop-words elimination, Part of Speech (POS) tagging, lemmatization) [46, 47], applies *Tf-Idf* and uses a distributed architecture [48, 50] to perform the Singular Values Decomposition. With regards to LDA, the parallel topics model used iterative Gibbs sampling over the training corpora [49] with 10,000 iterations and 100 topics, as recommended by [5]. Overall, in order to better grasp cohesion between textual fragments, we have combined information retrieval specific techniques, mostly reflected in word repetitions and normalized number of occurrences, with semantic distances extracted from ontologies or from LSA- or LDA-based semantic models.

In order to have a better representation of discourse in terms of underlying cohesive links, we introduced a *cohesion graph* [2, 51] (see Fig. 13.2) that can be seen as a generalization of the previously proposed utterance graph [52–54]. We are building a multi-layered mixed graph consisting of three types of nodes [55]: (1) a central node, the *document* that represents the entire reading material, (2) *blocks*, a generic entity that can reflect paragraphs from the initial text and (3) *sentences*, the main units of analysis, seen as collections of words and grammatical structures obtained after the initial NLP processing. The decomposition is applied to chat conversations or forum discussion threads where blocks are instantiated by utterances or interventions.

In terms of *edges*, *hierarchical links* are enforced by inclusion functions (sentences within a block, blocks within a document) and two types of links are introduced between analysis items of the same level: *mandatory* and *relevant links*. *Mandatory links* are set between adjacent blocks or sentences and are used for best modeling the information flow throughout the discourse, thus making possible the identification of cohesion gaps. Adjacency links are enforced between the previous block and the first sentence of the next block and, symmetrically, between the last sentence of the current block and the next block. Links ensure cohesiveness between structures at several levels within the cohesion graph, disjoint with regards to the inclusion function, and augment the importance of the first/last sentence of the current block, in accordance with the assumption that topic sentences are usually at the beginning/ending of a paragraph and ensure in most cases a transition from the previous paragraph [56].

Optional *relevant links* are added to the cohesion graph for highlighting fine-grained and subtle relations between distant analysis elements. In our experiments, the use as threshold of the sum of mean and standard deviation of all cohesion values from within a higher-level analysis element provided significant additional links for the proposed discourse structure. As cohesion can be regarded as the sum of semantic links that hold a text together and give it meaning, the underlying cohesive structure influence the perceived complexity level. In other words, the lack of cohesion may increase the perceived textual complexity as a text's proper understanding and representation become more difficult to achieve. In order to better highlight this

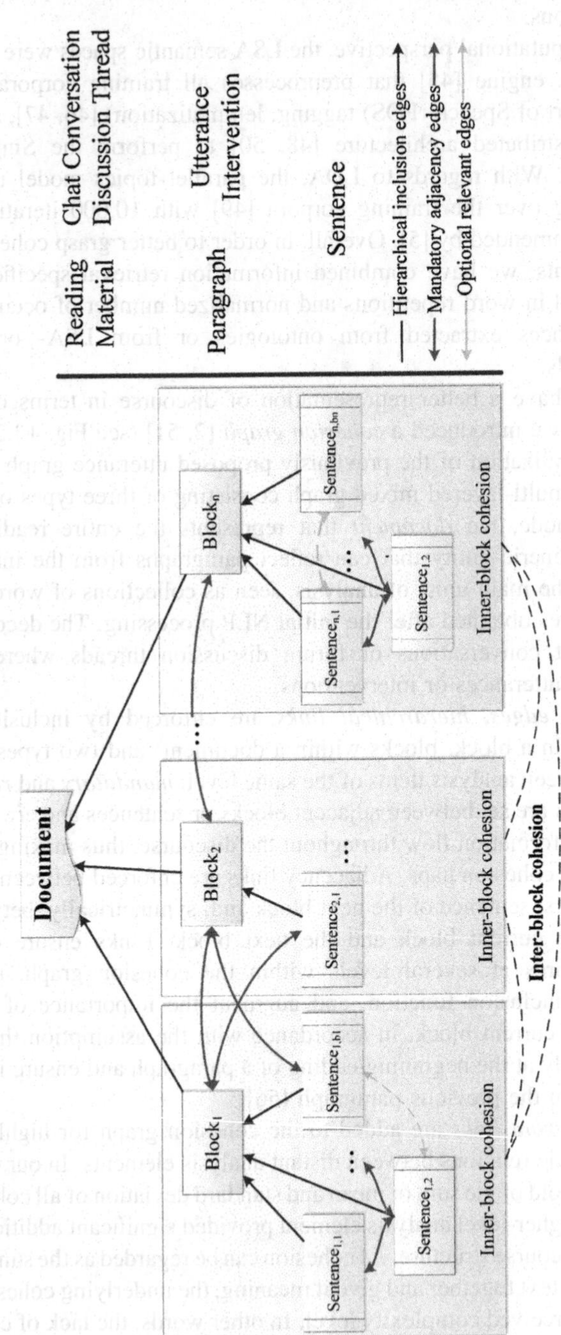

Fig. 13.2 The cohesion graph as underlying discourse structure

perspective, two measures for textual complexity were defined, later to be assessed: *inner-block cohesion* as the mean value of all the links from within a block (adjacent and relevant links between sentences) and *inter-block cohesion* that highlights semantic relationships at global document level.

13.5 Topics Extraction

The identification of covered topics or keywords is of particular interest within our analysis model because it enables us to grasp an overview of a document, but also in observing emerging points of interest or shifts of focus. Tightly connected to the cohesion graph, topics can be extracted at different levels and from different constituent elements of the analysis (e.g., the entire document or conversation, a paragraph or all the interventions of a participant). The relevance of each concept mentioned in the discussion and depicted by its lemma is defined by combining a multitude of factors:

1. *Individual normalized term frequency—1 + log(no. occurrences)* [57]; in the end, we opted for eliminating inverse document frequency, as this factor is related to the training corpora and we wanted to grasp the specificity of each analyzed text.
2. *Semantic similarities* through the cohesion function (LSA cosine similarity and inverse of LDA Jensen–Shannon divergence) with the analysis element and to the whole document for ensuring global resemblance and significance.
3. A *weighted similarity* with the corresponding *semantic chain* multiplied by the importance of the chain; semantic chains are obtained by merging lexical chains determined from the disambiguation graph modeled through semantic distances from *WordNet* and *WOLF* [58] through LSA and LDA semantic similarities and each chain's importance is computed as its normalized length multiplied with the cohesion function between the chain, seen as an entity integrating all semantically related concepts, and the entire document.

In addition, as an empirical improvement and as the previous list of topics is already pre-categorized by corresponding parts of speech, the selection of only nouns provided more accurate results in most cases due to the fact that nouns tend to better grasp the conceptualization of the document.In terms of a document's visualization, the initial text is split into paragraphs, cohesion measures are displayed in-between adjacent blocks and the list of sorted topics with their corresponding relevance scores is presented to the user, allowing him to filter the displayed results by number and by corresponding part of speech.

As an example, Fig. 13.3 presents the user interface of *ReaderBench* for a French story—*Miguel de la faim* [59]— highlighting the following elements: block scores (in square brackets after each paragraph), demarcation with bold of sentences considered most important according to the summarization facility and document topics and identified topics ordered by relevance. Although the block

ReaderBench - Document Visualization

Title: Miguel
Source: LSE URI: http://web.upmf-grenoble.fr/sciedu/

Contents

le village dormait , la maison était noire et paisible . [0.099]

Cohesion [Leacock-Chodorow=0.6; WU-Palmer=0.08; Path=0.04; cos(LSA)=0.04; sim(LDA)=0.34; dist=1]=0.15

miguel s' efforçait d' être calme et de ne pas écouter les douleurs qui lui fouettaient le ventre . il était tendu à force d' essayer de marcher sans bruit . [0.246]

Cohesion [Leacock-Chodorow=1.11; WU-Palmer=0.21; Path=0.1; cos(LSA)=0; sim(LDA)=0.45; dist=1]=0.22

arrivé près de l' arbre , il posa son argent plus une médaille bénite pour le cas où le prix des poulets excéderait la somme qu' il pouvait y mettre . [0.635]

Cohesion [Leacock-Chodorow=1.23; WU-Palmer=0.31; Path=0.13; cos(LSA)=0.08; sim(LDA)=0.27; dist=1]=0.22

puis , avec son couteau , il coupa doucement la ficelle qui tenait au tronc , s' avança , les deux mains prêtes à saisir le cou des volailles en même temps , afin de les étrangler avant qu' elles ne crient , et frappa . [0.517]

Cohesion [Leacock-Chodorow=1.11; WU-Palmer=0.37; Path=0.16; cos(LSA)=0.06; sim(LDA)=0.52; dist=1]=0.32

hélas , il était trop épuisé pour faire du bon travail . les poules crièrent et battirent des ailes comme des forcenées . un homme sortit sur le pas de la porte avec son fusil . [0.841]

Cohesion [Leacock-Chodorow=1.25; WU-Palmer=0.32; Path=0.12; cos(LSA)=0.04; sim(LDA)=0.37; dist=1]=0.24

- ne tirez pas , capitaine , s' écria miguel ! ce n' est que moi je venais prendre les poulets que j' ai achetés . [0.55]

Cohesion [Leacock-Chodorow=0.99; WU-Palmer=0.29; Path=0.09; cos(LSA)=0.05; sim(LDA)=0.6; dist=1]=0.31

- que tu as achetés , voleur , dit l' homme , sur un ton de colère . [0.931]

Cohesion [Leacock-Chodorow=0.56; WU-Palmer=0.17; Path=0.04; cos(LSA)=0.1; sim(LDA)=0.57; dist=1]=0.28

et il leva son fusil . [0.083]

Cohesion [Leacock-Chodorow=0.52; WU-Palmer=0.22; Path=0.04; cos(LSA)=0.15; sim(LDA)=0.7; dist=1]=0.36

- pose ces poulets . [0.268]

Cohesion [Leacock-Chodorow=1.46; WU-Palmer=0.42; Path=0.14; cos(LSA)=0.13; sim(LDA)=0.63; dist=1]=0.39

- je les pose , mais je les ai achetés , dit miguel , désespéré , sans lâcher son bien . j' ai mis par terre de l' argent et une médaille bénite . [0.871]

Advanced View | Visualize Multi-Layered Cohesion Graph | Select Voices | Display Voice Inter-animation

Topics

Nouns only 0 25 50

Topics	Relevance
homme	2.71
force	2.07
enfant	1.81
poulet	1.78
médaille	1.7
ventre	1.69
fils	1.6
chèvre	1.53
prix	1.42
fusil	1.34
bruit	1.03
maison	1.01
voix	0.99
argent	0.92
difficulté	0.89
somme	0.88
accent	0.88
cas	0.86
village	0.84
travail	0.82
couteau	0.81
poule	0.8
lune	0.79
main	0.78
porteur	0.78
cou	0.76

Generate network of concepts

Fig. 13.3 Reading material visualization

score can be elevated (e.g., *hélas* …), the presented values are a combination of individual sentence scores; therefore, underlying sentences might not be selected in the summarization process. The later scoring and summarization facilities are presented in the next sections.

An appealing extension to topics identification is the visualization of the corresponding semantic space that can also be enlarged with semantically similar concepts, not mentioned within the discourse and referred to in our analysis as *inferred concepts* (see Fig. 13.4). Therefore, an inferred concept does not appear in the document or in the conversation, but is semantically related to it.

From a computational perspective, the list of additional inferred concepts identified by *ReaderBench* is obtained in two steps. The first stage consists of merging lists of similar concepts for each topic, determined through synonymy and hypernymy relations from *WordNet*/*WOLF* and through semantic similarity in terms of LSA and LDA, while considering the entire semantic spaces. Secondly, all the concepts from the merged list are evaluated based on the following criteria: semantic relatedness with the list of identified topics and with the analysis element, plus a shorter path to the ontology root for emphasizing more general concepts.

The overall generated network of concepts, including both topics from the initial discourse and inferred concepts, takes into consideration the aggregated cohesion measure between concepts (LSA and LDA similarities above a predefined threshold) and, in the end, displays only the dominant connected graph of related concepts (outliers or unrelated concepts that do not satisfy the cohesion threshold specified within the user interface are disregarded). The visualization uses a Force Atlas layout from *Gephi* [60] and the dimension of each concept is proportional with its betweenness score [61] from the generated network.

Although the majority of displayed concepts make perfect sense and seem really close to the given initial text, in most cases there are also some dissonant words that appear to be off-topic at a first glimpse. In the example set in Fig. 13.4, *campaigner* might induce such an effect, but its occurrence in the list of inferred concepts is determined by its synonymy relationship from *WordNet* to *candidate*, a concept twice encountered in the initial text fragment that has a final relevance of 2.14. Moreover, the concept has only 7 occurrences in the TASA training corpus for LSA and LDA, therefore increasing the chance of making incorrect associations in the semantic models as no clear co-occurrence pattern can emerge.

In this context, additional improvements must be made to the previous identification method in order to reduce the fluctuations of the generated inferred concepts, frequent if the initial topics list is quite limited or the initial text is rather small, and to diminish the number of irrelevant generated terms by enforcing additional filters. All the previously proposed mechanisms were fine-tuned after detailed analyses on different evaluation scenarios and on different types of texts (stories, assigned reading materials and chat conversations), generating in the end an extensible and comprehensive method of extracting topics and inferred concepts.

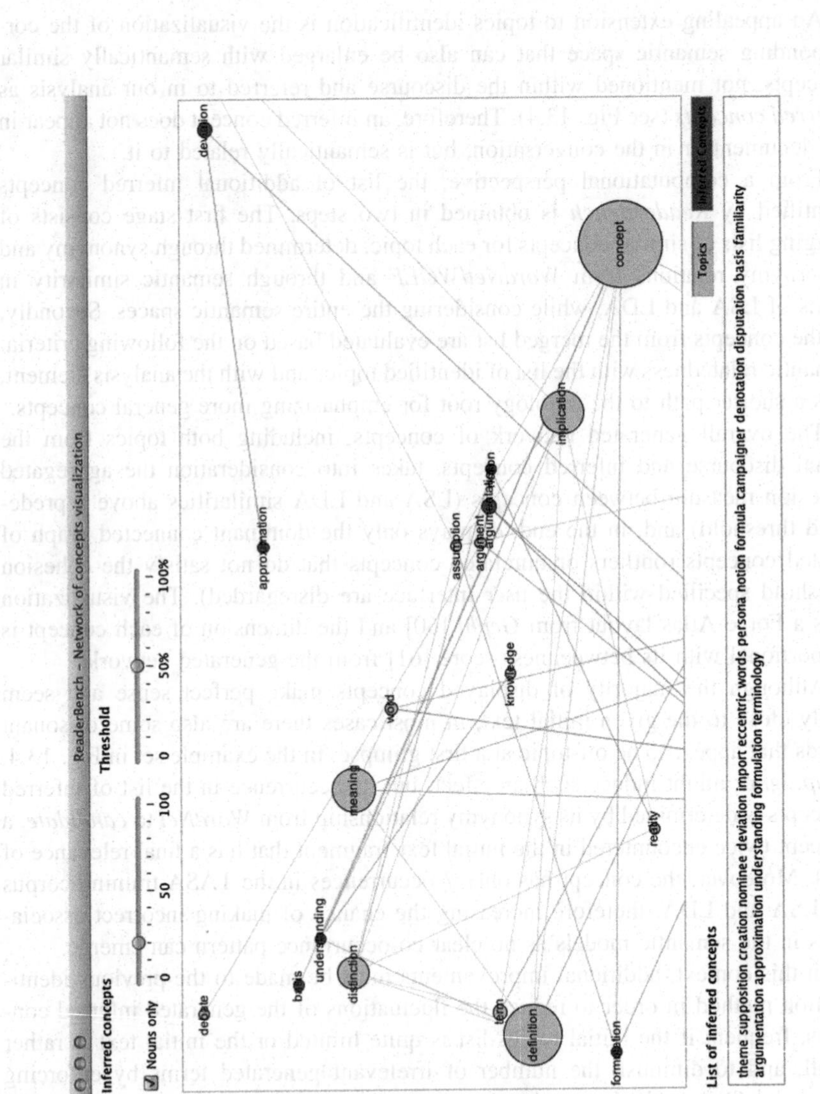

Fig. 13.4 Network of concepts visualization from and inferred from [62]

13.6 Cohesion-Based Scoring Mechanism of the Analysis Elements

A central component in the evaluation process of each sentence's importance is our bottom-up scoring method. Although tightly related to the cohesion graph [55] that is browsed from bottom to top and is used for augmenting the importance of the analysis elements, the initial assessment of each element is based on its topics coverage and their corresponding relevance, with regards to the entire document. Therefore, topics are used to reflect the local importance of each analysis element, whereas cohesive links are used to transpose the local impact upon other inter-linked elements.

In terms of the scoring model, each sentence is initially assigned an individual score equal to the normalized term frequency of each concept, multiplied by its relevance that is assigned globally during the topics identification process presented in the previous section. In other words, we measure to what extent each sentence conveys the main concepts of the overall conversation, as an estimation of on-topic relevance. Afterwards, at block level (utterance or paragraph), individual sentence scores are weighted by cohesion measures and summed up in order to define the inner-block score. This process takes into consideration the sentences' individual scores, the hierarchical links reflected in the cohesions between each sentence and its corresponding block and all inner-block cohesive links between sentences.

By going further into our discourse decomposition model (document > block > sentence), inter-block cohesive links are used to augment the previous inner-block scores, by also considering all block-document similarities as a weighting factor of block importance. Moreover, as it would have been a discrepancy in the evaluation in terms of the first and the last sentence of each block for which there were no previous or next adjacency links within the current block, their corresponding scores are increased through the cohesive link enforced to the previous, respectively next block. This augmentation of individual sentence scores is later on reflected in our bottom-up approach all the way to the document level in order to maintain an overall consistency, as each higher level analysis element score should be equal to a weighted sum of constituent element scores.

In the end, all block scores are combined at document level by using the block-document hierarchical links' cohesion as weight, in order to determine the overall score of the reading material. In this manner, all links from the cohesion graph are used in an analogous manner for reflecting the importance of analysis element; in other words, from a computational perspective, hierarchical links are considered weights and are characterized as a spread of information into subsequent analysis elements, whereas adjacency or relevant links between elements of the same level of the analysis are used to augment their local importance through cohesion to all inter-linked sentences or blocks.

In addition, starting from tutors' general observations that an extractive summarization facility, combined with the demarcation of the most important sentences, is useful for providing a quick overview of the reading material, we envisioned an *extractive summarization* facility within *ReaderBench*. This functionality can be considered a generalization of the previous scoring mechanism built on top of the cohesion graph and can be easily achieved by considering the sentence importance scores, in descending order, as we are enforcing a deep discourse structure, topics coverage and the cohesive links between analysis elements. Overall, the proposed unsupervised extraction method is similar to some extent to TextRank [63] that also used an underlying graph structure based on the similarities between sentences. Nevertheless, our approach can be considered more elaborate from two perspectives: (1) instead of simple word co-occurrences we use a generalized cohesion function and (2) instead of computing all similarities between all pairs of sentences, resulting in highly connected graph, inapplicable for large text, we propose a multi-layered graph that resembles the core structure of the initial texts in terms of blocks or paragraphs.

13.7 Identification Heuristics for Reading Strategies

Starting from the two previous studies and the five types of reading strategies used by [64], our aim was to integrate within *ReaderBench* automatic extraction methods designed to support tutors at identifying various strategies and to best fit the aligned annotation categories. The automatically identified strategies within *ReaderBench* comprise *monitoring*, *causality*, *bridging*, *paraphrase* and *elaboration* due to two observed differences. Firstly, very few predictions were used, perhaps due to the age of the pupils, compared to McNamara's subjects; secondly, there is a distinction in *ReaderBench* between causal inferences and bridging, although a causal inference can be considered a kind of bridging, as well as a reference resolution, due to their different computational complexities. Our objective was to define a fine-grained analysis in which different valences generated by both the identification heuristics and the hand coding rules were taken into consideration when defining the strategies taxonomy.

In addition, we have tested various methods of identifying reading strategies and we will focus solely on presenting the alternatives that provided in the end the best overall human–machine correlations. In ascending order of complexity, the simplest strategies to identify are *causality* (e.g., *parce que*, *pour*, *donc*, *alors*, *à cause de*, *puisque*) and *control* (e.g., *je me souviens*, *je crois*, *j'ai rien compris*, *ils racontent*) for which cue phrases have been used. Additionally, as *causality* assumes text-based inferences, all occurrences of keywords at the beginning of a verbalization have been discarded, as such a word occurrence can be considered a speech initiating event (e.g., *Donc*), rather than creating an inferential link.

Afterwards, *paraphrases*, that in the manual annotation were considered repetitions of the same semantic propositions by human raters, were automatically identified through lexical similarities. More specifically, words from the verbalization were considered paraphrases if they had identical lemmas or were synonyms (extracted from the lexicalized ontologies—*WordNet*/*WOLF*) with words from the initial text.

In addition, we experimented identifying paraphrases as the overlap between segments of the dependency graph (combined with synonymy relations between homologous elements), but this was inappropriate for French as there is no support within the Stanford Log-linear Part-Of-Speech Tagger [31].

In the end, the strategies most difficult to identify are *knowledge inference* and *bridging*, for which semantic similarities have to be computed. An inferred concept is a non-paraphrased word for which the following three semantic distances were computed: the distance from word w_1 from the verbalization to the closest word w_2 from the initial text (expressed in terms of semantic distances in ontologies, LSA and LDA) and the distances from both w_1 and w_2 to the textual fragments in-between consecutive self-explanations. The latter distances had to be taken into consideration for better weighting the importance of each concept, with regards to the entire text. In the end, for classifying a word as inferred or not, a weighted sum of the previous three distances is computed and compared to a minimum imposed threshold which was experimentally set at 0.4 for maximizing the precision of the knowledge inference mechanism on the used sample of verbalizations.

As bridging consists of creating connections between different textual segments from the initial text, cohesion was measured between the verbalization and each sentence from the referenced reading material. If more than 2 similarity measures were above the mean value and exceeded a minimum threshold experimentally set at 0.3, bridging was estimated as the number of links between contiguous zones of cohesive sentences. Compared to the knowledge inference threshold, the value had to be lowered, as a verbalization had to be linked to multiple sentences, not necessarily cohesive one with another, in order to be considered bridging. Moreover, the consideration of contiguous zones was an adaptation with regards to the manual annotation that considered two or more adjacent sentences, each cohesive with the verbalization, members of a single bridged entity.

Figure 13.5 depicts the cohesion measures with previous paragraphs from the story in the last column and the identified reading strategies for each verbalization marked in the grey areas, coded as follows: control, causality, paraphrasing [index referred word from the initial text], inferred concept [*] and bridging over the inter-linked cohesive sentences from the reading material. The grey sections represent the pupil's self-explanations, whereas the white blocks represent paragraphs from "*Matilda*" [65]. Causality, control and inferred concepts (that through their definition are not present within the original text) are highlighted only in the verbalization, whereas paraphrases are coded in both the self-explanation and the initial text for a clear traceability of lexical proximity or identity. *Bridging*, if present, is highlighted only in the original text for pinpointing out the textual fragments linked together through cohesion in the pupil's meta-cognition.

ReaderBench – Meta-cognition Processing

Document title: Matilda [config/LSA/lemonde_fr, config/LDA/lemonde_fr]

Verbalization: (MATILDA CM2 .xml)

View document

Contents

Text	Causality	Control	Paraphr...	Knowle...	Bridging	Cohesion
la mère[8] devint toute blanche . elle dit[5] à son mari il y a quelqu' un dans la maison[2] . ils arrêtèrent[9] tous de manger[10] . ils étaient tous sur le qui – vive . la voix[7] reprit[11] salut[6] , salut[6] , salut[6] . le frère[12] se mit à crier ça recommence[13] ! matilda se leva et alla éteindre la télévision[3] .						0.315
je ai compris[4] que c' est une famille[2] la famille[2] dans laquelle il ? suis qui dînent[1] devant la télé[3] . et qui . tout de un coup il z entendent[4] une voix[7] qui leur dit[5] salut[6] . et du coup ils ont peur donc parce que la mère[8] de matilda ? donc c' est qué je pense que ils ont peur . alors ils arrêtent[9] de manger[10] . puis le frère[12] commence à comprendre quelque cho quelque chose en disant ça recommence[13]	5	1	13	0	1	
la mère , paniquée , dit à son mari : henri , des voleurs[15] . ils sont dans le salon . tu devrais[14] y aller . le père , raide sur sa chaise ne bougea pas . il n' avait pas envie de jouer au héros . sa femme lui dit : alors , tu te décides ? ils doivent[14] être en train de faucher l' argenterie[16] !						0.294
alors je pense que c' est une famille[*] peut – être assez riche parce que il y a de l' argenterie[16] . et qui pensent que ceux qui doit[14] être riche ou que y a beaucoup de voleurs[15] dans notre dans leur maison donc	2	1	3	1	1	
monsieur verdebois s' essuya nerveusement les lèvres avec sa serviette et proposa d' aller[17] voir[18] tous ensemble . la mère attrapa un tisonnier au coin de la cheminée . le père[19] s' arma d' une canne de golf posée dans un coin . le frère attrapa un tabouret . matilda prit[9] le couteau avec lequel elle mangeait . puis ils se dirigèrent tous les quatre vers la porte du salon en marchant sur la pointe des pieds .						0.399
à ce moment – là , ils entendirent à nouveau la voix . matilda fit alors irruption dans la pièce en brandissant son couteau et cria haut[20] les mains[21] , vous êtes pris[9] ! les autres la suivirent en agitant leurs armes .						0.189
donc la c' est déjà comment s' appelle la famille . et puis que là vu que le père[19] veut pas y aller[17] tout seul . il est accompagné de toute sa famille pour aller[17] voir s' y a un voleur . et y a la le la parole[*] ça le bruit aussi ? qui recommence . et du coup elle , la petite fille[*] qui s' appelle matilda commence à avoir peur . donc elle lui dit haut[20] les mains[21] vous êtes pris[9]	4	2	5	2	1	

Fig. 13.5 Visualization of automatically identified reading strategies

13.8 Multi-Dimensional Model for Assessing Textual Complexity

Assessing textual complexity can be considered a difficult task due to different reader perceptions primarily caused by prior knowledge and experience, cognitive capability, motivation, interests or language familiarity (for non-native speakers). Nevertheless, from the tutor perspective, the task of identifying accessible materials plays a crucial role in the learning process since inappropriate texts, either too simple or too difficult, can cause learners to quickly lose interest.

In this context, we propose a multi-dimensional analysis of textual complexity, covering a multitude of factors integrating classic readability formulas, surface metrics derived from automatic essay grading techniques, morphology and syntax factors [66], as well as new dimensions focused on semantics [55]. In the end, subsets of specific factors are aggregated through the use of Support Vector Machines (SVM) [67], which has proven to be the most efficient for providing a categorical classification [68, 69]. In order to provide an overview, the textual complexity dimensions, with their corresponding performance scores, are presented in Table 13.1, whereas the following paragraphs focus solely on the semantic dimension of the analysis. In other words, besides the factors presented in detail in [66] that were focused on a more shallow approach, of particular interest is how semantic factors correlate to classic readability measures [55].

Firstly, *textual complexity* is linked to *cohesion* in terms of comprehension; in other words, in order to understand a text, the reader must first create a well-connected representation of the information withheld, a situation model [70]. This connected representation is based on linking related pieces of textual information that occur throughout the text. Therefore, cohesion reflected in the strength of inner-block and inter-block links extracted from the cohesion graph influences readability, as semantic similarities govern the understanding of a text. In this context, discourse cohesion is evaluated at a macroscopic level as the average value of all links in the constructed cohesion graph [2, 55].

Table 13.1 Textual complexity dimensions

Depth of metrics	Factors for evaluation	Avg. EA	Avg. AA
Surface analysis	Readability formulas	0.71	0.994
	Fluency factors	0.317	0.57
	Structure complexity factors	0.716	0.99
	Diction factors	0.545	0.907
	Entropy factors (words vs. characters)	0.297	0.564
	Word complexity factors	0.546	0.926
Morphology and syntax	Balanced CAF (complexity, accuracy, fluency)	0.752	0.997
	Specific POS complexity factors	0.563	0.931
	Parsing tree complexity factors	0.416	0.792
Semantics	Cohesion through lexical chains, LSA and LDA	0.526	0.891
	Named entity complexity factors	0.575	0.922
	Co-reference complexity factors	0.366	0.738
	Lexical chains	0.363	0.714

Secondly, a variety of metrics based on the *span* and the *coverage of lexical chains* [58] provide insight in terms of lexicon variety and of cohesion, expressed in this context as the semantic distance between different chains. Moreover, we imposed a threshold of minimum of 5 words per lexical chain in order to consider it relevant in terms of overall discourse; this value was determined experimentally after running simulations with increasing values and observing the correlation with predefined textual complexity levels.

Thirdly, *entity-density features* proved to influence readability as the number of entities introduced within a text is correlated to the working memory of the text's targeted readers. In general, entities consisting of general nouns and named entities (e.g., people's names, locations, organizations) introduce conceptual information by identifying, in most cases, the background or the context of the text. More specifically, entities are defined as a union of named entities and general nouns (nouns and proper nouns) contained in a text, with overlapping general nouns removed.

These entities have an important role in text comprehension due to the fact that established entities form basic components of concepts and propositions on which higher level discourse processing is based [71]. Therefore, the entity-density factors focus on the following statistics: the number of entities (unique or not) per document or sentence, the percentages of named entities per document, the percentage of overlapping nouns removed or the percentage of remaining nouns in total entities.

Finally, another dimension focuses on the ability to resolve *referential relations* correctly [38, 72] as *co-reference inference* features also impact comprehension difficulty (e.g., the overall number of chains, the inference distance or the span between concepts in a text, number of active co-reference chains per word or per entity).

13.9 Results

Of particular interest are the thorough cognitive validations performed with *ReaderBench* that were centered on providing a comparison to learners' performances. In terms of the presented functionalities, the validations for *ReaderBench* covered: (1) the aggregated cohesion measure by comparison to human evaluations of cohesiveness between adjacent paragraphs; (2) the scoring mechanism perceived as a summarization facility; (3) the identification of reading strategies by comparison to the manual scoring scheme and (4) the textual complexity model emphasizing morphology and semantics factors, compared to the surface metrics used within the Degree of Reading Power (DRP) score [19].

Firstly, for *validating* the aggregated *cohesion measure* we used 10 stories in French for which sophomore students in educational sciences (French native speakers) were asked to evaluate the semantic relatedness between adjacent paragraphs on a Likert scale of [1; 5]; each pair of paragraphs was assessed by more than 10 human evaluators for limiting inter-rater disagreement. Due to the subjectivity of the task and the different personal scales of perceived cohesion, the average values of intra-class correlations (ICC) per story were: ICC-average measures = 0.493 and ICC-single measures = 0.167. In the end, 540 individual cohesion scores were aggregated and then used to determine the correlation between different semantic measures and the gold standard. On the two training corpora used (Le Monde and TextEnfants), the correlations were: Combined-Le Monde (r = 0.54), LDA-Le Monde (r = 0.42), LSA-Le Monde (r = 0.28), LSA-TextEnfants (r = 0.19), Combined-TextEnfants (r = 0.06), Wu-Palmer (r = −0.06), Path Similarity (r = −0.13), LDA-TextEnfants (r = −0.13) and Leacock–Chodorow (r = −0.40).

All those correlations are non-significant, but the inter-rater correlations are on a similar range and are smaller than the Combined-Le Monde score. The previous results show that the proposed combined method of integrating multiple semantic similarity measures outperforms all individual metrics, that a larger corpus leads to better results and that Wu–Palmer, besides its corresponding scaling to the [0; 1] interval (relevant when integrating measurements with LSA and LDA), behaves best in contrast to the other ontology based semantic distances.

Moreover, the significant increase in correlation between the aggregated measure of LSA, LDA and Wu–Palmer, in comparison to the individual scores, proves the benefits of combining multiple complementary approaches in terms of the reduction of errors that can be induced by using a single method.

Secondly, for the preliminary *validation* of the *scoring mechanism* and of the proposed extractive summarization facility we have performed experiments on two narrative texts in French: *Miguel de la faim* [59] and *La pharmacie des éléphants* [73]. Our validation process used the measurements initially gathered by Mandin [74] in which 330 high school (9th to 12th grade) students and 25 tutors were asked to manually highlight the most important 3–5 sentences from the two presented stories [74]. The inter-rater agreement scores were rather low, as the ICC values were of 0.13, respectively 0.23, highlighting the subjectivity of the task at hand.

Afterwards, as suggested by [75], four equivalence classes were defined, taking into consideration the mean – standard deviation, mean and mean + standard deviation of each distribution as cut-out values. In this context, two measurements of agreement were used: *exact agreement* (EA) that reflects precision and *adjacent agreement* (AA) that allows a difference of one between the class index automatically retrieved and the one evaluated by the human raters.

By considering the use of the equivalence classes, we notice major improvements in our evaluation (see Table 13.2) as both documents have the best agreements with the tutors, suggesting that our cohesion-based scoring process entails a deeper perspective of the discourse structure reflected in each sentence's importance.

Table 13.2 Exact and adjacent agreement using equivalence classes

Text	Exact/adjacent agreement (EA/AA)					Avg. EA/AA
	9th grade	10th grade	11th grade	12th grade	Tutor	
Miguel de la faim	0.33/0.83	0.42/0.75	0.29/0.88	0.38/0.88	0.46/0.88	0.38/0.84
La pharmacie des éléphants	0.22/0.83	0.28/0.89	0.33/0.78	0.39/0.94	0.44/0.89	0.33/0.87

Moreover, our results became more cognitively relevant as they are easier to interpret by both learners and tutors—instead of a positive value obtained after applying the scoring mechanism, each sentence has an assigned importance class (1—less important; 4—the most important). In addition, we obtained 3 or 4 sentences per document that were tagged with the 4th class, a result consistent with the initial annotation task of selecting the 3–5 most important sentences.

Therefore, based on promising preliminary validation results, we can conclude that the proposed cohesion-based scoring mechanism is adequate and effective, as it integrates through cohesive links the local importance of each sentence, derived from topics coverage, into a global view of the discourse.

Thirdly, in the context of the *validation* experiments for the *identification* of *reading strategies*, pupils read aloud a 450 word-long story, "Matilda" [65], and stopped in-between at six predefined markers in order to explain what they understood up to that moment. Their explanations were first recorded and transcribed, then annotated by two human experts (PhD in linguistics and in psychology), and categorized according to scoring scheme.

Disagreements were solved by discussion after evaluating each self-explanation individually. In addition, automatic cleaning had to be performed in order to process the phonetic-like transcribed verbalizations.

Verbalizations from 12 pupils were transcribed and manually assessed as a preliminary validation. The results for the 72 verbalization extracts in terms of precision, recall and F1-score are as follows: causality (P = 0.57, R = 0.98, F = 0.72), control (P = 1, R = 0.71, F = 0.83), paraphrase (P = 0.79, R = 0.92, F = 0.85), bridging (P = 0.45, R = 0.58, F = 0.5) and inferred knowledge (P = 0.34, R = 0.43, F = 0.38). As expected, paraphrases, control and causality occurrences were much easier to identify than information coming from pupils' experience [76].

Moreover we have identified multiple particular cases in which both approaches (human and automatic) covered a partial truth that in the end is subjective to the evaluator. For instance, many causal structures close to each other, but not adjacent, were manually coded as one, whereas the system considers each of them separately. For example, "fille" ("daughter") does not appear in the text and is directly linked to the main character, therefore marked as an inferred concept by ReaderBench, while the evaluator considered it as a synonym.

Additionally, when looking at manual assessments, discrepancies between evaluators were identified due to different understandings and perceptions of pupil's intentions expressed within their metacognitions. Nevertheless, our aim was to support tutors and the results are encouraging (correlated also with the previous precision measurements and with the fact that a lot of noise existed in the transcriptions), emphasizing the benefits of a regularized and deterministic process of identification.

In the end, for training and *validating* our *textual complexity model*, we have opted to automatically extract English texts from TASA, using its DRP score, into six classes of complexity [19] of equal frequency (see Table 13.3). This validation scenario consisting of approximately 1,000 documents was twofold: on one hand,

Table 13.3 Ranges of the DRP scores as a function of defining the six textual complexity classes [after 19]

Complexity class	Grade range	DRP minimum	DRP maximum
1	K-1	35.38	45.99
2	2–3	46.02	51.00
3	4–5	51.00	56.00
4	6–8	56.00	61.00
5	9–10	61.00	64.00
6	11-CCR	64.00	85.80

we wanted to prove that the complete model is adequate and reliable and, on the other, to demonstrate that high level semantic features provide relevant insight that can be used for automatic classification.

In the end, *k*-fold cross validation [77] was applied for extracting the following performance features (see Table 13.1): precision or exact agreement (EA) and adjacent agreement (AA) [68], as the percent to which the SVM was close to predicting the correct classification.

By considering the granular factors, although simple in nature, readability formulas, the average number of words per sentence, the average length of sentences/words and balanced CAF [78] provided the best alternatives at lexical and syntactic levels; this was expected as the DRP score is based solely on shallow evaluation factors.

From the perspective of word complexity factors, the average polysemy count and the average word syllable count correlated well with the DRP scores. In terms of parts of speech tagging, nouns, prepositions and adjectives had the highest correlation of all types of parts of speech, whereas depth and size of the parsing tree provided also a good insight of textual complexity.

In contrast, semantic factors taken individually had lower scores because the evaluation process at this level is mostly based on cohesive or semantic links between analysis elements and the variance between complexity classes is lower in these cases. Moreover, while considering the evolution from the first class of complexity to the latest, these semantic features do not necessarily have an ascending and linear evolution; this can fundamentally affect a precise prediction if the factor is taken into consideration individually.

Only two entity-density factors had better results, but their values are directly connected to the underlying part of speech (noun) that had the best EA and AA of all morphology factors. Also, the most difficult classes to identify were the second and the third because the differences between them were less noteworthy.

Therefore *ReaderBench* enables tutors to assess the complexity of new reading materials based on the selected complexity factors and a pre-assessed corpus of texts, pertaining to different complexity dimensions. Moreover, by comparing multiple loaded documents, tutors can better grasp each evaluation factor, refine the model to best suit their interests in terms of the targeted measurements and perform new predictions using only their selected features (see Fig. 13.6).

ReaderBench – Document Complexity Evaluation

Path: in/corpus_complexity_tasa_en

Select complexity factors

Train SVM Model

Results

Factor	Wittgenstein, Mind and Mean... config/LSA/tasa_en config/LDA/tasa_en	A Walk With My Dog config/LSA/tasa_en config/LDA/tasa_en	The Hidden Treasure config/LSA/tasa_en config/LDA/tasa_en
Complexity prediction	6	1	3
Readability Flesh	-2.194	102.727	65.424
Readability FOG	19.522	1.365	6.439
Readability Kincaid	15.978	-0.221	5.447
Number of words per sentence	9.712	3.412	5.286
Average number of syllables per word	2.354	1.19	1.608
Percentage of complex words (>2 syllables)	39.092	0	10.811
Normalized number of commas	4.989	3.197	2.946
Normalized number of words	8.24	6.094	6.283
Normalized number of blocks	3.197	1	2.386
Average block size	782.111	592	197.25
Normalized number of sentences	5.06	3.773	3.708
Average sentence length	146.379	47.375	66.2
Average word length	121.362	37	52.6
Word entropy	5.055	4.447	4.276
Character entropy	2.876	2.941	2.85
Lexical Diversity	7.494	4.138	4.018
Lexical Sophistication	6.6	3.562	4.356
Syntactic Diversity	0.44	0.482	0.452
Syntactic Sophistication	10.678	4.294	5.438
Balanced CAF	12.915	6.488	7.12
Average number of nouns	5.621	2.562	2.467
Average number of pronouns	1.155	1.688	1.667
Average number of verbs	4.069	1.875	3.133
Average number of adverbs	1.414	0.562	0.333
Average number of adjectives	2.293	0.188	0.667
Average number of prepositions	3.207	1.188	1.333

Fig. 13.6 Document complexity evaluation

Table 13.4 *ReaderBench* versus *iStart* [79, 80, 82]

Benefits of *ReaderBench*	Benefits of *iStart*
Educational perspective	
Adaptation of the proposed methodology to the specificity of primary school pupils—elliptical expressions, pauses and repetitions in oral speech that impacted the transcription process	Initial methodology designed for assessing reading comprehension with high school/university students—adequate and coherent language, direct recording of textual representation
Refinement of the reading strategies in terms of the observed pupil's behavior (no prediction, elaboration was generalized to knowledge inference)	Initial taxonomy of reading strategies
Separate identification of reading strategies and a more fine-grained comparison to the gold standard, without a direct liaison to predicting learner comprehension	Assignment of an overall relevance score on a [1; 4] scale, easily linkable to comprehension
Retrospective view, with focus on accurate identification of different strategies	Proactive perspective, with emphasis on the impact of the system on students' comprehension
Tutor inquiry oriented analysis, with accent on the demarcation of different strategies	The use of different animated agents to present a warmer, more interactive and more user friendly perspective of the analysis
Technical perspective	
In-depth methods of extracting reading strategies using multiple heuristics (word- and LSA- heuristics were analyzed in the first two studies, later refined in ReaderBench)	Word-based and LSA centered extraction of strategies
French corpus, much more difficult to analyze in terms of natural language processing; moreover, the system enforces a NLP processing pipe to both French and English texts	English self-explanations analyzed within a web-form, with no NLP specific processing
Preprocessing and cleaning of verbalizations was required after manual phonetic transcription	

Table 13.5 *ReaderBench* versus *Coh-Metrix* [21, 83]

Benefits of *ReaderBench*	Benefits of *Coh-Metrix*
Educational perspective	
Explicit extraction of reading strategies and assessment of textual complexity using cohesion as a central measure (ingoing links with regards to cohesion as perspective of the analysis)	Emphasis on coherence from which multiple analysis dimensions emerge (outgoing links from coherence as method of building the evaluation model)
Technical perspective	
Multi-hierarchical analysis, integrating multiple natural language analysis techniques	Extensive use of LSA and of other relevant measures
Internal discourse structure built as the cohesion graph	Most commonly, similarity is expressed as LSA cosine similarity between adjacent analysis elements
Broader view, integrating factors identified as adequate within other studies	A more detailed analysis of possible factors, covering more scenarios
	Aggregation of results and visualization of multiple graphs

Two additional measurements were performed in the end. Firstly, an integration of all metrics from all textual complexity dimensions proved that the SVMs results are compatible with the DRP scores (EA = 0.779 and AA = 0.997), and that they provide significant improvements as they outperform any individual dimension precisions.

The second measurement (EA = 0.597 and AA = 0.943) used only morphology and semantic measures in order to avoid a circular comparison between factors of similar complexity, as the DRP score is based on shallow factors. This result showed a link between low-level factors (also used in the DRP score) and in-depth analysis factors, which can also be used to accurately predict the complexity of a reading material.

13.10 A Comparison of *ReaderBench* with Previous Work

As an overview, in terms of individual learning, *ReaderBench* encompasses the functionalities of both *CohMetrix* [21] and *iStart* [79, 80] as it provides teachers and learners information on their reading/writing activities: initial textual complexity assessment, assignment of texts to learners, capture of meta-cognitions reflected in one's textual verbalizations, and reading strategies assessment.

Nevertheless, *ReaderBench* covers a different educational purpose, as its validation was performed on primary school pupils, whereas *iStart* mainly targets high school and university students [81] (see Table 13.4 for a detailed comparison between *ReaderBench* and *CohMetrix*).

With regards to *Coh-Metrix* [21], ReaderBench integrates different factors, measurements, and uses SVMs [67, 68] for increasing the validity of textual complexity assessment [66] (see Table 13.5 for a detailed comparison).

Table 13.6 *ReaderBench* versus *Dmesure* [68, 84]

Benefits of *ReaderBench*	Benefits of *Dmesure*
Educational perspective	
Broad view covering multiple analysis levels, from surface analysis to semantics	Focalized analysis, granting a comprehensive view of lexical, syntactic and morphological factors
Technical perspective	
Integration of a complete NLP pipe for both french and english	Application of specific NLP techniques, but limited due to the use TreeTagger [85], a language independent parser
Integration of the most commonly used factors, plus a multitude of new factors extracted from the cohesion graph	Exhaustive analysis of possible factors (more than 300 factors), therefore enhancing the chance of accurately predicting the complexity class by combining multiple inputs; similar to some extent to Kukemelk and Mikk [86] regarding the spread of statistics; mostly surface, lexical and morphological factors, with only two factors derived from LSA
The use of solely SVMs for classifying documents as multiple studies consider them the most accurate classifiers, efficient also when addressing non-linear separable variables	A comprehensive analysis of multiple classification algorithms, including SVMs
Intuitive user interface, enabling the training and the evaluation of a new textual complexity model based on the factors selected by the user, plus a comparison of different document features	No visual interface
1,000 documents used for training the SVM; Drawback: the comparison was made using the DRP scores from TASA	French as a foreign language (FFL) corpus, manually annotated, which greatly improved the overall relevance of the analysis
Greater agreement values and near perfect adjacent agreement, as results are compared to automatic scores that induced a normalization of the initial documents classification; experiments performed on approx. 250 online reading assignments [66] proved that correlations dramatically decrease when using inconsistent initial classifications	Lower scores, meaningful nevertheless and completely justifiable while considering the used corpus and its specificity

Moreover, *ReaderBench* encompasses textual complexity measures similar to *Dmesure* [68, 84], but with emphasis on more in-depth, semantic factors. In other words, the aim of *ReaderBench* is to provide a shift of perspective towards demonstrating that high-level factors can be also used to accurately predict the complexity of a document (see Table 13.6 for a detailed comparison).

13.11 Conclusions

ReaderBench, a multi-lingual and multi-purpose system, supports learners and teachers to mine and analyze textual materials, learners' productions and identify reading strategies that enable an 'a priori' and an 'a posteriori' assessment of comprehension.

Our system allows computing a large range of measures that have been validated and compared to human ones. Moreover, *ReaderBench* infers data regarding the cognitive processes engaged in understanding and can be integrated in several pedagogical scenarios.

As a recall to Fig. 13.1, our system supports all the proposed learning activities from both perspectives, learner and tutor centered. On one hand, tutors can select learning materials by using the multi-dimensional textual complexity model, can compare the learners' productions to the automatically extracted features (topics, most important sentences or the strength of the cohesive links in-between adjacent paragraphs) and can evaluate self-explanations while addressing the identified reading strategies.

On the other, learners can take advantage of the document assessment facilities in order to better understand the structure, difficulty level and topics of the assigned material. Moreover, they can improve their own self-regulated processes through the system's feedback, especially in the case of their self-explanations in terms of the used reading strategies.

In addition, the potential of *ReaderBench*'s multi-lingual unified vision based on textual cohesion is confirmed by performing thorough validations on both analysis languages, English and French. Therefore, all the previous aspects make the integration of *ReaderBench* appropriate in various educational settings.

Further research will investigate the use of *ReaderBench* in classrooms by teachers and learners in order to validate the pedagogical scenarios. Moreover, the large range of raw data generated by *ReaderBench* will be subject to analysis in other educational data mining platforms, for example *UnderTracks* [87].

Acknowledgments This research was supported by an Agence Nationale de la Recherche (ANR-10-BLAN-1907) grant, by the 264207 ERRIC–Empowering Romanian Research on Intelligent Information Technologies/FP7-REGPOT-2010-1 and the POSDRU/107/1.5/S/76909 Harnessing human capital in research through doctoral scholarships (ValueDoc) projects. We also wish to thank Sonia Mandin, who kindly provided experimental data used for the validation of sentence importance. Some parts of this paper stem from [55].

References

1. Agrawal, R., Batra, M.: A detailed study on text mining techniques. Int. J. Soft Comput. Eng. **2**(6), 118–121 (2013)
2. Trausan-Matu, S., Dascalu, M., Dessus, P.: Textual complexity and discourse structure in computer-supported collaborative learning. In: Cerri, S.A., Clancey, W.J., Papadourakis, G., Panourgia, K. (eds.) ITS 2012. LNCS, vol. 7315, pp. 352–357. Springer, Heidelberg (2012)
3. Manning, C.D., Schütze, H.: Foundations of Statistical Natural Language Processing. MIT Press, Cambridge (1999)
4. Landauer, T.K., Dumais, S.T.: A solution to Plato's problem: the latent semantic analysis theory of acquisition, induction and representation of knowledge. Psychol. Rev. **104**(2), 211–240 (1997)
5. Blei, D.M., Ng, A.Y., Jordan, M.I.: Latent dirichlet allocation. J. Mach. Learn. Res. **3**(4–5), 993–1022 (2003)
6. Koedinger, K.R., Baker, R.S., Cunningham, K., Skogsholm, A., Leber, B., Stamper, J.: A data repository for the EDM community: the PSLC datashop. In: Romero, C., Ventura, S., Pechenizkiy, M., Baker, R.S. (eds.) Handbook of Educational Data Mining, pp. 43–55. CRC Press, Boca Raton (2010). (Chapman & Hall/CRC Data Mining and Knowledge Discovery Series)
7. Zou, M., Xu, Y., Nesbit, J.C., Winne, P.H.: Sequential pattern analysis of learning logs: methodology and applications. In: Romero, C., Ventura, S., Pechenizkiy, M., Baker, R.S. (eds.) Handbook of Educational Data Mining, pp. 107–121. CRC Press, Boca Raton (2010). (Chapman & Hall/CRC Data Mining and Knowledge Discovery Series)
8. Sheard, J.: Basics of statistical analysis of interactions data from web-based learning enviroments. In: Romero, C., Ventura, S., Pechenizkiy, M., Baker, R.S. (eds.) Handbook of Educational Data Mining, pp. 27–42. CRC Press, Boca Raton (2010). (Chapman & Hall/CRC Data Mining and Knowledge Discovery Series)
9. Tapiero, I.: Situation Models and Levels of Coherence. Lawrence Erlbaum Associates Inc, Mahwah (2007)
10. Schnotz, W.: Comparative instructional text organization. In: Mandl, H., Stein, N.L., Trabasso, T. (eds.) Learning and Comprehension of Text, pp. 53–81. Lawrence Erlbaum Associates Inc, Hillsdale (1984)
11. McNamara, D., Kintsch, E., Songer, N.B., Kintsch, W.: Are good texts always better? Interactions of text coherence, background knowledge, and levels of understanding in learning from text. Cogn. Instr. **14**(1), 1–43 (1996)
12. Oakhill, J., Garnham, A.: On theories of belief bias in syllogistic reasoning. Cognition **46**(1), 87–92 (1993)
13. O'Reilly, T., McNamara, D.S.: Reversing the reverse cohesion effect: good texts can be better for strategic, high-knowledge readers. Discourse Process. **43**(2), 121–152 (2007)
14. Cain, K., Oakhill, J.: Reading comprehension development from 8 to 14 years: the contribution of component skills and processes. In: Wagner, R.K., Schatschneider, C., Phythian-Sence, C. (eds.) Beyond Decoding: the Behavioral and Biological Foundations of Reading Comprehension, pp. 143–175. Guilford Press, New York (2009)
15. Kintsch, W.: Comprehension: a Paradigm for Cognition. Cambridge University Press, Cambridge (1998)
16. McNamara, D.S., O'Reilly, T.: Theories of comprehension skill: knowledge and strategies versus capacity and suppression. In: Colombus, A.M. (ed.) Progress in Experimental Psychology Research, pp. 113–136. Nova Science Publishers, Hauppauge (2009)
17. Winne, P.H., Baker, R.S.: The potentials of educational data mining for researching metacognition, motivation and self-regulated learning. J. Educ. Data Mining **5**(1), 1–8 (2013)
18. Eason, S.H., Goldberg, L., Cutting, L.: Reader-text interactions: how differential text and question types influence cognitive skills needed for reading comprehension. J. Educ. Psychol. **104**(3), 515–528 (2012)

19. McNamara, D.S., Graesser, A.C., Louwerse, M.M.: Sources of text difficulty: across the ages and genres. In: Sabatini, J.P., Albro, E. (eds.) Assessing Reading in the 21st Century: Aligning and Applying Advances in the Reading and Measurement Sciences, Rowman & Littlefield Publishing, Lanham (in press)
20. Nelson, J., Perfetti, C., Liben, D., Liben, M.: Measures of text difficulty. Technical Report, Gates Foundation (2011)
21. McNamara, D.S., Louwerse, M.M., McCarthy, P.M., Graesser, A.C.: Coh-metrix: capturing linguistic features of cohesion. Discourse Process. **47**(4), 292–330 (2010)
22. Millis, K., Magliano, J.: Assessing comprehension processes during reading. In: Sabatini, J. P., O'Reilly, T., Albro, E. R. (eds.) Reaching an understanding pp. 35–54. Lanham: Rowman & Littlefield (2012)
23. McNamara, D.S., Magliano, J.P.: Self-explanation and metacognition. In: Hacher, J.D., Dunlosky, J., Graesser, A.C. (eds.) Handbook of Metacognition in Education, pp. 60–81. Erlbaum, Mahwah (2009)
24. Millis, K., Magliano, J.: Assessing comprehension processes during reading. In: Sabatini, J.P., O'Reilly, T., Albro, E.R. (eds.) Reaching an Understanding, pp. 35–54. Rowman & Littlefield Publishing, Lanham (2012)
25. McNamara, D.S.: SERT: self-explanation reading training. Discourse Process. **38**, 1–30 (2004)
26. Nardy, A., Bianco, M., Toffa, F., Rémond, M., Dessus, P.: Contrôle et Régulation de la Compréhension: L'acquisition de Stratégies de 8 à 11 ans. In: David, J., Royer, C. (eds.) L'apprentissage de la Lecture: Convergences, Innovations, Perspectives. Peter Lang, Bern (2003) (in press)
27. Hayes, A.F.: Introduction to Mediation, Moderation, and Conditional Process Analysis: A Regression-Based Approach. The Guilford Press, New York (2013)
28. Budanitsky, A., Hirst, G.: Evaluating WordNet-based measures of lexical semantic relatedness. Comput. Linguist. **32**(1), 13–47 (2006)
29. Alias-i: LingPipe, http://alias-i.com/lingpipe
30. McCandless, M., Hatcher, E., Gospodnetic, O.: Lucene in Action (2nd ed.): Covers Apache Lucene 3.0. Manning Publications, Greenwich (2010)
31. Toutanova, K., Klein, D., Manning, C.D., Singer, Y.: Feature-rich part-of-speech tagging with a cyclic dependency network. In: Conference of the North American Chapter of the Association for Computational Linguistics on Human Language Technology, pp. 173–180. Association for Computational Linguistics, Stroudsburg (2003)
32. Toutanova, K., Manning, C. D.: Enriching the knowledge sources used in a maximum entropy part-of-speech tagger. In: Joint SIGDAT Conference on Empirical Methods in Natural Language Processing and Very Large Corpora, pp. 63–70. Association for Computational Linguistics, Stroudsburg (2000)
33. Klein, D., Manning, C.D.: Accurate unlexicalized parsing. In: 41st Annual Meeting of the Association for Computational Linguistics, pp. 423–430. Association for Computational Linguistics, Stroudsburg (2003)
34. Green, S., de Marneffe, M., Bauer, J., Manning, C.D.: Multiword expression identification with tree substitution grammars: a parsing tour de force with French. In: Conference on Empirical Methods in Natural Language Processing EMNLP 2011, pp. 725–735. Association for Computational Linguistics, Stroudsburg (2011)
35. Snowball, http://snowball.tartarus.org/
36. Centre National de Ressources Textuelles et Lexicales. le Lexique Morphalou, http://www.cnrtl.fr/lexiques/morphalou/LMF-Morphalou.php
37. Finkel, J.R., Grenager, T., Manning, C.D.: Incorporating non-local information into information extraction systems by gibbs sampling. In: 43rd Annual Meeting on Association for Computational Linguistics, pp. 363–370. Association for Computational Linguistics, Stroudsburg (2005)

38. Lee, H., Chang, A., Peirsman, Y., Chambers, N., Surdeanu, M., Jurafsky, D.: Deterministic coreference resolution based on entity-centric, precision-ranked rules. Comput. Linguist. **39**(4), 1–32 (2013)
39. Raghunathan, K., Lee, H., Rangarajan, S., Chambers, N., Surdeanu, M., Jurafsky, D., Manning, C.D.: A multi-pass sieve for coreference resolution. In: Conference on Empirical Methods in Natural Language Processing, pp. 492–501. Association for Computational Linguistics, Stroudsburg (2010)
40. Miller, G.A.: WordNet: a lexical database for english. Commun. ACM **38**(11), 39–41 (1995)
41. Sagot, B., Darja, F.: Building a free french WordNet from multilingual resources. In: 6th International Conference on Language Resources and Evaluation, Ontolex 2008 Workshop, pp. 14–19. ELRA, Marrakech (2008)
42. Wu, Z., Palmer, M.: Verb semantics and lexical selection. In: 32nd Annual Meeting on Association for Computational Linguistics, pp. 133–138. Association for Computational Linguistics, Stroudsburg (1994)
43. Leacock, C., Chodorow, M.: Combining local context and WordNet similarity for wordsense identification. In: Fellbaum, C. (ed.) WordNet: An Electronic Lexical Database, pp. 265–283. MIT Press, Cambridge (1998)
44. Denhière, G., Lemaire, B., Bellissens, C., Jhean-Larose, S.: A semantic space for modeling children's semantic memory. In: Landauer, T.K., McNamara, D.S., Dennis, S., Kintsch, W. (eds.) Handbook of Latent Semantic Analysis, pp. 143–165. Psychology Press, New York (2007)
45. Dascalu, M., Trausan-Matu, S., Dessus, P.: Utterances assessment in chat conversations. Res. Comput. Sci. **46**, 323–334 (2010)
46. Lemaire, B.: Limites de la Lemmatisation pour L'extraction de Significations. In: 9es Journées Internationales d'Analyse Statistique des Données Textuelles, pp. 725–732. Presses Universitaires de Lyon, Lyon (2008)
47. Wiemer-Hastings, P., Zipitria, I.: Rules for syntax, vectors for semantics. In: 23rd Annual Conference of the Cognitive Science Society. Lawrence Erlbaum Associates Inc, Mahwah (2001)
48. Low, Y., Bickson, D., Gonzalez, J., Guestrin, C., Kyrola, A., Hellerstein, J.M.: Distributed GraphLab: a framework for machine learning and data mining in the cloud. VLDB Endowment **5**(8), 716–727 (2012)
49. Mallet: A machine learning for language toolkit, http://mallet.cs.umass.edu/
50. Low, Y., Gonzalez, J., Kyrola, A., Bickson, D., Guestrin, C., Hellerstein, J.M.: GraphLab: a new parallel framework for machine learning. In: Grünwald, P., Spirtes, P. (eds.) 26th Conference on Uncertainty in Artificial Intelligence, pp. 340–349. AUAI Press, Catalina Island (2010)
51. Dascalu, M., Trausan-Matu, S., Dessus, P.: Cohesion-based analysis of CSCL conversations: holistic and individual perspectives. In: 10th International Conference on Computer-Supported Collaborative Learning, vol. 1, pp. 145–152. University of Wisconsin-Madison, Madison (2013)
52. Trausan-Matu, S., Stahl, G., Sarmiento, J.: Supporting polyphonic collaborative learning. E-service J. **6**(1), 58–74 (2007). (Indiana University Press)
53. Rebedea, T., Dascalu, M., Trausan-Matu, S., Chiru, C.G.: Automatic feedback and support for students and tutors using CSCL chat conversations. In: 1st International K-Teams Workshop on Semantic and Collaborative Technologies for the Web, pp. 20–33. Politehnica Press, Bucharest (2011)
54. Trausan-Matu, S., Rebedea, T.: A polyphonic model and system for inter-animation analysis in chat conversations with multiple participants. In: Gelbukh, A. (ed.) 11th International Conference Computational Linguistics and Intelligent Text Processing. LNCS, vol. 6008, pp. 354–363. Springer, Heidelberg (2010)
55. Dascalu, M., Dessus, P., Trausan-Matu, S., Bianco, M., Nardy, A.: ReaderBench: an environment for analyzing text complexity and reading strategies. In: Lane, H.C., Yacef, K.,

Mostow, J., Pavlik. P. (eds.) 16th International Conference on Artificial Intelligence in Education. LNCS, vol. 7926, pp 379–388. Springer, Heidelberg (2013)
56. Topic Sentences and Signposting. Harvard University, Writing Center, http://www.fas.harvard.edu/~wricntr/documents/TopicSentences.html
57. Manning, C.D., Raghavan, P., Schütze, H.: Introduction to Information Retrieval. Cambridge University Press, Cambridge (2008)
58. Galley, M., McKeown, K.: Improving word sense disambiguation in lexical chaining. In: 18th International Joint Conference on Artificial Intelligence, pp. 1486–1488. Morgan Kaufmann Publishers, San Francisco (2003)
59. Vidal, N.: Miguel de la Faim. Amitié-G.T. Rageot, Paris (1984)
60. Bastian, M., Heymann, S., Jacomy, M.: Gephi: An open source software for exploring and manipulating networks. In: 3rd International Conference on Weblogs and Social Media, pp. 361–362. AAAI Press, Menlo Park (2009)
61. Brandes, U.: A faster algorithm for betweenness centrality. J. Math. Sociol. **25**(2), 163–177 (2001)
62. Williams, M.: Wittgenstein, Mind and Meaning: Towards a Social Conception of Mind. Routledge, New York (2002)
63. Mihalcea, R., Tarau, P.: TextRank: bringing order into texts. In: Conference on Empirical Methods in Natural Language Processing, pp. 404–411. Association for Computational Linguistics, Stroudsburg (2004)
64. McNamara, D.S., O'Reilly, T.P., Rowe, M., Boonthum, C., Levinstein, I.B.: iSTART: a web-based tutor that teaches self-explanation and metacognitive reading strategies. In: McNamara, D.S. (ed.) Reading Comprehension Strategies: Theories, Interventions, and Technologies, pp. 397–420. Lawrence Erlbaum Associates Inc, Mahwah (2007)
65. Dahl, R.: Matilda. Gallimard, Paris (2007)
66. Dascalu, M., Trausan-Matu, S., Dessus, P.: Towards an integrated approach for evaluating textual complexity for learning purposes. In: Popescu, E., Li, Q., Klamma R., Leung, H., Specht, M. (eds.) 11th International Conference in Advances in Web-Based Learning. LNCS, vol. 7558, pp. 268–278. Springer, Heidelberg (2012)
67. Cortes, C., Vapnik, V.N.: Support-Vector Networks. Mach. Learn. **20**(3), 273–297 (1995)
68. François, T., Miltsakaki, E.: Do NLP and machine learning improve traditional readability formulas? In: 1st Workshop on Predicting and Improving Text Readability for Target Reader Populations, pp. 49–57. Association for Computational Linguistics, Stroudsburg (2012)
69. Petersen, S.E., Ostendorf, M.: A machine learning approach to reading level assessment. Comput. Speech Lang. **23**, 89–106 (2009)
70. van Dijk, T.A., Kintsch, W.: Strategies of Discourse Comprehension. Academic Press, New York (1983)
71. Feng, L., Jansche, M., Huenerfauth, M., Elhadad, N.: A comparison of features for automatic readability assessment. In: 23rd International. Conference on Computational Linguistics, pp. 276–284. Association for Computational Linguistics, Stroudsburg (2010)
72. Lee, H., Peirsman, Y., Chang, A., Chambers, N., Surdeanu, M., Jurafsky, D.: Stanford's multi-pass sieve coreference resolution system at the CoNLL-2011 Shared task. In: 15th Conference on Computational Natural Language Learning: Shared Task, pp. 28–34. Association for Computational Linguistics, Stroudsburg (2011)
73. Pfeffer, P.: Les Pharmacies des Éléphants. Vie et Mort d'un Géant: L'éléphant d'Afrique, Flammarion, Paris (1989)
74. Mandin, S.: Modèles Cognitifs Computationnels de L'activité de Résumer: Expérimentation d'un Eiah auprès D'élèves de Lycée. Laboratoire des Sciences de l'Éducation. PhD thesis. Université Grenoble (2009)
75. Donaway, R.L., Drummey, K.W., Mather, L.A.: A comparison of rankings produced by summarization evaluation measures. In: Workshop on Automatic Summarization, vol. 4, pp. 69–78. Association for Computational Linguistics, Stroudsburg (2000)
76. Graesser, A.C., Singer, M., Trabasso, T.: Constructing inferences during narrative text comprehension. Psychol. Rev. **101**(3), 371–395 (1994)

77. Geisser, S.: Predictive Inference: An Introduction. Chapman and Hall, New York (1993)
78. Schulze, M.: Measuring textual complexity in student writing. In: American Association of Applied Linguistics. AAAL 2010, Atlanta (2010)
79. McNamara, D.S., Boonthum, C., Levinstein, I.B.: Evaluating self-explanations in iSTART: comparing word-based and LSA algorithms. In: Landauer, T.K., McNamara, D.S., Dennis, S., Kintsch, W. (eds.) Handbook of Latent Semantic Analysis, pp. 227–241. Psychology Press, New York (2007)
80. Graesser, A.C., McNamara, D.S., VanLehn, K.: Scaffolding deep comprehension strategies through point & query, AutoTutor, and iStart. Educ. Psychol. **40**(4), 225–234 (2005)
81. Nardy, A., Bianco, M., Toffa, F., Rémond, M., Dessus, P.: Contrôle et Régulation de la Compréhension: L'acquisition de Stratégies de 8 à 11 ans. In: David, J., Royer, C. (eds.) L'apprentissage de la Lecture: Convergences, Innovations, Perspectives. Peter Lang, Bern (in press) (2003)
82. O'Reilly, T.P., Sinclair, G.P., McNamara, D.S.: iSTART: a web-based reading strategy intervention that improves students' science comprehension. In: Kinshuk, K., Sampson D. G., Isaías P. (eds.) IADIS International Conference Cognition and Exploratory Learning in Digital Age: CELDA 2004 pp. 173-180. IADIS Press, Lisbon (2004)
83. Graesser, A.C., McNamara, D.S., Louwerse, M.M., Cai, Z.: Coh-metrix: analysis of text on cohesion and language. Behav. Res. Meth. Instrum. Comput. **36**(2), 193–202 (2004)
84. François, T.: Les Apports du Traitement Automatique du Langage à la Lisibilité du Français Langue Étrangère. Centre de Traitement Automatique du Langage, PhD thesis. Université Catholique de Louvain, Faculté de Philosophie, Arts et Lettres, Louvain-la-Neuve (2012)
85. TreeTagger—A Language Independent Part of Speech Tagger, http://www.cis.uni-muenchen.de/~schmid/tools/TreeTagger/
86. Kukemelk, H., Mikk, J.: The prognosticating effectivity of learning a text in physics. Quant. Linguist. **14**, 82–103 (1993)
87. Bouhineau, D., Luengo, V., Mandran, N., Toussaint, B.M., Ortega, M., Wajeman, C.: Open platform to model and capture experimental data in technology enhanced learning systems. In: Workshop on Data Analysis and Interpretation for Learning Environments,Vienna University of Economics and Business, Vienna (2013)

Chapter 14
Maximizing the Value of Student Ratings Through Data Mining

Kathryn Gates, Dawn Wilkins, Sumali Conlon, Susan Mossing and Maurice Eftink

Abstract Student ratings of instruction are an important means of assessment within universities and have been the focus of much study over the last 50 years. Until very recently it has been difficult to perform meaningful analysis of student narrative comments given that most universities collected them as hand-written notes. This work uses statistical and text mining techniques to analyze a data set consisting of over 1 million student comments that were collected using an online process. The methodology makes use of positive and negative "category vectors" representing instructor characteristics and a domain-specific lexicon. Sentiment analysis is used to detect and gauge attitudes embedded in comments about each category. The methodology is validated using three approaches, two quantitative

K. Gates (✉)
Office of Information Technology, The University of Mississippi, 300 Powers Hall, University, Oxford, MS 38677, USA
e-mail: kfg@olemiss.edu

D. Wilkins
Department of Computer and Information Science, The University of Mississippi, 215 Weir Hall, University, Oxford, MS 38677, USA
e-mail: dwilkins@cs.olemiss.edu

S. Conlon
Department of Management Information Systems, The University of Mississippi, 247 Holman Hall, University, Oxford, MS 38677, USA
e-mail: conlons@bus.olemiss.edu

S. Mossing
Center for Excellence in Teaching and Learning, The University of Mississippi, 106 Hill Hall, University, Oxford, MS 38677, USA
e-mail: smossing@olemiss.edu

M. Eftink
Office of the Provost, The University of Mississippi, 137 Lyceum, University, Oxford, MS 38677, USA
e-mail: eftink@olemiss.edu

A. Peña-Ayala (ed.), *Educational Data Mining*,
Studies in Computational Intelligence 524, DOI: 10.1007/978-3-319-02738-8_14,

and one qualitative. While useful to individual instructors and administrators, it is only through data mining that student perceptions of teaching can be analyzed en masse to inform and influence the educational process.

Keywords Sentiment analysis · Text mining · Student ratings · Evaluation of instruction · Teaching effectiveness

Abbreviations

KWIC	Keyword in context
MOOC	Massively open online courses
NAR1	What are some positive characteristics or strengths of the course and/or instructor?
NAR2	What are some negative characteristics or weaknesses of the course and/or instructor?
Q10	How would you rate the difficulty level of this course, compared to other courses you have taken so far at Ole Miss?
Q11	How would you rate the instructor's overall performance in this course?
SEEQ	Students' evaluation of education quality

14.1 Introduction

Student evaluation of teaching is a standard practice for courses taught at universities in the US and many other countries [1–4]. These evaluations are performed for multiple purposes: to assess students' perceptions of the instructor's teaching effectiveness and to gather students' input about the characteristics of the courses and learning environment; for use of the assessment results by the instructors to improve their teaching and by their supervisors for summative decisions; and to provide other students with information for the selection of courses and instructors.

The standard design of a student evaluation instrument is a Likert-scale multiple choice survey, which may be administered as either a paper form in the classroom or as an online form [5]. In addition, students are frequently given the opportunity to provide written comments.

Written comments have usually been viewed only by the instructor and possibly by his or her supervisors, with these comments generally being thought to be private communications. Whether or not the written comments have been used to improve teaching effectiveness is uncertain, since written comments have a relatively unstructured nature and there is generally no formal framework for interpreting the comments. The impetus of this study is to develop a procedure for systematically analyzing written student course evaluation comments. In this work we explore the use of text mining and sentiment analysis in the realm of student comments about teaching.

Although similar approaches for text mining have been used to analyze social media and online reviews [6], we are not aware any large-scale applications involving student evaluations of instruction or any other applications of these techniques in higher education. Our goal is not only to apply sentiment analysis methods to student evaluation written comments, but also to develop a framework for analyzing these comments in terms of effective teaching characteristics, as interpreted by researchers in the higher education field [1–4].

The University of Mississippi converted to an online course evaluation process in 2003 and has been consistent in its administration of the system since 2006. We now have over 1 million individual evaluations in a digital database. In addition to a set of sixteen Likert-scale standard questions, our online course evaluation system also includes four narrative questions, two of which are targets of this study: "What are some positive characteristics or strengths of the course and/or instructor?" (referred to as NAR1) and "What are some negative characteristics or weaknesses of the course and/or instructor?" (NAR2). Because the two narrative questions are open-ended, students choose to discuss many topics (features), including everything from instructor characteristics and course assignments to the use of PowerPoint slides and the temperature in the classroom. Some of these topics are not captured in the Likert-scale questions [7]. The entry rates for the narrative questions have been 39 and 34 % for NAR1 and NAR2, respectively, with the average length of characters in the written comments being 84 and 73 characters, respectively, for these two questions. Thus we have available a sizeable database of student written comments. The primary goal of this research is to develop a process and framework for analyzing comments to enhance their value in evaluating instruction and courses.

The remainder of this document is organized as follows. Section 14.2, Description of the Data Set, explains the history of student evaluations at the University of Mississippi and provides details on the student comments database that we have available for this study. Section 14.3, The Methodology, describes the categories, the domain-specific lexicon, and the algorithm used to perform the sentiment analysis. Section 14.4, Assessment Results, provides details on the three assessment methods used to validate the work and summarizes the results of each assessment method. Section 14.5, Applications of the Methodology, describes applications within the University to enhance teaching and learning as well as other educational applications. Section 14.6, Future Work, summarizes our plans, both short and long term, for improving the performance of the methodology and expanding the focus of the study.

14.2 Description of the Data Set

14.2.1 The Process for Collecting Evaluations and Presenting Results

The University's process for collecting student ratings has evolved over a period of about 30 years using recommended practices such as collecting responses from at least two-thirds of the class, including several summary items, and not making results available to instructors until grades are turned in [8]. 10 years ago, the University moved to an online process for collecting evaluations. The online process was developed in-house using Java and JSP and resides on an Oracle database. Each evaluation contains a "booking id" that provides a link back to the University's student information system, SAP's Student Lifecycle Management module. This allows for the collection of related data about students, instructors, and courses at any point in time.

When moving to an online process for collecting evaluations, one of the biggest challenges is maintaining a satisfactory response rate [9]. Using a combination of no-cost incentives, we have achieved an average response rate of 78.08 %. Prior to the end of each term (several days to several weeks depending on the length of the term), students are invited to complete evaluations for the sections in which they are enrolled. The invitations are sent out as email and SMS text messages.

This is referred to as the "normal" window for completing evaluations. For fall and spring semesters, students are told that they will be able to view their grades early and register for classes early the following spring or fall semester, provided they complete 100 % of their evaluations in the normal window. The system is closed during final exams and then re-opened the following weekend before final grades are made available. In addition, students are given a "last chance" to complete their evaluations when viewing final grades. For the first few days of final grades viewing in fall and spring terms, students are blocked from viewing their grades until they have completed some percentage of their evaluations. Even with the incentives, 47 % of students complete their evaluations during the "last chance" opportunity.

We have created an expectation on campus of having selected results of student evaluations made viewable by students and faculty, so that the results are important to both faculty and students. Evaluation results are made available within the campus portal to students and employees several weeks after the term ends.

Before the evaluations are opened up for public viewing, analysts run a program to check for terms within comments that need to be replaced with "Expletive Deleted." The sixteen Likert-scale responses are available for viewing by all students and employees, whereas only one of the narrative responses is available, "What do you want other students to know about your experience in this class?"

All results are available to the instructor and those in the instructor's reporting line. Instructors have the ability to "opt out" of having their results available to others, but only a very few (less than ten out of thousands) have selected this option. Department chairs and deans have a separate portal interface for viewing more sophisticated reports such as comparisons by division or department, departmental summaries by question, and course GPAs.

14.2.2 Details About the Data Set

The corpus includes 7 years of evaluation results, spanning 2006–2012, with a total of 803,035 evaluations. We selected this window because, since 2006, we have had consistent elements of our online evaluation system for all course types (e.g., lecture, lab, seminar, studio), all delivery modes (e.g., face to face, online, regional campus), and all academic areas. There are 1,045,129 total narrative responses, including 312,423 non-null responses to NAR1 and 270,248 non-null responses to NAR2, yielding a total of 582,671 student comments targeted by this study. The evaluations and related details about students, courses, and instructors were migrated to a separate, consolidated relational database to aid in the analysis of the results.

14.2.3 Questions and Variables of Interest

In addition to NAR1 and NAR2, two Likert-scale questions are targeted by this study:

Q10: How would you rate the difficulty level of this course, compared to other courses you have taken so far at Ole Miss?
Easy—1; Average—2; Difficult—3; Very Difficult—4; Extremely Difficult—5
Q11: How would you rate the instructor's overall performance in this course?
Poor—1; Marginal—2; Good—3; Excellent—4; Superior—5

Five of the sixteen Likert-scale questions rotate in on a term-by-term basis and focus on specific topics such as textbooks, learning outcomes, and dignity and respect. Other questions are customized by academic area and course delivery mode. Several of the Likert-scale questions would be good candidates for inclusion in future sentiment analysis work but were not appropriate for the initial phase that focused on the full data set. For example:

- Was the instructor well organized and prepared for class sessions?
- Did the instructor speak clearly and distinctly?

Prior to beginning the text-mining phase of the project, we performed an extensive statistical analysis on the evaluations, focusing on the following variables of interest:

- Student: Grade in course; Overall GPA; Academic area; Classification; Age; Gender; Ethnicity.

- Instructor: Age; Gender; Ethnicity; Rank.
- Course: Term when course was taught; Campus where course was taught; Time of day when course was taught; Period within term when ratings were collected; Section size; Level; Category (lecture, laboratory, studio, etc.); Delivery mode (online, interactive video, traditional); Academic area.

The statistical analysis yielded very interesting results, some of which are summarized next.

14.2.4 Selected Results from the Statistical Analysis

Almost half (48.83 %) of the evaluations submitted during the normal window included a response to NAR1 versus slightly less than one third (31.27 %) during the "last chance" window. The average length of NAR1 responses submitted during the normal window was 101 characters, whereas the average length of NAR1 responses submitted during the "last chance" window was 67 characters. Although the response rates and comment lengths were lower in the "last chance" window, we were encouraged that a substantial volume of narrative data was still being collected when students were eager to view their grades. However, these results indicated that we should continue to seek ways to encourage early submission of evaluations.

In general, the higher the course level, the longer the response to NAR1, with doctoral level courses having an average NAR1 length of 135 characters. We saw differences in response rates for the narrative questions based on the academic area of the course and the student's program of study. Education majors were most likely to enter NAR1 responses at 49.94 %, and Military Science majors were least likely to enter NAR1 responses at 21.37 %. Law courses had the longest responses to NAR1 and NAR2 with averages of 122 and 117 characters respectively.

Students who made an "A" in the course had an average response rate of 84.18 %, whereas students who made an "F" in the course had an average response rate of 48.33 %. Students who made an "A" in the course entered the longest responses to NAR1 (101 characters) and NAR2 (76 characters). Students who made a "D" or "F" grade in the course entered the next longest responses to NAR2 at 76 and 73 average characters respectively. Among the 803,035 evaluations, there were 2,235 cases of a student making an "F" grade in a course and then submitting another evaluation for the repeated course with the same instructor. Given the variety of circumstances in which courses may be repeated and the relatively small number of occurrences, these evaluations were not given special treatment.

We observed a rather surprising result when we plotted the length of narrative responses against the Q11 rating (see Fig. 14.1). We found that students who gave a rating of "Poor" for Q11 wrote NAR2 responses that were almost three times the average length (213 characters).

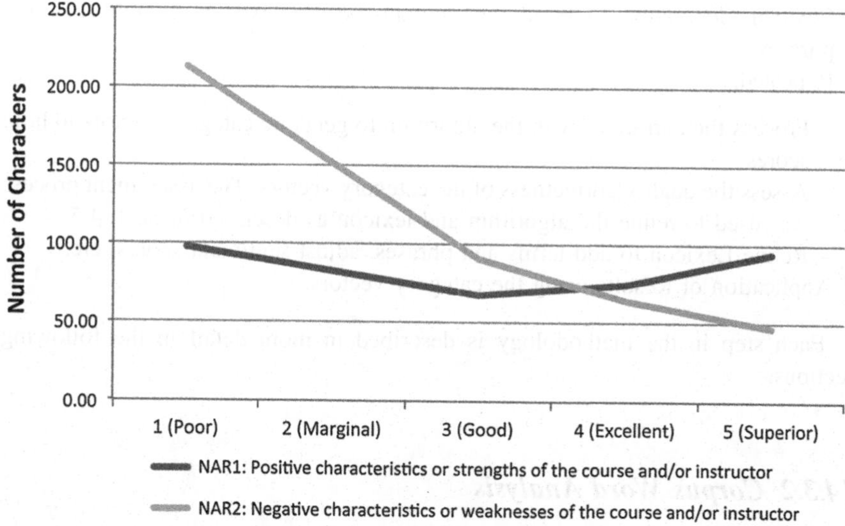

Fig. 14.1 Average length in characters of non-null responses by Q11 rating

We concluded that students write more when they have very negative experiences than when they have positive experiences. It would be easy to overlook this feedback using manual methods due to the volume of data; hence, one important aspect of the sentiment analysis work is the ability to detect "red flags" in student comments.

14.3 The Methodology

14.3.1 A High-Level View of the Process

Deriving sentiment from unstructured text is inherently a difficult problem. When the text is voluminous and noisy it is even more challenging. Given the number of narratives in the corpus, supervised techniques [10, 11], where each narrative has been previously classified, were not feasible. Instead we employ a form of aspect-based (or feature-based) sentiment analysis [12, 13] driven by a domain-specific database of terms and phrases. The design of the lexicon was influenced by recent work to analyze other opinion-based data sets such as movie reviews [14]. Lexicon-based methods using term scoring are standard in sentiment analysis [15, 16], and our technique uses this approach. The steps in the methodology are as follows:

- Corpus Word Analysis to understand the domain, build spelling correction mappings, and identify the universe of words.
- Category Selection to identify the domain-specific categories of interest.

- Lexicon Generation to build the domain-specific lexicon of keywords and phrases.
- Repeatedly …
  - Process the narratives with the algorithm to generate category vectors to hold scores.
  - Assess the quality/correctness of the category vectors. The assessment process was used to refine the algorithm and lexicon as described in Sect. 4.5.
  - Refine Lexicon to add terms and phrases, adjust sentiment scores, etc.
- Application of Results using the category vectors.

Each step in the methodology is described in more detail in the following sections.

14.3.2 Corpus Word Analysis

We knew that the data would be noisy, but it was not until we began the analysis that we fully realized just how noisy it was. We iteratively generated lists of all words occurring in the comments, ran spell-checkers to find unknown words, and created spelling correction mappings. This resulted in a table of almost six thousand spelling correction mappings. The ten words with the highest number of spelling variations are shown in Table 14.1. The interface to collect comments does not include an auto-correct or spell-check feature. Given the high number of spelling and typographical errors occurring in the comments, it may be beneficial to add spelling support and auto-correct features to the online form for collecting student comments in the future. The importance of the data cleansing process in domains involving student comments has been noted elsewhere [17, 18].

Table 14.1 Words occurring in student comments with the most variations in spelling

Word	Number of spelling variations	Number times misspelled	Total number of uses	Percent times misspelled (%)
Enthusiastic	67	1,077	17,602	6.12
Instructor	42	1,661	48,131	3.45
Enthusiasm	41	539	6,672	8.08
Knowledgeable*	38	3,816	10,787	35.38
Difficult	33	276	20,526	1.34
Material	33	302	57,068	0.53
Assignments	33	377	11,689	3.23
Understand	30	133	23,612	0.56
Interesting	29	715	26,415	2.71
Teacher	29	194	56,958	0.34

* Multiple correct spellings mapped to most common spelling

We made a decision to not stem words; otherwise, there was a potential of losing usage nuances for terms in this domain. An example is "understand" versus "understandable." The context must be considered in order to determine the intended meaning. In the table representing our universe of words, only words with a frequency count of two or more were retained, leaving a word list of approximately 21,000 words.

14.3.3 Category Selection

The selection of categories began with a survey of the literature. Marsh [19] developed the Students' Evaluation of Education Quality (SEEQ), which includes eight characteristics of effective teaching: learning, individual rapport, enthusiasm, examinations, organization, breadth, group interaction and assignments. This set of characteristics was a good starting point for developing our categories, but NAR1 and NAR2 did not focus the students on several of these categories, e.g., breadth and group interaction. As for sentiment, two of Marsh's characteristics, "examinations" and "assignments," were typically discussed in the context of course workload and difficulty.

As with many aspect-based opinion-mining methodologies [20], an early step was to mine the comments for features. Using the word table, we began to identify the most frequent sentiment words that fit with Marsh's characteristics, as well as those that did not. As a consequence of this manual process, some additional categories were identified and others were merged. The final list of categories was influenced by Marsh's traits and was adjusted based on the features found in the comments.

As noted by Kim and Calvo [7], difficulty/workload has a different sentiment structure than the other characteristics. It is clear for the other categories that a higher value is better. But for difficulty, the distinction is not so clear. Is the comment "a lot of tests" a positive or negative sentiment? What about "the homework was hard" or "too much homework"? We decided that a higher difficulty value would mean the student perceived the course as being more difficult, and a lower value in difficulty would mean the student perceived the course as being relatively easier.

Ultimately, two categories were created: "Learning" and "Difficulty." The "Learning" category contains words such as "learn," "think," "study," "teach," and "read." The "Difficulty" category contains words such as "hard," "difficult," "easy," "tough," "strict."

A category for "Accountability" was added to aid in identifying "red flag" situations, such as excessive class cancellations, students feeling threatened in class and profanity in the narratives.

In order to incorporate as much of the sentiment expressed in the narratives as possible, the "General" category was added. Many narratives had sentences such as "It was awesome" or "Terrible" where no subject or object was explicitly stated. The word "great" appeared in the narratives more than fifty thousand times.

Table 14.2 Categories for classifying narratives

Organized	Helpful
Knowledgeable	Fair
Engaging	Clear
Learning	General
Accountability	Difficulty

Often, these words are used in multiple ways, e.g., as a stand-alone general expression of approval or disapproval, and also as a modifier to other terms such as "has great enthusiasm." These sentiments are classified in the lexicon in a manner that supports awareness of context. Specifically, "great" as a stand-alone term contributes to the "General" category, whereas "great" when used as a modifier, such as "great enthusiasm," contributes to the category being modified, in this case "Engaging." The final set of categories/characteristics is shown in Table 14.2.

14.3.4 The Domain-Specific Lexicon

Once the categories were determined, a small set of seed words for each category was selected using the frequency count table and Keyword in Context (KWIC) indices [21]. The seed words were each manually assigned a sentiment score and category. The sentiment scores were influenced by SentiWordNet [22] ratings but were scaled for suitable use in algorithm calculations involving multiplicative factors. Most positive sentiment terms were given a value between +1 and +2, while negative sentiment terms were given values between −1 and −2. For example: atrocious (−1.875), incompetent (−1.75), worthless (−1.625), incoherent (−1.5), monotone (−1.375), inappropriate (−1.250), confused (−1.125), forgetful (−1), eye-opening (1), amusing (1.125), affirming (1.25), enjoyable (1.375), trustworthy (1.5), encouraging (1.625), brilliant (1.75), and exemplary (1.875).

A semi-automated approach using the thesaurus API at bighugelabs.com [23] was used to grow the set of keywords. In addition to sentiment, the seed words were marked with allowable parts of speech, e.g., noun and/or adjective, to narrow the results in the automated lookup. The online thesaurus resource was queried with the seed words, and synonyms were given the same sentiment score as the seed word while antonyms were given a sentiment score of −1 times the sentiment score of the seed word. This process was repeated (using recursion) for two rounds. Initially, we intended to fully automate the expansion of lexicon words, but we found that manual intervention was required. For the word, "growth," the thesaurus returned "development" and "maturation," which were appropriate for our domain, but also "botany" and "sickness"—words that occurred in our universe of words but which held entirely different meanings. To address this, two columns were added to the seed table, one to indicate that the word should not be expanded and the other to indicate not to use the word, allowing for partial automation. We set as an objective for this phase of the project to score all domain-relevant lexicon words that occurred at least twenty times in our universe of words.

To facilitate the identification of syntactical relationships among terms, the lexicon tags listed in Table 14.3 were established, effectively implementing simplified parts-of-speech tagging [20]. In order to distinguish the subject of a comment, the tags "sc", "ss", and "si" were used for courses, students and instructors, respectively.

The "sc" terms have subcategories to further disambiguate course assessment ("sc-a"), course materials ("sc-m"), and other course topics that were excluded from the study ("sc-e"). Typical words for each tag are provided in the table along with the count of the number of terms associated with each tag in the database.

The keywords ("k"), keyword modifiers ("km"), and intensifying modifiers ("m") [24] were grown using the semi-automated approach previously described [20]. The intensifying modifiers play a crucial role in the algorithm to accurately quantify the embedded sentiment. They are important for not only the intensity they convey but also because they occur so frequently in the corpus as shown in Table 14.4. Modifiers have sentiment scores ranging from +1 to +2. The sentiment of the word or phrase being modified is multiplied by the sentiment score of the modifier. The resulting value contributes to the category associated with the word or phrase.

A set of phrases was added to the lexicon during the refinement process. The use of phrases makes it possible to correctly score words that possess a unique meaning when grouped together, e.g., "hard to keep up," "straight from the book," or "good luck." Phrases are also used to nullify terms that are not likely to be scored correctly using simple methods. For example, "I would suggest," "I never thought," and "should have." Phrases have a sentiment score, optional category, and a priority. Nullifying a phrase can be accomplished by simply setting the sentiment score for the phrase to zero.

The phrases Jia et al. [25] exclude from negation, such as "not only," "not just," "not to mention," and "no wonder" are examples. A phrase can also have an optional set of rules [26] that influence the processing of the phrase, nearby terms, and sentiment score. The process of selecting phrases for inclusion in the lexicon was facilitated by a custom Java program designed to find the most commonly used phrases consisting of two to five terms. Candidate phrases were hand-pruned as part of the iterative refinement cycle during the qualitative assessment phase.

The final lexicon contains about 400 unique phrases that complement the approximately 2,100 unique words that have been tagged and assigned sentiment scores. Of the 8.28 million occurrences of the lexicon words in the comments targeted by this study, 1.64 million, or nearly 20 %, have been assigned sentiment scores. Table 14.5 shows the saturation of lexicon words in the comments by category.

14.3.5 The Assessment Process

A process of triangulation was used to iteratively refine the algorithm and assess its validity. First, teaching and learning specialists were asked to manually score the

Table 14.3 Tags with descriptions, examples, and counts

Tag	Title	Description	Examples	Count
Sc	Subject—course	The class, course, or university	Class, course	11
sc-a	Subject—course—assessment	Any type of assessment	Test, midterm, examination	15
sc-e	Subject—course—excluded	Words to exclude from processing due to not pertaining to course or instructor	Clicker, textbook, black board	29
sc-m	Subject—course—materials	Class materials	Podcast, homework, material	75
Ss	Subject—student	Students enrolled in the class	I, my, student	14
ss-i	Subject—student—inquiry	Inquiries from students for the purpose of gauging helpfulness	Email, question	5
si	Subject—instructor	Instructors	Dr, prof, she	22
A	Article		A, an, the	3
C	Conjunction		And, or, but	9
Hv	Helping verb		Is, am, are	24
K	Keyword	Words with inherent sentiment and relevance to this domain	Learn, ambiguous	1,830
Km	Keyword/modifier	Words that can serve as a general-purpose stand-alone keyword or as a modifier depending on the context	Great, awesome, inferior	157
M	Intensifying modifier	Words that increase or decrease the sentiment embedded in other words	Very, sometimes, many	128
N	Negator	Words that reverse the sentiment embedded in other words	No, not, never	10
P	Preposition		Above, to, of	36

Table 14.4 Five most frequently occurring intensifying modifiers

Intensifying modifier	Sentiment score	Total # of occurrences in comments	% of all lexicon words in comments
Very	1.25	120,847	1.46
Really	1.25	35,319	0.43
Always	1.5	30,217	0.37
Lot	1.25	29,887	0.36
More	1.125	28,741	0.35

Table 14.5 Saturation of scored lexicon words by category

Category	Total # of occurrences in comments	% of all lexicon words in comments
Organized	29,934	0.36
Helpful	116,305	1.40
Knowledgeable	29,884	0.36
Fair	16,773	0.20
Engaging	158,008	1.91
Clear	65,415	0.79
Learning	206,067	2.49
Difficulty	86,258	1.04
Accountability	4,288	0.05
General	349,366	4.22
Intensifying modifiers	574,955	6.95

results of the algorithm using a scale of "Very Accurate," "Accurate," "Inaccurate," and "Very Inaccurate." This process was repeated three times during the development of the algorithm and one more time after the algorithm was finalized.

In each case, five courses were selected representing varying academic levels and disciplines. Test files were generated for each course consisting of 50 NAR1 comments and 50 NAR2 comments along with the category and summary sentiment scores produced by the algorithm. Five reviewers each scored one hundred different comments. The reviewer comments were used to determine cases where the algorithm was performing well versus cases where further adjustments needed to be made.

The second method for assessing the algorithm's validity consisted of comparing the results of two Likert-scale questions: Q10 (difficulty) and Q11 (overall effectiveness) with the summary sentiment scores for the positive and negative category vectors. The summary sentiment scores (excluding difficulty) were plotted against Q11 responses to determine whether the sentiment expressed in the comments, as gauged by the algorithm, correlated with the Q11 (overall effectiveness) rating. Likewise, the sentiment scores for the "Difficulty" category were plotted against Q10 responses to determine whether sentiment expressed in the comments about difficulty, as gauged by the algorithm, correlated with the Likert-scale Q10 (difficulty) response.

The third method for assessing the algorithm's validity involved comparing the category and summary sentiment scores for recipients of a university-wide outstanding teaching award with all instructors.

Award recipients are selected through an extensive committee-based process involving personal recommendations from campus constituencies. Evaluations for faculty members who received the award at any point in time were included. While not a large comparison group, the expectation was that the average sentiment scores for award recipients should be better than average sentiment scores for all instructors.

14.3.6 Refining the Lexicon

In the first deployment of the algorithm, only keywords and a four-term window preceding keywords were considered. The algorithm performed well when students simply described the attributes of the instructor. However, analysis of feedback from teaching and learning specialists revealed that the algorithm performed poorly in three scenarios: (1) when students gave prescriptive advice on how to correct teaching or course inadequacies; (2) when words were combined to form entirely new meanings; and (3) when modifiers followed keywords. Examples showing where the first deployment of the algorithm performed poorly are as follows:

- "It would be nice if he spoke a little slower and wrote a little more clearly."
- "She was very willing to work with you or your group outside of class."
- "It was sometimes hard to understand the instructor."
- "I learned nothing."

In the first case, the student is stating how to correct inadequacies. When looking at keywords only, the term "nice" registers positive sentiment with the "Helpful" category. Moreover, the phrases, "spoke a little slower" and "wrote a little more clearly," are interpreted as the current behavior, not something the student is suggesting happen. In the second case, when considering keywords only, the phrase "willing to work with you" registers sentiment in the "Difficulty" category due to the keyword "work," rather than being associated with the "Helpful" category.

In the third case, when looking at keywords only, the phrase, "hard to understand," cannot be classified correctly as the instructor was hard to understand versus the material was hard to understand. Lastly, the phrase, "I learned nothing" reduces to just "I learned" when looking at only the terms preceding keywords rather than those preceding and following the keyword.

The feedback from teaching and learning specialists led to several crucial enhancements to the algorithm. First, the notion of static and dynamic phrases was added. Phrases can be used to treat several words as a single concept, to derive categories dynamically based on the presence of a nearby keyword, and to flexibly determine which words should be grouped as a phrase. With this enhancement, the algorithm can properly handle phrases such as "hard to" followed by a keyword. That is, the core phrase, "hard to," with one or more accompanying rules, can correctly assign "hard to hear," "hard to understand," and "hard to focus."

Although not currently planned, there is a potential for applications of the algorithm to influence personnel decisions. To minimize risk, we take a conservative approach in that we consider correctness to be a higher priority than completeness.

That is, misstating an instructor's negative characteristics is worse than simply omitting phrases that are beyond the capabilities of the algorithm. Accordingly, we added the ability to assign phrases a sentiment score of zero, thereby effectively eliminating them. We also added a warning to indicate that some phrases had been omitted from processing due to complexity. This enhancement provides a method for handling prescriptive advice, e.g., phrases starting with "should," "would," and "if."

Another important enhancement to the algorithm that came during the qualitative assessment phase was the addition of limited processing of modifying terms that follow, rather than precede, keywords. The algorithm handles terms following keywords more conservatively than those preceding keywords. This enhancement, even with its limited scope, allows for correct processing of phrases such as "I learned nothing."

The final issue that surfaced during the qualitative assessment phase was the need to be very clear in the definition of the categories. When first deployed, the lexicon grouped together in one category the notions of a course being challenging, i.e., the amount of learning that took place, and overall difficulty. These were ultimately separated out as two categories, "Learning" and "Difficulty," where a positive value in the "Difficulty" category means that the student perceives that the course is very difficult, not that the student feels good about the course being very difficult. For this reason, the "Difficulty" category is not included in summary sentiment scores.

As we converge on an effective process, at each iteration, the refinements become more about adjusting sentiment scores and adding phrases and less about structural changes to the algorithm and overall process.

14.3.7 The Algorithm

The algorithm makes use of the lexicon and sentiment scores stored in the database. The process begins by first extracting sentences as an array of strings using the LingPipe [27] sentence tokenizer method with context knowledge about possible and impossible sentence terminators, penultimates, common abbreviations, etc. Potentially interfering punctuation is addressed. Each sentence is then converted into an array of terms. Each term is checked against the table of spelling corrections and mapped to the correctly spelled term if necessary. The sentences are then reconstructed with correctly spelled terms.

Each narrative is processed a sentence at a time. Phrases are handled first in decreasing order of priority, and then keywords are handled. Some simplifying assumptions have been made. A window size of four is used throughout [25], both before and after a keyword or phrase. Once a phrase or keyword is processed, it cannot be matched again as a phrase or keyword, but its terms can be used in the pre- or post-window of another phrase or keyword.

The rules associated with phrases have the syntax, tag:position:scope. Tag (see Table 14.3) can be any of the tags supported in the lexicon, position can be "b" for before, "a" for after, or "e" for either, and scope can be "o" for optional or "r" for required. Multiple rules can be specified for each phrase. An additional rule of #EOL has been defined to indicate that the system should match on this phrase through the end of the current sentence.

The use of #EOL was helpful when students made prescriptive instead of descriptive comments, e.g., "it would be nice if he spoke a little slower." To explain the use of rules, consider the phrase "should have" with rule "si:b:r". For the rule to match the narrative, it is required that a "si" tagged term (subject instructor) appear before the phrase "should have." The position "before" (or "after") is limited to a window of size four, including the "si" term.

Therefore, this rule will match a portion of a narrative that consists of a subject instructor (si), followed by up to three optional words, then the phrase "should have". Some example matching phrases are: "he should have" or "Dr. Jones really should have." If the rule "k:a:o" was included in the "should have" phrase rules also, then an optional keyword ("k") can be identified following the target phrase, e.g., "Dr. Jones really should have been more organized." In this manner, the core phrase "should have" is expanded to the adjusted phrase, "Dr. Jones really should have been more organized." If a category key is not explicitly specified, then the category associated with the first keyword encountered in the adjusted phrase is used. If no keyword is found then the default category, "General," is used.

The result of processing each narrative is a positive and a negative category vector. That is, if a narrative indicates that the professor was smart (Knowledgeable, positive) and nice (Helpful, positive) but graded harshly (Fair, negative) and was boring (Engaging, negative), and the course was hard (Difficulty, positive), then the raw vectors may appear as shown in Table 14.6. Recall that "Difficulty" is analyzed separately.

In addition to raw scores, we also compute normalized scores. If the raw positive vector is P, and the raw negative vector is N, then T is the sum of the raw values over all categories. Then the normalized positive vector (NP) and the normalized negative vector (NN) are computed as shown in Eqs. (14.1)–(14.3) (when T is not zero):

$$T = \sum_i P[i] + \sum_j N[j] \tag{14.1}$$

$$NP[i] = \frac{P[i]}{T} \forall i \tag{14.2}$$

Table 14.6 A sample category vector with difficulty score

	Org	Help	Know	Fair	Engage	Clear	Learn	Gen	Safe	Dif
P: Positive	0	1.375	1.625	0	0	0	0	0	0	1.62
N: Negative	0	0	0	1.5	1.25	0	0	0	0	0

$$NN[j] = \frac{N[j]}{T} \forall j \tag{14.3}$$

We now present the pseudocode for the core algorithm.

Pseudocode for Algorithm score_narratives

```
Let hash cat{k} = c where c is category number for
keyword or phrase k
Let hash sent{k} = s where s is sentiment score  -2 … +2
for keyword, modifier, or phrase k

For each narrative nar {
  For each category c {
    P[c] = 0        // Positive vector
    N[c] = 0        // Negative vector
    }

  For each phrase p in nar in priority order {
    s = sent{p}
    If check_rules() returns true {
      p' = adjusted p using substr bounded by li and lr
      process_window(p')
      If s < 0 { N[c] +=  -1 * s }
      Else { P[c] += s}
      } // next phrase

  For each keyword kw tagged "k" or "km" in nar {
    If kw was not handled in phrase processing {
      If (kw is tagged "k") or (kw is tagged "km" and
        there is no "k" term in the pre- or post-window) {
        s = sent{kw}
        process_window(kw)
        c = cat{kw}
        If s < 0 { N[c] +=  -1 * s }
        Else P[c] += s
        }
      }
    } // next keyword

  Insert N[] and P[] for nar into database
  }   // next narrative
```

Pseudocode for process_window

```
For each word w in the pre-window of kw or p {
  If w is tagged "m" or "km" { s = s * sent{w} }
  Else if w is tagged "sc-e" {
    s = 0
    stop processing phrase or keyword
  }   }
For each word w in the post-window in L to R order {
  If w is not tagged one of ["m", "km", "a", "n", "hv"]
    stop processing post-window
  If w is tagged "m" or "km" { s = s * sent{w} }
  }
If the pre- and post-windows contain one negator {
  s = s * -1 }
```

Pseudocode for check_rules

```
  If no rules exist for phrase p { return true }
  Else {
    For each rule r for phrase p with a required tag {
      If p does not satisfy rule { return false }
      }   }
  For each rule r for phrase p {
    // adjust phrase boundaries
    li = starting index of p
    lr = ending index of p
    If rule position is "before" or "either" {
      li-tmp = left-most index in pre-window that
        matches tag specified in rule
      if li-tmp < li { li = li-tmp }
      }
    If rule position is "after" or "either" {
      lr-tmp = right-most index in post-window that
        matches tag specified in rule
      if lr-tmp > lr { lr = lr-tmp }
      }   } // next rule
  Set new phrase boundaries: li and lr // for p'
  c = cat{p}
  If c = 0 {
    c = cat{w} for first word w in p' tagged as "k"
    If no word w tagged as "k" is found {
      c = default category // "General"
      }   }
  return true
```

14.4 Assessment Results

14.4.1 Qualitative Validity Assessment of Category Scores by Teaching and Learning Specialists

In the final round of the qualitative validity assessment phase, five courses were selected representing a variety of academic levels and disciplines. Scoring forms were generated showing the original comment together with the categories and sentiment assigned by the system (see Fig. 14.2). Each of the five reviewers was asked to score one hundred comments for one of the courses as "Very Accurate," "Accurate," "Inaccurate," and "Very Inaccurate." The comments were split evenly among NAR1 and NAR2 question types.

Some of the scores of "Inaccurate" or "Very Inaccurate" were inconsistent with the directions provided. For example, confusion lingered about the meaning of the "Difficulty" category. Some reviewers scored the system as "Inaccurate" or "Very Inaccurate" when it registered positive sentiment in the "Difficulty" category for comments related to course difficulty as shown in Fig. 14.3.

```
Evaluation #18:  This was the most difficult class for me but he made it so
simple with the way he broke it down and explained it.  Knows a lot about
the course!  A whole lot!  Very funny!
Overall Sentiment Scores (Excluding Difficulty)
    Total -- Positive: 5.875  Negative: 0.000
    Normalized -- 100.000     Negative: 0.000
Category Scores
    Engaging: 1.875 (positive) 0.000 (negative)
    Clear: 2.500 (positive) 0.000 (negative)
    Learning: 1.500 (positive) 0.000 (negative)
    Difficulty: 1.688 (positive) 0.000 (negative)
```

Fig. 14.2 Sample evaluation with system-generated sentiment scores

```
Evaluation #9: it is just a hard class.
Overall Sentiment Scores (Excluding Difficulty)
    Total – Positive: 0.000   Negative: 0.000
    Normalized – Positive: 0.000  Negative: 0.000
Category Scores
    Difficulty: 1.625 (positive)  0.000 (negative)
____ Very accurate ____ Accurate ____ Inaccurate _X__ Very inaccurate
Comments (optional):
Difficulty is not positive
```

Fig. 14.3 Sample form with incorrect scoring for "Difficulty" category

The lexicon, through the use of the "sc-e" tag, was designed to exclude sentiment related to the book, the learning management system, and so on. Yet, in several cases, the reviewers marked the system as being inaccurate when it did not register sentiment for comments related to these items. Future rounds of qualitative validity assessment should involve multiple reviewers scoring the same narratives to ensure consistency.

Reviewers rated 88.41 % of the comments as very accurate or accurate. NAR1 comments exhibited a higher success rate than NAR2 at 95.63 and 77.70 % respectively. As noted earlier, the system performed better when students simply described instructor and course characteristics rather than giving suggestions for improvements. Some comments were not scored—17.20 % of NAR1 and 43.20 % of NAR2—due to the responses being sentiment-free, e.g., "none" or "n/a". Table 14.7 summarizes cases where the system did not perform well. As can be seen, many issues can be easily addressed through enhancements to the lexicon, whereas others are beyond the capabilities of the current approach.

14.4.2 *Quantitative Assessment Through the Comparison of Summary Scores with Overall Instructor Performance Ratings*

The data set provides a major benefit related to assessment in that each evaluation contains Likert-scale questions as well as free-form narrative responses, allowing for the comparison of responses to the Likert-scale question with the sentiment scores generated by the algorithm for the narrative responses. Specifically, we can compare the results of Q11 (overall effectiveness) with the summary sentiment scores generated for NAR1 and NAR2 where students are asked to describe positive and negative characteristics of the course and/or instructor. As well, we can compare the results of Q10 (difficulty) with the sentiment scores associated with the "Difficulty" category.

Unlike the qualitative validity assessment process where comments were checked for a subset of courses, with the quantitative assessment, all comments were analyzed through computational methods. Figure 14.4 shows the average of the summary sentiment scores for the positive and negative category vectors for NAR1 and NAR2 combined plotted against the responses to Q11 for all courses in the corpus (7 years worth of data).

The results match the expectation in that the more positive the response to Q11, the higher the average summary score for the positive vector, and the more negative the response to Q11, the lower the average summary score for the negative vector. Stated simply, the responses to Q11 correlate well with the sentiment scores gauged by the algorithm for NAR1 and NAR2. Figure 14.5 shows the normalized summary sentiment scores for the positive category vectors for NAR1

Table 14.7 Sampling of issues identified in final round of qualitative assessment

Comment fragment	Problem with System scoring	Explanation
"Pay attention in class and you will do well"	Registered as positive in the "Engaging" and "Clear" category	Currently not possibly to recognize commands; however, these probably can be addressed using phrases
"The professor was superior"	Did not register any sentiment	Term "superior" is used in both positive and negative ways in the corpus, e.g., "He acted as if he was superior to you," and therefore has not been assigned sentiment. Negative usages can be addressed with one or more phrases to determine context
"Didn't necessarily test the student's knowledge and understanding of the material"	Registered as positive in the "Learning" category	The negator "not" occurs more than four terms from keywords "knowledge" and "understanding." Difficult to address with current model
"I have never struggled like I did over her material"	Registered negative sentiment (easy) in the "Difficulty" category	Negator "never" was applied to keyword "struggled" reversing the intended meaning. Can be addressed with a new phrase in the lexicon, i.e., "I have never …"
"[Instructor] is the man"	Did not register any sentiment	System not able to interpret some forms of student vernacular. Possibly can be addressed with phrases.
"[Other Instructor] should be teaching this class instead of [Instructor]…. She talks slower, and is sure that we understand the material before she moves on"	Registered as 100 % positive in "Helpful" and "Learning" categories	Student talked about the characteristics of the teacher he/she would like to teach the class, not the one who actually taught the class. First "should" occurrence resulted in text being omitted through "…" but rest was processed. Difficult to address

(continued)

Table 14.7 (continued)

Comment fragment	Problem with System scoring	Explanation
"Professor was nice enough to only put 11 chapters on one test that was not a final"	Registered as positive in the "Helpful" category, yet student probably intended to be sarcastic	Difficult to address
"The instructor can sometimes get easily side-tracked during lecture"	Did not register any sentiment	The term, "side-tracked" occurs in the comments seventeen times and fell beneath the threshold of twenty. Can be addressed by adding sentiment to more words
"Talks to softly"	Did not register any sentiment	Student used "to" instead of "too." This happens frequently, making accurate processing difficult

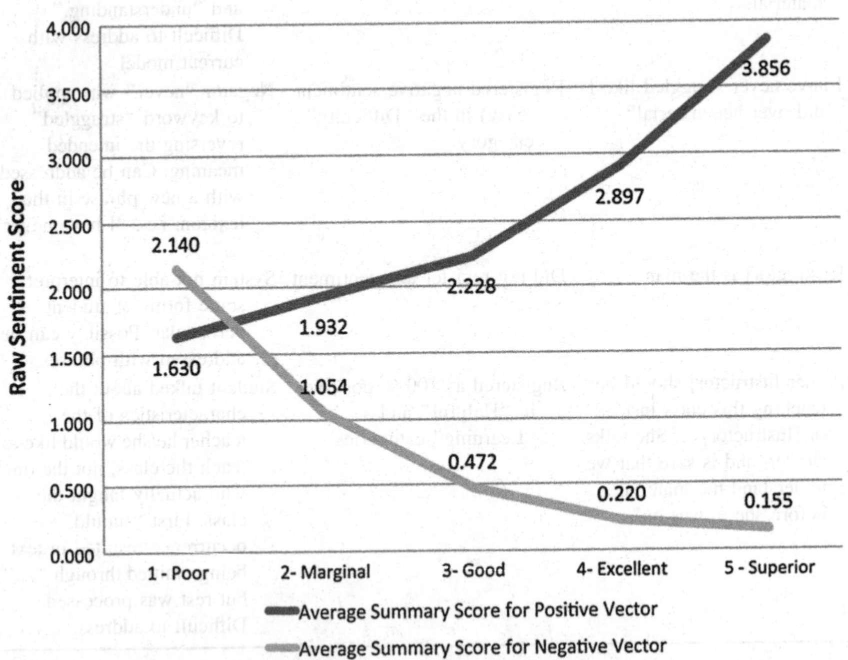

Fig. 14.4 Average raw summary sentiment scores for positive and negative category vectors derived from NAR1 and NAR2 combined plotted against responses to Q11, "How would you rate the instructor's overall performance in this course?"

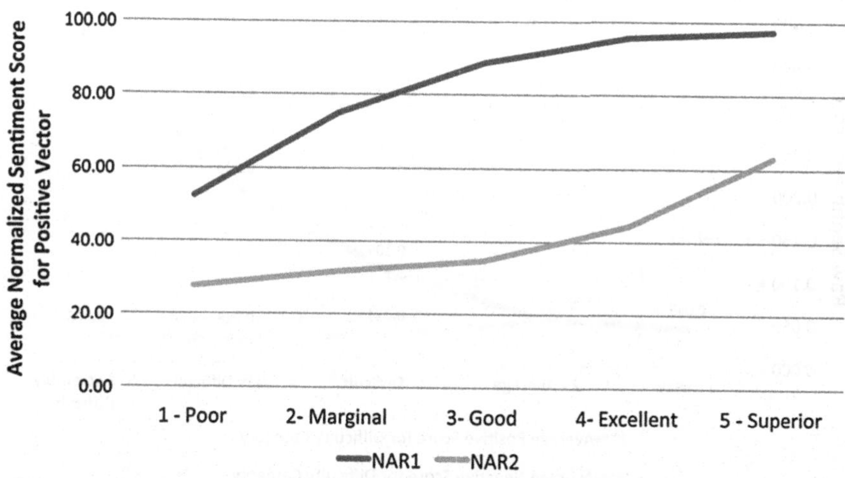

Fig. 14.5 Average normalized sentiment scores for positive category vectors for NAR1 and NAR2 plotted against responses to Q11, "How would you rate the instructor's overall performance in this course?"

and NAR2 plotted against the responses to Q11, further confirming the correlation between sentiment expressed in the comments, as gauged by the algorithm, and responses to Q11.

It is worth noting that in cases where the Q11 responses are very low, the algorithm registers some amount of positive sentiment even for NAR2 when students were asked to name negative characteristics or weaknesses of the course and/or instructor. This is partially due to the "correct versus complete" decision, i.e., phrases in the narrative responses deemed too complicated for the algorithm are simply omitted. Other studies [28] have reported similar results, i.e., students are not overly negative when entering comments online.

Figure 14.6 shows the average positive and negative sentiment scores for the "Difficulty" category for NAR1 and NAR2 combined plotted against the responses to Q10. The results match the expectation in that the more positive the response to Q10, i.e., the course is very or extremely difficult, the higher the average of the positive sentiment scores for the "Difficulty" category.

Likewise, the more negative the response to Q10, i.e., the course is easy, the lower the average of the negative sentiment scores for the Difficulty category. Figure 14.7 shows the normalized summary sentiment scores for the positive category vectors for NAR1 and NAR2, together with the average normalized values for Q11, all plotted against the responses to Q10. (The Q11 responses that ordinarily range from 1 to 5 have been adjusted for a 100-point scale.) The three sets of values exhibit virtually identical trends with the peak positive sentiment occurring for courses perceived as having "Average" difficulty but only small amounts of variation overall.

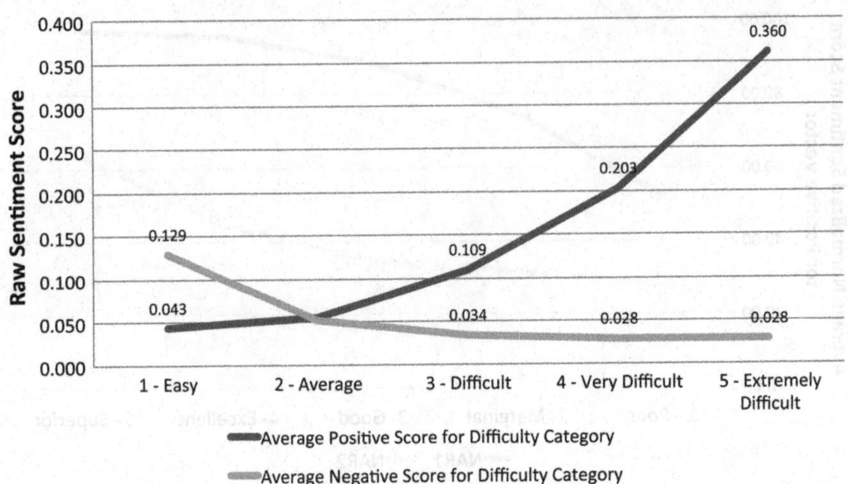

Fig. 14.6 Average raw sentiment scores for the "Difficulty" category derived from NAR1 and NAR2 combined plotted against responses to Q10, "How would you rate the difficulty level of this course, compared to other courses you have take so far at Ole Miss?"

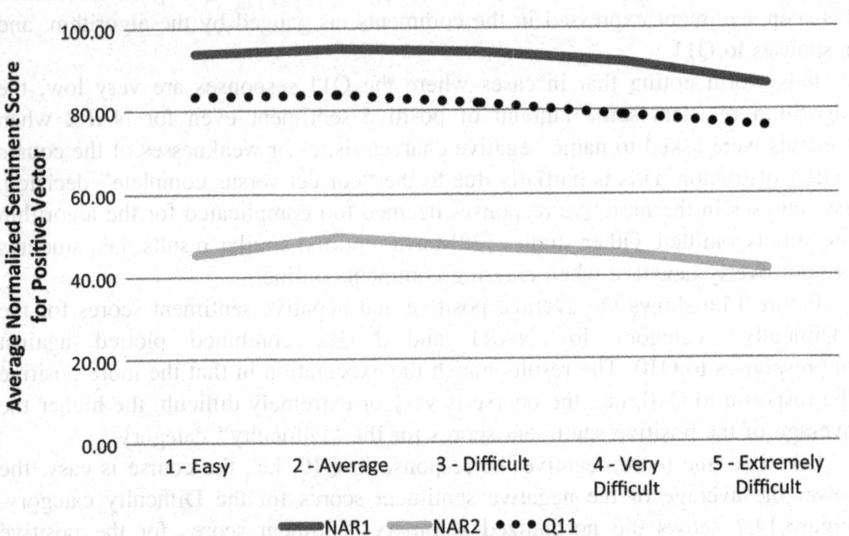

Fig. 14.7 Average normalized sentiment scores for positive category vectors for NAR1 and NAR2, together with average normalized values for Q11, plotted against responses to Q10, "How would you rate the difficulty level of this course, compared to other courses you have take so far at Ole Miss?"

14.4.3 Quantitative Assessment Through the Comparison of Category and Summary Scores for Teaching Award Winners with All Instructors

In the third assessment method, the evaluations for instructors in the data set who had been named as recipients of the Elsie M. Hood Outstanding Teacher Award [29] were partitioned into a group that could be compared with all evaluations. This set consisted of 17 instructors with a total of 16,366 evaluations, including 6,585 NAR1 responses and 5,369 NAR2 responses. The average raw summary sentiment score of the positive vector for Hood award winners was 3.311 (84.94 % normalized) versus 2.660 (77.61 % normalized) for all instructors.

The average raw summary sentiment score of the negative vector for Hood award winners was 0.456 (15.07 % normalized) versus 0.647 (22.39 % normalized) for all instructors. When looking at the raw scores, the Hood award winners outperformed their peers in all categories except for positive sentiment expressed for the "Helpful" category; however, they had less negative sentiment expressed in the "Helpful" category than their peers.

Figures 14.8 and 14.9 show the average normalized positive and negative category sentiment scores for Hood award winners together with that of all instructors. When looking at normalized sentiment scores for both positive and negative vectors, the Hood award winners outperformed their peers by the largest margins in the "General" (8.05 % difference), "Engaging" (5.33 % difference), "Organized" (1.41 % difference), and "Clear" (1.32 %) categories. Courses taught by Hood award winners were perceived as more difficult than all courses.

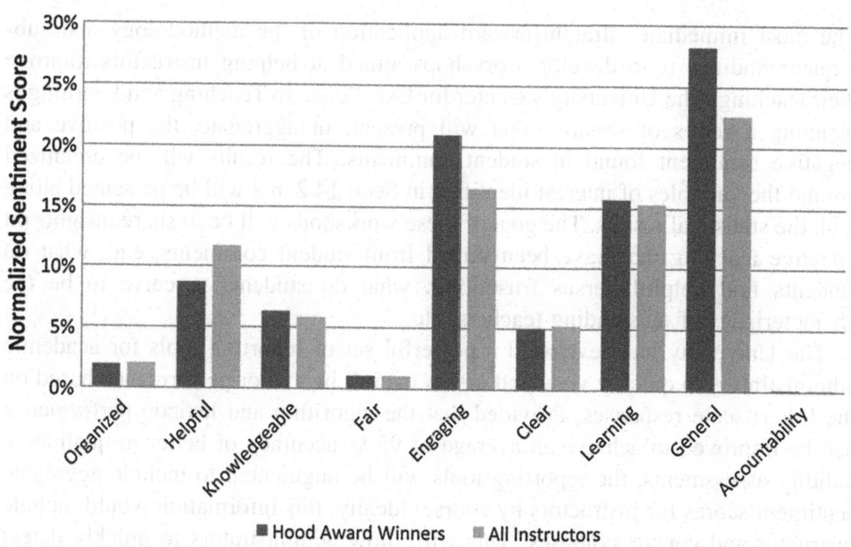

Fig. 14.8 Normalized positive category sentiment scores for Hood award winners versus all

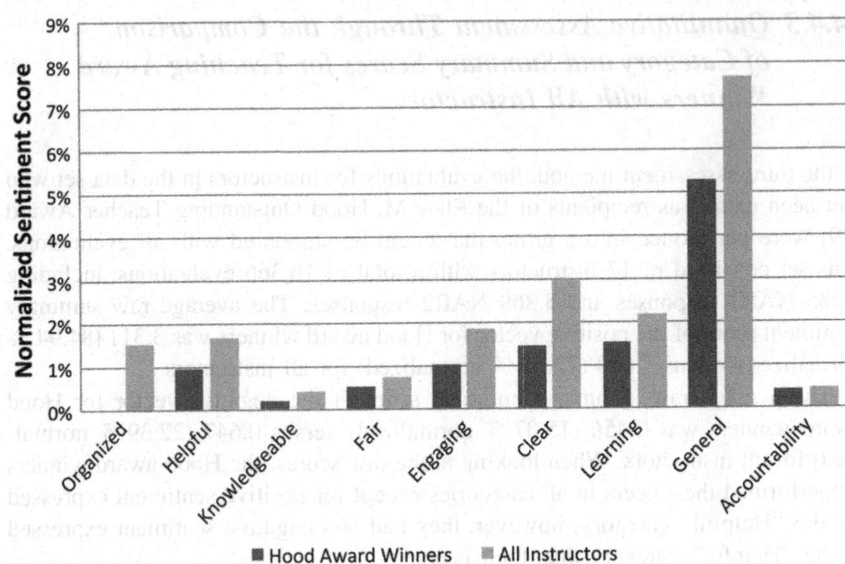

Fig. 14.9 Normalized negative category sentiment scores for Hood award winners versus all

14.5 Applications of the Methodology

14.5.1 Evaluation of Instruction at the University of Mississippi

The most immediate, straightforward application of the methodology and subsequent findings is to develop workshops aimed at helping instructors improve their teaching. The University's Center for Excellence in Teaching and Learning is planning a series of sessions that will present, in aggregate, the positive and negative sentiment found in student comments. The results will be organized around the variables of interest identified in Sect. 14.2 and will be presented along with the statistical results. The goal of these workshops will be to share insights on effective teaching that have been mined from student comments, e.g., what do students find helpful versus frustrating, what do students perceive to be the characteristics of outstanding teachers, etc.

The University has developed a powerful set of reporting tools for academic administrators to quickly view and assess overall instructor performance based on the Likert-scale responses. Provided that the algorithm and lexicon performance can be improved to achieve an average of 95 % accuracy or better in qualitative validity assessments, the reporting tools will be augmented to include aggregate sentiment scores for instructors by course. Ideally, this information would include instructor and course synopses. This will allow administrators to quickly detect very high performing and very low performing instructors based on student comments like they can currently do with Q11.

One academic department chair has asked for a summary of student comments about instructors that he can view when completing faculty performance evaluations. That is, it is not practical for him to drill down and read every comment for every instructor in his department. He would like to see a summary organized by instructor with links to the most representative comments. Prior to this project, we considered this request to be difficult and perhaps impossible. The category vectors, aggregated by instructor and course, should satisfy this need nicely.

As described in Sect. 14.2, one of the most surprising findings from the statistical analysis of the corpus was that students write extensively—about three times the average length—when they rate instructors very poorly. The concern is that students may be sharing actionable information, but, due to the volume of data, those in a position to act are not listening. We plan to address this with a new report that generates administrative alerts for comments with highly negative scores in the "Accountability" category as well as highly negative summary sentiment scores. Recall that the "Accountability" category is designed to help identify cases of high institutional risk, e.g., harassment, bigotry, negligence, and so on.

As the performance of the algorithm is improved, we may consider making aggregate results available to students to increase awareness about the nature of certain courses, including the requirements for success, as has been done in recent course recommendation systems [18]. From a retention perspective, it is important for students to have a good understanding about the commitment required to succeed. For example, by showing students the overall "Difficulty" rating, as expressed by other students, they can be more informed and perhaps avoid the risk of failure, e.g., enrolling in too many courses at one time. Also, some students may respond better to certain instructor profiles such as "Helpful" or "Clear" than others. By giving students access to trustworthy information about the course and instructor, they will be in a stronger position to direct their own academic futures.

Lastly, we plan to use the methodology and results to identify courses that regularly register sentiment falling below acceptable thresholds. The intent is to determine what students find unsatisfactory and feed this back into the course planning process to make improvements. We anticipate that many other ideas for applications of the methodology will emerge from our internal eco-system of instructors, students, support personnel, and administrators.

Our plan is to handle these as we would with other projects, evaluating proposed applications for suitability in our environment, balancing overall benefit to the University with fairness, privacy, and other factors.

14.5.2 Other Educational Applications

Just as there are numerous applications of the methodology within the University specifically designed to enhance teaching and learning, there are numerous other educational applications. The following list highlights some of the ideas that we are considering:

- Some of our outreach programs, e.g., enrichment programs to increase minority enrollment in STEM fields, include components where students blog [30] about their academic experiences. The methodology could be used to summarize student perceptions for the purposes of improving the program and recruiting students.
- Many online learning environments, including Massively Open Online Courses (MOOCs), employ discussion groups to help learners connect with their peers. These discussion groups can get unwieldy very quickly, making it difficult to get to the information being sought. This is especially true for MOOCs where thousands of students may be enrolled simultaneously. The methodology could be used to summarize and organize the free-form responses in an orderly way for more productive information retrieval.
- Many applications to academic programs, scholarships, special cohorts, and residential colleges require essays or writing samples. We envision that the methodology could be used to categorize submissions by theme and determine frequency of comments related to issues of interest, such as civility on campus or student debt. The results of this data mining might then inform actions by the University.
- Students, as well as other campus constituencies, are active on social media, generating huge amounts of freely accessible, unstructured content with each passing minute. It may be important to monitor what they are saying from an institutional reputation and risk management perspective. Paltoglou and Thelwall [31] motivate their sentiment analysis work using information retrieval methods in this very domain. The volume of data makes manual monitoring impractical. The methodology could be used to gauge sentiment and assess risk in social media in an automated manner similar to what has been done here with student comments.
- Most universities conduct annual performance evaluations of faculty and staff, and these evaluations include narrative comments from supervisors as well as responses from employees. Another possible application of the methodology is to mine these text comments for use in improving employee-supervisor relationships and detecting overall patterns. As with many of the previously mentioned application areas, care must be taken to maintain confidentiality.
- In some settings, evaluations have been performed in real time using SMS texts or tweets [17]. This is becoming popular in conferences where immediate attention is needed, e.g., to address audiovisual issues. The methodology could be used to quickly process these streams of information and issue alerts and summary reports.

The methodology, with its straightforward steps of corpus analysis, category selection, lexicon generation, and continuous application and refinement, can be applied to virtually any domain. In this study, we focused on student comments about teaching, but the process can be applied to any scenario that involves unstructured or semi-structured feedback and the analysis of user experiences.

14.6 Future Work

The most immediate next step for this project will be to continue to refine the lexicon and algorithm so that the system is ready for use in a production environment. Because the application areas have the potential to influence institutional decisions and reflect on individual performance, it is critical that the system be as accurate as possible. The first priority is to accurately gauge overall sentiment, and the second priority is to accurately classify comments by category. A supervised learning method such as those used in [32, 33] may be helpful. As part of this work, we will set up a process to add new comments to the existing corpus as evaluations are collected for current and future terms. We also plan to incorporate comparative analysis [12] and other more complicated sentence structures.

Before we began the text-mining phase of this project, we performed extensive statistical analysis of the Likert-scale questions and comments. As described in Sect. 14.2, we identified variables of interest such as section size, course level, grade in class, GPA, student gender, instructor age, instructor rank, and more. We plan to augment the previous work with a similar analysis using the category vectors. For example, we would like to see whether and how category vectors differ when aggregated by academic area of the student and course.

Our expectation is that there will be major differences as we found with the statistical analysis. Likewise, we would like to perform a thorough analysis of sentiment scores relative to overall GPA and grade in class as others have done [34], especially given the differences that we observed in the statistical analysis.

In the statistical analysis, we found that students wrote longer comments during the normal window than in the "last chance" window when they were eager to view their grades. As is shown in Fig. 14.10, stronger sentiment—both positive and negative—was registered in the normal window than in the "last chance" window. However, the normalized values were comparable. The normalized positive summary sentiment score for NAR1 in the normal window was 92.43 versus 93.21 % in the "last chance" window. The normalized positive summary sentiment score for NAR2 in the normal window was 50.10 versus 47.37 % in the "last chance" window. Although stronger sentiment was registered in the normal window, the balance between positive and negative sentiment did not vary much by window.

This work focused on two Likert-scale questions only, Q10 and Q11. We plan to extend the analysis by looking at other Likert-scale questions such as those listed in Sect. 14.2. We started with Q10 and Q11 for two reasons: (1) they are broad in scope; and (2) they are used consistently in all evaluation types. The evaluations include questions that rotate in by semester as well as questions appropriate for specific academic areas and course formats, e.g., online courses,

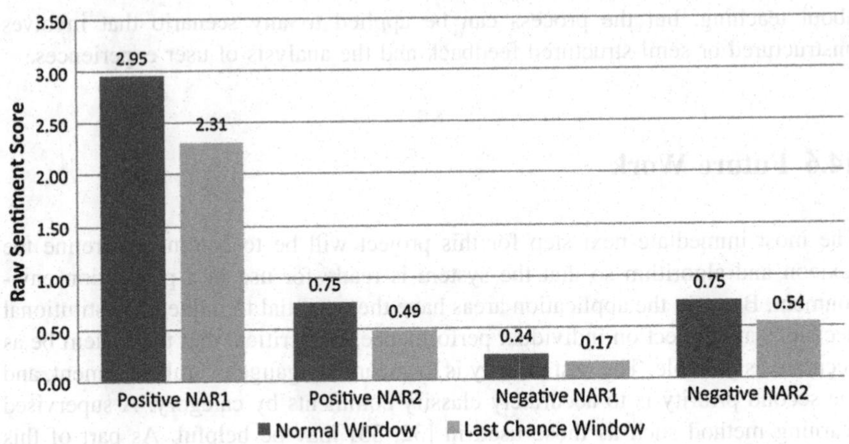

Fig. 14.10 Average raw summary sentiment score by submission window

labs, and Law courses. Including additional questions increases complexity but should yield very interesting, useful results. Once the lexicon and algorithm are in production, we plan to extend the statistical analysis to include the text mining results and add in more Likert-scale questions.

Up until this point, all code has been written in Java and the work has been performed using a relational database without the benefit of sophisticated front-end reporting tools. We imagine that other visualizations of the data, e.g., tag clouds and dashboards, would make analysis by end users more meaningful and would help with trend reporting such as faculty performance over time.

Ideally, we would have a tool that academic administrators and learning specialists could use directly to perform queries and manipulate the data. The system currently hosting the database is a virtualized Solaris server running on an Oracle T series processor architecture. Using this platform, we are able to process about 600,000 comments in 4 hours. As funds permit, we would like to move the database to a system designed for handling big data such as SAP's HANA [35] in-memory platform.

We are encouraged by the progress that we were able to make in this first effort to mine student comments. Using relatively simple methods, we were able to achieve 88.41 % accuracy in our qualitative validity assessment. The quantitative assessment results further validated the effectiveness of the methodology. For several decades, the analysis of student comments has been mostly limited to instructors and, in some cases, supervisors reading individual responses. We are optimistic about future directions for this work, specifically the opportunities for gaining insights and extracting actionable information to improve teaching and learning in a much more strategic manner. Moreover, the methodology resulting from this work offers great promise for mining voluminous, unstructured and semi-structured data sets in educational domains beyond teaching and learning.

Acknowledgments This research was made possible through the support of several University of Mississippi units: The Office of the Provost; The Center for Excellence in Teaching and Learning; and The Office of Information Technology. The authors wish to thank Sarah Hill for her help in coordinating the qualitative assessment process.

References

1. Feldman, K.A.: Effective college teaching from the students' and faculty's view: matched and mismatched priorities. Res. High. Educ. **28**(4), 291–329 (1988)
2. Marsh, H.W., Bailey, M.: Multidimensional students' evaluations of teaching effectiveness. J. Higher Educ. **64**(1), 1–18 (1993)
3. Cashin, W.E.: Student ratings of teaching: the research revisited. IDEA paper No. 32. center for faculty evaluation and development, Kansas State University, Manhattan, 1995
4. Rindermann, H.: Generalizability of multidimensional student ratings of university instruction across courses and teachers. Res. High. Educ. **42**(4), 377–399 (2001)
5. Abrami, P.C., d'Apollonia, S., Rosenfield, S.: The dimensionality of student ratings of instruction: what we know and what we do not. In: Perry, R.P., Smart, J.C. (eds.) The Scholarship of Teaching and Learning in Higher Education: An Evidence-Based Perspective, pp. 385–456. Springer, The Netherlands (2007)
6. Jin, W., Ho, H., Srihar, R.: OpinionMiner: a novel machine learning system for web opinion mining and extraction. In: Proceedings of the 15th ACM SIGKDD International Conference on Knowledge Discovery and Data Mining, pp. 1195–1204. ACM, New York (2009)
7. Kim, S.M., Calvo, R.A.: Sentiment analysis in student experiences of learning. In: Baker, R.S.J.D., Merceron, A., Pavlik Jr., P.I. (eds.) Proceedings of 3rd International Conference on Educational Data Mining, pp. 111–120. International Educational Data Mining Society, Pittsburgh (2010)
8. Cashin, W.E.: Student ratings of teaching: recommendations for use. IDEA paper No. 22. Technical report, Center for Faculty Evaluation and Development, Kansas State University, Manhattan, 1990
9. Stowell, J., Addison, W., Smith, J.: Comparison of online and classroom-based student evaluations of instruction. Assess. Eval. High. Educ. **37**(4), 465–473 (2012)
10. Avancini, H., Lavelli, A., Magnini, B., Sebastiani, F., Zonoli, R.: Expanding domain-specific lexicons by term categorization. In: ACM Symposium on Applied Computing, pp. 793–797. ACM, New York (2003)
11. Sahlgren, M., Coster, R.: Using bag-of-concepts to improve the performance of support vector machines in text categorization. In: Proceedings of 20th International Conference on Computational Linguistics, Geneva, pp. 487–493. Association for Computational Linguistics, Stroudsburg (2004)
12. Liu, B.: Sentiment analysis and subjectivity. In: Indurkhya, N., Damerau, F.J. (eds.) Handbook of Natural Language Processing, pp. 627–666. CRC Press, Boca Raton (2010). (Chapman and Hall/CRC Machine Learning and Pattern Recognition series)
13. Liu, B. (ed.): Web Data Mining: Exploring Hyperlinks, Contents, and Usage Data, 2nd edn. Springer, Heidelberg (2011)
14. Simmons, L., Conlon, S., Mukhopadhyay, S., Yang, J.A.: Computer aided content analysis of online reviews. J. Comput. Inf. Syst. **52**(1), 43–55 (2011)
15. Taboada, M., Brooke, J., Tofiloski, M., Voll, K., Stede, M.: Lexicon-based methods for sentiment analysis. Comput. Linguist. **37**(2), 267–307 (2011)
16. Neviarouskaya, A., Prendinger, H., Ishizuka, M.: SentiFul: a lexicon for sentiment analysis. IEEE Trans. Affect. Comput. **2**(1), 22–36 (2011)
17. Leong, C.K., Lee, Y.H., Mak, W.K.: Mining sentiments in SMS texts for teaching evaluation. Expert Syst. Appl. **39**(3), 2584–2589 (2012)

18. Vialardi, C., Bravo, J., Shafti, L., Ortigosa, A.: Recommendation in higher education using data mining techniques. In: Barnes, T., Desmarais, M., Romero, R., Ventura, S. (eds.) Proceedings of 2nd International Conference on Educational Data Mining, pp. 190–199. International Educational Data Mining Society, Cordoba (2009)
19. Marsh, H.W.: SEEQ: a reliable, valid, and useful instrument for collecting students' evaluations of university teaching. Br. J. Educ. Psychol. **52**, 77–95 (1982)
20. Hu., M., Liu, B.: Mining and summarizing customer reviews. In: Proceedings of Tenth ACM SIGKDD International Conference on Knowledge Discovery and Data Mining, pp. 168–177. ACM, New York (2004)
21. Luhn, H.P.: Keyword-in-context index for technical literature. Am. Doc. **11**(4), 288–295 (1960)
22. SentiWordNet, http://sentiwordnet.isti.cnr.it/
23. Big Huge Thesaurus, http://words.bighugelabs.com/
24. Dey, L., Haque, M.: Opinion mining from noisy text data. Int. J. Doc. Anal. Recogn. **12**(3), 205–226 (2009)
25. Jia, L., Yu, C., Meng, W.: The effect of negation on sentiment analysis and retrieval effectiveness. In: Proceedings of 18th ACM International Conference on Information and Knowledge Management, pp. 1827–1830. ACM, New York (2009)
26. Qiu, G., Liu, B., Bu, J., Chen, C.: Opinion word expansion and target extraction through double propagation. Comput. Linguist. **37**(1), 9–27 (2011)
27. LingPipe, http://alias-i.com/lingpipe/
28. Hardy, N.: Online ratings: fact or fiction. New Dir. Teach. Learn. **2003**(96), 31–38 (2003)
29. Elsie, M.: Hood outstanding teaching award at the University of Mississippi, http://www.olemiss.edu/news/hood_award/
30. Melville, P., Gryc, W., Lawrence, R.: Sentiment analysis of blogs by combining lexical knowledge with text classification. In: Proceedings of 15th ACM SIGKDD International Conference on Knowledge Discovery and Data Mining, pp. 1275–1284. (2009)
31. Paltoglou, G., Thelwall, M.: A study of information retrieval weighting schemes for sentiment analysis. In: Proceedings of 48th Annual Meeting of the Association for Computational Linguistics, pp. 1386–1395. Association for Computational Linguistics, Stroudsburg (2010)
32. Agarwal, B., Mittal, N.: Optimal feature selection for sentiment analysis. In: Gelbukh, A. (ed.) Computational Linguistics and Intelligent Text Processing, vol. 7817, pp. 13–24. LNCSSpringer, Heidelberg (2013)
33. Elberrichi, Z., Rahmoun, A., Bentaalah, M.: Using WordNet for text categorization. Int. Arab. J. Inf. Technol. **5**(1), 16–24 (2008)
34. Campagni, R., Merlini, D., Sprugnoli, R., Verri, M. C.: Comparing examination results and course evaluation: a data mining approach. In: Didamatica 2013, pp. 883–892. Scuola Superiore Sant'Anna, Pisa (2013)
35. Sap Hana, http://www.saphana.com/

Chapter 15
Data Mining and Social Network Analysis in the Educational Field: An Application for Non-Expert Users

Diego García-Saiz, Camilo Palazuelos and Marta Zorrilla

Abstract With the increasing popularity of social networking services like Facebook, social network analysis (SNA) has emerged again. Undoubtedly, there is an inherent social network in any learning context, where teachers, learners, and learning resources behave as main actors, among which different relationships can be defined, e.g., "participate in" among blogs, students, and learners. From their analysis, information about group cohesion, participation in activities, and connections among subjects can be obtained. At the same time, it is well-known the need of tools that help instructors, in particular those involved in distance education, to discover their students' behavior profile, models about how they participate in collaborative activities or likely the most important, to know the performance and dropout pattern with the aim of improving the teaching–learning process. Therefore, the goal of this chapter is to describe our E-learning Web Mining tool and the new services that it provides, supported by the use of SNA and classification techniques.

Keywords Data mining · Educational data mining · Social network analysis · Learning analytics

Abbreviations

API	Application programming interface
DM	Data mining
EDM	Educational data mining

D. García-Saiz · C. Palazuelos · M. Zorrilla (✉)
Department of Mathematics, Statistics, and Computer Science, University of Cantabria, Avenida de los Castros s/n 39005 Santander, Spain
e-mail: marta.zorrilla@unican.es

D. García-Saiz
e-mail: diego.garcia@unican.es

C. Palazuelos
e-mail: camilo.palazuelos@unican.es

A. Peña-Ayala (ed.), *Educational Data Mining*,
Studies in Computational Intelligence 524, DOI: 10.1007/978-3-319-02738-8_15,

ElWM	E-learning web miner
KDD	Knowledge discovery in databases
LA	Learning analytics
LMS	Learning management system
MOOC	Massive open online course
SNA	Social network analysis
SOA	Service-oriented architecture
SOAP	Simple object access protocol
UC	University of Cantabria
WSDL	Web services description language
WS	Web service
XML	eXtended Markup Language

15.1 Introduction

Since the late 1990s, the use of computer-based technologies has drastically changed learning and teaching processes in all academic levels, from elementary school to university. Nowadays, it is very frequent that teachers include in their subjects activities that require the use of Web 2.0 technologies in order to develop contents and social and communication skills.

Collaborative activities, e.g., content search [1, 2], collaborative writing [3], and discussion forums [4], appear in many curricula independently of the educational field and level of the studies. Other tools frequently used, regardless of whether teaching is face-to-face or virtual, are the learning management systems (LMS), e.g., Moodle [5], Blackboard [6], or Shakai [7], which offer different modules, e.g., blogs, wikis, or forums, to develop collaborative activities that enable students to adapt to new environments and work in heterogeneous teams. This new scenario, where the degree of interaction among different actors, e.g., learners, educators, and resources, is very high, poses new situations and needs to instructors.

They need to know the students' level of cohesion, their degree of participation in forums, the identification of the most influential ones, which students help their classmates, and so on. This information might be helpful for teachers to organize team-works with different social profiles, grade the activities performed by their students according to their contribution, or spread news or relevant explanations through the most influential students. Especially, the analysis of social interaction might help teachers to better understand their students' social behavior and, as a consequence, assist them to improve their skills, as well as their results in the subjects involved.

Hence it is necessary to develop applications that help teachers to extract and analyze interaction data produced in the different teaching activities and their impact on student performance. This application must fulfill some requirements with the aim of being useful for non-expert users in the learning analytics field.

The 2013 Horizon Report [8] describes learning analytics (LA) as the "Field associated with deciphering trends and patterns from educational big data, or huge sets of student-related data, to further the advancement of a personalized, supportive system of higher education." This is a very wide field in which different techniques and tools are used by educators for gaining insights into student interaction with online texts and courseware and, consequently, being able to take actions to improve the teaching process.

This field comprises, among others, techniques from the educational data mining (EDM) field, which deals with the development and application of computational methods to detect patterns in large collections of educational data. Its main goal is to better understand how students learn and identify the settings in which they learn to improve educational outcomes and gain insights into and explain educational phenomena [9]. Techniques from the SNA field are also employed in the academic context since the analysis of the structure and composition of relationships in the network provides useful information on the cohesion of individuals, their level of participation, or which individuals are the most active or influential.

Regrettably, both research fields use techniques and algorithms that make them unsuitable for people outside the fields of mathematics or computer science. Therefore, these algorithms and the associated processes for their execution must be wrapped in such a way that the end users should only worry about interpreting the results.

Furthermore, another key feature of this analytic tool is that it is independent from specific web applications or any other resource available that can be used in a collaborative activity. In such a way, it can be easily extended and enhanced as well as be used in different scenarios.

For example, LMS or massively open online course (MOOC) platforms are tools in which its inclusion would be very valuable since educators can design all the teaching process inside them. Furthermore, these platforms generally collect users' interaction—when students connect, how often, or when they write a post or perform a test—in databases that make it easier its utilization for the analysis processes. Although these platforms offer some monitoring tools, these are limited enough and, as stated by Macfadyen et al. [10], instructors in the new world of education are in need of new tools and strategies that will allow them to quickly identify students at risk and devise ways of supporting their learning.

In order to contribute to fill this gap, this chapter describes our enhanced version of E-learning Web Miner (ElWM) [11] and, more specifically, the new options that allow educators to gain insights into interaction, social behavior, and performance. Among other services, ElWM offers the generation of predictive models of students' performance based on classification techniques (in particular, decision trees), descriptive models for the characterization of learners from a social perspective based on SNA, graphs showing students' cohesion and those ones who are the most influential by using graph mining techniques, specifically FRINGE [12]. Likewise, we show its usefulness and simplicity through the analysis of different virtual courses and collaborative activities developed in both Moodle and Blackboard platforms at the University of Cantabria (UC).

This chapter is organized as follows. In Sect. 15.2, we provide a brief introduction to the context of our work, the theoretical foundation to understand our work and relate works published in the field of SNA applied to the educational context. We also cite the most relevant works focused on the generation of models of students' performance and dropout, and briefly describe other tools similar to our ElWM and discuss the main differences found. Section 15.3 describes the purpose of our tool, its architecture and the new services provided. Section 15.4 presents and discusses several case studies with the aim of showing the usefulness, simplicity, and added value that ElWM provides to the educational context. Finally, we summarize the contents of this chapter and discuss our future work.

15.2 Background and Related Work

EDM is an interdisciplinary area in which methods and techniques from computer science, education, and statistics are combined. LA is another field very related to EDM and with which it shares some goals and interests. However, LA integrates a broader array of academic disciplines, e.g., computer science, information science, learning sciences, psychology, sociology, and statistics.

Although there is no hard and fast distinction between these two fields [13], EDM focuses on developing methods and applying techniques from statistics, machine learning, and data mining (DM) to analyze data collected during teaching and learning with the aim of answering questions related to the educational practice, e.g., "What sequence of topics is the most effective for a specific student?" "What student actions are associated with more learning?" "What features of an online learning environment lead to better learning?" or "What will predict student success?" To accomplish its goals, EDM mainly uses methods based on prediction, classification, clustering, and association.

On the other hand, LA emphasizes measurement and data collection as activities that institutions need to undertake and understand, and focuses on the analysis and reporting of data. That means that LA, unlike EDM, does not develop new algorithms for data analysis but addresses the application of known methods and models in order to answer important questions that affect student learning and organizational learning systems.

Hence LA does not only focus on student performance, but it is also used to assess curricula, programs, and institutions. LA uses techniques related to concept, discourse, influence, and sentiment analyses, as well as sense-making models, SNA, statistics, and visualization [13].

Thus, our work fits in both fields since our tool uses exploratory and analytical techniques, as well as provides patterns generated with DM techniques. Specifically, we show the utility of SNA and its contribution to predict students' performance.

In this section, we provide some background on classification and SNA techniques and relate some of the most important works that apply DM techniques to discover prediction models of students' performance and dropout and SNA techniques to understand the interactions of students in e-learning courses. Furthermore, we include a section about tools developed for these purposes.

15.2.1 Social Network Analysis

SNA is the methodical study of the relationships present in connected actors from a social point of view. SNA represents both actors and relationships in terms of network theory, depicting them as a graph or network, where each node corresponds to an individual actor within the network, e.g., a person or an organization, and each link symbolizes some form of social interaction between two of those actors, e.g., friendship or kinship.

Although social networks have been studied for decades [14, 15], the recent emergence of social networking services like Facebook or Twitter has been the cause of the unprecedented popularity that this field of study has now. Since then, an extraordinary variety of SNA techniques has been developed, allowing researchers to model different types of interactions, e.g., movie actors [16] or sexual contact networks [17], and giving solution to very diverse problems, e.g., detection of criminal and terrorist patterns [18] or identification of important actors in social networks [19].

In order to estimate the prominence of a node in a social network, many centrality measures have been proposed. The research devoted to the concept of centrality addresses the question "Which are the most important nodes in a social network?" Although there are many possible definitions of importance, prominent nodes are supposed to be those that are extensively connected to other nodes. Generally, in social networks, people with extensive contacts are considered more influential than those with comparatively fewer contacts. Perhaps, the most simple centrality measure is the *degree* of a node, which is the number of links connected to it, without taking the direction of the links into consideration. If we consider that direction of the links, a node has both *indegree* and *outdegree*, which are the number of incoming and outgoing links attached to it, respectively. There are more complex centrality measures, such as the *betweenness* [20] of a node, which is equal to the number of shortest paths from all nodes to all others that pass through such a node.

Mathematically, let g_n^{pq} [1] be 1 if node n lies on the shortest path from p to q and 0 otherwise. Then the betweenness centrality of a node is given by (15.1).

$$b_n = \sum_{pq} g_n^{pq} \qquad (15.1)$$

Authorities and *hubs* [20] are also two examples of more complex centrality measures; a node is an authority if its incoming links connect it to nodes that have a large number of outgoing links, whereas a node is a hub if its outgoing links connect it to nodes that have a large number of incoming links. Mathematically [see Eq. (15.2)], the authority centrality of a node is defined to be proportional to the sum of the hub centralities of the nodes that point to it, where α is a constant and A_{nm} is an element of the adjacency matrix of the network.

$$a_n = \alpha \sum_m A_{nm} h_n \tag{15.2}$$

Similarly, the hub centrality of a node [see Eq. (15.3)] is proportional to the sum of the authority centralities of the nodes it points to, where β is another constant.

$$h_n = \beta \sum_m A_{mn} a_m \tag{15.3}$$

From a node-level point of view, centrality measures constitute a very useful tool for the inference of the importance of nodes within a network. Due to their own nature, some of them, e.g., betweenness, cannot be trivially calculated, so that network-level metrics—which can be computed more easily and provide helpful information by considering the network as a whole—can be used for complementing the aforementioned centrality measures.

One of these network-level metrics is the density of the network, which measures the number of links within the network compared to the maximum possible number of links. The diameter of the network is also a useful network-level metric; it is defined as the largest number of nodes that must be traversed in order to travel from one node to another. Other meaningful network-level metric is the number of connected components of the network, i.e., the number of subnetworks in which any two nodes are connected to each other by paths without taking the direction of their links into consideration. Finally, the last metric to be mentioned is reciprocity; it occurs when the existence of a link from one node to another triggers the creation of the reverse link.

SNA techniques do not just concentrate on social networks, but also focus on other fields, such as marketing (customer and supplier networks) or public safety. Another field of application is education, although it is not deeply explored yet. There are some case studies in the literature, for instance: Brewe et al. [21] used a multiple regression analysis of the Bonacich centrality to evaluate the factors that influence participation in learning communities, e.g., students' age or gender. Crespo and Antunes [22] proposed a strategy to quantify the global contribution of each student in a team-work through adaptations of the PageRank algorithm.

Cuéllar et al. [23] proposed a method for the formulation and interpretation of learning management platforms as social networks with the aim of making further studies about the social structure among learners, teachers, and learning resources, and discovering useful relationships to improve the learning process. Rabbany et al. [24] built Meerkat-ED, a tool designed to assess the students' participation in asynchronous discussion forums of online courses.

SNA has also been used for extracting relevant information that can be used in EDM tasks. For example, Dawson et al. used SNA to monitor the learners' creative capacity [25], to detect and encourage students at risk [26]. Obsivac et al. [27] used several centrality measures for the prediction of dropouts. A similar work was performed by Palazuelos et al. [28] but, in this case, the goal was to evaluate whether SNA attributes are useful to build more accurate classification models.

15.2.2 Classification Applied to the Educational Context: Students' Performance and Dropout

DM is the process of automatically discovering useful information in large data repositories. DM is an integral part of Knowledge Discovery in Databases (KDD) [29] which is the overall process of converting raw data into useful information. This process comprises several steps: data selection and preparation, technique selection, testing and result evaluation. This process is not trivial; whenever a very accurate model is needed, one must turn to an expert in DM. Our tool will not achieve the most accurate model, but a reasonably good one at a very low cost that allows non-expert users to take advantage of using data mining towards its democratization.

DM techniques are generally divided into two major categories: predictive and descriptive tasks, being the former the one that arouses most interest. The prediction task consists of extracting relevant features from labeled training data to build a model that discriminates between classes to classify unlabeled observed objects. Prediction methods are, in turn, divided into classification and regression techniques. The former are used when the predicted variable is a categorical value and the latter, when the predicted variable is a continuous value or a probability density function. In EDM, classification techniques are more used, particularly, to forecast student performance [30–33] and detect anomalous students' behaviors, such as dropout [27, 34, 35].

Classification is the task of learning a target function f that maps each attribute set x to one of the predefined class labels y. There are different techniques to identify the model that best fits the relationship between the attribute set and the class label of the input data. Some examples are: decision tree classifiers, rule-based classifiers, neural networks, support vector machines, and naïve Bayes classifiers. The decision of which one to use depends on the DM goals that decision makers pursue. In the EDM field, we find some works in the scientific literature in which the authors try to establish which algorithms are the best to predict the students' performance and dropout.

Dekker et al. [34] presented a case study to predict student dropout in which demonstrated the effectiveness of several classification techniques and the cost-sensitive learning approach on several datasets with about 500 instances with numerical and nominal attributes that corresponded to pre-university and university characteristics.

Their experimental results showed that rather simple and intuitive classifiers, referring to decision trees, give a useful result with accuracies between 75 and 80 %. Kotsiantis et al. [35] also compared six classification algorithms to predict dropouts. Their comparison showed the Naïve Bayes algorithm [36] was the most appropriate.

Regarding predicting performance, a similar study was performed by Hämäläinen et al. [37]. In this paper, the authors compared five classification methods: multiple linear regression and support vector machine techniques to predict the numerical mark and three variations of the naïve Bayes classifier to predict the categorical qualification (pass/fail), concluding that all methods achieve about the same accuracy. Zafra et al. [33] compared several algorithms based on multi-instance learning to predict student's performance using data from a Moodle platform and the method that they proposed, G3P-MI [38], got the best results.

Finally, two of the authors of this work [38] compared five classification algorithms in order to determine which was the most suitable for educational datasets from e-learning platforms. Their experimentation concluded that there was no algorithm that achieved significantly better accuracy. When the dataset was very small (less than 100 instances) and had numeric attributes, naïve Bayes performed adequately; on the other hand, when the data set was bigger, BayesNet TAN [39] was a better alternative. However, J48 (implementation of the C4.5 [40] algorithm in Weka [41]) was suitable for datasets with more instances and/or with the presence of nominal attributes, being also the most interpretable. Thus, J48 was the algorithm chosen to be wrapped in ElWM. The process of building the classifier comprises two steps: training and test. When the dataset size is reduced, a good practice is to evaluate the performance of the classifier by means of cross-validation. Finally, the model must be evaluated. This task is based on the number of test records correctly and incorrectly predicted by the model. The most frequently used metrics are *accuracy*, *specificity* and *sensitivity*. Accuracy measures the number of correct predictions made by the model divided by the total of predictions. Sensitivity (true positive rate) measures the proportion of actual positives which are correctly identified as such. And specificity (true negative rate) measures the proportion of negatives that are correctly identified.

15.2.3 Data Mining Tools for Non-expert Users

The EDM field provides a large quantity of techniques and tools to further understand students and the settings which they learn in. In the last decade, many works have been carried out, as can be read in this survey [42]. Regrettably, most of these methods and tools are not directly used by non-experts in data mining, e.g., teachers. To fill this gap, our research group developed ElWM [43].

It aims to help instructors involved in distance education to discover their students' behavior profiles and models about how they navigate and work in their virtual courses. There are a few tools with a similar purpose, e.g., TADA-Ed [44] and

Moodle Data Mining Tool [45]. Both of them provide different techniques like ElWM but instructors must have certain knowledge about data mining concepts in order to use them since they are responsible for doing the phases of selection and pre-processing of attributes and the selection of algorithms and their parameter setting.

There are more specific tools, focused on a kind of problem. For example, García et al. [46] described a collaborative EDM tool based on association rule mining for the ongoing improvement of e-learning courses, allowing teachers with similar course profiles to share and score the discovered information. Kotsiantis [31] developed a decision support tool to predict students' mark from the analysis of a list of attributes defined by the user as a spreadsheet in the CSV (Comma-Separated Value) file format. The tool builds a regressor using the M5-rules algorithm [47] and ranks the influence of each attribute according to a statistical measure named RRELIEF, whose goal is to estimate the quality of attributes according to how well their values distinguish between the instances that are close to each other.

Other tools have been developed and embedded in an LMS. This is the case of the recommender integrated in the adaptive web-based educational system Aha! [48] in order to propose the most appropriate links and Web pages to students. It also includes a web mining tool to help instructors to fulfill the whole web mining process but, unlike our proposal, its use is limited to the interaction produced in the system.

15.3 E-Learning Web Miner

In this section, we describe our web-based tool, its enhanced architecture and the new services that it offers as well as its internal mode of working.

15.3.1 Description of E-Learning Web Miner

ElWM [43] is designed to help instructors to improve the teaching–learning process since it offers models and patterns that allow them to gain insight into the activity performed by their students in their virtual courses, analyze their course design and also discover their students' behavior profile and the kind of sessions they performed. In such a way, they make informed decisions driven by data. Its characteristic is that, despite using DM, end users do not require DM knowledge for its use, since the knowledge discovery process is automated and hidden. In the previous version [11], teachers only had to send a data file according to one of the templates provided by the application and request the results. As teachers considered that the task of preparing the data file was cumbersome, in the current version a new service for Moodle and Blackboard has been built, in such a way that the instructor only has to choose the question and indicate the course under study, as it is stated in Sect. 15.3.2.

Regarding DM techniques, the tool defines a template for each question and chooses the algorithm and its parameter setting according to the data file and kind of problem (association rules, clustering, classification and so on).

The techniques chosen [30, 38, 49–51] are easier to interpret and represent their results since our tool is addressed to non-expert data miners. All that means that we prefer, for instance, yacaree [52] as a rule associator rather than the well-known apriori (for instance, Borgelt implementation [53]) since the number of rules that it offers is more reduced; or kmeans or kmedoids [54] for clustering instead of a hierarchical or density-based method; and decision trees instead of neural networks, vector support machines or genetic techniques because they are more interpretable and more suitable for the size of educational datasets that, most of times, collect categorical attributes. In general, we configure these algorithms with default parameters since our experimentation shows that it is suitable for this kind of datasets.

15.3.2 General View of the E-Learning Web Miner Architecture

Our tool has been designed following a service-oriented architecture (SOA). The term SOA describes a concept to align an enterprise's IT environment with its business processes. This is achieved by providing loosely coupled atomic services that can be flexibly combined with others. A SOA can be implemented with the help of any arbitrary service-based architecture, but web services (WS) are most commonly used. We must say that our tool has been designed as a service that makes use of other services offered by our tool, though in the future, it could be orchestrated with other external services to offer a more powerful functionality. Figure 15.1 depicts this architecture.

As can be observed, we have created a web application with an easy-to-use application programming interface (API) that puts in communication the different WSs implemented. The web application, as previously mentioned, inquires the instructor about what kind of query he or she wants to make, e.g., predicting students' performance according to the global activity carried out by the students in a course or discovering who the leaders in a class are through their interaction in the forum. The API, supported by a set of configuration files where the KDD process is set up, requests its participation to each WS. Next, we describe each WS developed. The WS-ElWM Service exposes the services that can be consumed by a client application or component software that wants to include the functionality of ElWM, for example, a LMS. This service is found, when it is publicly available, or be statically bound, and possibly, choreographed into a composite service. It must be said that ElWM is an adaptation of this architecture to the e-learning context but is easily configurable for other contexts. The service is stated in the web services description language (WSDL). The WSDL file describes four data:

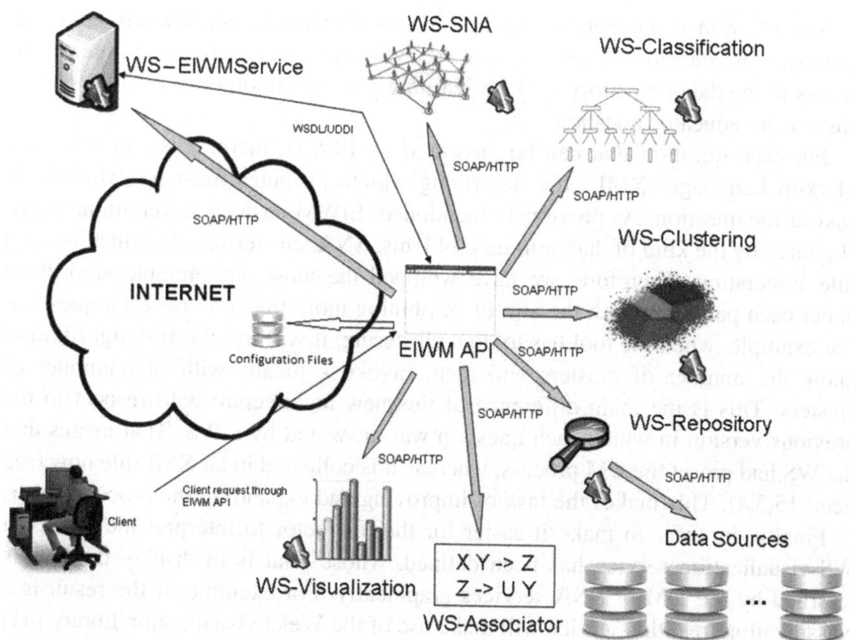

Fig. 15.1 EIWM SOA architecture

(1) interface information describing all publicly available functions (<definitions>); (2) data type information for all message requests and message responses (<types>); (3) binding information about the transport protocol to be used (<binding>), in this case Simple Object Access Protocol (SOAP); (4) address information for locating the specified service (<service>). An example can be seen in the WSDL file example presented as in the next program code.

```
<definitions xmlns:wsu"http://docs.oasis-open.org/...>
        ...
<types>
    <xsd:schema>
        <xsd:import namespace=
        "http://server/" schemaLocation="http://.../WServices?xsd=1"/>
        ...
    </xsd:schema>
</types>
        ...
<binding name="WS_RepositoryPortBinding" type="tns:WS_Repository">
    <service name="WS_Repository">
        ...
```

[WSDL file example]

Since ElWM is independent of the LMS used (Moodle, Blackboard, etc.), it is necessary to develop a service, named WS-Repository, which configures the access to the data repository of the e-learning platform in order to read the data to answer the educator question.

For each question that can be answered by ElWM, there exists an eXtended Markup Language (XML) file describing which attributes must be extracted to answer the question. As previously mentioned, ElWM answers different questions classified by the kind of data mining problems: SNA, clustering, classification, and rule association. Therefore, we have wrapped the most interpretable algorithms under each paradigm with the aim of combining more than one for each question. For example, when the tool has to do a clustering, it will use the EM algorithm to know the number of clusters and then, invoke *k* means with that number of clusters. This is the main difference of this new architecture with respect to the previous version in which each question was answered by a WS. That means that the WS had preset the DM process, whereas it is collected in an XML file now (see Sect. 15.3.4). This makes the task of improving and expanding the process easier.

Finally, in order to make it easier for the instructor to interpret the results, a WS-visualization service has been defined, whose goal is to display the results returned by the DM or SNA services graphically. For example, if the result is a classification tree, this service will make use of the Weka visualization library [41] to represent the tree graphically. Currently, we represent clusters by means of spiders and bar diagrams, and we display association rules using a MATLAB library.

15.3.3 New Services Provided

Initially, the set of models that we proposed and implemented in ElWM used only descriptive techniques, e.g., clustering and association rules, because these easily allow instructors to gain an insight into students' characteristics and depict students' learning patterns [43].

Now, we have added new services focused on classification and SNA techniques. In particular, we include statistical data about the degree of collaboration produced among learners, educators, and resources applying SNA techniques. This offers instructors the possibility of building social networks from the activity performed in forums or blogs to discover which students are more active, how the resources are used, detect isolated learners, and so on. It also provides the possibility of detecting the social communities defined in the virtual course.

Likewise, we offer the generation of models to predict the students' performance and the students' dropout because they are very useful for educational context and are very demanded by educators. ElWM allows educators to choose which type of activity performed in the LMS must be considered to build the model, e.g., getting a model based exclusively on the global activity performed in the virtual course measured by means of the time spent on each tool, the number of sessions performed, the number of messages written the replied in mail or forums,

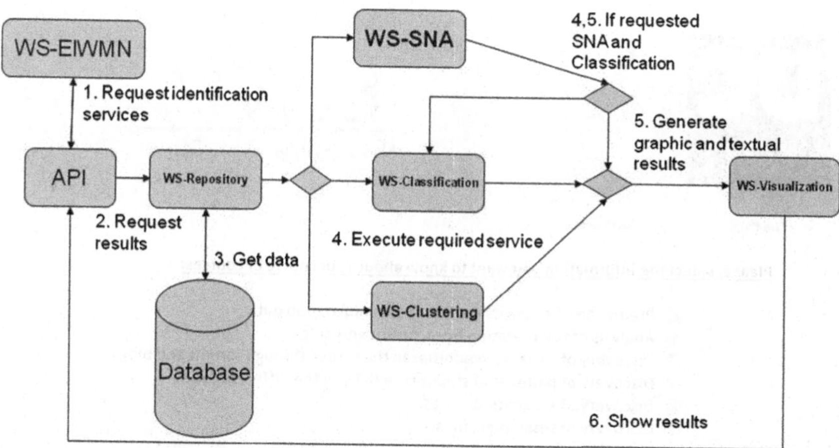

Fig. 15.2 ElWM workflow

and the number of pages edited in a wiki; or getting a model that only considers the activity carried out on certain resources. Thanks to adding SNA in ElWM, these predictive models can include and take advantage of SNA features in such a way that educators can assess the effectiveness of collaboration within the group for the students to pass the course.

Currently, ElWM implements prediction models using the J48 classification algorithm, which is not only easy to interpret but also return the most accurate results, according to our experimentation [55].

In short, we have added new services focused on classification and SNA techniques. The questions that can be answered by ElWM are shown in Fig. 15.2. The new ones are "Prediction of students' performance and/or dropout" (classification task), "Analysis of collaboration from forums and blogs" (SNA), and "Discovery of social communities in the course through forums and blogs" (SNA). When an instructor selects a query, e.g., "Prediction of students' performance and/or dropout," ElWM inquires information related to this process, as can be observed in Fig. 15.3.

15.3.4 Mode of Working

Figure 15.2 depicts how ElWM coordinates the call of the different services from the user request to the model generation and the presentation of results. The instructor through a web form (see Figs. 15.3, 15.4), will select what question he or she wants to inquire. After that, the WSs have to be identified by WSDL and then they will exchange SOAP messages between them. The first service to be called is WS-Repository, which will return the concrete data needed, obtained by querying the LMS database.

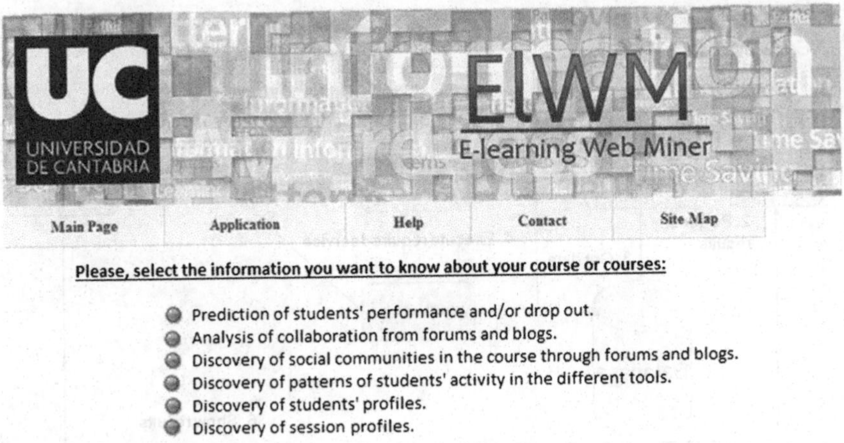

Please, select the information you want to know about your course or courses:

- Prediction of students' performance and/or drop out.
- Analysis of collaboration from forums and blogs.
- Discovery of social communities in the course through forums and blogs.
- Discovery of patterns of students' activity in the different tools.
- Discovery of students' profiles.
- Discovery of session profiles.

Fig. 15.3 ElWM questions

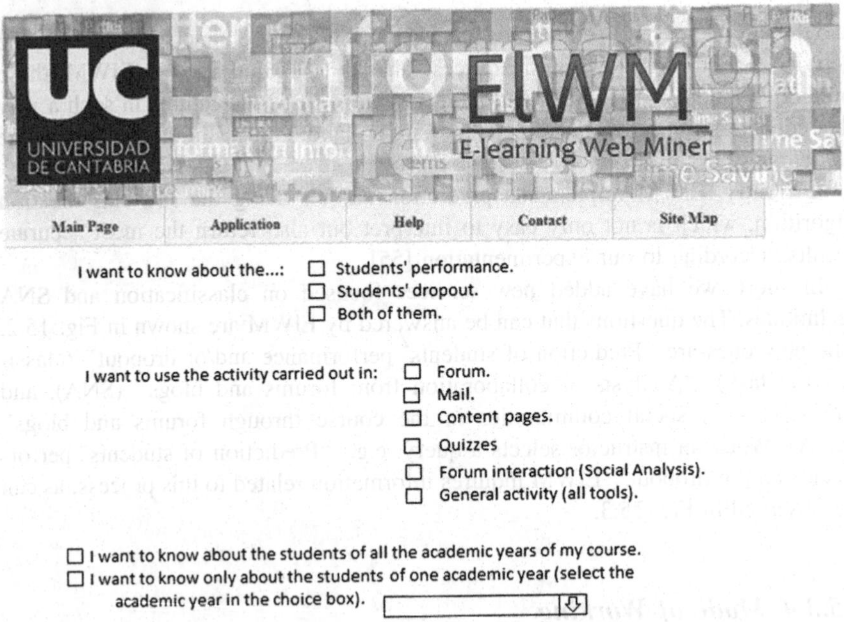

UC UNIVERSIDAD DE CANTABRIA

ElWM E-learning Web Miner

Main Page | Application | Help | Contact | Site Map

I want to know about the...: ☐ Students' performance. ☐ Students' dropout. ☐ Both of them.

I want to use the activity carried out in: ☐ Forum. ☐ Mail. ☐ Content pages. ☐ Quizzes ☐ Forum interaction (Social Analysis). ☐ General activity (all tools).

☐ I want to know about the students of all the academic years of my course.
☐ I want to know only about the students of one academic year (select the academic year in the choice box).

Fig. 15.4 Classification options provided by ElWM

In the next step, the API will call one of the WSs among the following: WS-Classification, WS-Clustering and WS-SNA, depending on what information the instructor has requested. Note that these services are not exclusive: for example, to answer the question about the students' performance using the SNA information, ElWM has to first use the WS-SNA to generate this information and send it to the WS-Classification service in order to make the prediction task.

This last selected service will return the textual results and send them to the WS-Visualization service, which will return the same result graphically. Finally, both graphical and textual results will be shown to the instructor by the web interface (see Figs. 15.6, 15.8).

After the user has selected the options, the ElWM API has to access an XML configuration file that records, for each question and its options, which WSs are needed to get the right results. Both WS-Repository and WS-Visualization are always invoked because the former is needed to get the data from the database and the latter to visualize the results. The rest of WS are invoked depending on the questions to be answered. A piece of its content is shown in the program code XML configuration.

Such a program code depicts that, if the user selects the question to obtain the students' performance and/or dropout (*<question id="Q1"*) and he or she only wants to consider the activity carried out by the students in mail tool (*<option name="Mail"*>), then the needed WS is WS-Classification. However, if the user requests the model that requires to take the forum interaction data into account (*<option name="SNAClassification"*>), then the ElWM API must first invoke WS-SNA, and next WS-Classification.

```
<question id="Q1">
        <option name="Forum">
                <webservice n="1">WS_Classification</webservice>
        </option>
        <option name="Mail">
                <webservice n="1">WS_Classification</webservice>
        </option>
        <option name="ContentPages">

        <option name="SNAClassification">
                <webservice n="1">WS_SNA</webservice>
                <webservice n="2">WS_Classification</webservice>
        </option>
</question>
<question id="Q2">
                        ...
```

[XML configuration file]

The code SOAP messages to get the data from Moodle displays an example of the SOAP messages exchanged between the ElWM API (the requester) and the WS-Repository (requested). The first SOAP message contains the parameters (see *Body* tag) that the method *getData* of the WS-Repository needs to know in order to perform the operation of accessing the data in Moodle and returning to the API.

The parameters indicate that the data requested is for a prediction task whose class attribute is the students' performance ("*<performance>*"). Furthermore it indicates that, this task must consider the activity carried out by the students in all the tools ("<tool>") from the course selected, in this case *Course1* ("<course>"). Finally, the parameters related to the user login are provided to connect to Moodle database.

The second message is the response that the WS-Repository sends to ElWM API after accessing the Moodle database and executing the predefined SQL sentence to get the requested data. WS-Repository returns the *courseData.xml* file that presents the format shown in the code XML data from Moodle. This process is repeated between the ElWM API and the needed WSs to get the final results.

15.4 Case Study

This section aims to show the usefulness of ElWM for the educational context. We configure different queries on several virtual courses in the same way any teacher would do and we also discuss how educators can take advantage of its use. We organize the experiments in two sections: SNA in e-learning context and prediction of students' performance and dropout.

15.4.1 Courses

First, we briefly describe the courses used for this purpose. We work with four virtual courses imparted at the UC, three of them hosted in the Moodle platform and another one in Blackboard.

The first course, hosted in Blackboard and entitled "Introduction to multimedia methods," offered in three academic years (2007–2010) with an average of 70 students enrolled from different degrees (economics, engineering, and sciences). The second course, a computer science course taught in the 2007–2008 academic year with a total of 432 enrolled students from the Computer Science degree. A course oriented to train transversal skills named "Creativity and Innovation" imparted during the first semester of 2013 with 28 learners enrolled and, finally, a congress named "Congress of Learning Styles[1]" where four discussion forums were organized.

The reasons to choose these courses were: (1) the students enrolled had different demographic profiles, (2) these learners were enrolled in degrees from different branches of knowledge, (3) the design of these courses was completely virtual and (4) we had a good relationship with the teachers to discuss the results later on.

[1] See http://congresoestilosdeaprendizaje.blogspot.com..

```
<!-- Soap Request -->
<S:Envelope xmlns:S="http://schemas.xmlsoap.org/soap/envelope/">
    <S:Header>
        <To xmlns="http://www.w3.org/2007/08/addressing">
            http://localhost:8080/ElWM/WS_Repository</To>
        <Action xmlns="http://www.w3.org/2005/08/addressing">
            http://server/WS_Repository/getData</Action>
        <ReplyTo xmlns="http://www.w3.org/2007/08/addressing">
                ...
    </S:Header>
    <S:Body>
        <ns2:getData xmlns:ns2="http://server">
            <prediction>performance</prediction>
            <tool>all</tool>
            <course>Course1</course>
            <loginData>
                <user>User1</user>
                <password>pass</password>
                <datasource>
                    http://localhost:3535/Moodle
                </datasource>
            </loginData>
        </ns2:getData>
    </S:Body>
</S:Envelope>

<!-- Soap Response -->
<S:Envelope xmlns:S="http://schemas.xmlsoap.org/soap/envelope/">
    <S:Header>
        <To xmlns="http://www.w3.org/2007/08/addressing">
            http://www.w3.org/2007/08/addressing/anonymous</To>
        <Action xmlns="http://www.w3.org/2005/08/addressing">
            http://server/WS_Repository/getData</Action>
                ...
    </S:Header>
    <S:Body>
        <ns2:getDataResponse xmlns:ns2="http://server">
            <return>/.../courseData.xml</return>
        </ns2:getDataResponse>
    </S:Body>
</S:Envelope>
```

[SOAP messages to get the data from Moodle]

```
<numberOfRows>675</numberOfRows>
  <classValueName>class</classValueName>
  <row id="1">
       <attribute name="totalTimeSpent">17239</attribute>
       <attribute name="numberOfSessions">178</attribute>
                        ...
       <attribute name="class">Pass</attribute>
  </row>
  <row id="2">
                        ...
```


[XML data from Moodle]

15.4.2 Social Network Analysis in E-Learning Courses

ElWM offers educators two questions to be answered using SNA: (1) analysis of collaboration from forums and blogs and (2) discovery of social communities in the course through forums and blogs. Our experimentation begins with the study of the course "Introduction to Multimedia Methods," with which we made a general SNA of the students and instructors interaction in the forum. In particular, Fig. 15.5 shows the social networks for the academic year 2008–2009. For this purpose, ElWM automatically extracts the data with all the interactions of the students in the forum querying the Moodle Database at the UC. In this course, we found a single connected component and a diameter of 11, as well as low values of density (0.07) and reciprocity (7 % of the links were reciprocal).

In the previous and the subsequent courses (2007–2008 and 2009–2010), we found three and one connected components, and diameters of 19 and 16, respectively, as well as low values of density (0.05 and 0.06) and reciprocity (12 and 8 %). A possible explanation for the low values detected of both density and reciprocity is that the instructor answered the questions in the forum faster than students, preventing them from helping one another. As can be observed, the node with more links is the one corresponding to the main instructor (node 1). With this information extracted by ElWM, the instructor can know that, in this course, most interactions in the forum occur between the instructor and the students, whereas it is less frequent that those interactions that occur among the students themselves. Thus, the forum is mainly used in two different ways: (1) students make questions about the contents or the organization of the course that should be answered by the instructor and (2) the instructor makes important announcements.

These conclusions can be better understood by analyzing the node centrality values exposed in Table 15.1. As can be observed in Fig. 15.5, the instructor (node 1) has the highest values of degree and outdegree. Moreover, the difference in outdegree between the instructor and the second and the third ranked users is very high. This also happens to the betweenness and hub centrality measures. Thus, we

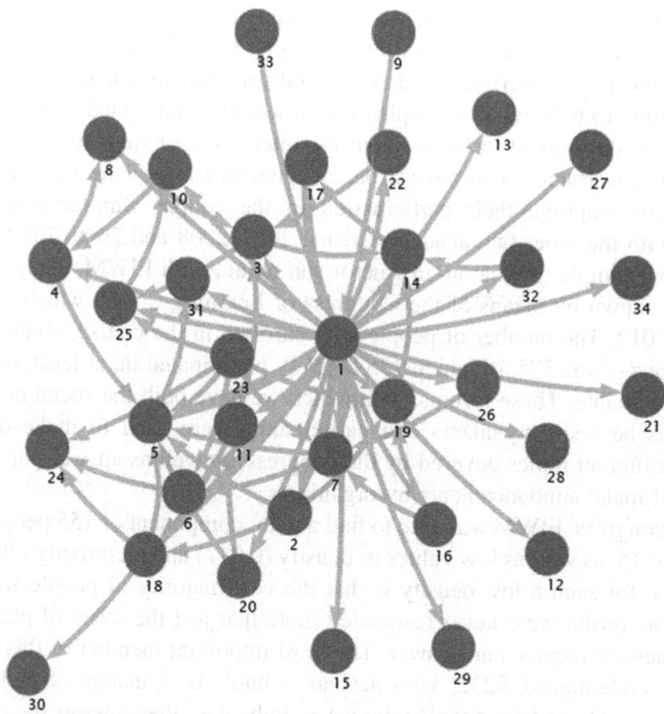

Fig. 15.5 Network of interactions between the instructor and the students of the course "Introduction to Multimedia Methods" taught in 2008–2009 at the UC

Table 15.1 Ranking of top 3 nodes for different centrality measures

	First ranked		Second ranked		Third ranked	
	Node ID	Value	Node ID	Value	Node ID	Value
Degree	1	166	3	39	5	35
Indegree	3	36	5	33	6	17
Outdegree	1	157	2	6	23	6
Betweenness	1	505	17	151	14	152
Authority	3	0.93	5	0.76	6	0.41
Hub	1	1.41	23	0.03	2	0.03

can conclude that the instructor is the user that answered the great majority of messages posted by the students in the forum.

On the other hand, the highest indegree and authority values correspond to nodes 2, 3, and 6. These students are the users that posted more messages in the forum. As a matter of fact, these three students scored the best in the course. Thus, with this analysis, we can conclude that students with a high number of interactions in the forum are likely to get good scores, a fact to be analyzed using DM techniques.

The instructor, thanks to ElWM, can obtain these results easily, with both textual and graphical visualizations. For example, ElWM does not only show the rank with the degree attribute, but it also informs the instructor about what it means. Figure 15.6 shows an example of what would be presented to the instructor in case he or she wanted to know about the interaction of students in the forum.

With this, instructors can have a better understanding of the students' social behavior and improve their participation in the course. Similar results were obtained with the other two academic years, 2007–2008 and 2009–2010.

Another example of what an instructor can obtain with ElWM, using the SNA service, is shown by means of the Congress of Learning Styles, which was celebrated in 2012. The number of people who enrolled in the course—including the 13 organizers—was 375, of which 155 actively participated in, at least, one of the 4 forums available. These forums, with which we have built the social network of interactions between organizers and participants, were used to make questions about the different topics covered by the congress, as well as answer participants' doubts and make announcements by organizers.

In this congress, ElWM was able to find a giant component of 155 people with a diameter of 15, as well as low values of density (0.017) and reciprocity (29 %). An explanation for such a low density is that the vast majority of people who intervened in the forum were never responded; note that just the 29 % of people who asked a question received an answer. The most important member of this forum is a student, code-named S232, who acts as a hub. As a matter of fact, in this congress, we can see how people who act as hubs, i.e., they answer the questions of people who are highly responded, are considered by SNA to be the most important actors in the network.

Although this behavior might seem logical, we have just seen how, in an educational course context, authority nodes can also turn out to be the most important ones. Also, it is noticeable that no organizer is considered to be important, i.e., most interactions in the forum occur among participants, whereas it is less frequent that those interactions occur between them and organizers.

Finally, community detection algorithms, which are able to identify clusters of vertices with a high internal density of edges, were also considered to check whether the forum structure of the congress is independent and recognizable enough to be recovered by such a type of algorithms.

For this purpose, FRINGE [12], an overlapping community detection algorithm with a recent approach based on the dynamics of social networks, was wrapped in ElWM and used. It was able to recover four overlapping community members, i.e., S232, S303, S296, and S220, as leading members of every community, respectively. As can be observed in Fig. 15.7, these leading members are the same as those obtained in the four forums of the congress.

This result demonstrates the power of community detection algorithms and justifies their use under this kind of scenarios. The educational context would allow instructors to discover the students' degree of interaction, which is paramount to

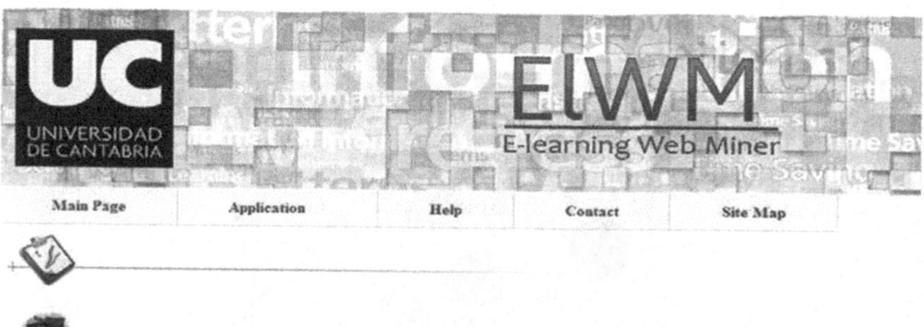

Number of students evaluated: 67
Number of post in forum: 368
Number of discussions: 157
Number of topics considered: 1

The user "instructor" (1) has answered most of the discussions. The rest of users have a lower activity and interaction.

The users with id "2", "3" and "6" have posted or initialized a lot more discussions than the other users. These 3 users are the ones with better performance in the course.

	First Ranked		Second Ranked		Third Ranked	
	Node ID	Value	Node ID	Value	Node ID	Value
Degree	1	166	3	39	5	35
Indegree	3	36	5	33	6	17
Outdegree	1	157	2	6	23	6
Betweenness	1	505	17	151	14	142
Authority	3	0.93	5	0.76	6	0.41
Hub	1	1.41	23	0.03	2	0.03

How can I interpret these results? click here for help.

Interactions between students and instructors in the forum:

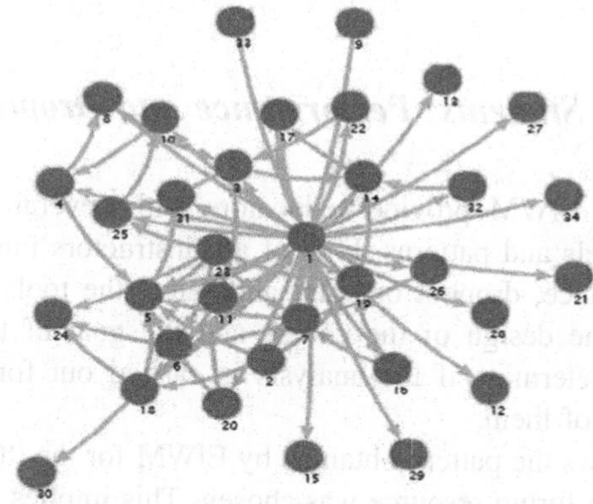

How can I interpret these results? click here for help.

Fig. 15.6 ElWM results applying SNA to the course "Introduction to Multimedia Methods" taught in 2008–2009 at the UC

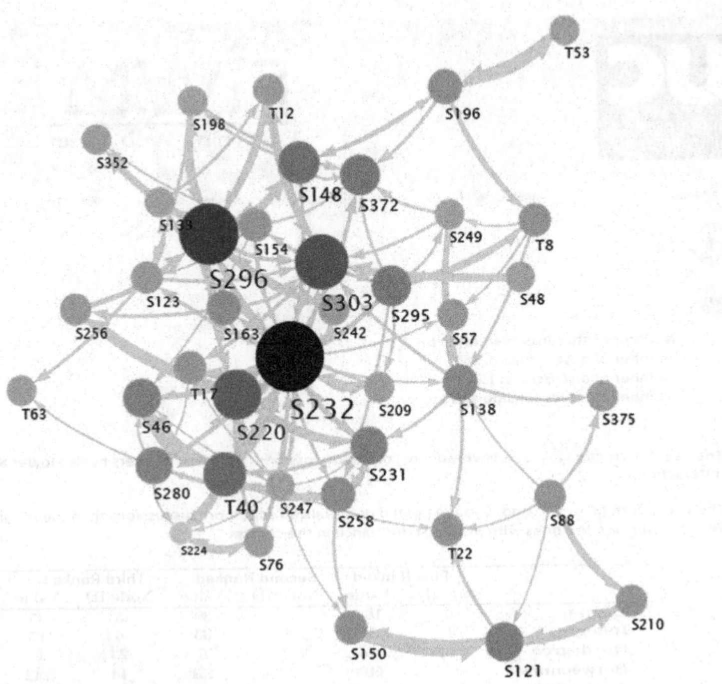

Fig. 15.7 Network of interactions among the participants in "Congress of Learning Styles"

create team-works with different social profiles. Also, it could be used to score the contribution of each student to a certain forum's discussion. Thus, it would take an objective variable of participation into consideration.

15.4.3 Prediction of Students' Performance and Dropouts

As shown in Fig. 15.3, ElWM provides educators with several configuration options to generate models and patterns. First of all, instructors must choose the kind of model: performance, dropout or both, and select the tools that must be analyzed according to the design of the course and the goal of their analysis. Furthermore, they can determine if the analysis is carried out for one specific academic year or for all of them.

Our first example shows the pattern obtained by ElWM for the "Creativity and Innovation" course when forum resource was chosen. This implies the use of the following attributes: number of posts read, number of posts initiated, and number of messages posted by each student in order to discover whether learners will pass. Figure 15.8 displays the result shown to the end user.

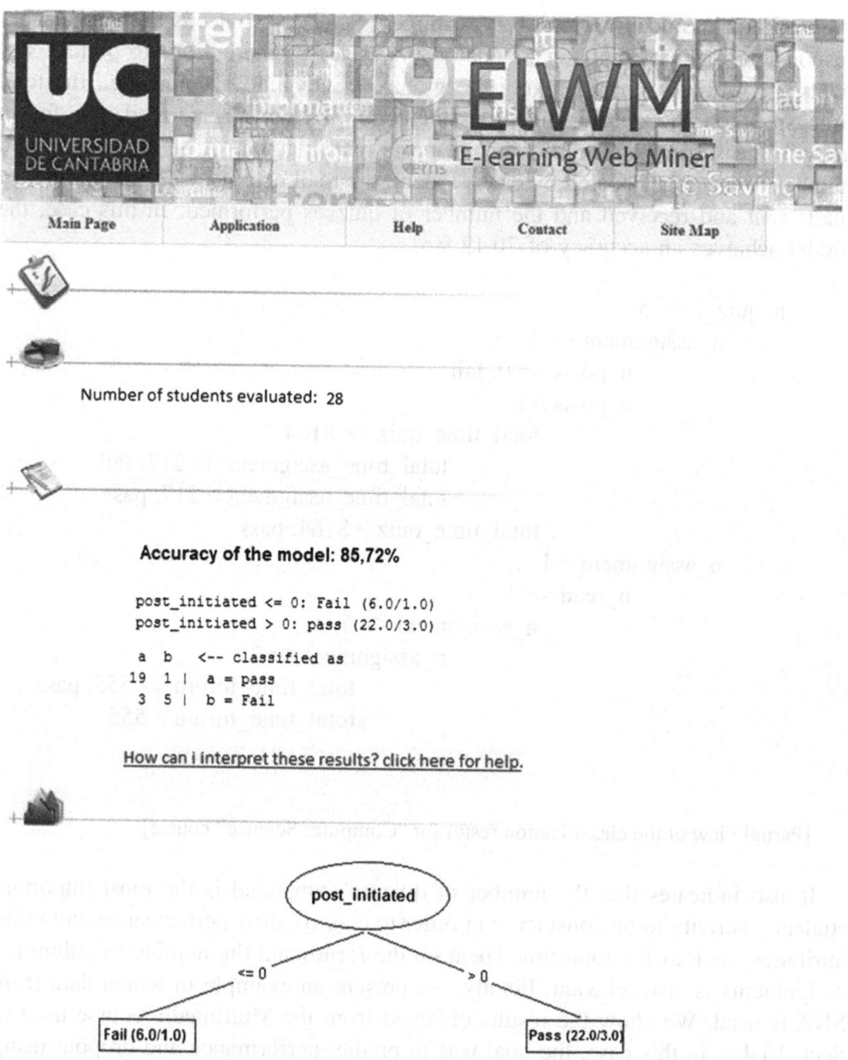

Fig. 15.8 ElWM classification results for "Creativity and Innovation" course

This is a tree, obtained with J48 algorithm from Weka set up with default parameters, which says that students, who initialize 1 or more posts, would pass the course with 86 % of accuracy and students who do not initialize any posts would fail with an accuracy of 95 %. The global accuracy of the model is 85.72 %. Hence, the instructor could encourage their learners to participate in forums because the information written was useful to pass the course.

A more complex and rich output would be obtained if the number of students and resources to be analyzed grew. It is partially shown in program code for the view of the classification result for "Computer Science" course, where the teacher responsible for the Computer Science course selected the options to build a students' performance model taking the global activity into account, i.e., the total time spent on the course and total number of sessions performed, the forum activity measured from the number of posts read, number of posts initiated and number of messages posted by each student, the mail activity with the number of mails sent and received and the number of quizzes performed. In this case, the model achieves an accuracy of 70.12 %.

```
n_quiz_a <= 5
    n_assignment <= 1
        n_posts <= 0: fail
        n_posts > 0
            total_time_quiz <= 8164
                total_time_assigment <= 217: fail
                total_time_assigment > 217: pass
            total_time_quiz > 8164: pass
    n_assignment > 1
        n_read <= 1
            n_assignment <= 6
                n_assignment <= 2
                    total_time_forum <= 555: pass
                    total_time_forum > 555
                        ...
```

[Partial view of the classification result for "Computer Science" course]

It also indicates that the number of quizzes completed is the most important students' activity to be considered in order to classify their performance, but other attributes, such as the total time spent on the forum and the number of submitted assignments is also relevant. Finally, we present an example in which data from SNA is used. We show the results obtained from the Multimedia course used in Sect. 15.4.2. In this case, the goal was to predict performance and dropout using the following global activity attributes (in this case, since this course was imparted in Blackboard, the number of global attributes that ElWM can get is higher): total time spent, total number of sessions, average time spent per week and average number of sessions per week and the activity in forum and mail.

The model generated is shown in the code for Partial view of the classification model with SNA attributes. By reading this code, the teacher discovers that students with an average time per week lower than 63 min are likely to dropout, and also finds out that the interaction among students in the forum is important for the classification task: the students with an average time per week equal or higher than 63 min are considered to fail or pass the course depending, for example, on whether they are an authority in the forum or if they have written a high quantity of posts. By using ElWM and having these last results, the instructor can know not only about the interactions of the students, but also how these interactions affect the prediction of students' performance and dropout.

15.5 Conclusions

In this chapter we describe the architecture and functionality of an educational tool that assists educators to monitor, analyze, and better understand the behavior of their students in the development of collaborative activities using Web 2.0 resources.

```
average_time_per_week <= 63: dropout
average_time_per_week > 63
    number_of_messages_written_in_the_forum <= 0
        average_number_of_sessions_per_week <= 3
            number_of_messages_read_in_the_forum <= 52: pass
            number_of_messages_read_in_the_forum > 52: dropout
        average_number_of_sessions_per_week > 3
            total_time <= 1962: pass
            total_time > 1962
                total_time <= 2000: dropout
                total_time > 2000: pass
    number_of_messages_written_in_the_forum > 0
        number_of_messages_written_in_the_forum <= 8
            in_degree_centrality <= 0.013
                number_of_messages_read_in_the_forum <= 87: pass
                number_of_messages_read_in_the_forum > 87
                    number_of_messages_read_in_the_mail <= 24
                    number_of_messages_written_in_the_forum <= 4
            in_degree_centrality_unbalanced <= 1
                authority_centrality <= 0.029
                    number_of_messages_read_in_the_mail <= 7: fail
```

[Partial view of the classification model with SNA attributes]

For instances forums, blogs, or wikis, as well as, achieving course objectives. This tool, called E-learning Web Miner, relies on the application of DM and SNA techniques on data from logs of the services used to develop the teaching–learning process (e-learning platforms, social networks, wiki spaces, etc.).

Its architecture, based on WSs, makes it an easily extensible and embeddable one in any tool, as for example in an LMS. Its main feature is that its use is oriented to non-expert educators in analytics since it hides the knowledge discovery processes from the user. Its mode of working is simple; instructors have to connect to ElWM and choose the question to solve by pointing out the virtual courses under study.

ElWM carries out the mining process without user's interaction and displays the results in textual and graphical mode. This new version of ElWM includes three new questions: "Prediction of students performance and/or dropout," "Analysis of collaboration from forums and blogs," and "Discovery of social communities in the course through forums and blogs" using classification techniques and social analysis techniques for these tasks.

Currently, our research is focused on meta-learning [30] to build a recommender that automatists the choice of the most suitable classification algorithm for each dataset at hand. Furthermore, we are adapting our FRINGE algorithm to directed and weighted networks with the aim of finding out the leaders (our most active learners) in a network according to their weight (e.g., messages answering doubts in a forum can have less value than contributions in a wiki from an instructional point of view).

In addition to the direction of the edges, i.e., the relationship between learners (both support themselves or the help is always in one sense). At the same time, we are working in the phase of tool testing and we hope to deploy it in the coming months. Next, we will study new and interesting questions for educators and look for the best answers to these questions. This will lead us to add new algorithms and visualization tools in our web services.

Acknowledgements The authors would like to thank the anonymous referees for their constructive comments, which led to a significant improvement of this paper. The authors are also deeply grateful to CEFONT, the department of the UC that is responsible for LCMS maintenance, for their help and collaboration. Likewise, the authors gratefully acknowledge the valuable collaboration of the instructors involved in the courses analyzed. This work is partially supported by the Ministry of Education, Culture, and Sport of the Government of Spain, under grant Beca de colaboración (2012–2013), and by the UC, under a Ph.D. studentship (2011–2015).

References

1. Capra, R., Arguello, J., Chen, A., Hawthorne, K., Marchionini, G., Shaw, L.: The results space collaborative search environment. In: 12th ACM/IEEE-CS Joint Conference on Digital Libraries, pp. 435–436. ACM, New York (2012)
2. Lin, C.C., Tsai, C.C.: Participatory learning through behavioral and cognitive engagements in an online collective information searching activity. Int. J. Comput. Support. Collaborative Learn. **7**(4), 543–566 (2012)
3. McNely, B.J., Gestwicki, P., Hill, J.H., Parli-Horne, P., Johnson, E.: Learning analytics for collaborative writing: a prototype and case study. In: Dawson, S., Haythornthwaite, C., Shum, S.B., Gasevic, D., Ferguson, R. (eds.) Second International Conference on Learning Analytics and Knowledge, pp. 222–225. ACM, New York (2012)
4. Joubert, M., Wishart, J.: Participatory practices: lessons learnt from two initiatives using online digital technologies to build knowledge. Comput. Educ. **59**(1), 110–119 (2012)
5. Rice, W.: Moodle E-learning Course Development. A Complete Guide to Successful Learning Using Moodle. Packet Publishing, Birmingham (2006)

6. Southworth, H., Cakici, K., Vovides, Y., Zvacek, S.: Blackboard for Dummies. Wiley, New York (2006)
7. Korcuska, M., Berg, A.M.: Sakai Courseware Management: The Official Guide. Packt Publishing, Birmingham (2009)
8. Johnson, L., Adams-Becker S., Cummins, M., Estrada, V., Freeman, A., Ludgate, H.: NMC Horizon Report: Higher Education Edition. Report. The New Media Consortium (2013)
9. Romero, C., Ventura, S.: Data mining in education. Wiley Interdiscip. Rev.: Data Min. Knowl. Disc. **3**(1), 12–27 (2013)
10. Macfadyen, L.P., Dawson, S.: Mining LMS data to develop an early warning system for educators: a proof of concept. Comput. Educ. **54**(2), 588–599 (2010)
11. Zorrilla, M., García-Saiz, D.: A service oriented architecture to provide data mining services for non-expert data miners. Decis. Support Syst. **55**(1), 399–411 (2013)
12. Palazuelos, C., Zorrilla, M.E.: FRINGE: a new approach to the detection of overlapping communities in graphs. In: Murgante, B., Gervasi, O., Iglesias, A., Taniar, D., Apduhan, B.O. (eds.) ICCSA 2011. LNCS, vol. 6784, pp. 638–653. Springer, Heidelberg (2011)
13. Bienkowski, M., Feng, M., Means, B.: Enhancing teaching and learning through educational data mining and learning analytics: an issue brief. U.S. Department of Education, Office of Educational Technology (2012)
14. Scott, J.: Social Network Analysis: A Handbook. SAGE Publications, London (2000)
15. Wasserman, S., Faust, K.: Social Network Analysis: Methods and Applications. Structural Analysis in the Social Sciences. Cambridge University Press, Cambridge (1994)
16. Watts, D., Strogatz, S.: Collective dynamics of small-world networks. Nature **393**(6684), 440–442 (1998)
17. Klovdahl, A., Potterat, J., Woodhouse, D., Muth, J., Muth, S., Darrow, W.: Social networks and infectious disease: the colorado springs study. Soc. Sci. Med. **38**(1), 79–88 (1994)
18. Krebs, V.: Mapping networks of terrorist cells. Connections **24**(3), 43–52 (2002)
19. Freeman, L.: A set of measures of centrality based on betweenness. Sociometry **40**(1), 35–41 (1977)
20. Kleinberg, J.: Authoritative sources in a hyperlinked environment. J. ACM **46**(5), 604–632 (1999)
21. Brewe, E., Kramer, L.H., Sawtelle, V.: Investigating student communities with network analysis of interactions in a physics learning center. Phys. Rev. Spec. Top. Phys. Educ. Res. **8**(1), 1–9 (2012)
22. Crespo, P.M.T., Antunes, C.: Social networks analysis for quantifying students performance in teamwork. In: Yacef, K., Zaïane, O., Hershkovitz, A., Yudelson, M., Stamper, J. (eds.) 5th International Conference on Educational Data Mining, pp. 234–235. International Educational Data Mining Society, Chania (2012)
23. Cuéllar, M.P., Delgado, M., Pegalajar, M.C.: Improving learning management through semantic web and social networks in e-learning environments. Expert Syst. Appl. **38**(4), 4181–4189 (2011)
24. Rabbany, R., Takaffoli, M., Zaïane, O.R.: Social network analysis and mining to support the assessment of on-line student participation. SIGKDD Explor. **13**(2), 20–29 (2011)
25. Dawson, S., Tan, J.P.L., McWilliam, E.: Measuring creative potential: using social network analysis to monitor a learners' creative capacity. Australas. J. Educ. Technol. **27**(6), 924–942 (2011)
26. Dawson, S.: Seeing the learning community: an exploration of the development of a resource for monitoring online student networking. Br. J. Educ. Technol. **41**(5), 736–752 (2010)
27. Obsivac, T., Popelinsky, L., Bayer, J., Geryk, J., Bydzovska, H.: Predicting drop-out from social behaviour of students. In: Yacef, K., Zaïane, O., Hershkovitz, A., Yudelson, M., Stamper, J. (eds.) 5th International Conference on Educational Data Mining, pp. 103–109. International Educational Data Mining Society, Chania (2012)
28. Palazuelos, C., García-Saiz, D., Zorrilla, M.: Social network analysis and data mining: an application to the e-learning context. In: International Conference on Computational Collective Intelligence Technologies and Applications (2013, in press)

29. Fayyad, U., Piatetsky-Shapiro, G., Smyth, P.: From data mining to knowledge discovery in databases. J. Artif. Intell. **17**(1), 37–54 (1996)
30. García-Saiz, D., Zorrilla, M.E.: Towards the development of a classification service for predicting students' performance. In: D'Mello, S.K., Calvo, R.A., Olney, A. (eds.) 6th International Conference on Educational Data Mining, pp. 318–319. International Educational Data Mining Society, Memphis (2013)
31. Kotsiantis, S.B.: Use of machine learning techniques for educational proposes: a decision support system for forecasting students' grades. J. Artif. Intell. **37**(4), 331–344 (2012)
32. Romero, C., Ventura, S., Espejo, P.G., Hervás, C.: Data mining algorithms to classify students. In: Baker, R.S.J.D., Barnes, T., Beck, J.E. (eds.) 1st International Conference on Educational Data Mining, pp. 8–17. International Educational Data Mining Society, Montreal (2008)
33. Zafra, A., Romero, C., Ventura, S.: Predicting academic achievement using multiple instance genetic programming. In: Ninth International Conference on Intelligent Systems Design and Applications, pp. 1120–1125, IEEE, Washington (2009)
34. Dekker, G., Pechenizkiy, M., Vleeshouwers, J.: Predicting students drop out: a case study. In: Barnes, T., Desmarais, M., Romero, C., Ventura, S. (eds.) 2nd International Conference on Educational Data Mining, pp. 41–50. International Educational Data Mining Society, Cordoba (2009)
35. Kotsiantis, S.B., Pierrakeas, C., Pintelas, P.E.: Preventing student dropout in distance learning using machine learning techniques. In: Palade, V., Howlett, R.J., Jain, L.C. (eds.) KES. LNCS, vol. 2773, pp. 267–274. Springer, Heidelberg (2003)
36. John, G., Langley, P.: Estimating continuous distributions in bayesian classifiers. In: Eleventh Conference on Uncertainty in Artificial Intelligence, pp. 338–345. Morgan Kaufmann, San Francisco (1995)
37. Hämäläinen, W., Vinni, M.: Comparison of machine learning methods for intelligent tutoring systems. In: Ikeda, M., Ashley, K., Chan, T.W. (eds.) Intelligent Tutoring Systems. LNCS, vol. 4053, pp. 525–534. Springer, Heidelberg (2006)
38. Zafra, A., Ventura, S.: G3P-MI: a genetic programming algorithm for multiple instance learning. Inf. Sci. **180**(23), 4496–4513 (2010)
39. Friedman, N., Geiger, D., Goldszmidt, M., Provan, G., Langley, P., Smyth, P.: Bayesian network classifiers. Mach. Learn., 131–163. Kluwer Academic Publishers, Boston (1997)
40. Quinlan, J.R.: C4.5: Programs for Machine Learning. Morgan Kaufmann Publishers, San Francisco (1993)
41. Witten, I.H., Frank, E.: Data Mining: Practical Machine Learning Tools and Techniques, 2nd edn. Morgan Kaufmann, San Francisco (2005)
42. Romero, C., Ventura, S.: Educational data mining: a review of the state of the art. IEEE Trans. Syst. Man Cybern. Part C Appl. Rev. **40**(6), 601–618 (2010)
43. García-Saiz, D., Zorrilla, M.E.: E-learning web miner: a data mining application to help instructors involved in virtual courses. In: Pechenizkiy, M., Calders, T., Conati, C., Ventura, S., Romero, C., Stamper, J. (eds.) 4th International Conference on Educational Data Mining, pp. 323–324. International Educational Data Mining Society, Eindhoven (2011)
44. Benchaffai, M., Debord, G., Merceron, A., Yacef, K.: TADA-ED, a tool to visualize and mine students' online work. In: McKay, E., Collis, B. (eds.) International Conference on Computers in Education, pp. 1891–1897. RMIT, Melbourne (2004)
45. Romero, C., Ventura, S., García, E.: Data mining in course management systems: Moodle case study and tutorial. Comput. Educ. **51**(1), 368–384 (2008)
46. García, E., Romero, C., Ventura, S., de Castro, C.: A collaborative educational association rule mining tool. Internet High. Educ. **14**(2), 77–88 (2011)
47. Holmes G., Hall, M., Frank, E.: Generating rule sets from model trees. In: 12th A.J.C. on Artificial Intelligence. LNCS, vol. 1747, pp. 1–12. Springer, Heidelberg (1999)
48. Romero, C., Ventura, S., Zafra, A., de Bra, P.: Applying web usage mining for personalizing hyperlinks in web-based adaptive educational systems. Comput. Educ. **53**(3), 828–840 (2009)

49. Balcázar, J.L., Tîrnauca, C., Zorrilla, M.E.: Filtering association rules with negations on the basis of their confidence boost. In: International Conference on Knowledge Discovery and Information Retrieval, pp. 263–268, INSTICC, Valencia (2010)
50. Zorrilla, M.E., García, D.: A data mining service to assist instructors involved in virtual education. In: Zorrilla, M., Mazón, J., Ferrández, Ó., Garrigós, I., Daniel, F., Trujillo, J. (eds.) Business Intelligence Applications and the Web: Models, Systems and Technologies, pp. 222–243. Business Science Reference, Hershey (2012)
51. Zorrilla, M.E., García-Saiz, D., Balcázar, J.L.: towards parameter-free data mining: mining educational data with Yacaree. In: Pechenizkiy, M., Calders, T., Conati, C., Ventura, S., Romero, C., Stamper, J. (eds.) 4th International Conference on Educational Data Mining, pp. 363–364. International Educational Data Mining Society, Eindhoven (2011)
52. Balcázar, J.L.: Parameter-free association rule mining with Yacaree. In: Khenchaf, A., Poncelet, P. (eds.) Extraction et Gestion des Connaissances, pp. 251–254. Hermann, Brest (2011)
53. Borgelt, C.: Efficient implementations of Apriori and Eclat. In: Goethals, B., Zaki, M.J. (eds.) ICDM Workshop of Frequent ItemSet Mining Implementations. CEUR-WS, Melbourne (2003)
54. Park, H.S., Jun, C.H.: A simple and fast algorithm for K-medoids clustering. Expert Syst. Appl. **36**(2), 3336–3341 (2009)
55. García-Saiz, D., Zorrilla, M.: Comparing classification methods for predicting distance students' performance. In: Diethe, T., Balcázar, J.L., Shawe-Taylor, J., Tîrnauca, C. (eds.) Journal of Machine Learning Research, Workshop and Conference Proceedings. 2nd Workshop on Applications of Pattern Analysis, vol. 17, pp. 26–32 (2011)

Chapter 16
Collaborative Learning of Students in Online Discussion Forums: A Social Network Analysis Perspective

Reihaneh Rabbany, Samira Elatia, Mansoureh Takaffoli and Osmar R. Zaïane

Abstract Many courses are currently delivered using Course Management Systems (CMS). Discussion forums within these systems provide the basis for collaborative learning. In this chapter, we present the use of Social Network Analysis (SNA) to analyze the structure of interactions between the students in these forums. Various metrics are introduced for ranking and determining roles, while clustering and temporal analysis techniques are applied to study the student communications, the forming of groups, the role changes, as well as scrutinizing the content of the exchanged messages. Our approach provides the instructor with better means to assess the participation of students by (1) identification of participants' roles; (2) dynamic visualization of interactions between the participants and the groups they formed; (3) presenting hierarchy of the discussed topics; and (4) tracking the evolution and growth of these patterns and roles over time. The applicability of the proposed analyses are illustrated through several case studies.

Keywords Social network analysis · Student participation assessment · Student monitoring · Content summarization · Discussion forums

Abbreviations

CSCL Computer supported collaborative learning
CMS Course management systems

R. Rabbany (✉) · M. Takaffoli · O. R. Zaïane
Department of Computing Science, University of Alberta, Edmonton, Canada
e-mail: rabbanyk@ualberta.ca

M. Takaffoli
e-mail: takaffol@ualberta.ca

O. R. Zaïane
e-mail: zaiane@ualberta.ca

S. Elatia
Campus Saint Jean, University of Alberta, Edmonton, Canada
e-mail: selatia@ualberta.ca

A. Peña-Ayala (ed.), *Educational Data Mining*,
Studies in Computational Intelligence 524, DOI: 10.1007/978-3-319-02738-8_16,

MOOC Massive open online course
SNA Social network analysis

16.1 Introduction

There is a growing number of courses delivered using e-learning environments both using computer-supported collaborative learning (CSCL) tools: such as Moodle, WebCT and Blackboard, or massive open online course (MOOC) delivery systems, such as Coursera, Udacity, and EdX. Online asynchronous discussions in these environments play an important role in collaborative learning processes of students. Through interaction, students become more actively engaged in sharing information and perspectives with each other [1]. These e-learning course adds-on environment provide a fertile ground for independent learning and a wealth of information that teachers can use to enhance teaching and learning.

More than four decades now, several studies have investigated and emphasized the benefits of collaborative learning in general. CSCL, in particular, offers a unique media for collaborative learning activities, where peer and independent learning as well as peer feedback are thriving, i.e. threaded discussion forums.

Consequently, there is a theoretical emphasis in CSCL literature on the role of threaded discussion forums for collaborative learning activities [2]. Even basic CSCL tools enable the development of these threads where the learners could access text, revise it or reinterpret it; which allow them to connect, build, and refine ideas, along with stimulating deeper reflection [2].

From a teacher's perspective, these types of activities provide insight into the quality of learning and teaching. By being able to assess on a general level without much intervention from the teacher, s/he can actually have a better grasp of what has been learned and what challenges the students are still having during the course. The teacher then can benefit from the information accumulated through the students' interactions to build a diagnostic assessment model that would allow for directing both teaching and learning.

By working independently on these assignments, students engage in a self/peer-learning process that helps demonstrate what they have acquired in the course. The element of pressure from being observed by the teacher is lessened, and students are more comfortable seeking help from each other on matters they are still struggling with. It is in a way, a great tool for inductive learning/teaching where students come to grasp with the concepts being studied.

However, a large amount of messages/entries are generated within few weeks within these forums, often containing lengthy discussions bearing many interactions between students. This amount of generated data can be overwhelming to teachers who want to monitor and assess these interactions in these forums. Given this situation, CSCL tools, should provide means to facilitate the instructors' task of evaluating students input. It would be time consuming and even impossible for teachers to manually analyze this data. Moreover, the magnitude of this information deluge is even more accentuated with the advent and quick popularity of MOOCs where thousands of learners can take a course at the same time [3].

On the other hand, current CSCL tools do not provide much information regarding the participation of students and structure of interactions between them in discussion threads. In many cases, only some statistical information is provided such as frequency of postings, which is not a sufficient or even useful measure for interaction activity [1]. This means that the instructors, who are using these tools, do not have access to convenient indicators that would allow them to evaluate the participation and interaction in their classes [4]. Instructors usually have to monitor the discussion threads manually which is hard, time consuming, and prone to human error and in the case of MOOCs, manual monitoring is hardly possible.

There exists a large body of research on studying the participation of students in such discussion threads using traditional research methods: content analysis, interviews, survey observations and questionnaires [5].

These methods try to detect the activities that students are involved in while ignoring the relations between students. For example, content analysis methods, as the most common traditional methods, provide deep information about specific participants. However, they neglect the relationships between the participants while their focus is on the content, not on the structure [4].

In order to fully appreciate the participation of students, we need to understand their patterns of interactions and answer questions like who is involved in each discussion, who is the active/peripheral participant in a discussion thread [5].

The practicality of social network analysis methods in CSCL is demonstrated in [6–8], as methods for obtaining information about relations, fundamental structural and collaborative patterns. Moreover, there is a recent line of work on applying social network analysis techniques for evaluating the participation of students in online courses e.g. [1, 4, 5, 9–11].

The major challenges these attempts tried to tackle are: extracting social networks from asynchronous discussion forums (might require content analysis), finding appropriate indicators for evaluating participation (from education's point of view) and measuring these indicators using social network analysis. As clarified in the background and related works section, none of these works provides a complete or specific mechanism or framework for analyzing discussion threads. However, they attempted to address one of these challenges to some extent.

In the rest of this chapter, we elaborate on the importance of social network analysis for mining structural data in the field of computer science and its applicability to the domain of education specifically for evaluating collaborative learning of students within the media of discussion forums. This chapter is an extension of our earlier works in [12, 13], where we first introduced Meerkat-ED. Meerkat-ED is a specific and practical toolbox for analyzing interactions of students in asynchronous discussion forums of online courses.

Through our case studies we present how Meerkat-ED analyzes both the structure and content of these interactions using social network analysis techniques including community mining. Which gives the instructor a quick view of what is discussed in these forums, what are the topics, and how much each student has participated in these topics and how they collaborated on each discussion.

In the following, we first introduce some basic concepts in social network analysis. We then illustrate two information networks that can be extracted from an on-line course and discuss different structural analyses that can be performed on these networks. Finally we present Meerkat-ED a comprehensive social network analysis toolbox specific for analysis of online courses and illustrate its practicality on our case study data.

16.2 Background and Related Works

In this section, we first overview the two recent paradigm shifts in Education: the first one in the method of learning shifted toward collaborative learning and the second on in the mode of delivery shifted toward e-learning. After educational background on collaborative learning and e-learning, we then we review the background on social networks analysis, its major techniques. Finally we survey relevant research that use these techniques to assess collaboration of students and individuals' level of participation in (discussion threads i.e. a means for collaborative learning) a course.

16.2.1 On Collaborative Learning and E-Learning: An Educational Perspective

In the last fifty years, we witnessed two major shifts in education: the first one in the relationship between students and teacher and second on in the mode of delivery. In the first one, higher education is no longer one way method of learning/teaching, where a teacher lectures and students listen, and learn individually, and where the focus is mostly on the lower levels of the cognitive domains. It is now a two-way approach to learning, where the classical teacher is replaced with a facilitator and where the focus is on working in groups collaboratively in problem solving, finishing various tasks and projects, and on creating and innovating. The second shift, deals with the mode of delivery. We are no longer bound with the classical classroom and library. E-learning and social networking as tools of interaction and as platform to working in groups are present more in our lives than ever in history. Both two shifts are interconnected and we are witnessing a major paradigm shift in education where the two converge into one major change in education.

Collaborative learning fosters learning and e-learning. When a student works in collaborative situation, s/he is able to perform intellectually at a much higher level of thinking[1] [15]. As Gokhale [16] advocates, "the peer support system makes it possible for the learner to internalize both external knowledge and critical thinking skills." When working in groups, we are faced with different perspectives and interpretations that each member in the group brings into the work. Gokhale [16] contends that collaborative learning yields significantly better learning results, particularly at higher levels of critical of thinking. Bruner [17] further argues, collaborative and/or cooperative learning methods improve problem-solving strategies. Pantiz [18] lists many benefits to collaborative learning that could be categorizes into social, psychological, and academic level. For the academic benefits group, Laal and Ghodsi [19] make two distinction: the first group deals with the student's own learning benefit; in the second group, collaborative learning brings in a new ways for assessment and interaction between teacher and students.

Threaded discussion forums offer a unique media for collaborative learning activities, where peer and independent learning as well as peer feedback are thriving. Analyzing the structure and quality of these messages gives indications of higher-order thinking and a clearer insight into collaborative learning [20].

Therefore there are many literature with a theoretical emphasis on the role of threaded discussion forums in collaborative learning activities in CSCL [2], however the empirical studies are rather limited [21]. Many visualization approaches have been investigated to support analysis of these forums, mainly focusing on one discussion or one thread, a comparison of these could be found in [20]. Here we investigate monitoring and assessing the collaborative activities of students within these forums using the power of social networks.

16.2.2 Social Networks: A Data Mining Perspective

Social networks are first introduced in social and behavioural sciences and focus on relations between entities and patterns of these relations. Social Networks are formally defined as a set of actors which are tied by one or more types of relations [21]. The actors are most commonly persons or organizations; however, they could be any entities such as web pages, countries, proteins, documents, etc. There could also be many different types of relationships, to name a few, collaborations, friendships, web links, citations, information flow, etc. These relations are represented by the edges in the network connecting the actors and may have a direction indicating the flow from one actor to the other; and a strength denoting how much, how often, or how important the relationship is. See Fig. 16.1[2] as an example.

[1] We are using the levels of the learning domains from Bloom's taxonomy [14]. According to Bloom, the cognitive domain is divided into the following six levels of thinking/learning: Knowledge, comprehension, application, analysis, synthesis and evaluation.

[2] The figures presented in this chapter illustrate the structure of the graphs; where, the figure tags label nodes represent different students, but their text is not relevant to be extended.

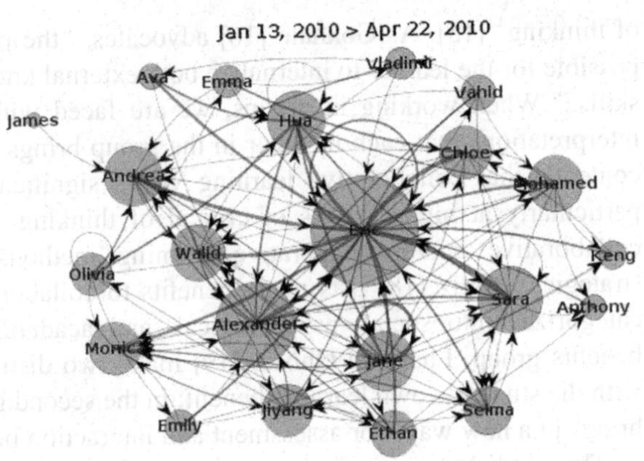

Fig. 16.1 Social network of students interacting in an online discussion forum. Nodes represent actors or students, while an edge from a student to the other summarizes messages sent in that direction and the thickness of that edge corresponds to the number of messages sent

Social Network Analysis Unlike proponents of attribute based social sciences, social network analysts argue that causation is not located in the individuals, but in the social structure [22]. Social network analysis is the study of this structure. Rooted in sociology, nowadays, social network analysis has become an interdisciplinary area of study, including researchers from anthropology, communications, computer science, education, economics, criminology, management science, medicine, political science, and other disciplines [22]. For example in medicine, it is used to understand the progression of the spread of an infectious disease [23]; in criminology, it is an important part of a conspiracy investigation and identifying the nature and extent of conspiratorial involvement [24]; or in education it is helpful in monitoring interactions and participation of students in online courses [7].

Social network analysis examines the structure and composition of ties in a given network and provides insights into its structural characteristics. There are several analyses that could be done on social networks. The most common analysis is ranking individuals based on different centrality measures to find actors with the most prestige, influence, prominence or to detect the outlier actors. The statistics of the network itself could also be insightful, such as, the density i.e. proportion of possible ties that actually exist in the network, or the clustering coefficient, i.e. how much actors tend to group together. The actors that are communicating more often with each other are called communities and could be detected using a communing mining approach.

The other upcoming trend is dynamic analysis of networks to examine the evolution of networks over time, which is useful to predict changes or to make recommendations based on the social structure of actors. Based on the application in hand, one could also perform many further analyses, for example, to examine the flow and diffusion of information within the network and to find actors that are involved in passing information through the network. In the following, we elaborate more on some of these analyses that are useful in our analysis of educational discussion forums.

Centrality of a node in a network/community measures its relative importance within that network/community. There are many measures defined for measuring centrality in social networks. The three most common ones are namely degree centrality, betweenness centrality, and closeness centrality. The degree centrality of a node simply measures the number of edges incident upon that node which implies to some extent the popularity of that node in the network.

In Fig. 16.1 size of nodes corresponds to their degree centralities. The betweenness centrality, represents the control of a node over communication within its community which measures the number of shortest paths between any other nodes that have to pass through this node. Actors/nodes with high betweenness centrality tend to be the hubs in the network connecting different groups of actors, i.e. communities.

On the basis of which, they are often called mediators. The closeness centrality, on the other hand, ranks nodes based on their position in the network—how fast they can spread the information to the whole network, which can be estimated by averaging shortest paths from this node to all the other nodes.

Detecting communities Densely connected actors have been pursued by sociologists for many decades. More recently, it has also attracted attention from physicists, applied mathematicians and computer scientists [25] as a result of its significant practical importance. The availability and growth of large datasets of information networks makes community detection a very challenging research topic in social networks analysis. This line of research resembles well-studied clustering methods in machine learning. However, clustering approach in machine learning is closer to individualist approach in social sciences, as they both focus on the attributes of data entities. This interest resulted in the emergence of a variety of different community detection approaches, e.g. Clique percolation [26], FastModularity [27] and Local [28]; refer to [29] for a recent survey.

For instance, [26] proposed clique Percolation method, known as CFinder, to partition networks into overlapping communities. Based on the observation that edges within communities are likely to form cliques, they defined a community as the union of adjacent cliques.

Modularity optimization based approaches are the most prominent family of community detection methods. The modularity Q is proposed as a measure of the quality of a particular division of a network. The basic idea is to compare the division to a randomized network with exactly the same vertices and same degrees, in which edges are placed randomly without regard to community structure [30]. Figure 16.2 illustrates the communities found by this approach in the network of Fig. 16.1.

Dynamic Social Network Analysis is studying evolution of networks over time, which provides insight into how the characteristics of network and the flow of information in the network changes over time. Figure 16.3 illustrates the changes of the network of Fig. 16.1 over three snapshots. This approach of converting an evolving network into a series of static network snapshots is the basis of many dynamic analysis approaches [31]. Analysis of these changes helps detecting structural events and transitions patterns that occurs in the network.

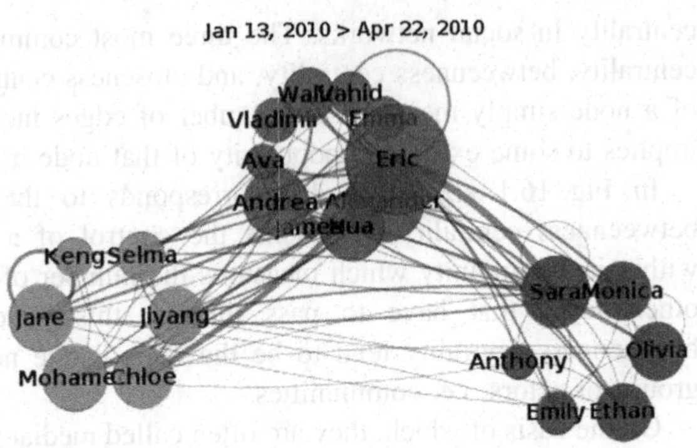

Fig. 16.2 Communities detected by fast modularity approach in the social network of students interacted in the online discussion forum. Different colours represent the three different communities of students that communicated mostly within themselves throughout the course

16.2.3 Social Network Analysis of Online Educational Forums: Related Works

In the following we review the related work specific to social network analysis of asynchronous discussion forums in online courses offered using e-learning environments particularly to assess participation of students. We first overview the related work to how to extract the social network from the e-learning forums, and then we summarize different measures they defined for assessing effective participation.

CSCL tools, used to provide e-learning environments, usually record log files that contain the detailed actions made by learners. These log files include information about the activity of the participants in the discussion forums [1, 4–6, 11] used the log files to *extract the social network* underneath of discussion threads. Laghos and Zaphiris [11] stated that they considered each message as directed to all participants in that discussion thread while others considered it as only directed to the previous message. Figure 16.1 shows a network extracted using the latter approach, which we used in our analysis. Gruzd and Haythornthwaite [32], Gruzd [33] proposed an alternative and more complicated way of extracting social networks, called named network.

They argue that using this common method (connecting a poster to the previous poster in the thread) would result in losing much of the connections. Their approach briefly is: first using named entity recognition to find the nodes of the network, then counting the number of times that each name is mentioned in posts by others to obtain the ties, and finally weighting these ties by the amount of information exchanged in the posts. However, their final reported results are not that promising and even obtaining those results required many manual corrections during the process.

More recently, Dawson [9], Dawson et al. [34] developed a cross platform toolbox called SNAPP which is able to capture the discussion threads of different CSCL platforms from their content in the web browser. However this crawling process is very time consuming compared to reading an input log file. Regarding

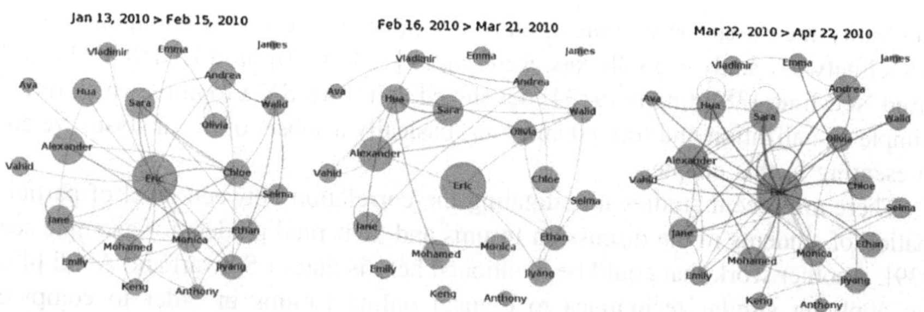

Fig. 16.3 Changes in the network of students over three time snapshots: beginning, middle, and end of the course

what we should consider as the participation in extracting the social network, Hrastinski [35] suggested that apart from writing, there are other indicators of participation like accessing the e-learning environment, reading posts or the quantity and quality of the writing.

Particularly, one might construct the network by linking the author of a message and all other participants whom read that message. However, all of methods mentioned above extract networks just based on posts by student—writing level. For *measuring the effectiveness of participation*, Daradoumis et al. [36] defined high level weighted (showing the importance) indicators to represent collaboration learning process; task performance, group functioning, social support, and help services.

They further divided these indicators to skills and sub-skills, and assigned every sub-skill to an action. For example, group functioning is divided into: active participation behavior, task processing, communication processing. On the other hand, communication processing is divided into more sub-skills: clarification, evaluation, illustration, etc. and clarification is then mapped to the action of changing description of a document.

In the education context, Calvani et al. [2] defined 9 indicators for measuring the effectiveness of participation to compare different groups within a class; extent of participation (number of messages), proposing attitude (number of messages with proposal label), equal participation (variance of messages for users), extent of role (portion of roles used), rhythm (variance of daily messages per day), reciprocal reading (portion of messages that have been read), depth (average response depth), reactivity to proposal (number of direct answers to messages with proposal label) and conclusiveness (number of messages with conclusion label); all summarized for the group interactions and compared relatively to the mean behaviour of all groups.

Similarly, Nandi et al. [21], review the necessity of evaluating the interaction of students in the discussion forums, proposing a set of criteria for evaluating the interactions, including use of social cues or emotions to engage, and the consistency of participation. However, for measuring the effectiveness of participation, most of the previous works simply use general social network measures (different centrality measures, betweenness, etc.), available in one of the common generic social network analysis toolboxes. References [1, 4, 5, 10] used UCINET [11, 37] used NetMiner [38]; finally [9, 34] developed their own SNA toolbox which offers simple visualization and limited analyses, basically a subset of the analyses we are presenting in this chapter.

There are recent studies investigating the correlation between level of participation of students in the discussion forums and their final grades, for example see [39]. Another work that could be mentioned here is that of Stewart and Abidi [40] on applying similar techniques to clinical online forums in order to compare participation of institutions and professions.

16.3 Network Analysis in E-Learning

In this section, we illustrate the practicality of social network analysis in analyzing information networks underlying e-learning environments. Particularly we focus on two types of networks, the network of interactions between students in a course and the network of terms they have used in their interactions.

These networks could be extracted from any e-learning environment and analyzing these networks is helpful in monitoring the students, evaluating their participation in the course, detecting peripheral and central students, etc. Here we describe different analysis on each of these two networks.

16.3.1 Students Interaction Network

The network of interactions between students summarizes all the interactions that occurred during the course. Visualizing this network provides an easy way for the instructor to monitor the structure of these interactions, examine which students are the leaders, and who are the peripheral students. Here, we first describe how the network is extracted based on the information from the discussion threads.

Then, we continue by bringing an analysis of leadership of the students, the collaborative groups of students, and also dynamic analysis of these aspects. In the students' interaction network, each node stands for a student and edges between nodes encode interactions between the students. Figure 16.4 illustrates an example of a student interaction network.

Here, we extracted the network from the discussion forums recorded in the e-learning environment. Consequently, edges correspond to exchange written messages. These edges are weighted by the number of messages passed between the two incident students. This network could be built either directed or undirected.

In the directed model, each message is considered connecting the author of the message to the author of its parent message, while in the undirected network, each edge contains all the correspondences between the two students, see Fig. 16.5.

Fig. 16.4 Visualization of network of students interacting in an online discussion forum. Nodes represent actors or students, edges the interactions, and the thickness of an edge corresponds to the volume of the interaction

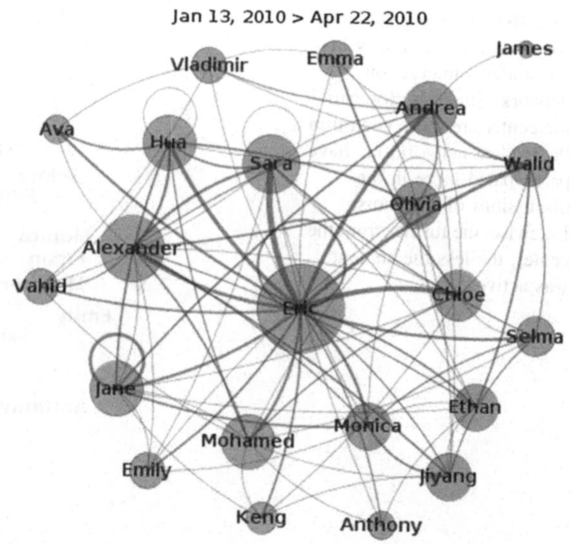

Each interaction edge incorporates several messages. This *visualization of the structure of the interactions* in the course provides an overall view of the underlying structure of the communications in the course, and an apparent way to examine them.

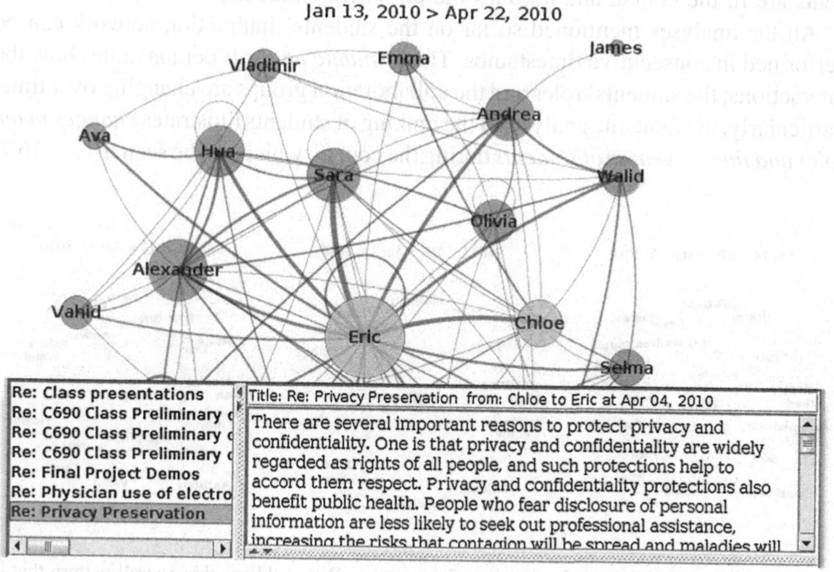

Fig. 16.5 Each interaction edge consists of several messages. This figure investigates the content of messages in the interaction edge between two students—Chloe and Eric

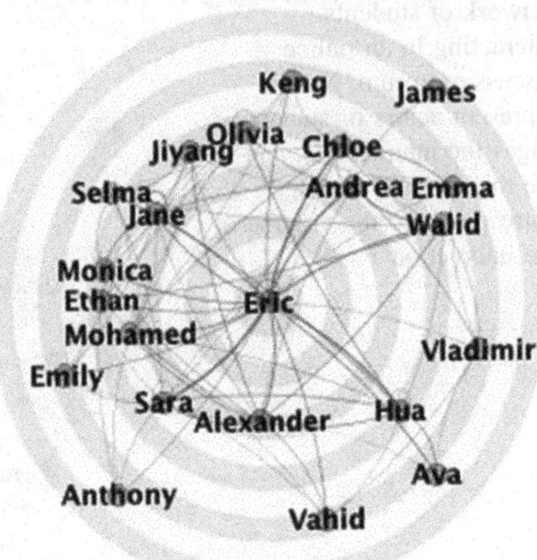

Fig. 16.6 Ranking students based on their centrality in the students interaction network. Students closer to the center are more central in the student network, i.e. have participated more in the discussions of the course. Likewise, the further from the center, the less the student was active

Size of the nodes in Fig. 16.4 corresponds to their centrality in the network. Since centrality measures the relative importance, the leadership and influence of students in the discussions could be compared by analyzing the centrality of nodes corresponding to them in the network.

Consequently, students could be ranked explicitly in a concentric centrality graph in which the more central/powerful the node is, the closer it is to the center, see Fig. 16.6. This *ranking of the students* clearly distinguishes how active students are in the course and outlines the peripheral students.

All the analyses mentioned so far on the students' interaction network can be performed in consecutive timestamps. This *dynamic analysis* demonstrates how the interactions, the students' roles and the collaboration groups are changing over time. Particularly, the dynamic analysis of the ranking of students illustrates *changes in the roles and the activeness of students* during the course; which can be seen in Fig. 16.7.

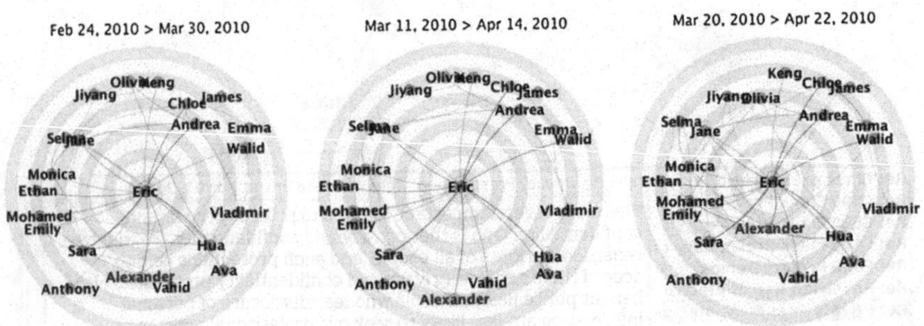

Fig. 16.7 Changes of students' roles during the course. We could see this overall pattern that in the middle snapshot, students tend to be less active. We could also focus in each of the students and monitor how he/she is changing his behavior throughout the course

This systematic monitoring can produce an effective vision for the instructor and/or students about the flow of the course, which can be used to recommend and implement necessary changes. For instance this monitoring can alert the instructor about the unusual low participation of some students in some periods of time. Also it can be used to detect students that are losing interest in the course and recommend and intervene to motivate them to engage more.

Detecting communities in the students interaction network effectively outlines groups of students that collaborated more with each other, refer to Fig. 16.2 for an example. This *Identification of collaborative groups* has an emphasized practical importance for the understanding of academic collaborations as discussed in [7, 41].

We could detect patterns and events in students' collaborations by a dynamic analysis of the communities, such as transition of students between these groups, formation of new discussion groups and much more. Such events can affect participation and engagement of students in the course and detecting these events could be used to make proper recommendation to modify students' behaviors affected by these changes. For instance a community split is detected in Fig. 16.8, and it can be predicted that the participation level of the detached students would drop, accordingly, they could be recommended and/or invited by one of the remaining collaboration groups to join and engage in their discussions.

16.3.2 Term Co-Occurrence Network

In addition to the different analyses on the structure of interactions between the students, content analysis of these interactions could yield to more profound realization about the essence of engagement of students. With this in mind, we have extracted the network of terms used by students in their interactions.

In this network, nodes are terms and edges are their co-occurrence in the same context, i.e. same sentence. Building such network has previously been discussed for different purposes such as improving the results of a search engine [42]. In fact it has been observed that different sets of terms are used to discuss different topics

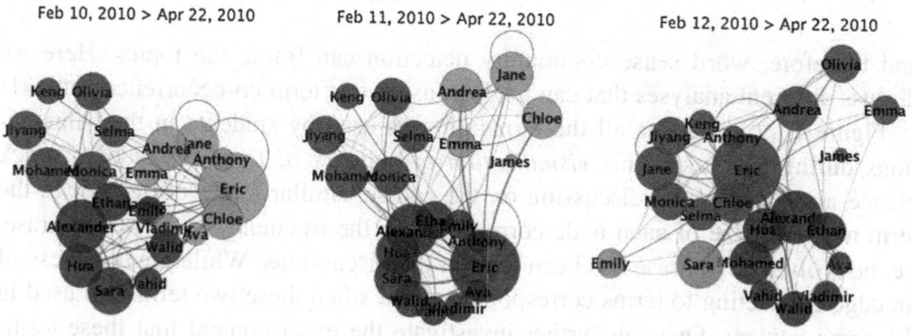

Fig. 16.8 Changes in the collaborative student groups over time, one could observe how the Cyan community first loses some of its members and then splits in the next time stamp

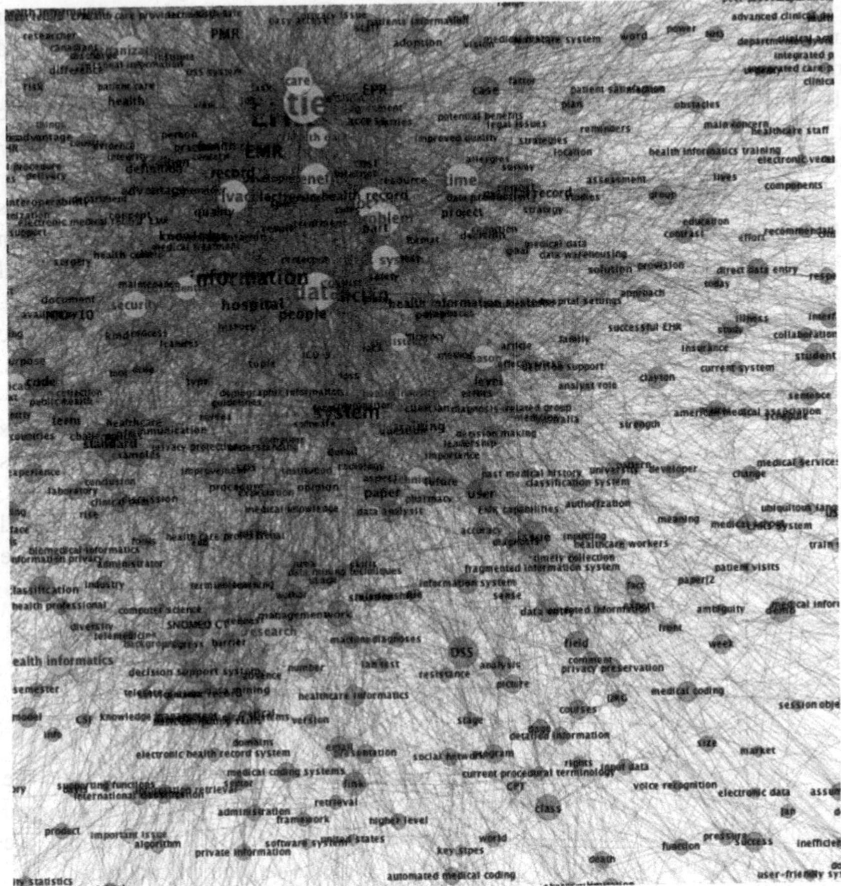

Fig. 16.9 Visualization of network of terms used in an online discussion forum. Nodes represent noun phrases, edges their co-occurrence, and the thickness of an edge corresponds to the fraction of times they have been used in the same sentence. Filtering this graph on term frequency and/or topics can weed out irrelevant terms

and therefore, word sense community detection can frame the topics. Here we discuss different analyses that can be done using this term co-occurrence network.

Figure 16.9 visualizes all the terms phrases used by students in their interactions during a course. This *visualization of network of terms* provides a quick glance at what is under discussion on the course, similar to a word cloud. In the term network, size of each node corresponds to the frequency of its noun phrase, i.e. how often this phrase has been used in the discussions. While the thickness of an edge connecting to terms corresponds to how often these two terms are used in the same context. One can further investigate the exact context that these terms have been used together, as it can be seen in Fig. 16.10.

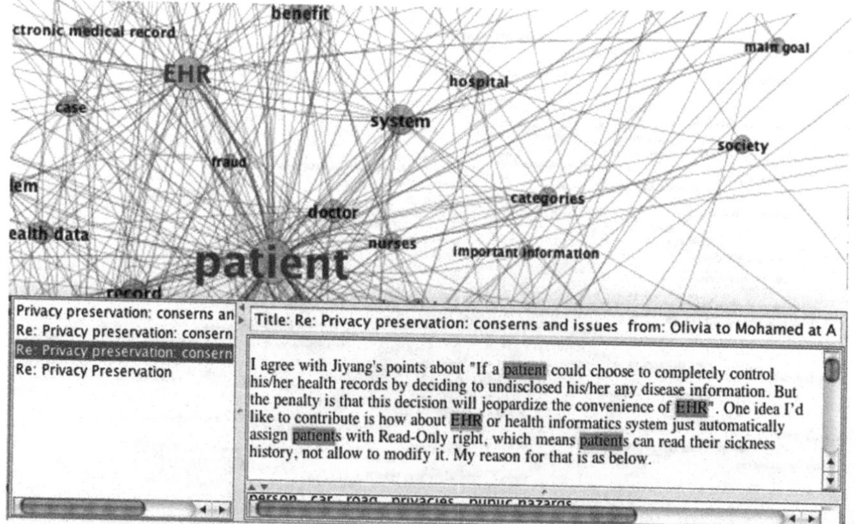

Fig. 16.10 Each co-occurrence edge connects two terms and consists of the messages that these two terms appeared together

The term co-occurrence network can also be extracted per each discussion thread. This provides a straightforward *thread content comparison* of content of different threads, i.e. key words used in each of these discussions? Moreover, one can filter words that have been used by a particular student(s) to inspect and compare the points different students made in the discussions. Figure 16.11 for instance, highlights the terms used by one of the students, among all terms used in a discussion threads. This *student term usage comparison* is helpful in comparing the contextual engagement of the students and determining their points and topics of interest.

These term co-occurrence networks include large amount of information and are not very easy to interpret. They require to be further analyzed for more clear patterns. Clustering is one of the most appropriate analysis when dealing with large correlated data. Thus, we cluster the words into words that co-occurred more often using a hierarchical community mining algorithm. The detected communities summarize the *hierarchy of the discussed topics* in the course. This analysis can also be performed per specific discussion thread to obtain *thread topic comparison*. Figure 16.12 illustrates the structure of hierarchical topics discovered for the discussion thread of Fig. 16.11.

Additionally, these topics could be filtered for a specific student(s) to outline his/her topics of interest and involvement. This *student topic involvement comparison* is illustrated in Fig. 16.12. Using this approach, we could compare the range of participation of different students and detect students who participated in a wider range of topics. In like manner, we could also filter and rank students by their participation in a particular topic. This *Ranking of students by their level of*

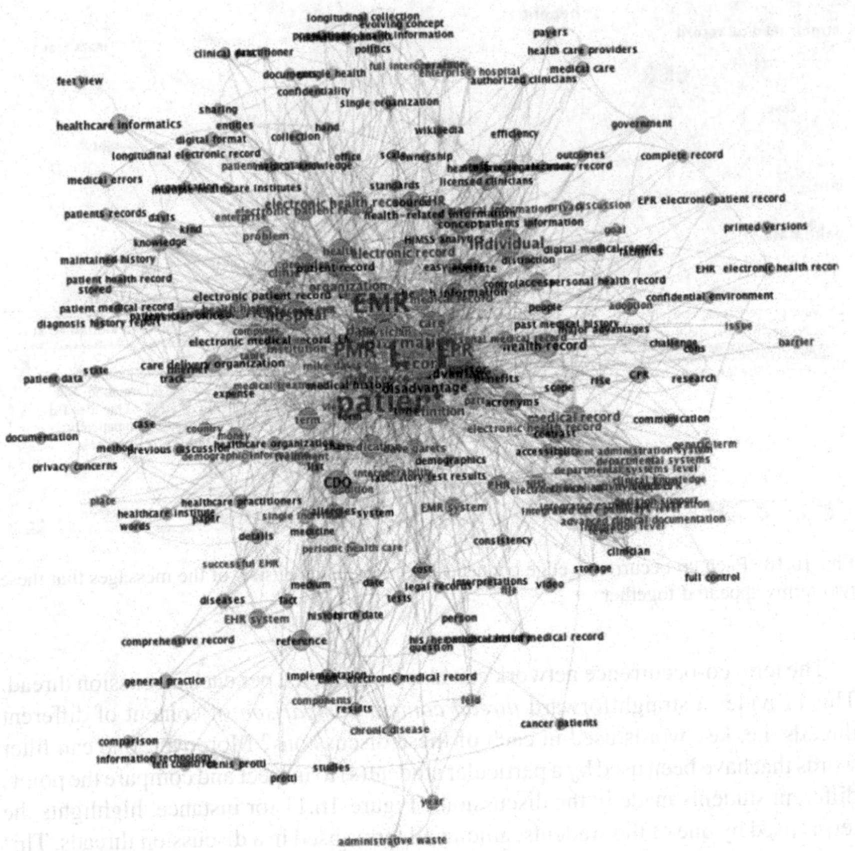

Fig. 16.11 Examining words used by a particular student in one of the discussion threads in an e-learning course. This can be used to determine and compare interest of different students as well as their level of participation

engagement in each topic can be used to recommend the less active students to engage more on that particular topic.

In the next section, we illustrate the practicality of the proposed analyses in e-learning environments by applying them on our two case study courses. We first describe our dataset, and the tool we have used for analyzing these courses. Then we present the result of our analyses and the patterns we found in these courses.

16.4 Case Studies

In this section, we illustrate the practicality of the proposed analyses on two case study courses. The data set we have used is obtained from a post-secondary course offered in two consecutive years. The course titled Electronic Health Record and

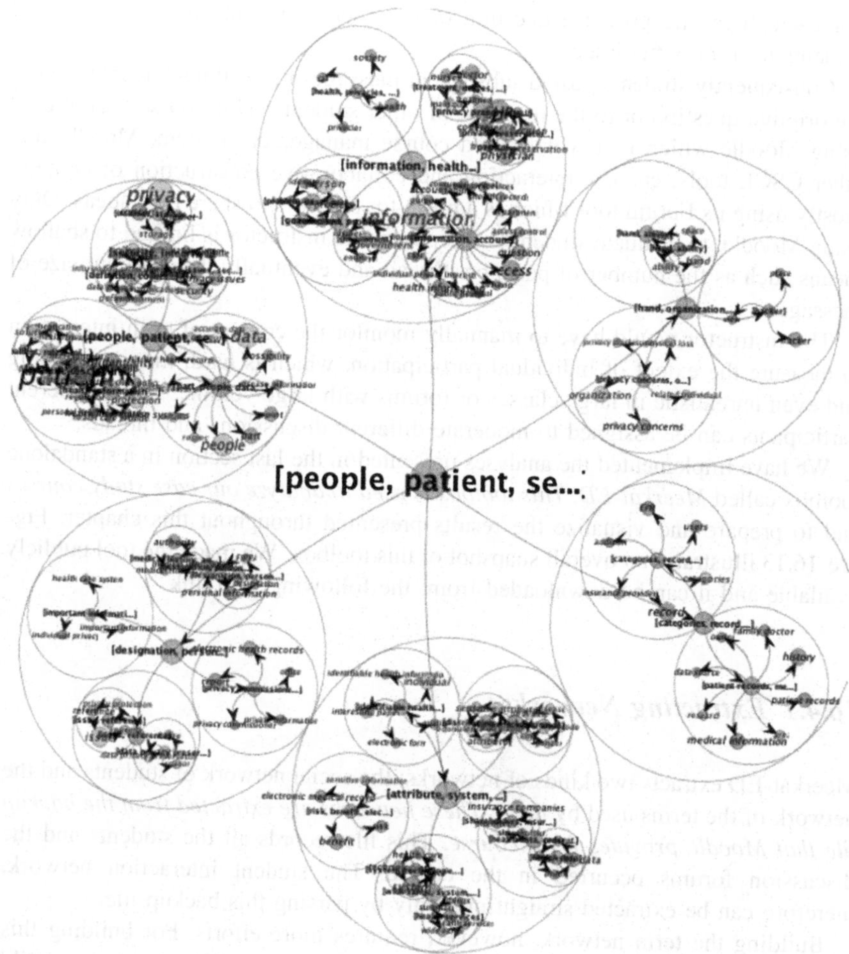

Fig. 16.12 Hierarchical topic clustering of a discussion thread. In which one can also examine topics that a particular student involved with, the content of these topics and the range of participation of the student

Data Analysis, was offered in winter 2010 and also winter 2011 at the University of Alberta. The permission to use the anonymized course data for research purposes was obtained from all the students registered in the course, at the end of the semester so as not to bias the communications taking place.

This data is further anonymized by assigning fake names to students and replacing any occurrence of first, last or user name of the students in the data (including content of the messages in discussion forums) with the assigned fake name. We also removed all email addresses from the data. In the chosen course, as is also usual in other courses, the instructor initiated different discussion threads.

For each thread he posted a question or provided some information and asked students to discuss the issue.

Consequently students posted subsequent messages in the thread, responding to the original question or to the response of other students. This course was offered using Moodle which is a widely-used course management system. Moodle like other CSCL tools, enables interaction and collaborative construction of content, mostly using its Forum tool which is a place for students to share their ideas. Only using Moodle, to evaluate student participation the instructor is limited to shallow means such as the number of posts per thread and eventually the apparent size of messages.

The instructor would have to manually monitor the content of each interaction to measure the extent of individual participation, which is hard, time consuming and even unrealistic in large classes or forums with large volume, where different participants can be assigned to moderate different discussions and threads.

We have implemented the analyses presented in the last section in a standalone toolbox called *Meerkat-ED. This toolbox is used to analyze our case study courses* and to prepare and visualize the results presented throughout this chapter. Figure 16.13 illustrates an overall snapshot of this toolbox. We made the tool publicly available and it can be downloaded from: the following web link.[3]

16.4.1 Extracting Networks

Meerkat-ED extracts two kinds of networks: the social network of students and the network of the terms used by them. *These networks are extracted from the backup file that Moodle provides for a course*. This file records all the students and the discussion forums occurred in the course. The student interaction network, therefore can be extracted straightforwardly by parsing this backup file.

Building the term network, however, requires more efforts. For building this network, we need to first extract the noun phrases from the discussions, then build the network by setting the extracted phrases as nodes and checking their co-occurrence in all the sentences of every message in order to create the edges.

We have used the OpenNlp toolbox (http://opennlp.apache.org/) for extracting noun phrases out of discussions. OpenNlp is a set of natural language processing tools for performing sentence detection, tokenization, pos-tagging, chunking, parsing, and etc. Using sentence detector in OpenNlp, we first segmented the content of messages into sentences. The tokenizer was used to break down those sentences to words. Having the tokenized words, we used the Part-Of-Speech tagger to determine their grammatical tags—whether they are noun, verbs, adjective, etc.

3 http://http://webdocs.cs.ualberta.ca/~rabbanyk/MeerkatED/.

Then using the chunker, we grouped these words into phrases, and picked the detected noun phrases, which are sequences of words surrounding at least one noun and functioning as a single unit in the syntax. For obtaining better sets of terms to depict the content of the discussions, pruning on the extracted noun phrases was necessary.

We removed all the stop words, and split the phrases that have stop word(s) within into two different phrases. For example the phrase "privacy and confidentiality" is split into two terms: "privacy", and "confidentiality". To avoid having duplicates, the first characters were converted to lower case (if the other characters of the phrase are in lowercase) and plurals to singular forms (if the singular form appeared in the content). For instance "Patients" would be "patients" then "patient". As a final modification, we removed all the noun phrases that only occurred once.

16.4.2 Interpreting Students Interaction Network

We have already reported partial results of the analyses on the 2010 course in the last section for illustrations. Figure 16.4 shows the structure of interaction between students in the 2010 course, while Fig. 16.3 represents the dynamics of these interactions over tree snapshots—beginning, middle and end of the course. Using these visualizations we can overview the structure of interactions between students and detect the interesting patterns. For example, the snapshot of Fig. 16.3 clearly outlines a less engagement from students in the middle of the course, compared to

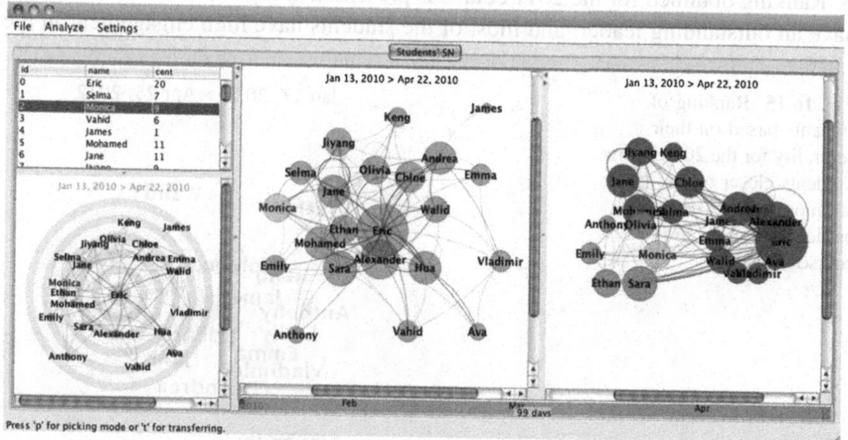

Fig. 16.13 A snapshot of Meerkat-ED toolbox for the 2010 course over the whole 99 days period. We can see a list of the students in the course (*top left*), ranking of leader to peripheral students (*bottom left*), the structure of their interactions (*middle*), and their collaborative groups (*right*)

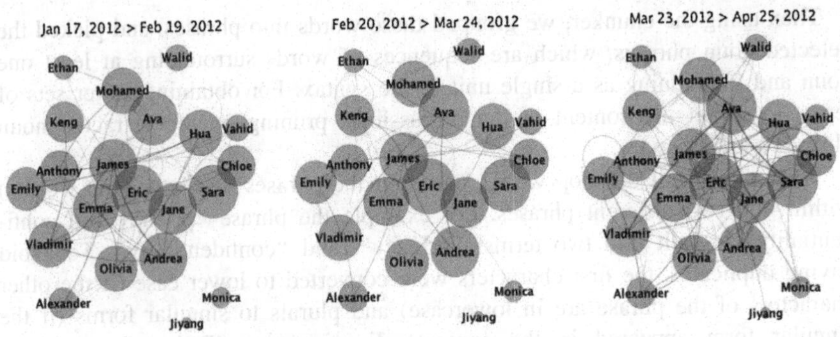

Fig. 16.14 Changes in the network of students over three time snapshots: beginning, middle, and end of the course offered in the 2011

the higher activity at the beginning and in the end of the course. A similar trend is true for the 2011 course as it can be seen in Fig. 16.14.

In these figures, size of the nodes corresponds to their degree centrality in the network—the number of incident edges. This means that the bigger a node is, the more messages the student represented by that node sent and received. Moreover, students are ranked more explicitly in a concentric centrality graph in which the more central/powerful the node is, the closer it is to the center, as presented in Fig. 16.6. From this graph we can see that Eric is notably the most active member in the course, whom is in fact the instructor. Besides the instructor, the rest of students have about the same participation overall, except for the few that have very low activity—James, Antony, Vahid and Ava. The dynamics of this concentric view determines the role change of students in different time periods.

Ranking obtained for the 2011 course is presented in Fig. 16.15. Here we do not have an outstanding leader, and most of the students have high engagement in the

Fig. 16.15 Ranking of students based on their centrality for the 2011 course. Students closer to the center are more central i.e. have participated more in the course

Fig. 16.16 Changes of students' roles during the 2011 course

course while there are few salient outliers that relatively have a very low participation in the course—Jiyang, Vahid, Ethan, Walid and Monica.

We can also monitor how the roles are changing during the course by dynamic analysis of this concentric graph. An example is given in Fig. 16.7 for the 2010 course where roles are pretty much preserved, as Eric always is the main leader and the rest of the class have about the same activity with minor changes.

This however is not the case for the 2011 course. As we can see in Fig. 16.16, the leader changes during the course, from Emma at the beginning to James in the middle of the course and Eric in the end.

Furthermore, we can monitor the changes in collaborative groups of students and detect events and patterns. For example in Fig. 16.8, we have seen a community split that have occurred during the 2010 course.

A community growth can also be detected proceeding that community split as illustrated in Fig. 16.17. Where the Red community recruits new members while at the same time the Green community dissolves into Purple.

We can also see the effect of the leader move between communities that clearly has triggered most of these events. In Fig. 16.8 for example, moving Eric from the Cyan community to Purple community caused the Cyan one to split, while in

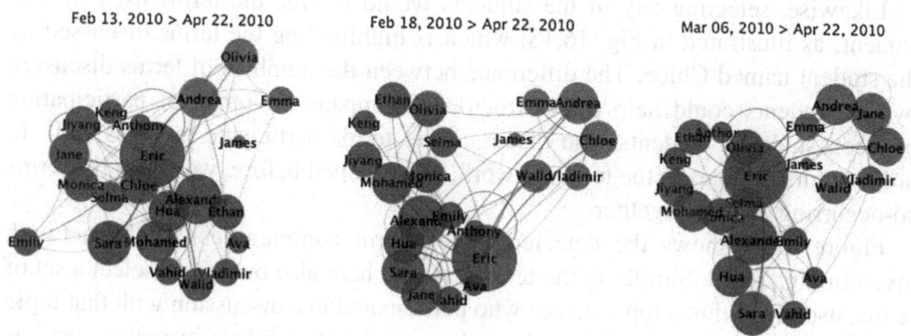

Fig. 16.17 The effect of leader move in collaborative groups, (1) the Green community follows leader into the Purple community, (2) Purple members leave the community after the leader moves to the Blue community

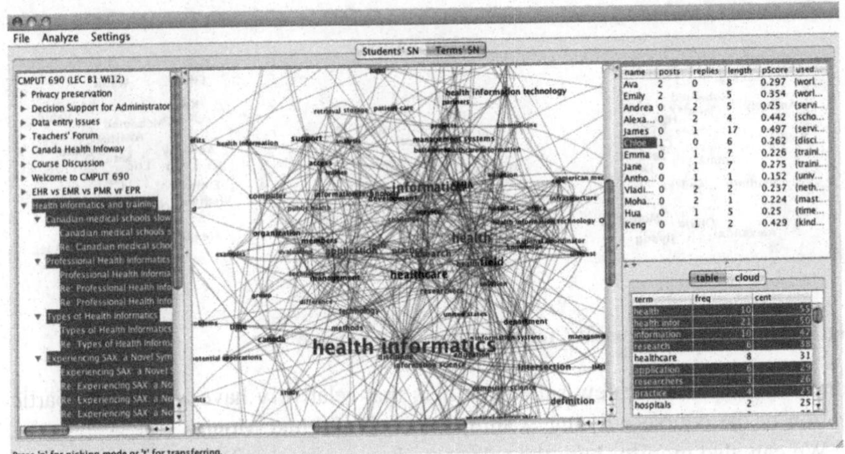

Fig. 16.18 A snapshot of Meerkat-ED for the 2011 course; network of the terms used in "Health Informatics and Training" discussion is visualized and the terms used by Chole are highlighted

Fig. 16.17 his next move from Purple to Blue, helped the Red community to enlist some of the Purple members.

16.4.3 Interpreting Term Co-Occurrence Network

Meerkat-ED also extracts and analyzes the term co-occurrence network. In this visualization the instructor would see a list of the discussion threads in the course while selecting any set of those discussions/messages would bring up the corresponding term network, along with the list of terms occurring in them and the list of students that participated in these selected set of discussions/messages. Selecting any of these terms would show the students that used that term.

Likewise, selecting any of the students would outline the terms used by that student, as illustrated in Fig. 16.18; which is highlighting the terms discussed by the student named Chloe. The difference between the numbers of terms discussed by the students could help the instructor to compare the range of participation between students: students who discuss more terms participate more as well. In order to further analyze the term Network, as explained before, we group the terms co-occurring mostly together.

Figure 16.19 shows the detected topics (term communities) in the network given in Fig. 16.18. Similar to the term network, here also one could select a set of terms, usually within a topic, to see who participated in a discussion with that topic and to what extent. For example here the topic of "school, informatics, etc." is selected and we can see that most of the students have participated in this topic.

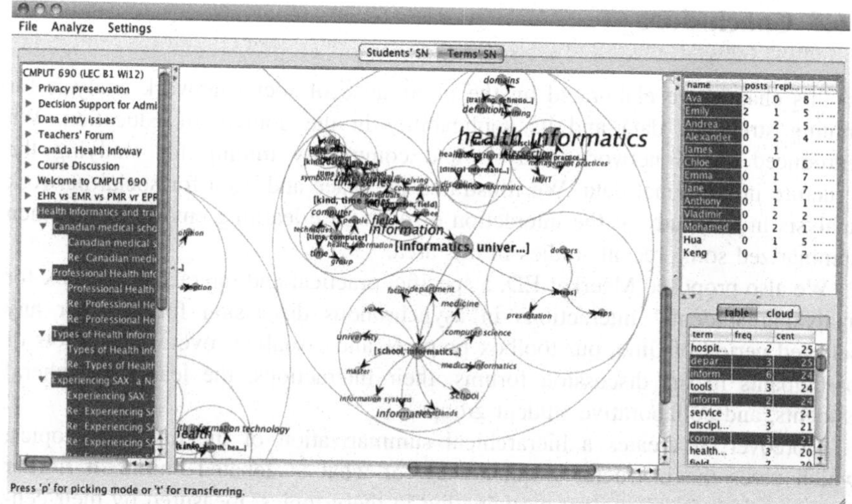

Fig. 16.19 A snapshot of Meerkat-ED for the 2011 course; topics discussed in "Health Informatics and Training" discussion are outlined and the students participated in the selected sub topic of "school, informatics" are highlighted

16.4.4 Objective Evaluation

The instructor of the course denoted the usefulness of the results of these analyses in evaluating the participation of students in the course. Like in [10] where the authors noted that using SNA it was easy to identify the "workers and the lurkers" in the class, in this case study, the instructor reported that *using Meerkat-ED it was easy to have an overview of the whole participation and it was possible to identify influential students in each thread as well as identify quiet students or unvoiced opinions*, something that would have been impossible with the simple statistics provided by Moodle.

More importantly, focusing on the relationships in the graph one can identify the real conduit for information rather than simply basing assessment of participation on message size or frequency of submissions. Learners who place centrally in the network as conduit for the information control and can cause more knowledge exchange which is desirable in an online class.

Regardless of the frequency of messages, their size or content, if they do not have influence, their authors remain marginal and sit on the periphery of the network (See Fig. 16.7). This role of conduit of information versus marginal students can change during the course of the semester or from one discussed thread to the other. *The systematic analysis of centrality of participants per topic discussed provided by Meerkat-ED allowed a better assessment of the participation of learners* at each discussion topic level.

16.5 Conclusions

In this chapter we elaborated on the importance of social network analysis for mining structural data and its applicability in the domain of education. We introduced social network analysis and community mining for studying the structure in relational data. We illustrated the place and need for social network analysis in the study of the interaction of users in e-learning environments; then summarized some recent studies in this area.

We also proposed Meerkat-ED, a specific, practical and interactive toolbox for analyzing students' interactions in asynchronous discussion forums. For any selected period of time, our toolbox prepares and visualizes overall snapshots of participants in the discussion forums, their interactions, the leaders/peripheral students, and collaborative student groups.

Moreover, it creates a hierarchical summarization of the discussed topics, which gives the instructor a quick view of what is debated online. It further illustrates individual student participation in these topics, measured by their centrality in the discussions on that topic, their number of posts, replies, and the portion of terms used by them. We believe exploiting the mining abilities of this toolbox would facilitate fairer evaluation of students' participation in online courses.

References

1. Erlin, E., Yusof, N., Rahman, A.: Students' interactions in online asynchronous discussion forum: a social network analysis. In: International Conference on Education Technology and Computer, pp. 25–29. IEEE Computer Society, Washington (2009)
2. Calvani, A., Fini, A., Molino, M., Ranieri, M.: Visualizing and monitoring effective interactions in online collaborative groups. Br. J. Educ. Technol. **41**(2), 213–226 (2010)
3. Mak, S., Williams, R., Mackness, J.: Blogs and forums as communication and learning tools in a MOOC. In: Dirckinck-Holmfeld, L., Hodgson, V., Jones, C., de Laat, M., McConnell, D., Ryberg, T. (eds.) 7th International Conference on Networked Learning, pp. 275–285. Aalborg University, Denmark (2010)
4. Willging, P.A.: Using social network analysis techniques to examine online interactions. US-Chin. Educ. Rev. **2**(9), 46–56 (2005)
5. de Laat, M., Lally, V., Lipponen, L., Simons, R.J.: Investigating patterns of interaction in networked learning and computer-supported collaborative learning: a role for social network analysis. Int. J. Comput. Support. Collab. Learn. **2**(1), 87–103 (2007)
6. Nurmela, K., Lehtinen, E., Palonen, T.: Evaluating CSCL log files by social network analysis. In: Hoadley, C.M., Roschelle, J. (eds.) Conference on Computer Support for Collaborative Learning, p. 54. International Society of the Learning Sciences, Palo Alto (1999)
7. Haythornthwaite, C.: Building Social Networks Via Computer Networks: Creating and Sustaining Distributed Learning Communities, pp. 159–190. Cambridge University Press, Cambridge (2002)
8. Cho, H., Gay, G., Davidson, B., Ingraffea, A.: Social networks, communication styles, and learning performance in a CSCL community. Comput. Educ. **49**(2), 309–329 (2007)

9. Dawson, S.: Seeing the learning community: an exploration of the development of a resource for monitoring online student networking. Br. J. Educ. Technol. **41**(5), 736–752 (2010)
10. Sundararajan, B.: Emergence of the most knowledgeable other (MKO): social network analysis of chat and bulletin board conversations in a CSCL system. Electron. J. e-Learn. **8**, 191–208 (2010)
11. Laghos, A., Zaphiris, P.: Sociology of student-centred e-learning communities: a network analysis. In: International Conference on Cognition and Exploratory Learning in Digital Age, e-Society, IADIS Press, Barcelona (2006)
12. Rabbany, R., Takaffoli, M., Zaïane, O.R.: Analyzing participation of students in online courses using social network analysis techniques. In: Pechenizkiy, M., Calders, T., Conati, C., Ventura, S., Romero, C., Stamper, J. (eds.) 4th International Conference on Educational Data Mining, pp. 21–30. International Educational Data Mining Society, Eindhoven (2011)
13. Rabbany, R., Takaffoli, M., Zaïane, O.R.: Social network analysis and mining to support the assessment of online student participation. ACM SIGKDD Explor. Newsl. **13**(2), 20–29 (2011). ACM, New York
14. Bloom, B.S., Englehart, M.D., Furst, E.J., Hill, W.H., Krathwohl, D.R.: Taxonomy of Educational Objectives: Handbook 1. Cognitive Domain. David McKay Company (1956)
15. Vygotskiĭ, L.L.S.: Mind in Society: The Development of Higher Psychological Processes. Harvard University Press, Cambridge (1978)
16. Gokhale, A.A.: Collaborative learning enhances critical thinking. J. Technol. Educ. **7**(1) (1995)
17. Bruner, J.: Vygotsky: a historical and conceptual perspective. In: Wertsch, J.V. (ed.) Culture, Communication, and Cognition: Vygotskian Perspectives, pp. 21–34. Cambridge University Press, Cambridge (1985)
18. Panitz, T.: Benefits of cooperative learning in relation to student motivation, In: Theall, M. (ed.) Motivation from Within: Approaches for Encouraging Faculty and Students to Excel, New Directions for Teaching and Learning, Josey–Bass publishing, San Francisco (1999)
19. Laal, M., Ghodsi, S.M.: Benefits of collaborative learning. Procedia Soc. Behav. Sci. **31**, 486–490 (2012)
20. Jyothi, S., McAvinia, C., Keating, J.: A visualisation tool to aid exploration of students' interactions in asynchronous online communication. Comput. Educ. **58**(1), 30–42 (2012)
21. Nandi, D., Hamilton, M., Harland, J.: Evaluating the quality of interaction in asynchronous discussion forums in fully online courses. Distance Educ. **33**(1), 5–30 (2012)
22. Scott, J.G., Carrington, P.J. (eds.): Handbook of Social Network Analysis. Sage, London (2011)
23. Keeling, M.J., Eames, K.T.: Networks and epidemic models. J. Roy. Soc. Interface **2**(4), 295–307 (2005)
24. Davis, R.H.: Social network analysis: an aid in conspiracy investigations. FBI Law Enforc. Bull. **50**(12), 11–19 (1981)
25. Newman, M.E.J.: Detecting community structure in networks. Eur. Phys. J. B **38**, 321–330 (2004)
26. Palla, G., Derenyi, I., Farkas, I., Vicsek, T.: Uncovering the overlapping community structure of complex networks in nature and society. Nature **435**, 814–818 (2005)
27. Clauset, A., Newman, M.E.J., Moore, C.: Finding community structure in very large networks. Phys. Rev. E. **70**(6) (2004)
28. Chen, J., Zaïane, O.R., Goebel, R.: detecting communities in large networks by iterative local expansion. In: International Conference on Computational Aspects of Social Networks, CASoN, pp. 105–112. IEEE Computer Society, Washington (2009)
29. Lancichinetti, A. Fortunato, S.: Community detection algorithms: a comparative analysis. Phys. Rev. E. **80**(5) (2009)
30. Newman, M.E.J., Girvan, M.: Finding and evaluating community structure in networks. Phys. Rev. E. **69**(2) (2004)

31. Berger-Wolf, T.Y., Saia, J.: A framework for analysis of dynamic social networks. 12th ACM SIGKDD International Conference on Knowledge Discovery and Data Mining, pp. 523–528. ACM, New York (2006)
32. Gruzd, A., Haythornthwaite, C.A.: The analysis of online communities using interactive content-based social networks. In: American Society for Information Science and Technology Conference (ASIS&T), pp. 523–527. Association for Information Science and Technology, Columbus (2008)
33. Gruzd, A.: Automated discovery of social networks in online learning communities. PhD thesis, University of Illinois (2009)
34. Dawson, S., Bakharia, A., Heathcote, E.: Snapp: realising the affordances of real-time SNA within networked learning environments. In: Dirckinck-Holmfeld, L., Hodgson, V., Jones, C., de Laat, M., McConnell, D., Ryberg, T. (eds.) 7th International Conference on Networked Learning, pp. 125–133. Aalborg University, Denmark (2010)
35. Hrastinski, S.: What is online learner participation? A literature review. Compute. Educ. **51**(4), 1755–1765 (2008)
36. Daradoumis, T., Martínez-Monés, A., Xhafa, F.: A layered framework for evaluating on-line collaborative learning interactions. Int. J. Hum. Comput. Stud. **64**(7), 622–635 (2006)
37. Borgatti, S.P., Everett, M.G., Freeman, L.C.: Ucinet for Windows: Software for Social Network Analysis (2002)
38. CYRAM: Netminer Software. Available from http://www.netminer.com/
39. Lopez, M.I., Luna, J.M., Romero, C., Ventura, S.: Classification via clustering for predicting final marks based on student participation in forums. In: Yacef, K., Zaïane, O., Hershkovitz, A., Yudelson, M., Stamper, J. (eds.) 5th International Conference on Educational Data Mining, pp. 148–151. International Educational Data Mining Society, Chania (2012)
40. Stewart, S.A., Abidi, S.S.R.: Applying social network analysis to understand the knowledge sharing behaviour of practitioners in a clinical online discussion forum. J. Med. Internet Res. **14**(6), e170 (2012)
41. Reffay, C., Chanier, T.: How social network analysis can help to measure cohesion in collaborative distance-learning. In: Wasson, B., Ludvigsen, S., Hoppe, U. (eds.) Designing for Change in Networked Learning Environments. Computer Supported Collaborative Learning, vol. 2, pp. 343–352. Springer, The Netherlands (2003)
42. Chen, J., Zaïane, O.R., Goebel, R.: An unsupervised approach to cluster web search results based on word sense communities. 2008 IEEE/WIC/ACM International Conference on Web Intelligence and Intelligent Agent Technology, vol. 01, pp. 725–729. IEEE Computer Society, Washington (2008)

Author Index

A. Peña-Ayala (ed.), *Educational Data Mining*,
Studies in Computational Intelligence 524, DOI: 10.1007/978-3-319-02738-8,

Printed in the United States
By Bookmasters